T0073831

*Reading the Book of Nature*

# *Reading the Book of Nature*

## HOW EIGHT BEST SELLERS RECONNECTED CHRISTIANITY AND THE SCIENCES ON THE EVE OF THE VICTORIAN AGE

Jonathan R. Topham

*The University of Chicago Press* CHICAGO AND LONDON

The University of Chicago Press, Chicago 60637
The University of Chicago Press, Ltd., London

For more information, contact the University of Chicago Press, 1427 E. 60th St., Chicago, IL 60637.
Published 2022
Printed in the United States of America

31 30 29 28 27 26 25 24 23 22 1 2 3 4 5

ISBN-13: 978-0-226-81576-3 (cloth)
ISBN-13: 978-0-226-82080-4 (e-book)
DOI: https://doi.org/10.7208/chicago/9780226820804.001.0001

Published with the support of the Susan E. Abrams Fund.

Library of Congress Cataloging-in-Publication Data

Names: Topham, Jonathan R., author.
Title: Reading the book of nature : how eight best sellers reconnected Christianity and the sciences on the eve of the Victorian age / Jonathan R. Topham.
Other titles: How eight best sellers reconnected Christianity and the sciences on the eve of the Victorian age
Description: Chicago ; London : The University of Chicago Press, 2022. | Includes bibliographical references and index.
Identifiers: LCCN 2021061572 | ISBN 9780226815763 (cloth) | ISBN 9780226820804 (ebook)
Subjects: LCSH: Bridgewater treatises on the power, wisdom, and goodness of God, as manifested in the creation. | Religion and science—England—History—19th century. | Books and reading—England—History—19th century.
Classification: LCC BL245 .T67 2022 | DDC 261.5/5094109034—dc23/eng20220218
LC record available at https://lccn.loc.gov/2021061572

♾ This paper meets the requirements of ANSI/NISO Z39.48-1992 (Permanence of Paper).

*For my parents,*
*Joan and Keith Topham,*
*with much love and gratitude*

# *Contents*

*Note for Teachers* * ix

INTRODUCTION
Reading the Book of Nature * 1

PRELUDE
Trouble over Bridgewater * 19

**PART I**
Writing

CHAPTER 1
Becoming a Bridgewater Author * 53

CHAPTER 2
Writing God into Nature * 107

**PART II**
Publishing

CHAPTER 3
Distributing Design * 169

CHAPTER 4
Science Serialized * 227

## PART III
## Reading

CHAPTER 5
Science and the Practice of Religion * 289

CHAPTER 6
Preachers and Protagonists * 331

CHAPTER 7
Being a Christian "Man of Science" * 375

CHAPTER 8
Religion and the Practice of Science * 431

CONCLUSION
"The Fashionable Reign of the Bridgewater Treatises" * 471

*Acknowledgments* * 489
*Appendix A: Note on Currency and the Value of Money* * 493
*Appendix B: British Editions of the Bridgewater Treatises* * 495
*Appendix C: British Reviews of the Bridgewater Treatises, 1833–38* * 499
*Works Cited* * 511
*Index* * 555

# *Note for Teachers*

This book offers a narrative history of the writing, publishing, and reading of the Bridgewater Treatises, but it has been written with a particular intention of being useful in undergraduate and postgraduate teaching in several fields, and its subdivisions (parts, chapters, and sections) have been deliberately adapted to allow them to be assigned independently of the whole. For many courses, it will be helpful to assign the short introduction alongside whatever other chapters are assigned, in order to introduce the subject matter of the book and the general approach taken.

This is a book that draws on book history and the history of reading to offer a new way of thinking about the topic of science and religion, exploring the practical interconnectedness of these two aspects of culture through an analysis of the production, circulation, and use of printed materials. In addition to the introduction and conclusion, chapters that are likely to be of most relevance to science and religion courses are chapter 2, which offers the first comprehensive overview of the purposes and contents of the Bridgewater Treatises; chapters 5 and 6, which provide an account of how reading about the sciences became integrated with the practice of Protestant Christianity; and chapters 7 and 8, which concern how scientific readers integrated their religious concerns into their scientific practice. Also relevant is chapter 4, which gives unparalleled treatment of the culture of reviewing in relation to science and religion. Several of these chapters (especially 2, 5, and 6) will be of use in more general courses on the history of Christianity.

The book will also be of value to teachers of book history and the history of reading. Since it has been structured around themes of writing, publishing, and reading, those teaching courses on the history of the book should be able to identify chapters relevant for assigning in relation to these different aspects of their subject readily enough. All of the main chapters are well adapted for such courses, with the possible exception of chapter 2, which is more narrowly focused on the purposes and claims of the Bridgewater Treatises. Chapter 1 (on scientific authorship), chapter 4 (on science and journalistic culture), and chapters 5 and 6 (on scientific reading and religion) would also be valuable for courses on nineteenth-century literature or on science and literature.

Many of those who use this book in teaching will be historians of modern science, and the Bridgewater Treatises obviously constitute an important episode in that history. They are, of course, of particular relevance to the history of Darwinism, and teachers of courses on that topic will probably find chapter 2 (on the Bridgewaters themselves) and chapter 8 (on scientific readers, including especially Darwin) to be of most value. Those teaching more general courses on nineteenth-century science might find several of the other chapters useful in addition, including the prelude (on the reform controversy in British science), chapter 1 (on scientific careers and authorship), chapter 3 (on scientific publication, including illustration), and chapter 4 (on science and journalistic culture). Chapter 7 contains a uniquely wide-ranging survey of British university education in the sciences at a key moment of transition.

Teachers of nineteenth-century history courses wishing to include a session on Victorian science and religion will find the account of the Bridgewater Treatises in chapter 2 and the account of science and the practice of religion in chapters 5 and 6 especially relevant. Chapter 6 includes two sections on science, religion, and political radicalism ("Science in the Spiritual Battleground" and "Bringing Christianity into Disrepute").

✷ INTRODUCTION ✷

# *Reading the Book of Nature*

*But with regard to the material world, we can at least go so far as this—*
*we can perceive that events are brought about not by insulated*
*interpositions of Divine power, exerted in each particular case, but by*
*the establishment of general laws.*

CHARLES DARWIN, On the Origin of Species, *1859*[1]

First impressions matter. Charles Darwin understood as much when he sketched the title page of his nearly completed book on species in the spring of 1859. He already had an informative draft title, his substantial credentials as author, and the respectable name of his publisher, but he wanted to include an epigraph to set the tone and jotted down the name "Whewell" under a blank set of quotation marks (fig. I.1). The quotation that duly appeared opposite the title page of the published work was the sentence quoted above from the "Bridgewater Treatise" of William Whewell, one of his undergraduate mentors in Cambridge.[2] But why did Darwin choose to start his revolutionary book in this way? What were the Bridgewater Treatises and why did he expect that quoting from one of them would help in getting a fair hearing for his carefully crafted case for evolution?

The answer lies a generation earlier, before Victoria came to the throne. Darwin grew up in a country that was witnessing unprecedented change both in the sciences themselves and in their place in society. In a development that has sometimes been described as a second scientific revolution, the study of nature was rapidly becoming the preserve of new specialist societies, with aspiring researchers having to master increasingly arcane

1. Darwin 1859, [ii], quoting from Whewell 1833b, 356.

2. CUL, DAR 205.1:70r. The spring date is inferred from Murray's name, which began to be discussed at the start of 1859. For a presumably later draft see also Darwin to C Lyell, 28 and 30 Mar [1859], in *CCD*.

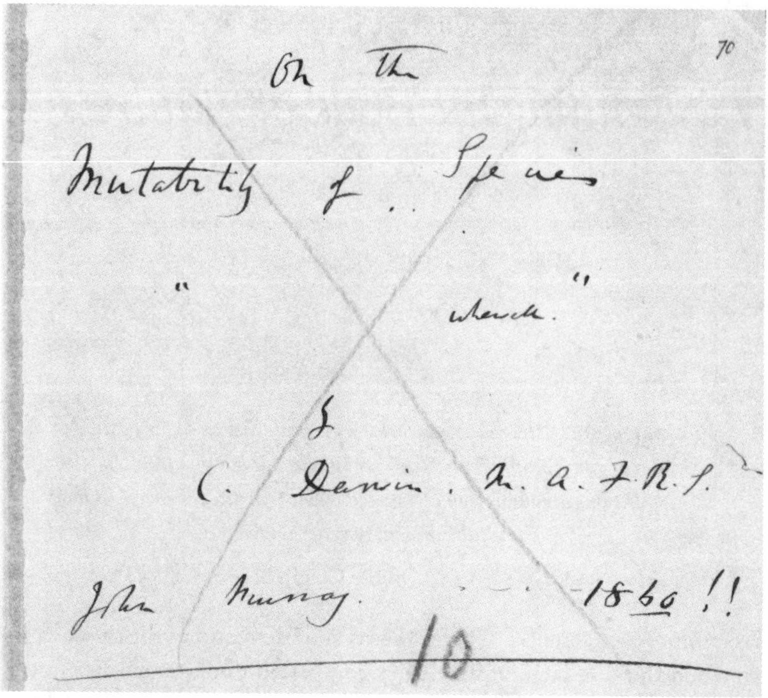

FIGURE I.1 Draft title page for *On the Origin of Species* by Charles Darwin, ca. Spring 1859. Detail from MS DAR 205.1:70r. Image: reproduced by kind permission of the Syndics of Cambridge University Library.

concepts and practices (often involving laboratories or advanced mathematics) before they could make a contribution. At the same time, a cadre of (notably gendered) "men of science" was gaining unprecedented public renown, especially through new kinds of "popular" publications aimed at an emerging industrialized mass market. Increasingly, these savants came to be seen as heroic discoverers who could change the world by unveiling surprising new phenomena and reducing them to laws. Moreover, they were ever-more confident in answering questions about the causes and history of the natural world that had not long before seemed to require theological answers. In postrevolutionary France, Pierre-Simon Laplace had reportedly declared that in offering an account of the development of the solar system, the hypothesis of a creator God was surplus to requirements.[3]

3. Antommarchi 1825, 1:265. On the "second scientific revolution"—or, more tellingly, the "invention of science"—see, for example, Cohen 1985, 91–101, Schaffer 1986, and Cunningham and Williams 1993.

Natural philosophers in Britain were scarcely so audacious, but whether they were speculating about the natural origin of new species, like Darwin's grandfather Erasmus, or advocating a reinterpretation of Genesis in the light of geological findings, they brought into question the long-established relationship between Christianity and the sciences.

This was the context in which the Bridgewater Treatises came to be of such defining importance. Written by some of the foremost scientific figures of the age, this series of eight works on "the Power, Wisdom, and Goodness of God, as manifested in the Creation" provided authoritative reassurance that, whether they were offering a law-governed account of the universe or a progressive history of the creation, the rapidly developing sciences would nevertheless continue to support rather than undermine Christianity. Moreover, their rebaptism of the sciences at this key moment of transition set the tone for the Victorian age. As one reviewer saw it in 1837, the Bridgewater authors had demonstrated that the various sciences could be shown to "establish the truths" on which Christianity was based. With the question likely to have been "set at rest for ever," correct views of the subject would henceforth pervade the "popular mind" and would become "part of the birth-right of every Englishman." In future, the reviewer was sure, educated or reflective people would be unable to think about the natural world without recognizing the creator who lay behind it.[4]

Such was the effect of the Bridgewater Treatises that a couple of decades later Darwin still considered a quotation from the series to be the perfect way of signaling the religious safety of his scientific views. For a generation the Bridgewaters had been an emblem of the widespread view that the rapidly developing scientific disciplines of nineteenth-century Britain were congruent with and supportive of Christianity. This book addresses how they came to be viewed in that light and considers the role that they played in reconnecting the sciences with Christianity on the eve of the Victorian age. Rooted in the history of print, it examines not only the purposes and activities of the books' authors but also the processes by which they came to be so widely familiar and intensely read. Above all it focuses on practices of reading, considering how and why readers engaged with the Bridgewaters in the months and years following their publication. In the process, it reorients the history of science and religion away from a central focus on theology and belief and more toward practical and experiential concerns.

4. "Yarrell's *History of British Fishes,*" *Eclectic Review,* 4th ser., 2 (1837): 598–601, on 598–99.

## *Science, Print, and Christianity in a Revolutionary Era*

The success of the Bridgewater Treatises took contemporaries by surprise. Their publication had been prompted by an extraordinary bequest through which the eighth Earl of Bridgewater had commissioned the president of the Royal Society to appoint an author (or authors) to write a work on God's attributes as manifested in nature. In the event, eight highly respected authors were appointed who between them filled twelve substantial volumes, published between 1833 and 1836, that would have cost a skilled London craftsman at least a month or two's pay.[5] Yet while one prospective publisher pulled out for fear that the series would be a financial failure, it proved to be a publishing sensation (the notion of "best sellers" was not introduced until much later in the century). More importantly, the treatises were read, reviewed, excerpted, discussed, and quoted across the land. From the richest aristocrat to the most reviled street radical, the Bridgewater Treatises were on everybody's lips.

The unexpected success of the Bridgewater Treatises owed much to a profound and ongoing transformation in the production and use of printed matter. Since the inception of moveable-type printing in the fifteenth century, books and other publications had been manufactured expensively by hand, and the great bulk of them had been the preserve of a small wealthy minority. Starting in the last decade of the eighteenth century, however, the rapid rise in demand for print in Britain's increasingly urban and industrial society led to the manufacture of books being progressively mechanized. One trade after another became part of a factory process—from paper making to printing and from printing to binding. In the 1820s these developments began to be exploited to a significant extent in the mass production of cheap but quality literature. Britain was entering a new world of print.

Contemporaries portrayed the 1820s as the era of the "march of intellect" (fig. I.2), dominated by an ambition to bring knowledge to the masses. Moreover, a significant proportion of the new industrially produced cheap literature that tumbled from the press related to the natural sciences. Never before had so many had access to such a body of scientific information. Especially notable were the cheap publications of the wonderfully named Society for the Diffusion of Useful Knowledge (SDUK). This organization was founded by reformers in 1826 with the intention of using the new technologies of cheap print to secure the availability of educational

5. The value of money in early nineteenth-century Britain is discussed in appendix A.

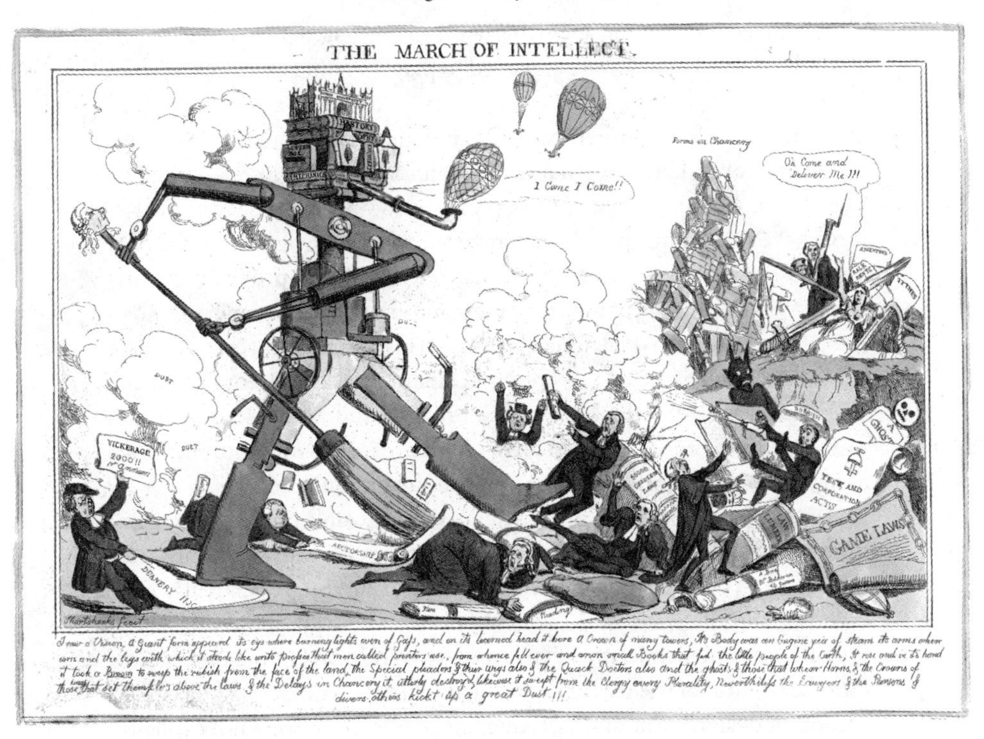


FIGURE 1.2 "The March of Intellect." Hand-colored etching by Robert Seymour, ca. 1828. The caption reads, "I saw a Vision, A Giant form appeard [ . . . ] and on its learned head it bore a Crown of many towers [London University], Its Body was an Engine yea of steam [ . . . ] and the legs with which it strode like unto presses [ . . . ] from whence fell over and anon small Books that fed the little people of the earth, It rose and in it's [*sic*] hand it took a Broom [Brougham] to sweep the rubish from the face of the land [ . . . ]." Image: © The Trustees of the British Museum. All rights reserved.

materials for workers. The need for such pamphlets seemed clearly to have been demonstrated by the numbers of workers who packed the burgeoning mechanics' institutions that were being founded in towns and cities across the country. Now it became possible for workers to buy authoritative introductions to the modern sciences for as little as sixpence a time. Moreover, commercial publishers were also soon busy using the new technologies to spread scientific knowledge far and wide.[6]

These developments in print reflected a wider sense of the growing social and cultural importance of scientific knowledge. Many of the Whig reformers who founded the useful knowledge society were in the same

6. Secord 2014; Topham 1992, 2007.

year involved in founding an unchartered and self-styled London University, where the sciences were expected to play a significant role in educating middle-class youth. This was, contemporaries felt, above all an "age of reform," and for reformers rooted in Enlightenment values, knowledge of the natural and human sciences was foundational for human progress. The central, defining event of the age of reform was the passage of the Reform Act of 1832—an episode that involved a titanic struggle yielding only a modest extension of the franchise among the urban property-owning classes and the ending of the worst electoral abuses. More generally, however, it was an age in which long-established institutions and customs were subject to public scrutiny, resulting in significant social reorganization. The sciences were at the heart of the process, and one of the outcomes was that the emergent "gentlemen of science" came to public prominence as figures with a significant role to play in the nation's progress.[7]

The growing prevalence of the sciences in accessible publications, combined with these new claims for cultural authority and autonomy, provoked anxiety about their association with irreligion. A particular concern was that so many of these new publications and bodies discussed the sciences independently of Christianity. While one of the earliest and most successful of the useful knowledge society's treatises had discussed "proofs of design in the animal frame," its publications were not explicitly Christian in orientation, and most made no reference to religion at all. In similar vein, most of the mechanics' institutions—like the middle-class literary and philosophical societies that had been founded in many provincial towns over recent decades—sought to avoid sectarian and political disputes by excluding theological and political subjects from their premises. Likewise, in establishing London University as a nonsectarian institution, its backers had agreed that the doctrines of Christianity should not be taught there, in order to avoid theological disagreements. As a result, many conservative Christians—notably High Church Anglicans and evangelicals of various casts—saw the increasing output of books on the sciences and the growing standing of the sciences in British society as undermining the national religion.[8]

Such anxiety was not new. In the aftermath of the French Revolution, British commentators had been quick to identify the close connection in France between "philosophers," "infidelity," and revolution. Similar alignments could easily be identified at home. Unitarian and chemist Joseph Priestley, for instance, lost his laboratory, library, and very nearly his life to

7. Morrell and Thackray 1981; [Hamilton] 1831, 384.
8. [Bell] 1827–29; Topham 1992.

"church and king" riots in 1791. In the new century British natural philosophers were widely perceived to be religiously neglectful if not explicitly unorthodox. Moreover, certain incidents provoked a new sense of "conflict" in which what were viewed as the erroneous views of particular philosophers were attacked by their orthodox peers. Especially notable in this regard was the "controversy concerning life and organization" triggered in the late 1810s by the surgeon William Lawrence, whose lectures at the Royal College of Surgeons prompted charges of materialism. Lawrence ultimately suppressed his published lectures in order to save his career, but the incident contributed to an ongoing disquiet.[9]

Matters were not helped when Lawrence's lectures were reprinted in cheap form by working-class radicals, for whom his materialism offered a means of attacking the power of the Anglican hierarchy. As political radicalism became resurgent after the end of the war with France in 1815, the earlier association of philosophers with religious and political danger acquired a new aspect. In the hands of journalist Richard Carlile, Lawrence became a hero of working-class radicalism, and in one pamphlet Carlile urged other "men of science" to join the surgeon in standing forward to "vindicate the truth from the foul grasp and persecution of superstition" (fig. I.3). On Carlile's account, science was fundamentally materialist and would destroy supernatural religion together with all the political and social abuses that were foisted on the public in the name of religion. With "very few exceptions," he wrote, "the medical and surgical professions" in the Metropolis had "discarded from their minds all the superstitious dogmas which Priestcraft hath invented." Men of science were well aware of the revolutionary potential of their work, but cowardice and self-interest had intervened at a time when revolution had been suppressed by "fixed bayonets and despotic laws."[10]

In such a context, it is not surprising that the surge of cheap scientific publications made possible in the 1820s by the new technologies of book production caused concern among more conservative Christians. It was not, generally, that they considered there to be an inherent conflict between scientific inquiry and Christianity. It was, rather, that they felt that the sciences were increasingly falling into the hands of those indifferent or even hostile to Christianity. Unlike philosophers of the seventeenth

9. "Somatopsychonoologia," *British Critic* 22 (1824): 225–45, on 225; Esther Houghton suggested that this review might have been written by a young William Whewell (Curran and Simons 2004–17). On the Lawrence affair, see Goodfield-Toulmin 1969, Jacyna 1983, and Desmond 1989b, esp. 117–21.

10. Carlile 1821, 19, 21.

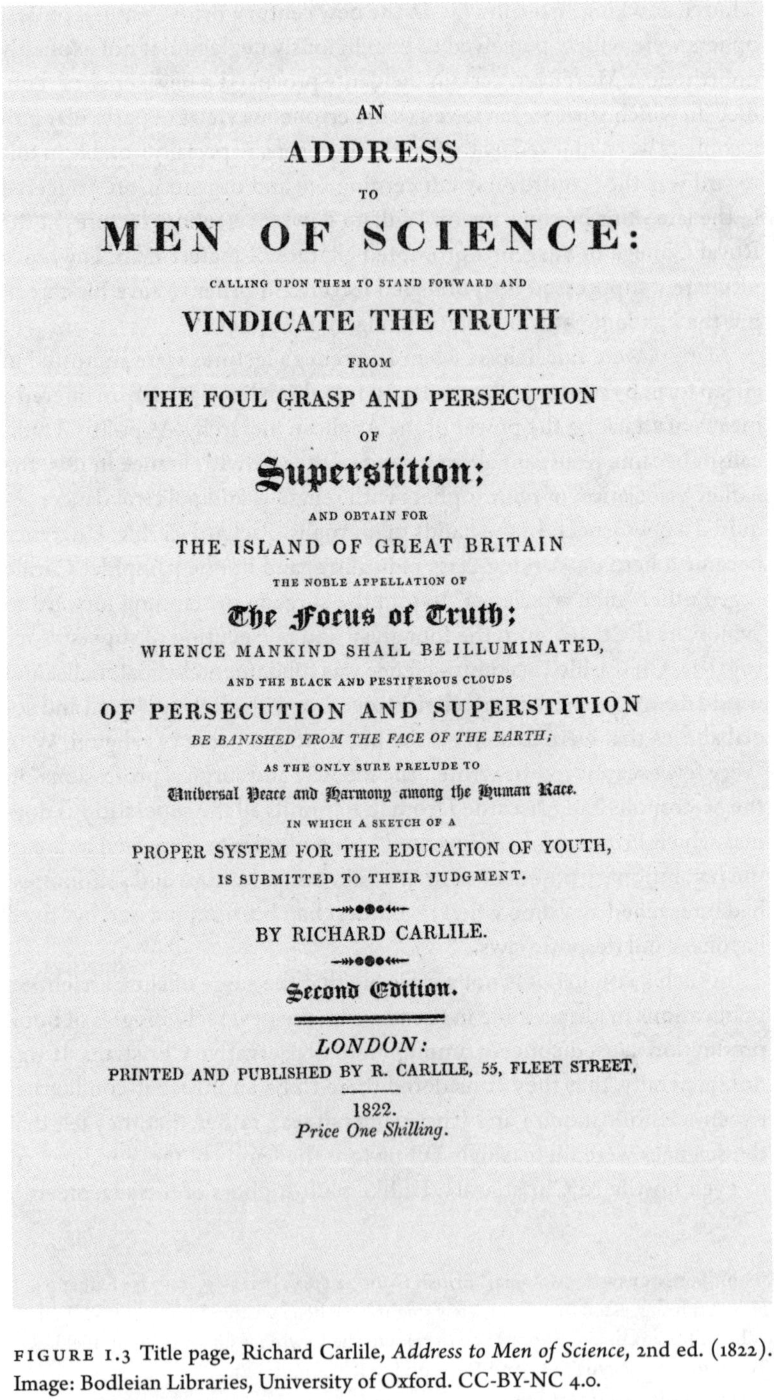

AN

ADDRESS

TO

MEN OF SCIENCE:

CALLING UPON THEM TO STAND FORWARD AND

VINDICATE THE TRUTH

FROM

THE FOUL GRASP AND PERSECUTION

OF

Superstition;

AND OBTAIN FOR

THE ISLAND OF GREAT BRITAIN

THE NOBLE APPELLATION OF

The Focus of Truth;

WHENCE MANKIND SHALL BE ILLUMINATED,

AND THE BLACK AND PESTIFEROUS CLOUDS

OF PERSECUTION AND SUPERSTITION

*BE BANISHED FROM THE FACE OF THE EARTH;*

AS THE ONLY SURE PRELUDE TO

Universal Peace and Harmony among the Human Race.

IN WHICH A SKETCH OF A

PROPER SYSTEM FOR THE EDUCATION OF YOUTH,

IS SUBMITTED TO THEIR JUDGMENT.

BY RICHARD CARLILE.

Second Edition.

*LONDON:*

PRINTED AND PUBLISHED BY R. CARLILE, 55, FLEET STREET,

1822.

*Price One Shilling.*

FIGURE I.3 Title page, Richard Carlile, *Address to Men of Science*, 2nd ed. (1822). Image: Bodleian Libraries, University of Oxford. CC-BY-NC 4.0.

century, one prominent Methodist minister wrote in 1824, the great majority of modern philosophers had "given no indication whatever of a devout spirit." It was, he continued, "notorious" that scientific books generally avoided "every observation or allusion" that "might expose the writer to a sneer as a religionist or a fanatic."[11] Some considered that the philosopher's task of tracing out the natural causes by which the creator made and sustained his creation was inherently unfavorable to religion. Most, however, considered that such dangers could be circumvented effectively if the philosopher were thoroughly imbued with a Christian piety rooted in the Bible and the teachings and practices of the church.

This concern was well expressed in 1823 by Thomas Dick, a Scottish minister turned teacher and self-styled "Christian philosopher." The problem, he claimed, was that the sciences had become separated from theology and were treated as "so many branches of secular knowledge." As a result, natural philosophy and religion were frequently arrayed against each other, with the ensuing "combats" being "equally injurious to the interests of both parties." On the one hand, philosophers occasionally investigated nature without reference to God and more frequently criticized the Bible. On the other, theologians became so zealous against infidel philosophers as to "declaim against the study of science, as if it were unfriendly to religion." It was, Dick claimed, "high time that a complete reconciliation were effected between these contending parties."[12]

As the sciences transformed and became more prominent in the early decades of the nineteenth century, there was thus a growing sense of religious fear and embattlement within a highly charged and rapidly changing social and political context. Such perceptions in no way imply a timeless conflict between science and Christianity, but they do help to explain how, in drawing together authoritative accounts of the modern disciplinary sciences with Protestant orthodoxy, the Bridgewaters came to be among the most widely known books of the age.[13]

## *Rebaptizing the Sciences in the Age of Reform*

That scientific inquiry supported Christianity was a claim that had, of course, been widely restated in Europe since the Middle Ages. Under-

11. [Watson] 1824, 33.

12. Dick 1824, 178–79.

13. On the problems inherent in seeing science and religion as timeless entities necessarily locked in conflict, see, for example, Brooke 1991b, Harrison 2015, and Hardin, Numbers, and Binzley 2018.

neath his epigraph from Whewell, Darwin also quoted a familiar passage from Francis Bacon, the seventeenth-century figurehead of the "experimental philosophy," asserting that God's two books—his word (the Bible) and his works (the creation)—should both be explored in full without fear or favor. This emphasis on the religious value of reading the book of nature was often expressed in the early nineteenth century by professors at Britain's universities, by the authors of scientific books, and, memorably, by the Anglican theologian William Paley in his often-reprinted *Natural Theology* of 1802. Nevertheless, the rapid transformation of the sciences in the early nineteenth century required the relation between God's two books to be revisited, and the Earl of Bridgewater's bequest offered an opportunity to rebaptize the sciences that was extremely timely.

Across an array of emerging disciplines, new developments raised questions about the relation of the sciences to Christianity. In the wake of Laplace's advances, the physical sciences promised to explain ever-more phenomena in relation to natural laws using analytical mathematics that involved formulas impenetrable to the uninitiated.[14] In particular, Laplace had opened the door to a universe that unfolded progressively through eons according to nature's laws, and the nebulae observed by astronomers now began to appear as transient stages in that progressive history. Nor was this the only way in which the history of creation was being reconfigured. The emerging science of geology increasingly made clear that the conventional chronology of earth history derived from Genesis could not accommodate the evidence from the strata. The earth was altogether more ancient than previously thought, and it seemed likely that it had begun as a ball of fire, arising from the astronomical processes described by Laplace. In addition, the fossils that were characteristic of the earth's various strata seemed to most geologists to indicate that progressively more complex forms of life had appeared over the course of its history.

The phenomena of life themselves were more intractable. Nevertheless, the burgeoning science of comparative anatomy (which explored the extraordinary continuities to be found between different organisms, living and fossil) was becoming increasingly "philosophical" as naturalists searched for the underlying laws of life. Philosophical or transcendental anatomists looked especially to a fundamental law of the "unity of composition," associated with the French naturalist Étienne Geoffroy Saint-Hilaire, which seemed to account for anatomical configurations that were otherwise unaccountable. In public at least, British naturalists repudiated

14. [Carlyle] 1829, 445.

the "transmutationist" notion that one species could develop into another, but the notorious theory of Jean-Baptiste Lamarck left no doubt about where lawlike accounts of the history and variety of living things might lead. Moreover, the sciences of chemistry and physiology were also promising new laws of life in which distinct "vital" agencies were displaced by such physical agencies as electricity or the complex organization of matter. Even the phenomena of mind and society were not off limits. The new and controversial science of phrenology claimed to reveal laws of mental function based on the configuration of the brain, while a developing science of political economy was offering an increasingly confident account of the laws governing industrial society.

In such a context, and with new developments gaining unheard of audiences through the expanding press, the relation of the sciences to Christianity seemed increasingly uncertain. Theological writers were clear that the rapidly changing sciences needed to be confronted. One High Church writer in 1834, for example, observed that Christianity had "passed through many ordeals," including "the ordeal of historical research, the ordeal of critical scholarship, [and] the ordeal of logical and metaphysical investigation," but that it now had to pass through "the ordeal of *physical science*." It must, he continued, "be submitted to the test of astronomical and geological, of chemical and physiological, phenomena," adding, "we make no complaint." There was "no shadow of doubt" that the "ultimate result" would be a triumph for Christianity, but the questions had to be asked and answered.[15] The Bridgewater Treatises did just that in a detailed and authoritative manner, with high-profile authors appointed to write across the full range of scientific subjects—from astronomy to physiology and from geology to political economy (fig. I.4).

For the "gentlemen of science" seeking to carve out a new role for themselves and their sciences at the heart of British society and culture, addressing these concerns was essential. Their headquarters, figuratively speaking, lay in the newly founded and peripatetic British Association for the Advancement of Science—one of a number of broadly based associations founded in the 1830s that strove to heal the rifts of party strife in both politics and religion. Together, moderate Whigs and liberal Tories, Anglicans and Protestant dissenters, used the association to build a vision of the future of science at the heart of a progressive Christian nation. This new "vision of science" also found expression in a number of publications aimed at the growing market for accessible works on the sciences, but the Bridgewater Treatises became a central component of that program, of-

15. [Bowden] 1834c, 233.

NOTICE.

THE series of Treatises, of which the present is one, is published under the following circumstances:

The RIGHT HONOURABLE and REVEREND FRANCIS HENRY, EARL of BRIDGEWATER, died in the month of February, 1829; and by his last Will and Testament, bearing date the 25th of February, 1825, he directed certain Trustees therein named to invest in the public funds the sum of Eight thousand pounds sterling; this sum, with the accruing dividends thereon, to be held at the disposal of the President, for the time being, of the Royal Society of London, to be paid to the person or persons nominated by him. The Testator further directed, that the person or persons selected by the said President should be appointed to write, print, and publish one thousand copies of a work *On the Power, Wisdom, and Goodness of God, as manifested in the Creation; illustrating such work by all reasonable arguments, as for instance the variety and formation of God's creatures in the animal, vegetable, and mineral kingdoms; the effect of digestion, and thereby of conversion; the construction of the hand of man, and an infinite variety of other arguments; as also by discoveries ancient and modern, in arts, sciences, and the whole extent of literature.* He desired, moreover, that the profits arising from the sale of the works so published should be paid to the authors of the works.

The late President of the Royal Society, Davies Gilbert, Esq. requested the assistance of his Grace the Archbishop of Canterbury and of the Bishop of London, in determining upon the best mode of carrying into effect the intentions of the Testator. Acting with their advice, and with the concurrence of a nobleman immediately connected with the deceased, Mr. Davies Gilbert appointed the following eight gentlemen to write separate Treatises on the different branches of the subject as here stated:

THE REV. THOMAS CHALMERS, D. D.

Professor of Divinity in the University of Edinburgh.

ON THE POWER, WISDOM, AND GOODNESS OF GOD AS MANIFESTED IN THE ADAPTATION OF EXTERNAL NATURE TO THE MORAL AND INTELLECTUAL CONSTITUTION OF MAN.

JOHN KIDD, M. D. F. R. S.

Regius Professor of Medicine in the University of Oxford.

ON THE ADAPTATION OF EXTERNAL NATURE TO THE PHYSICAL CONDITION OF MAN.

vi

THE REV. WILLIAM WHEWELL, M.A. F.R.S.

Fellow of Trinity College, Cambridge.

ASTRONOMY AND GENERAL PHYSICS CONSIDERED WITH REFERENCE TO NATURAL THEOLOGY.

SIR CHARLES BELL, K.G.H. F.R.S. L.& E.

THE HAND: ITS MECHANISM AND VITAL ENDOWMENTS AS EVINCING DESIGN.

PETER MARK ROGET, M.D.

Fellow of and Secretary to the Royal Society.

ON ANIMAL AND VEGETABLE PHYSIOLOGY.

THE REV. WILLIAM BUCKLAND, D.D. F.R.S.

Canon of Christ Church, and Professor of Geology in the University of Oxford.

ON GEOLOGY AND MINERALOGY.

THE REV. WILLIAM KIRBY, M.A. F.R.S.

ON THE HISTORY, HABITS, AND INSTINCTS OF ANIMALS.

WILLIAM PROUT, M.D. F.R.S.

CHEMISTRY, METEOROLOGY, AND THE FUNCTION OF DIGESTION, CONSIDERED WITH REFERENCE TO NATURAL THEOLOGY.

HIS ROYAL HIGHNESS THE DUKE OF SUSSEX, President of the Royal Society, having desired that no unnecessary delay should take place in the publication of the above mentioned treatises, they will appear at short intervals, as they are ready for publication.

FIGURE I.4 Introductory notice attached to each of the Bridgewater Treatises. Reproduced from the first US edition of William Prout's Bridgewater Treatise (1834).

fering confirmation that science's growing place in the nation would not jeopardize its Christian integrity.[16]

Even among the supporters of the British Association, however, there were somewhat different visions concerning what science was and what its significance should be, not least in its relation to Christianity. Not everyone shared the widespread view that while scientific investigation should not be curtailed by Christian considerations it would ultimately be found congruent with Protestant orthodoxy. Indeed, there were some scientific men, for instance in the medical schools and universities of London and Edinburgh, who plainly rejected supernatural religion, or at least orthodox Christianity. Yet in an age of heightened sensitivity to the possible consequences of religious heterodoxy, such voices were mostly muted. The

16. Secord 2014, esp. 239–40; Morrell and Thackray 1981.

professional and social consequences of publicly confronting Christianity in the name of science were clear to see, so that dissident visions were typically only indirectly or indistinctly expressed. In such a culture, the Bridgewater Treatises sometimes proved useful even for the heterodox as public markers of the religious wholesomeness of the sciences.[17]

The symbolic status of the series depended in part on the fact that the authors themselves offered somewhat different perspectives. The treatises ran to more than five thousand pages of text, and one contemporary suggested that it would be a "literary miracle" for anyone to have read them all.[18] For many readers, an individual Bridgewater that had appealed to them and consequently been made a particular focus of study came to stand for the whole. Thus, some readers took the Bridgewaters to claim that scientific pursuits should be conducted independently of the Bible while others concluded that the series indicated that such pursuits could only properly be conducted with Bible in hand. Similarly, the series might be taken to suggest that detailed knowledge of the creator could be inferred from the natural world independently of the Bible or, on the contrary, that it was only Christians rooted in knowledge of the Bible who could discover intelligible testimony of God's being and characteristics in the creation. With their disparate authors, and extending over many pages, the Bridgewaters had multiple faces.

This was nowhere more important than in relation to the subject of "natural theology" (strictly speaking, that form of theology in which knowledge is sought by reason alone, independently of God's self-revelation through miracles, prophecies, and scriptures). Many contemporaries wrote of the Bridgewaters as works of natural theology, and several of the titles actually included the phrase. Yet these references usually betrayed a degree of imprecision that was characteristic of the period, when natural theology was decidedly contested. There were, one reviewer pointed out, "few subjects on which a wider variety of opinion" had existed. Thus, some considered that God could only be known by reason, while others claimed that God could not be known by unaided reason at all and that the attempt was dangerous to religion. Underlying this was a lack of agreement about what natural theology really was, "so that what one man has condemned as natural theology" had "often been a very different thing from that which another has defended under the same name."[19] In short,

17. Desmond 1989b; Winter 1997.

18. [Willmott] 1837, 48.

19. *Edinburgh Christian Instructor*, 2nd ser., 2 (1833): 767. Full references for reviews of the Bridgewater Treatises are provided in appendix C.

some approved of what they called natural theology while using the term in an imprecise way to describe how the Christian enlightened by revelation could trace God's existence and attributes in nature, but others attacked it as erroneous and even dangerous while using the term more strictly to describe knowledge acquired independently of revelation.

That the Bridgewater authors took a range of positions on this controversial question probably perversely helped to secure the reputation of the series. Indeed, perhaps more significant still was the extent to which they sidestepped the question, offering relatively little by way of explicit theological analysis. It is thus unhelpful to think of the Bridgewater Treatises primarily as works of natural theology. Rather, they offered readers detailed, authoritative, and up-to-date accounts of the several sciences, showing how both the latest findings and current approaches enhanced rather than undermined Christian views of nature. Two themes were especially prominent. First, where the critics of "infidel philosophers" had been concerned about the replacement of God's agency with natural laws, several of the Bridgewater authors emphasized how knowledge of the laws by which the creator governed the universe should enhance the Christian's appreciation of God's role. Second, where some conservative Christians had felt the vast extension of geological time to be a threat to biblical accounts, several of the Bridgewater authors not only urged the consistency of deep time with the Bible record but also argued that the lengthened history of creation augmented the evidence of divine action.

In tackling these sources of concern, the authors consolidated and publicized principles that came to define the public face of the sciences in the middle years of the nineteenth century. The "theistic science" that was dominant in early Victorian Britain depended on a strong commitment to the uniformity of nature, the laws of which were underpinned by a divine legislator.[20] But to a significant extent, it was the Bridgewaters that had baptized such a commitment in the 1830s, as Darwin's opening quotation from Whewell suggests. At the same time, however, the series presented a vigorous riposte to scientific attempts to account for the origin of new species solely in lawlike terms, specifically through the process of transmutation. This, of course, was precisely the point for which Darwin's book would later contend. As Darwin's own reading demonstrated, the mixed messages that the Bridgewaters conveyed were open to such reinterpretation right from the start, and the use made of them by other authors helps to make clear how far their rebaptism of the sciences served to authenticate a lawlike vision of the history of creation.

20. Stanley 2015.

## *Reading the Book of Nature*

The story of how the Bridgewater Treatises came to have the effects that they did is fundamentally rooted in the changing world of print that was at the heart of the age of reform. Their authors faced significant challenges in mastering and manipulating the medium of print as new forms of publication emerged conveying exciting "visions of science" to wide audiences.[21] They were not alone in their task, however. They worked alongside publishers, illustrators, and the immensely powerful periodical press in engaging with readers through printed objects that looked different almost week by week. Without readers, of course, the Bridgewaters would have had no discernible effect, and in the 1830s the ways in which readers encountered and engaged with books were also rapidly changing. This study examines in turn how the Bridgewater Treatises were created, distributed, and used in order to uncover how, in practice, the series came to reconnect the sciences and Christianity on the eve of Victoria's reign. The account it offers reveals how refocusing on the dynamics of print communication can deepen and modify our understanding of familiar historical developments and of the social and cultural processes that underpin them.

The act of private patronage that led to the Bridgewater Treatises seemed, in a reformist era, to smack dangerously of the old guard. The eccentric Earl of Bridgewater was a hugely wealthy aristocrat and absentee clergyman—the very epitome of privilege—and some saw the handling of his bequest by Davies Gilbert, the Tory president of the Royal Society, as a continuation of entrenched interests. Nevertheless, it made possible the appointment of eight authors capable of writing about the sciences in ways that reached out to new audiences, offering an inspiring vision of how they might contribute to the future of the nation. These exclusively male authors were characteristic of the rising generation of scientific gentlemen, numbering five university professors (three of whom were clerical and two medical), two other physicians, and a country parson-naturalist. Between them, they possessed a significant amount of experience as authors, and most had already addressed larger audiences concerning the wider significance of the sciences. Two, indeed, were active in the useful knowledge society.

The approaches that the several authors took to their task and the resources that they drew on in doing so were strikingly diverse. The treatises were very uneven in length, ranging from just over three hundred pages to more than 1,250. They were also very disparate in writing style as well as in

21. Secord 2014.

the use they made of the rapidly proliferating technologies of illustration. The detailed analysis offered here of the authorial labor and craft involved in producing these eight books reveals their rich and varied roots. Reflective treatises on the sciences offered the primary vehicle for competing visions of science in the 1830s, but such works emerged out of a myriad of experiments, and a detailed analysis of the writing of the Bridgewaters yields important insights into that process. Exploring the task of authorship also makes much clearer and more intelligible the divergent visions that the authors offered of the proper connection between Christianity and the sciences. The distinctive tone of the various treatises reflected not only the authors' different theological perspectives but also the processes by which they repurposed text from sermons, lectures, and manuals, seeking to give coherence, clarity, and charm to what they wrote as they responded to their unexpected commission.

The Bridgewater authors brought considerable if varied expertise to their task. Writing a manuscript, however, is not the same as publishing a book. In the ferment of the 1830s, navigating the market for printed matter involved myriad uncertainties, and the story of the Bridgewaters reveals many of them. Their publication entailed a curious combination of conscious intent, misfortune, and good luck, yielding books that spoke in their form, as well as in their content, of a combination of traditional values and authority with progress and modernity (fig. I.5). Important as this was in securing the reputation of the series, it entailed a price ticket that was beyond the pocket of most—a point of no small significance in the age of reform. In fact, however, the expanding apparatus of the reform movement, including the libraries at mechanics' institutes, literary and philosophical societies, and other new institutions, and, above all, the mighty machinery of the periodical press, ensured that the price tag was not an insuperable barrier to readers gaining access to the contents of the Bridgewaters. Rather, the series was effectively relayed and recast to readers of many different classes, its meaning and importance being repeatedly redefined in the process. In understanding how the Bridgewaters came to be an important symbol of the harmony of the sciences with Christianity, it is important to recognize that many experienced them only at second hand.

Whether books are only encountered at one remove in this way or, indeed, merely have their titles read on a shelf or a table, they need in some broad sense to be "read" in order for their existence to have any causal efficacy. The largest part of this study thus focuses on how readers engaged with the Bridgewaters. Histories of science and religion have often focused on the printed or, less frequently, manuscript texts of well-educated men, examining their theological concerns in regard to the sciences. Yet while

FIGURE I.5 Two cloth-bound volumes of the Bridgewater Treatises sitting between a more traditional gentlemanly publication (Alexander Crombie's *Natural Theology*, 1829) and three state-of-the-art cheap publications (Dionysius Lardner's *Treatise on Hydrostatics and Pneumatics*, 1831, from his Cabinet Cyclopædia; James Rennie's *The Faculties of Birds*, 1835, from the SDUK's Library of Entertaining Knowledge; and David Brewster's *The Life of Sir Isaac Newton*, 1831, from John Murray's Family Library).

theology is an important aspect of Christianity, it is nevertheless a limited one. Indeed, while writers in the early nineteenth century sometimes expressed concern about the theological implications of scientific findings, they also frequently wrote about the effect that the sciences could have on religious feelings and habits. In an era when Christianity in Britain was increasingly dominated by evangelicalism, alongside High Anglicanism, the sense that true religion was rooted in such feelings and habits was notably strong. For many of the religious readers of the Bridgewaters, the key question was not so much whether scientific findings had undermined religious truths as whether reading about those findings was liable to strengthen or weaken the believer's sense of religious devotion.

The Bridgewaters were not only significant in enabling Christians to read about the sciences in ways that sustained the life of faith; they also assisted a generation of men in imagining and living out the respectable though gendered role of the Christian philosopher or man of science. Again, this was not simply, or perhaps even primarily, about theological commitments. At a time when the supposed irreligious arrogance of philosophers had been a potential barrier to their respectable social advancement, the Bridgewaters offered a banner behind which men of science could learn to articulate their pious humility. Moreover, these moral qualities were widely enjoined in scientific education at Britain's expanding universities, where readings of the Bridgewaters supported a continuing connection between Christianity and the practice of science. As the series demonstrated, that connection took many forms. The nature of the scientific project, the centrality of the principle of uniformity, and its relation to religious concerns all loomed large in many scientific readers' engagements with the Bridgewaters. So also did the hugely practical question of the role that perceiving design should play in the scientific study of living organisms. It was in the management of such practical concerns that the early Victorian rebaptism of the sciences also took place.

❋

In bringing the perspectives of the history of the book to bear on the Bridgewaters in this way, this study offers a new and revealing account of the historical interactions between Christianity and the sciences. Faced with the question of how the series came to have the symbolic value that Darwin sought to exploit at the start of his *Origin of Species*, we find that the answer lies in an understanding of the entanglement of Christianity and the sciences that goes well beyond the theological. It is rooted in such grand transformations as the industrialization of print production, the weakening and partial replacement of traditional forms of political power, and the rise to cultural authority of the increasingly disciplinary sciences. But it is also rooted in changes in the daily lives of Britain's majority of professing Christians and in the routine practice of those engaged in researching and teaching the sciences. It was in the lived experience of emotional response, spiritual quest, family life, and moral self-fashioning and in the practical challenges of scientific teaching, theory formation, and observation that the Bridgewater Treatises gained their value as guides to the interconnectedness of the sciences and Christianity.

✸ PRELUDE ✸

# *Trouble over Bridgewater*

Royal Society.—*This society was chartered expressly for the purpose of improving* Natural *Science, in the expectation of lessening the influence of* super-*natural science, which at the time when the society was founded, had become alarmingly extensive. As we are upon the subject of the Royal Society, we may mention that we some time ago inquired on behalf of a respectable correspondent, in what manner the late Earl of Bridgewater's legacy of* 8000l. *for two essays had been disposed of. We now learn that the affair has been snugly managed between Mr. Charles Bell, Dr. Roget, and Professor Buckland.*

Monthly Review, *March 1831*[1]

When the Reverend Francis Henry Egerton first signaled his intention in 1813 of asking the president of the Royal Society to administer a bequest to produce a work on "the Power, Wisdom, and Goodness of God, as manifested in the Creation," he can little have anticipated how controversial such a request would prove to be. Egerton had been elected a fellow of the Royal Society in 1781, fewer than three years into the long presidential reign of Sir Joseph Banks, and he was precisely the sort of well-connected individual that the new president considered critical to the society's success. As one later fellow put it, Banks's vision was for a society that combined "the working men of science" with "those who, from their position in society or fortune, it might be desirable to retain as patrons of science."[2] In addition to Egerton being the son of a prominent bishop, his father's cousin was one of the foremost improving aristocrats of the age, by whose pioneering canal ventures Banks was mightily impressed. Ad-

1. "Royal Society," *Monthly Review*, n.s., 1 (1831): 478.
2. B Brodie to CR Weld, 7 April 1848, in Weld 1848, 2:153–54.

mittedly, there was little prospect at the time of his election that the young Shropshire rector would inherit either his cousin's wealth or his titles, but Banks did not have a narrow conception of patronage. Moreover, while the new fellow's interests lay in classical and historical scholarship, this was no difficulty, since Banks was keen to make the society the hub of a "Learned Empire" in which the antiquarian, agricultural, and other learned interests of the aristocracy and landed gentry were incorporated.[3]

Egerton had devised elaborate plans for posthumous patronage long before the unexpected death of his brother made him the eighth Earl of Bridgewater, increasing his annual income to a princely £40,000. His plans were designed to uphold his view that the Egerton family—in whose patronage since the time of Bacon he took inordinate pride—was an ongoing instrument of divine providence. By the time of Egerton's death in 1829, however, Banks was almost nine years dead, and the Royal Society was in the midst of a crisis over reform. Banks's commitment to a society combining working men of science with aristocrats and others who had learned interests was being challenged by a new generation. Particularly problematic was the concern of some of the reformers that the way the society handled patronage was retarding the advancement of science in Britain or even causing its decline. Matters reached a head with the publication in April 1830 of *Reflections on the Decline of Science in England* by Cambridge's Lucasian Professor of Mathematics, Charles Babbage, followed by the election of a new president in November.

From the perspective of the reformers, Egerton's bequest epitomized the problem. When one of them, the astronomer John Herschel, was invited to take a part in the intended work, he declined. He was unhappy, he reported, with the idea of writing such a work for "pecuniary reward," but he was even unhappier about how the society was handling so valuable a "windfall." To Herschel, it was

> an opportunity of calling forth to something like lucrative exertion the talents of men, who with *real science* and *irreproachable character*, have their zeal chilled and their sphere of utility contracted by the "res angusta domi." To such persons [ . . . ] who live, or rather starve on their science, but who prefer hunger in that good cause to competency in a less dignified calling, a thousand pounds [ . . . ] would indeed be a more material and noble assistance.[4]

3. Miller 1981; Malet 1977, 100–104.
4. Herschel to D Gilbert, 1 July 1830, in Enys 1877, 6–7.

If Bridgewater's bequest were to be allowed at all, it must be used to support the working men of science on the basis of merit and need rather than as a species of patronage with which to curry favor or reward service. Of course a similar desire for accountability was abroad in the nation more generally. As politicians debated the Reform Act through 1831, newspapers and magazines were not slow to question the character of Bridgewater or the management of his bequest.

## *The Trouble with Bridgewater*

It was in October 1823, following his unexpected accession to the title of Earl of Bridgewater, that Egerton became for the first time well known to the British public in the character of an "eccentric." Widely reprinted columns in the newspapers of France and Britain reported sensational accounts of his oddities. Frail, and having apparently suffered a stroke, his carriage had to follow behind him the whole way whenever he walked through the streets of Paris, where he lived. He had a pair of boots for each day of the year. Having had a dream that the devil had possessed him, he had commissioned a hundred-foot effigy of Satan for his servants to attack. By December 1823 a West End printmaker had produced a satirical "View near Bridgewater," depicting the new earl as a young man with a large nose and protruding jaw, his long hair tied in a plait beneath a French-style hat and his riding habit buttoned beneath a fashionable Spencer. The image contrasted with the sober engraving reproduced the previous year from what was reported to be one of Baron Gérard's best portraits (fig. P.1).[5] To radicals such as Richard Carlile, Egerton appeared as an outrageous "Specimen of a Pampered Aristocrat." To others he was something to enter into the annals of eccentricity. As a paragraph that appeared in many of the newspapers put it, his actions were "well calculated to enrich the history of human oddities." Journalist and compiler John Timbs later obliged, using these newspaper reports to give the earl lasting notoriety in his *English Eccentrics and Eccentricities* (1866).[6]

A discourse of eccentricity was something new in late Georgian Britain, and its defining characteristic was the transgression of the boundaries by

5. George 1952, 383; *Morning Post*, 27 December 1821, 2b, and 6 March 1822, 2d; *The Times*, 5 November 1823, 3b, and 7 November 1823, 3a.

6. Timbs 1866, 1:111–13; Falk 1942, 185–218; "The Earl of Bridgewater," *The Times*, 18 October 1826, 2c (and widely reprinted); "Specimen of a Pampered Aristocrat," *Republican*, 20 October 1826, 477.

FIGURE P.1 Two contrasting portraits of Francis Henry Egerton. "A View near Bridgewater" (hand-colored etching by William Heath, 1823) and "Francis Henry Egerton, Earl of Bridgewater" (stipple engraving after Gérard, 1824). Image: © The Trustees of the British Museum. All rights reserved.

which social and cultural life was ordered. But to conclude from the eccentricity of Egerton's dotage that his bequest was a caprice of no larger significance would be a mistake. While his transgressive behavior was well suited to titillate newspaper readers, the English press conceded that reports about the elderly Egerton were "probably overcharged." Moreover, his highborn life of scholarly dilettantism, collecting, and posthumous patronage was quite characteristic of a certain kind of fellow of the Royal Society under Banks's command. While his distant cousin Sir Egerton Brydges was well aware of the unpalatable eccentricities of his later years and found him "vain, insolent, ostentations, and insufferably proud," he nevertheless considered Egerton to have had "many faculties of intellect; a vast memory, and much erudition."[7] His bequest reflected a culture of aristocratic privilege, learned interests, and lavish patronage.

Egerton's family were deeply rooted in the patronage networks of eighteenth-century England so that, having been educated at Eton and

7. [Brydges] 1832, 298; "The Earl of Bridgewater," *The Times*, 18 October 1826, 2c. On eccentricity, see Carroll 2008, 7.

Oxford, he soon had his pick of ecclesiastical plums. In 1780, at the age of just twenty-four, he became a Canon of Durham Cathedral through the influence of his father, who had become Bishop of Durham. This was a sinecure that brought him around £1,900 annually, and two years later his father's cousin, the third Duke of Bridgewater, presented him to one of the "fattest and most productive" church livings in England, worth almost £1,500. Declining a "High Station" in the church in 1796, he soon added a further family living, worth £1,000 per year. These appointments together made him a very rich man. Indeed, when rumors of his death began to circulate early in 1829, it was reported that they had "thrown the Church into a state of ferment and bustle" on account of his "rich pluralities."[8]

For Egerton, this handsome patronage did not represent an invitation to earnest parish duty. Rather, he employed curates to perform his clerical duties while devoting himself to the literary pursuits of a leisured gentleman. He was elected to the Royal Society in 1781 on the basis of his broad scholarly interests and wealthy connections and to the Society of Antiquaries a decade later.[9] In subsequent years he produced a trickle of unremarkable works of historical, classical, and literary scholarship as well as writings on the practical arts. First came a biography of one of his illustrious ancestors, the Elizabethan Lord Chancellor Thomas Egerton, that was initially written for a biographical dictionary but went through successive private printings over following decades, becoming bloated by extraneous documentary materials. Then there was an opulent but derivative edition of the *Hippolytus* of Euripides, printed by Oxford University Press. Changing tack, he published a literal translation into French and Italian of an important manuscript copy of Milton's masque *Comus* that he had found in family papers and saw as evidence of the family's importance as literary patrons.

Egerton's most valuable publication was arguably an account of the innovative inclined plane built for his father's cousin, the third Duke of Bridgewater, to carry barges between the different levels of the underground canals in his mines under Walkden Moor in Lancashire. The duke had come to be seen as the "father of Inland Navigation" on account of his scheme to link the coal mines of his manor of Worsley with Manchester and Liverpool using a groundbreaking summit-level canal. Egerton for a spell became his chief companion, living at his London mansion in the

8. "The Mirror of Fashion," *Morning Chronicle*, 26 January 1829, 3a; Egerton [1821], 88–89; *The Times*, 1 November 1826, 2c; *Report into Ecclesiastical Revenues* 1835, 29–30, 68–69, 514–15, 532–33.

9. Election Certificate, RSL, EC/1781/16.

heart of St. James and accompanying him to the ancestral seat of Ashridge, Hertfordshire. With the technical assistance of the duke's chief mining engineer, Egerton drew up an account of the inclined plane and submitted it to the Royal Society of Arts in 1800. It was published in the society's transactions and reprinted in the *Journal of Natural Philosophy* and in the *Annales des arts et manufactures*, establishing Egerton's credentials as an authority on the duke's canal works. This gave the cleric confidence, in later years, to pen increasingly eccentric tirades concerning the extent to which British industrial success depended on aristocratic patronage.[10]

The canal duke's death in 1803 brought Egerton a huge legacy of £40,000, although that was dwarfed by the fortune inherited by his older brother, who now became the Earl of Bridgewater and one of the largest landowners in the country.[11] Shortly afterward, however, Egerton's life took another unexpected turn. Like many wealthy Britons, he had traveled to Paris following the Treaty of Amiens in 1802, and the resumption of hostilities left him stranded there. Unlike most, he retained his liberty, apparently through the favor of Napoleon's wife Josephine, whom he had flattered with the gift of the first English barouche in Paris. In 1806 the Royal Society president Sir Joseph Banks, who was "a particular friend" of Egerton's brother, triumphantly brokered his return to Britain. Egerton was grateful enough to remain in London for a year or two, taking the opportunity to issue an anonymous millenarian pamphlet associating Napoleon and his campaigns with the "end times."[12] However, he soon returned to France, claiming a dispensation from his clerical duties on the grounds of ill health. In truth, he had become romantically embroiled in Paris with a twenty-year-old Catholic girl "of a great family," and he ultimately fathered several illegitimate children there. He also devoted himself to collecting historical manuscripts, traveling widely on the Continent even as the war raged. Settling to a comfortable life in Paris, he entertained leading scientific and literary figures while privately issuing a series of increasingly eccentric elaborations on his earlier scholarly productions.[13]

Following the Bourbon restoration, Egerton purchased the magnificent

10. Egerton [1819–20], 47–57; Egerton [1809], 1; [Carey] 1795, 13.

11. "Value of Estates," *Caledonian Mercury*, 18 October 1813, 2e; *Morning Post*, 7 April 1814, 3b.

12. [Egerton] 1808 (for the attribution, see Egerton [1819–1820], 89); "Anecdote," *Morning Post*, 11 November 1823, 3b; Banks to Lacépède, 24 May 1806, in De Beer 1960, 151.

13. Coult 1980, 176; Egerton [1828?], 5. See also Alger 1904, 191, and Falk 1942, 186. Five volumes of letters in the BL (Eg.60–64) detail his remarkable social life at this period.

Hôtel de Noailles in rue Saint-Honoré for the colossal sum of £26,000, renaming it the Hôtel Egerton. Moreover, while most of his brother's vast fortune passed elsewhere following his death in 1823, the new earl inherited a life interest in a portion of the estates as well as an annuity of £18,000, giving him an enormous annual income of approximately £40,000.[14] A man now grown as wealthy as he had grown odd and infirm, it is not surprising that he attracted widespread newspaper attention. His public notoriety was such that in 1831, one delusional optician's son, arrested late at night in Palace Yard, opposite the Houses of Parliament, fancied himself to be the Earl of Bridgewater on his way to see "his cousin the Archbishop of Canterbury." Yet the reports that greeted Egerton's death in February 1829 were divided. While some reminded readers of his "singularities," others followed the French government newspaper, the *Moniteur universel*, in reporting that his publications had "acquired him a reputation throughout Europe."[15]

Whatever his reputation, it was Egerton's bequests that gave him lasting recognition. His death was met with immediate newspaper speculation concerning his will. While reports related that his wealth had been exaggerated, Egerton's estate was nevertheless valued at £70,000. Alongside the £8,000 bequeathed for a work on the "Power, Wisdom, and Goodness of God," he left a colossal £4,000 for a monument to his own memory at the family vault, depicting a female figure surrounded by a stork, a dolphin, and an elephant, her right hand resting on a book inscribed "the works of creation." Much of his will was devoted to the bequest of his treasured manuscript collection to the British Museum, where it became one of the foundation collections of its manuscripts department, along with the especially valuable bequest of £12,000 and additional property for its upkeep and extension. There were various bequests for All Souls College, Durham Cathedral, and his former parishes, but most of the remainder of the fortune was to be devoted to the aggrandizement of his ancestors, with £8,500, together with the residue of his estate, left for the construction of obelisks in memory of his parents and the canal duke and for the embellishment of the tomb of Lord Chancellor Egerton.[16]

Egerton's will echoed the preoccupations of his life in its self-

14. Coult 1980, 183–87.

15. *Hampshire Telegraph*, 23 February 1829, 3c; "French Papers," *Morning Chronicle*, 17 February 1829, 3a; "Police," *The Times*, 4 November 1831, 4c.

16. Edwards 1870, 446–59; "The Late Earl of Bridgewater," *Standard*, 27 April 1829, 4d; *Gentleman's Magazine* 99, pt. 1 (1829): 560. Egerton's will is in the NA, PROB 11/1754/3. Documents relating to the executorship are in the HA, AH2487–2693.

aggrandizing commitment to scholarship, the practical arts, and the role of aristocratic patronage in fostering both. His self-conception as the latest in a long line of illustrious aristocrats led him to use his wealth to maintain a particular social vision, namely, that God's providential plan for human society entailed a hierarchy in which each rank was ordained to discharge certain duties. His fervor in making this claim had grown ever stronger as he experienced events in postrevolutionary France. In 1808, for instance, his millenarian pamphlet *John Bull* began by arguing that the hierarchical arrangement of human society was clear testimony to divine providence. Later, other publications dwelt on how far France had been damaged by turning its back on God's providential provision while elaborating on how his own family's history of patronage was the fulfilment of its divinely instituted role.[17] In the same spirit, Egerton's acts of testamentary patronage were intended in themselves to offer a vindication of the divinely ordained social hierarchy.

Egerton hoped that his bequest to the president of the Royal Society would ensure that his symbolic gesture resulted in an explicit statement of the providential character of a hierarchical social system. Such views were, of course, common in the British reaction to the French Revolution, and they received a conspicuous airing in Paley's lastingly popular *Natural Theology*. As an active clergyman in the industrializing North of England, Paley had found himself in close quarters with working people stirred to political action by events in France, and soon after the revolution he issued a pamphlet outlining *Reasons for Contentment, Addressed to the Labouring Part of the British Public* (1792). Egerton's experience was naturally rather different. By the time his mind turned to a work on natural theology, he was living in France as a leisured "milord."[18] He was thus rather insulated from the growing tide of anticlerical or even atheist street literature that emanated from British working-class radicals in the troubled years after the Battle of Waterloo in 1815. Nevertheless, his literary bequest was clearly motivated by a similar concern with defending Britain's social hierarchy from the fallout of the French Revolution.

Alongside Egerton's wider social and political agenda lay other motivations. A significant factor was probably a desire to make some recompense for his long and disgraceful career of clerical absenteeism. For while he showed little compunction in neglecting his parochial duties, he sometimes exhibited a sense of clerical vocation in his scholarly projects. Such a sense was on display in the 1790s, when his response to the possibility

17. Egerton [1821], 86n–87n; [Egerton] 1808, col. 3n.

18. *Republican*, 20 October 1826, 477; Turner 1993, 105–9.

of further ecclesiastical patronage had been to produce an edition of the *Hippolytus* of Euripides, purportedly with the intention of calling away the minds of "talented young men" from the vices of an "untrustworthy and heedless age." He observed,

> For how long am I to fill my mind with pleasure [ . . . ] if noble and honest young men from boyhood are often indoctrinated by unwholesome and unsuitable nebulous doctrines, by the foul and misinformed ravings of licentious poets; how am I to turn them away then, from the daring, impious, and delirious ravings of counterfeit philosophers and turn them in the direction of those healthier teachings of true knowledge which have been handed down to us by the auspices of Greek records and disciplines?

There was more than a hint of autobiography about this. As he later recalled, he had read a great deal about "ancient atomic physiology" in his youth. "Stumbling about and floundering in this 'dreary wild,'" he continued, he had been rescued by a "beaming rule of light" that illuminated a path out of the "joyless labyrinth," leading him "to a First Cause."[19] His bequest for a work on the "Power, Wisdom, and Goodness of God" reflected both this firsthand experience of religious skepticism and Egerton's consequent sense that such a work would be helpful to subsequent generations of wealthy young men.

Egerton had, in fact, put pen to paper on the subject himself, but the results hardly amounted to what the *Gentleman's Magazine* referred to after his death as a "splendid work." Rather, his observations concerning the proper means of answering atheism appeared in a fragmentary forty-page footnote to one of the eccentric addenda to his edition of *Hippolytus* that he issued while in Paris. This was "an unfinished Copy, requiring much revision," he reported, and he was already looking for a "favorable opportunity" for others to develop his arguments further by means of a bequest. Since the proposed work was intended to inculcate religious principles in Britain's cultured elite, Egerton was already planning for it to be administered by the president of the Royal Society, and he initially had in mind that a second work would be administered by the Institut de France. His approach to the subject was naturally through the ancient Greek atomism that had troubled his own youth. If the ancient atomists had been opposed by pertinent questions, Egerton suggested, they must

19. Egerton [1819–20], 88; Egerton [1821], 150; Egerton 1796, vi (translation courtesy of Sophie Weeks).

ultimately have had recourse to a first cause. This was the path by which the young Francis Egerton had originally escaped the "Labyrinth," and it was the pearl of wisdom that he sought to pass on to others. However, he had little to say about modern atheists, whom he regarded as merely "subalterns [ . . . ] who only hashed and served up again, and badly too, what better cooks have dressed."[20]

While Egerton's own focus was thus rather narrowly on answering Greek atomism, he hoped that those employed on his behalf would range further afield in their argumentation. When he first made provision in his will for a work of natural theology in 1813, the wording was very similar to the later rubric of his bequest. He prescribed a work on

> The Power, Wisdom, and Goodness of God, as manifested in the Creation, illustrating such work by all reasonable arguments; as, for instance, the variety and formation of God's creatures in the animal, vegetable and mineral kingdoms; the effect of digestion, and thereby of conversion; the construction of the hand of man; and an almost infinite number of other rational arguments; and also by discoveries ancient and modem; and generally by whatever may serve the purpose of the writer, or writers, in arts, in science, through the whole extent of literature, erudition, and *omne scientibile,* in setting forth, displaying, evidencing, manifesting, illustrating, [and] proving the Power, Wisdom, and Goodness of God in the Creation.[21]

The commission thus included arguments that drew on the natural sciences, but as in many eighteenth-century works of natural theology these were to sit alongside arguments drawn from history and literature, in keeping with Egerton's own scholarly interests. Moreover, such a conception was entirely consistent with his Banksian sense of the wide remit of the Royal Society.

Egerton provided further hints in his own writings as to the kinds of additional arguments he hoped to see developed in response to his bequest. Indeed, he had first outlined the argument from "the variety and formation of God's creatures" in 1808 in his millenarian pamphlet *John Bull,* where its social implications were fully evident. The infinitude of God's power was evident in the great variety of created things, he claimed, and that of God's goodness in its consequences for the "Moral Government

20. Egerton [1821], 38–43, 147–88; *Gentleman's Magazine* 99, pt. 1 (1829): 560. See also Egerton [1819–1820], 87n–89n.

21. Egerton [1821], 39; Egerton [1821?], 10.

of the World." Humans were all "unequal, unlike, and different," and from their different relations arose their different "moral and civil duties." Elaborating on the argument in the addenda to his edition of *Hippolytus,* he made clear its social consequences. He had heard much, he observed, of the word *égalité,* but while he was happy to accept the notion that all men were equal before the law, he considered that if the term were "tortured" into denying variety, it was not only "absurd" and "contrary to the constitution of things" but also "in opposition to the will of God."[22] God's goodness was evident, Egerton argued, in the stratified social system of which he was a leading beneficiary.

The second suggested topic, the "effect of digestion, and thereby of conversion," was intended to provide an argument for the immortality of the soul. However, Egerton admitted that his own account of it was no more than a cursory hint on which others might expand if they wished. God, he observed, had "imposed upon nature a law whereby all body should die and suffer conversion and digestion," but no such law seemed to apply to the soul. In short, the process of digestion emphasized the disanalogy of matter and spirit and the eternity of the latter. This argument was not, on the whole, scientifically informed, although Egerton referred readers to Alexander Hunter's lengthy disquisition on plant physiology in his edition of John Evelyn's *Silva; or, A Discourse of Forest Trees* (1776). He did, however, have one complaint to make about the scientific study of digestion, namely, that too many were inclined to attribute it to a purely mechanical cause: "One, to the one cause only of solution by the gastric juices; another to trituration and the compression of the coats of the stomach."[23]

A similar complaint of partial reasoning among scientific men attended the subject of Egerton's third recommended topic, "the construction of the human hand." Egerton agreed with Galen that the hand gave humanity dominion over the world and argued that its extraordinary construction clearly suggested design rather than chance. However, he complained that some, "finding a similar Anatomical configuration in the hand of the Oran-Outang" had "confounded [ . . . ] comparative with human anatomy." The reference is too imprecise to be sure what modern target Egerton had in mind, but he was clear that it was foolishness to think of the hand in Epicurean terms as being the chance product of a "Hurry-Scurry of Nimble atoms."[24]

Egerton's final example, "discoveries ancient and modern," derived

22. Egerton [1821], 186 (see also 148–149); [Egerton] 1808, cols 2n–4n.

23. Egerton [1821], 38, 40.

24. Egerton [1821], 38–39, 157. On the orangutan in eighteenth-century debates about humans, see Sebastiani 2019.

from his interest in the role his family had played in the history of inland navigation. In the fragmentary third part of his *Letter to the Parisians and the French Nation upon Inland Navigation*, he described how the ancients had considered inventive genius to be of "Super-Human" origin and had "deified many who were supposed to have been the inventors of useful arts." According to Egerton, however, "Extra-Ordinary Talent" was "*given* by God *our Father* in compassion to Fallen man, for the purpose of bettering the condition of the Whole Race of man, in this state of Probation." Thus, technological discoveries were to be seen as evidence of divine goodness rather than as grounds for human pride. Those who thought too highly of their own intellectual abilities were inclined to fall into error, with some even being led to boast that "the Light of Reason, Alone, is All-Sufficient, without That, Too, of Revelation!"[25] By contrast, Egerton considered that a proper appreciation of the divine origin of discoveries like those that his family had been involved in patronizing would foster true religion.

As a classically educated clergyman keen to answer the Epicurean atheism that had disquieted him as a young man and to repel critics of the social inequality of Christian Europe, Francis Egerton thus offered suggestions for his work on the "Power, Wisdom, and Goodness of God, as manifested in the Creation" that had relatively little direct bearing on the developing sciences. He was, in many ways, out of kilter with the spirit of the age. On one point, however, his concerns exactly matched those of a growing number of contemporary critics of science. "Certain philosophers," he noted,

> ascending into the bright regions of Science, have resolved their inquiries into natural causes; and, having thus far mounted, they stop there: so it is, they say, in nature. GOD is the Author of nature, they will reply; and therefore, when we speak of nature, we mean only to speak of the Author of nature [ . . . ]. Why must it remain to be gathered by implication? Does not every thing centre in the Author of nature, in the one great First Cause? Does it not behove the creature, man, to finish in his CREATOR, and his GOD?[26]

As we have seen, this sense that scientific savants were lacking in Christian piety was a key concern in the age of reform, and it was one that Egerton's

25. Egerton [1823?b], 103; Egerton [1823?a], 152–55. See also Egerton [1821], 37–46, 171.

26. Egerton [1821], 40–41.

bequest was to have the effect of addressing through the appointment of leading men of science who contributed to rebaptize the sciences. In the process, moreover, it brought his name more prominently and honorably before the public than it had ever been in life. But this outcome was not achieved without some difficulties along the way.

## *Patronage and the Royal Society*

When the news of Lord Bridgewater's bequest reached Royal Society president Davies Gilbert in February 1829, it cannot have been very welcome. A Tory Member of Parliament who for many years had been heavily involved in scientific administration, Gilbert had been Banks's chosen successor as president but had himself supported the election of his protégé Sir Humphry Davy. During Davy's seven-year presidency, Banks's "learned empire" had started to crumble and, while the new man had broadly supported attempts to make the society more strictly scientific, his final year had witnessed concerted pressure for reform from those who favored a more meritocratic approach to science. Taking over from Davy at the end of 1827, the conservative Gilbert had encountered similar agitation. In particular, his handling in 1828 of the award of the Royal Medals that had been created three years before by George IV suggested to reformers that the society was more interested in honoring old friends than in rewarding scientific distinction. As tensions mounted concerning the direction of the society and its role in fostering scientific excellence, the Earl of Bridgewater's bequest can only have appeared to Gilbert as a hostage to fortune. Above all, it was made to the president ex officio "without any control" rather than to the society's council, and to some it thus seemed redolent of the former autocratic regime (fig. P.2).[27]

The president acted immediately to consult Bridgewater's nephew by marriage, the politician and former member of the Royal Society council Charles Long, Lord Farnborough, who ensured the family's concurrence in the matter. After the will was proved in April, the newspapers reported that the money was "to be appropriated as a reward for the best written Essay on the Creation, on the Anatomy of Man, and especially on the powers, formation, and properties of the Hand." Gilbert immediately began to receive applications from prospective authors. By that date, however, he had made an astute move, designed to forestall criticism of his management of the patronage. He had secured the assistance of the Archbishop of Canter-

27. "Royal Society," *Literary Gazette*, 5 March 1831, 153–54, on 153; Hall 1984, chap. 2.

FIGURE P.2 Meeting of the Royal Society at Somerset House in the Strand. Engraving by H. S. Melville, 1844, after F. W. Fairholt, 1843. Image: Wellcome Collection. Public Domain Mark.

bury William Howley and the Bishop of London Charles James Blomfield with a view to "placing the whole transaction above even the suspicion of favouritism or partiality." Involving the bishop and the archbishop clearly emphasized that the matter had nothing to do with the Royal Society's council. Indeed, the archbishop's domestic chaplain George D'Oyly advised Gilbert, "whatever members of the Royal Society you may consult with, you ought to reserve entirely to yourself the disposal of the bequest according to the will of the donor."[28]

28. Gilbert to JFW Herschel, 29 June 1830, D'Oyly to Gilbert, 14 May [1829], and Long to Gilbert, 14 February [1829], in Enys 1877, 3–6; *Philosophical Magazine*,

Having done his best to defuse the issue, Gilbert now ignored it for a year, perhaps awaiting the release of funds. Meanwhile, discontent in the society reached new heights, with the publication in April 1830 of Babbage's *Reflections on the Decline of Science in England*. Mixing matters of principle with those of the author's personal pique, the work laid most of the blame for the supposed decline of English science at the door of the Royal Society. Babbage's central claim, that the Society was retarding science by admitting nonscientific fellows and that government support for science was preferable to that of private patrons, was one with which many of the younger fellows agreed. His other claims, particularly those respecting the abuse of patronage, were more debatable, and many reforming fellows considered Babbage's attacks unnecessarily intemperate. Nevertheless, the book provoked a significant amount of public linen washing, and by August rumors were beginning to circulate that the new king's brother—Augustus Frederick, Duke of Sussex—was likely to become the society's new president. Gilbert sought to smooth the transition to the new presidency, but Herschel was persuaded to stand as a scientific candidate, sparking further controversy and a close election late in November in which the duke's succession was finally secured. This seemed to many scientific reformers a retrograde step, but under the liberally minded duke the management of the society nevertheless underwent a process of gradual reform.[29]

It was in the midst of these tumultuous events, between May and September 1830, that Gilbert organized the disbursement of Bridgewater's bequest. By August, the likelihood of his imminent resignation was giving a particular sense of urgency to proceedings. Fortunately for Gilbert, however, none of the public critics of the society's handling of patronage extended their observations to this new and handsome prize, allowing Gilbert and his advisers to proceed according to their own lights.

Bridgewater's will had stipulated that the president of the Royal Society should nominate a "person or persons" to write—in the singular—"a work." Responding in May 1830 to an application for a share in the project from the young Irish mathematician John Thomas Graves, Archbishop Howley told Gilbert that he did not think Bridgewater had intended "to give a prize to an ingenious young man" but rather "to encourage the publication of a work or works which might be really useful to the

---

2nd ser., 9 (1831): 202; *Standard*, 27 April 1829, 4d. See also Brock 1966. The bulk of the original correspondence reproduced in Enys 1877 is in ESRO, GIL4/74–84, although some is elsewhere, including in a private collection.

29. Hall 1984; Miller 1981; MacLeod 1983; Babbage 1830; Secord 2014, chap. 2.

public." This, he considered, might be achieved by Gilbert "selecting a certain number of eminent persons, and desiring them to form outlines of a plan, the several parts of which might be filled up by them respectively, as they might agree amongst themselves." Gilbert initially considered the appointment of three authors but by the end of June had settled on "dividing the Work into eight Treatises by separate Authors under their respective names." As the Scottish natural philosopher and critic of the Royal Society David Brewster later observed in a caustic review, the number of the authors was doubtless suggested by the magnitude of the reward—each author to receive a round £1,000. The initial plan was that these would be published as a single work in two reasonably sized volumes, and it was much later that the question of their separate publication arose.[30]

The only difficulty the archbishop anticipated was that he was little acquainted with "the names of persons qualified for such a work." One point that they readily agreed on was that the authors should have significant scientific as well as literary credentials. This became clear during the course of the summer, when names were suggested to the selectors by those involved in the executorship of the will. Neither Lord Farnborough's nomination of his friend the artist Edward Hawke Locker nor banker and executor Jean-Charles Clarmont's nomination of High Church cleric Samuel Wix were considered favorably, despite the respectability of the individuals concerned. Locker was a fellow of the Royal Society and a commissioner of Greenwich Hospital and had edited with publisher Charles Knight one of the earliest cheap magazines, the *Plain Englishman* (1820–23). Yet as Howley remarked to Gilbert, while he was "a very useful writer for the lower classes," it was questionable "whether his depth of knowledge is such as to qualify him for the production of a book which would stand the test of professional criticism." Similar objections applied to Wix, the vicar of St. Bartholomew the Less in London, whom the archbishop described as "an ingenious and most respectable man." He was not, Howley indicated, someone whom, "all partiality apart," he would select "on account of his talents, or scientific acquirements," and his opinion was shared by the bishop.[31] By contrast, the authors who were finally chosen were all well established in their respective scientific fields, and most were genuinely "eminent."

30. *Edinburgh Review* 58 (1834): 425. Howley to Gilbert, 8 June 1830, Gilbert to Herschel, 29 June 1830, and PMR to Gilbert, 17 August 1830, in Enys 1877, 4–6, 8.

31. Howley to Gilbert, 26 August and 11 September 1830, Blomfield to Gilbert, 28 August 1830, and Farnborough to Gilbert, 7 and 10 August 1830, in Enys 1877, 7–8, 10–12, 15.

FIGURE P.3 Peter Mark Roget. Lithograph by W. Drummond, ca. 1836, after E. U. Eddis. Image: Wellcome Collection. Public Domain Mark.

Gilbert had early sent the archbishop a work by Peter Mark Roget (fig. P.3), a London physician and his close associate and supporter as secretary of the Royal Society. This was probably Roget's *Introductory Lecture on Human and Comparative Physiology*, which had been published by Longmans in 1826, and Howley was enthusiastic. On his own account, the archbishop reported that his only close scientific acquaintance was the geologist and surgeon John MacCulloch, and that he had been impressed by reading part of the manuscript of MacCulloch's *System of Geology*. However, this was a work written before 1824, and by the time MacCulloch managed to publish it in 1831, its focus on mineral analysis had been rendered distinctly outmoded by the new emphasis on fossil-based stratigraphy.[32] These early discussions were typical of the haphazard proceedings by which Gilbert and his advisers put together the list of eight names and subjects. Sometimes what came first were the names of individuals they thought might be suitable authors or who perhaps they wished to reward. In other cases, ideas for subjects prompted a search for appropriate authors. This was especially the case with Bridgewater's nominated topics of "the hand" and "the function of digestion" but also applied in developing areas of modern science where it seemed especially desirable to have theological reflection. However, there was no very carefully worked out plan that would ensure systematic or integrated treatment.

Gilbert evidently had Roget in mind as an author from the start, and it is difficult not to see the nomination as reflecting his desire to reward his loyal lieutenant. Nevertheless, Roget, who was now in his fifties, was a highly active and respected figure in scientific London. He had lectured on physiology at a range of scientific institutions (and since 1826 at the private Aldersgate Street medical school) and had in 1823 written the lengthy article on physiology for the supplement to the *Encyclopædia Britannica* as well as successful treatises for the Society for the Diffusion of Useful Knowledge. He was thus a reasonable choice for what Gilbert called "Physiological or Comparative Anatomy." The second invitation went to another individual prominent in the Royal Society debacle of 1830. Roget's immediate predecessor as secretary, the astronomer John Herschel, had resigned at the time of Gilbert's election partly because he disapproved of the new president's views respecting patronage.[33] While Gilbert must have hoped that his nomination would ease the tensions with this prominent reformer, the whole exercise only served to confirm Herschel's sense of Gilbert's failings. He was nevertheless a natural enough choice. Aged thirty-eight, he was one of the country's foremost astronomers and natural philosophers and had manifested his ability to write for a wide audience in his book-length contributions on "Physical Astronomy," "Light," and "Sound" to the *Encyclopædia Metropolitana* (a project that Howley had fostered as a counterblast to the secularity of the *Encyclopædia Britannica*). Indeed, not only was he at this moment preparing his *Preliminary Discourse on the Study of Natural Philosophy* for Dionysius Lardner's Cabinet Cyclopædia, he had already also agreed to write the *Treatise on Astronomy* (1833) for the same series, which may well have influenced his response.

In view of the strong tradition of Newtonian natural theology, astron-

32. Howley to Gilbert, 8 June 1830, in Enys 1877, 4–5.
33. Todd 1967, 228.

FIGURE P.4 William Whewell. Lithograph, 1835, after E. U. Eddis. Image: Wellcome Collection. Public Domain Mark.

omy was an obvious topic for inclusion in the planned work. However, the Astronomical Society was a hotbed of discontent, and it was perhaps unlikely that such stalwart critics of the Royal Society as the astronomers Sir James South or Francis Baily would be invited to contribute. Herschel's own suggestion—his penurious protégé, the Scottish natural philosopher and school rector William Ritchie, who had just begun lecturing at the Royal Institution—was not sufficiently eminent in either scientific or literary terms. Instead, it was Bishop Blomfield's suggestion that was quickly adopted. Blomfield was a talented classical scholar who had maintained close ties with Cambridge friends, and his mind went to a fellow of his old college, the professor of mineralogy William Whewell (fig. P.4). Aged thirty-six, Whewell was the youngest of the appointed authors, but he had a growing reputation as the author of mathematical textbooks and mineralogical publications, as an active participant in London's scientific societies, and as a clerical commentator on the sciences. As one new student reported in 1828, he might have a name "more easily whistled" than pronounced (leading some students to dub him "Billy Whistle"), but he was a "very consummate man" whose sermons were considered "magnificent" on all sides, and he was adept at verse and classical scholarship as well as physical science.[34]

The archbishop's first nomination, the Oxford reader in geology and mineralogy William Buckland, was likewise an Anglican clergyman. Perhaps Gilbert's response to the archbishop's initial mention of MacCulloch had been discouraging, but the triumvirate nevertheless considered that the recent rise to prominence of geology required the subject's inclusion in the planned work. In his midforties, Buckland was one of the country's foremost geologists, and he had won the Royal Society's Copley Medal in 1822 for his groundbreaking reconstruction of fossil remains in a Yorkshire cave. As Oxford's first geological reader, his highly engaging lectures had attracted the attention of many of the university's senior members. He was, moreover, a canon of Oxford's Christ Church cathedral and the author of two works that sought to explain "the connexion of geology with religion," and his appointment was readily agreed.

Next came Gilbert's suggestion of the distinguished surgeon Charles Bell to write on "Human Anatomy including of course Lord Bridgewater's favorite topic the Human Hand." Bell had, Gilbert reported, "eminently distinguished himself by writing on the Nervous System." Indeed, his pioneering work on its functional anatomy had led to his being awarded a Royal Society Royal Medal the previous year. He was also a seasoned medical lecturer, having long taught at one of London's leading private medical schools before becoming one of the founding professors at the new London University. Moreover, he had recently written a popular two-part treatise on "proofs of design in the animal frame" for the useful knowledge society.[35]

With four authors and topics thus agreed, Gilbert now pondered "whether Light including all the recent discoveries of Polarization; with the other imponderable Fluids might not form a proper division." Two authors came to mind, but the first was not known to him by name. Gil-

34. AH Hallam to WE Gladstone, November 1828, in Kolb, 1981, 244; Tennyson 1897, 1:39; Herschel to Gilbert, 1 July 1830, in Enys 1877, 6–7.

35. [Bell] 1827–29; Gilbert to Howley, 23 August 1830, in Enys 1877, 8–9.

bert had heard that a "gentleman of Cambridge" who had distinguished himself in the *Transactions of the Cambridge Philosophical Society* would be "well qualified to execute the task." The candidate in question was George Biddell Airy, who had recently been appointed Plumian Professor of Astronomy and director of the Cambridge observatory. Airy's optical papers were already a matter of discussion in the precincts of Somerset House and earned him the Royal Society's Copley Medal the following year. However, he was only twenty-nine and was not yet a fellow of the society (perhaps because of his limited funds), so he was probably insufficiently eminent. In any case, after the Bishop of London consulted with Whewell concerning his identity, the suggestion was quietly dropped.[36]

Gilbert's other suggestion, David Brewster, was altogether more proven. In his late forties, Brewster's internationally renowned researches in optics had been financially sustained by his editorship of the *Edinburgh Encyclopædia* (1808–30) and by his journalism, including the editing of his own philosophical journals. It was Brewster who contributed the *Treatise on Optics* (1831) to Lardner's Cabinet Cyclopædia, and he was awarded the society's Royal Medal for his optical researches later in the year. Yet he was not in the end nominated to the Bridgewater project, and one might wonder whether Gilbert, in the midst of his Royal Society crisis, had heard that Brewster was penning a lengthy essay expanding on Babbage's *Reflections on the Decline of Science in England* for publication in the October number of the *Quarterly Review*.

It seems, however, that the shifting organization of the subject matter was the issue. While Gilbert's ecclesiastical advisers approved of light as a theme, they also had other suggestions, including chemistry, natural history, and "the adaptation of the *Physical constitution* of man to his intellectual and moral faculties, his social propensities, and the provision made for his wants in the works of nature." Blomfield suggested that this last subject would be best divided into two departments—the intellectual and the moral—and Gilbert was thus left trying to shoehorn light and chemistry together, suggesting "Chemistry chiefly in reference to the Etherial or Imponderable Fluids or especially to Light, including Optics with the recent discoveries of Polarization." In this new form, the topic suggested the name of the physician William Prout who, in his midforties, was regarded by some as "the first chemist of the metropolis." Blomfield thought

36. Gilbert to Howley, 23 August 1830, and Blomfield to Gilbert, 18 September 1830, in Enys 1877, 8–9, 15–16.

him very able but confessed that he had not seen his publications.[37] This is not surprising since Prout's only significant publication other than his rather abstruse memoirs and papers was a successful professional work on urinary pathology. However, it was perhaps his work on metabolic chemistry, which earned him the Copley Medal in 1827, that gave him the advantage over Brewster, since it meant that he was able to deal with the Earl of Bridgewater's theme of "the function of digestion."

The bishop's suggested theme of the human constitution came with names attached, and it may be that it was inspired in part by the interests of authors who were considered desirable. For the physical discussion, the suggested choice was between Oxford's Regius Professor of Physic John Kidd and the London surgeon Herbert Mayo, who had just been appointed the inaugural professor of anatomy and physiology at King's College London. Both Howley and Blomfield were heavily involved in the founding of this new Anglican establishment, and the archbishop had reported early on that he had been impressed with Mayo's *Outlines of Human Physiology* (1827). The surgeon was a former pupil of Charles Bell, and his treatise was based on physiological lectures that he had delivered at Bell's Great Windmill Street medical school. Yet while he had conducted important investigations into the physiology of the nervous system, he was at the age of thirty-four still only moderately distinguished. In any event, Howley came to prefer Kidd, a man twenty years Mayo's senior.[38] Kidd had played a major role in reviving scientific education in Oxford, and in his inaugural lecture as Regius Professor of Physic he had sought to show how comparative anatomy could provide additional evidence of design. He was thus someone who could be trusted to write authoritatively on the topic.

For the "*intellectual* and *moral*" aspects of the human constitution, Blomfield suggested the prominent Scottish clergyman and Edinburgh's professor of divinity Thomas Chalmers. Howley had noted, overlooking Bell (who had been in London for nearly three decades) that it would be advisable to include at least one author who was Scottish, and he and Blomfield had a preference for Chalmers.[39] The Scot was a well-regarded communicator, and his immensely successful discourses on astronomy

37. Gilbert to Blomfield, 2 September 1830, Blomfield to Gilbert, 1 October 1830, and Blomfield to Gilbert, 28 August 1830, in Enys 1877, 11–13, 16; Morrell and Thackray 1981, 120.

38. Howley to Gilbert, 8 June 1830, Gilbert to Blomfield, 2 September 1830, and Blomfield to Gilbert, 18 September 1830, in Enys 1877, 4–5, 12–13, 15–16; Hearnshaw 1929, 87.

39. Blomfield to Gilbert, 28 August 1830, in Enys 1877, 11–12.

and Christianity had clearly demonstrated his ability to write on the religious bearings of the sciences. Moreover, Chalmers had established a considerable reputation as a political economist through his practical application of laissez-faire principles to parish affairs in *The Christian and Civic Economy of Large Towns* (1821–26). The bishops clearly expected that he would introduce into the project a discussion of providence in the mechanism of society and its associated moral features.

Preoccupied with Royal Society affairs, Gilbert readily agreed to these suggestions and to Blomfield's proposal that the subject of natural history should be entrusted to Leonard Knapp, the anonymous author of a "very ingenious and pleasant work," *The Journal of a Naturalist* (1829). Knapp was a gentleman botanist, and his *Journal* was modeled on Gilbert White's *Natural History of Selborne*, giving diary descriptions of the natural history and rural life of his Gloucestershire home. Published by the fashionable John Murray, the work proved very popular, passing through four editions in a decade. However, when Blomfield came to look at it again, he doubted Knapp's "competency to the task." By now it was October, and, as the bishop was one of the trustees of the British Museum, he consulted there with the assistant keeper of natural history. A former secretary of the Royal Society, John George Children was happy to support Blomfield's new suggestion of the septuagenarian naturalist William Kirby.[40] Kirby's reputation as one of Britain's foremost entomologists had been enhanced by his publication with William Spence of an *Introduction to Entomology* (1815–26), and he thus brought to the subject scientific and literary credentials far in advance of Knapp's. He was also an assiduous Anglican clergyman in rural Suffolk who shared Blomfield's High Church proclivities.

After five months of rather haphazard proceedings, the subjects and authors of the Bridgewater Treatises were thus finalized by 18 November 1830. Five days later, Gilbert issued the authors formal letters of appointment, just one week before his resignation as president of the Royal Society became effective.[41] From that date, Gilbert and his episcopal associates ceased to have any significant involvement in the management of the project. Nor, indeed, did the Royal Society's new president, the Duke of Sussex, take a close interest in the affair, except occasionally to urge the authors to bring their labors to a conclusion. Lord Farnborough contin-

40. Blomfield to Gilbert, 1 October 1830, in Enys 1877, 16; Children to WK, 29 November 1829, in Freeman 1852, 437–38.

41. Blomfield to Gilbert, 18 November 1830, in Enys 1877, 24–25; Gilbert to WW, 23 November 1830, TC, Add.Mss.a.63[1].

ued to act on behalf of his late uncle's executors partly because, as an active fellow of the Royal Society and a trustee of the British Museum, he was in regular contact with several of the authors. However, it fell to the authors themselves to organize all matters pertaining to the production of their treatises.

Early in December Roget, who became their secretary for practical purposes, organized a meeting of the three London-based authors together with Buckland and Kirby. While Roget still viewed them as "joint authors of a work 'on the power, goodness, and wisdom of God as manifested in the creation,'" they resolved among themselves that they would each "write a separate work, forming one or more octavo volumes of not less than 300 pages." It was also agreed that while they would be produced uniformly by the same publisher, the works would be sold independently. Although Gilbert appears to have acquiesced in these arrangements, it was the authors who, in the absence of external pressure, definitively turned the plan from being a unitary work into a series of separate treatises. By the following March, Roget had organized for leading fashionable publisher John Murray to publish the series before the end of 1832.[42] However, with no one to take the authors in hand, the writing dragged on, and the treatises appeared piecemeal between March 1833 and September 1836.

## *Trouble over Bridgewater*

In view of the huge sums involved, it is not surprising that the management of Bridgewater's bequest soon became a matter of public discussion. Patronage was a highly controversial topic in the age of reform. Its abuse was rampant in public bodies that notoriously included the Church of England but also hospitals and the medical Royal Colleges. More pointedly, between March 1831 and June 1832 it took three attempts to pass a bill to reform a parliament that still included members returned from "pocket boroughs" under the control of wealthy patrons. The administration of Bridgewater's bequest was inevitably seen as part of this larger picture.

Trouble started in February 1831, when the weekly *Literary Gazette* carried a report of the "conversazione" that had occurred after the Linnean Society's latest meeting. Several of the nominated authors, including Roget, Bell, and Buckland, were reported to have "nearly completed their works, as competitors for the legacy left by the late eccentric Duke of Bridgewater, for the best essay on the structure of the earth and the hu-

42. PMR to TC, 11 December 1830 and 9 April 1831, NC, CHA 4.147.33, 55; WB to PMR, 26 August 1832, BO, Mss.Eng.Lett.b.35, fols. 30–31.

man hand." This reference to "competitors" was freely elaborated on in a paragraph that appeared in several newspapers in which it was reported that the president of the Royal Society was intended to adjudicate between submissions. Essays had been expected from the Continent, it was claimed, alarming the council of the Royal Society, "who imagined that a considerable run would be made upon their funds for postage, so numerous and distant were the applications anticipated to turn out." The scurrilous Tory weekly *Age*, wishing to discomfort the liberal Duke of Sussex, suggested that he, "or any other person having any regard to the respectability of the Royal Society," should look into the matter. Was it right, the newspaper asked, given that the earl had prescribed a single essay, that it should be "made a matter of jobbing for three *philosophers*"? In particular, one of the three "Joint Stock Candidates," whose work was to be adjudicated by the council of the society, was its own secretary, Roget.[43]

William Buckland was alarmed and told Gilbert that the report would "go the round of all the papers unless contradicted." Gilbert accordingly prepared a statement concerning the bequest, its administration, and the names of the authors. It was published in the *Philosophical Magazine* at the start of March and was widely reprinted in the periodical and newspaper press. However, it did not appear in time to stop the March numbers of several monthlies from carrying the story. The idea that the secretary was one of the candidates when the council was responsible for making the decision was particularly shocking. The *Monthly Magazine* exclaimed, "If the matter be not altogether a jest passed upon the public, we should think it a very curious specimen of the new administration of the Royal Society."[44] By the following month Gilbert's statement, and perhaps especially the news of the involvement of the bishop and archbishop, had succeeded in stemming public criticism in more moderate quarters. Even then, the liberal *Monthly Review* suggested that "it would have been a far more useful and liberal plan to have admitted the whole body of scientific persons in this country to a general *concursus*" to determine the appropriate plan.

The criticisms were renewed the following spring. In March 1832 the publisher John Murray began advertising what he was calling the "Theology of Natural History; or, Treatises on the Power, Goodness, and Wis-

43. "Fashion and Table Talk," *Globe*, 5 February 1831, 2d; "Royal Society," *Age*, 6 February 1831, 44b; "Linnæan Society," *Literary Gazette*, 5 February 1831, 88; see also "Royal Society," *Standard*, 10 February 1831, 3c.

44. "Royal Society," *Monthly Review*, n.s., 1 (1831): 626; "Notes of the Month on Affairs in General," *Monthly Magazine*, n.s., 11 (1831): 318–28, on 323–24; Gilbert 1831; WB to Gilbert, 8 February 1831, in Enys 1877, 20–21.

dom of God, as Manifested in the Creation." This included the full list of titles and authors, information that was soon picked up by the monthly magazines. Perhaps as a result, the reformist weekly *Spectator* ran an article on the "Bridgewater bequest" that damned the "decided tendency to jobbing in those ranks and classes" that held a "monopoly in the disposal of public money." The bishop and archbishop were now singled out for criticism, and the writer declared, "Oh, shame upon such jobbery! If Lord BRIDGEWATER had wished to sink his money in paying parsons and college monks, the means were ready enough—he need not have gone to the President of the Royal Society for aid." The complaint was not that the authors were not "able men" but that the earl's expectation that his bequest would be used to "stimulate the genius of the country to the production of a great and lasting work of utility" and "bring forth some young PALEY" had been thwarted. The money had been given instead to established men who would produce, the *Spectator* supposed, large volumes costing four guineas each. What was wanted was a pocket book like the classic *Imitation of Christ* of Thomas à Kempis, not an "'Encyclopædia moralized,' in eight quarto volumes, price thirty-two guineas."[45]

The *Spectator*'s article was picked up the following day by the widely read weekly *Observer* in a much reduced form that quickly did the round of the daily and weekly newspapers and some of the magazines. The bequest "might have called forth another Paley," the *Observer* echoed, but it had instead been "rendered comparatively useless to the people." Reprinted in *The Times*, the paragraph prompted a letter from "Justinus," suggesting that the eight works should be adjudicated by a committee of "men, distinguished by theological and scientific attainments" to avoid "sanctioned mediocrity." The bequest was "a public interest," the letter continued; its "appropriation" was therefore "a matter of public concernment."[46] In the age of reform, there should be no parceling out of patronage in such a way as to favor private interests.

The full tide of public criticism was only reached once the several treatises began to be published at the end of March 1833. As reviews of the works themselves appeared over the next three and a half years, every aspect of the administration of Bridgewater's bequest was subjected to ex-

45. "The Bridgewater Bequest," *Spectator*, 31 March 1832, 301–2; "New Works Announced for Publication," *Gentleman's Magazine* 102, pt. 1 (1832): 315. Murray's announcement appeared in his advertising leaves for March 1832.

46. "The Bridgewater Bequest," *Observer*, 1 April 1832, 4e; "The Bridgewater Bequest," *The Times*, 2 April 1832, 2a; Justinus, "To the Editor of *The Times*," *The Times*, 3 April 1832, 3d.

amination. Indeed, the very fact of the bequest was an issue of political contention. According to the Tory *Quarterly Review*, while the abundant "charitable institutions" of Britain afforded "ample proof of the benevolent spirit that pervades the opulent orders of our community," Bridgewater's bequest offered a particularly inspiring instance that was devoted to an especially noble object. But the radical *Westminster Review* predictably saw Bridgewater's bequest in a diametrically opposite light. "It is not universally necessary to recall the failings of individuals" wrote the review's Benthamite editor, "but when the Quarterly Review chuses to be diffuse upon the case, *to the honour and glory of the clerical nobility*, it becomes a fraud not to state the whole." Whereas the *Quarterly* had dwelt on "the ardour of his faith as a Christian" and "the truth of his perceptions as a philosopher," the *Westminster* portrayed him as being "distinguished by the possession of immense wealth, and by leading a most eccentric and unclerical life at Paris, as far as can well be from anything like the odour of sanctity." Bridgewater, whose "general character and conduct as a clergyman, was not," even in the opinion of the more moderate *Metropolitan Magazine*, "altogether such as to mark him, in public estimation, for a champion of the church or of orthodoxy," was thus not ideally suited as the example with which to defend the reputation of aristocratic patronage.[47]

Whatever they thought of the earl, most commentators approved of the object of his bequest even if, as the liberal weekly *Spectator* put it, Paley had "in a single treatise of moderate bulk, long since forestalled the Bridgewater project." The *Examiner*, edited by the philosophical radical Albany Fonblanque, thought Bridgewater's purpose altogether "vain." Yet it still welcomed the bequest, declaring,

> No matter—Eight thousand pounds diverge from the pockets of a lord into those of eight able and deserving men: Lord Bridgewater thought he was doing good on a large scale: it is something to have succeeded even on a small one. We wish dying lords would recollect that Britain boasts a thousand such men as Prout, Roget, Buckland, Chalmers, who would be equally glad, and have far more need of a few thousand pounds.

The *United Kingdom*, a reformist weekly much more sympathetic to Bridgewater's object, likewise lamented, "Happy would it be for the country, and the cause of religion and morality, if we could find many a noble

47. *Metropolitan Magazine* 7 (1833): 32; *Westminster Review* 20 (1834): 2; *Quarterly Review* 50 (1833): 1.

and reverend lord so disposing of his property." The full irony of this remark becomes clear when we consider how far the paper's campaign against church abuses could have been applied to Bridgewater. He was a notable beneficiary of that "system of corruption, by which many noble families are benefited in the promotion of their children to situations almost useless, and which they but too frequently fill with either little credit to themselves or much credit to the community."[48]

If the bequest itself was welcomed, its administration was emphatically not. On the whole, the charge was one of incompetence more than impropriety, as might perhaps be expected given the involvement of the bishop and archbishop and the obvious eminence of the appointed authors. Yet the sense that matters could have been handled better was widespread, and such views were often expressed in relation to a wider critique of public administration. This was particularly notable with some of those involved in agitating for the improved organization of British science and medicine. Brewster, writing anonymously in the *Edinburgh Review*, was characteristically formidable, observing,

> The injudicious appropriation of a sum of nearly ten thousand pounds, which might have been made truly useful to science and religion, is of itself a sufficient evil to demand the censure of public criticism; but we have been induced to notice it more particularly at present, because many instances have recently occurred, in which public bodies have unnecessarily and injuriously thwarted the obvious intentions of testators.

Brewster's disenchantment with the system of scientific patronage in England is well known, and he claimed that news of the bequest had filled him with "extreme anxiety respecting its judicious application" from the start. Here, however, his comment was probably prompted by an incident north of the border, which he regarded as "the most scandalous job that the history of science records." The financially straightened Brewster was still smarting from the election in January 1833 of his former protégé, the Tory James David Forbes, as professor of natural philosophy at the University of Edinburgh, in preference to himself, on patently political grounds. In addition, however, Brewster's caustic remarks on the management of the Bridgewater bequest were no doubt colored by the fact that he had been omitted by Gilbert and his associates when he was so obviously a suitable candidate. Brewster was evidently speaking from the

48. "Literature," *United Kingdom*, 21 July 1833, 3d (compare the editorial at 3a); *Examiner*, 9 June 1833, 357a; *Spectator*, 14 June 1834, 566.

heart when he declared that "the most eminent of our philosophers would have felt themselves honoured by contributing even a line to that mighty Anthem."[49]

As with most of the critics, Brewster's complaint was not with the "eight eminent individuals" who had been selected. Rather, he objected to the "capricious" division of the work into "a disjointed assemblage of treatises" in which there would be omissions, repetitions, discrepancies, and possibly contradictions. The attempt to honor Bridgewater's favorite topics of the hand and the function of digestion had led to an especially ridiculous division of labor. Brewster was in sarcastic mood: "A system of Natural Theology which begins with the *Mind* and ends with the *Stomach* is not likely to be devoured *seriatim* in the order of its eight courses." What had been called for, Brewster averred, was for the charge to be entrusted to a scientific individual possessed of "the power of eloquent writing and lucid exposition" who might have sought the aid of "intelligent auxiliaries who could have collected from each branch of science its tributary stores, and submitted them to the assimilating power of the master-spirit who was to direct the whole." In this way a great classic of natural theology might have been written that "would have been translated into every written language."[50] It is easy enough to see that Brewster, whose editorship of the *Edinburgh Encyclopædia* had until recently kept him solvent, would have taken kindly to being the editor of such an encyclopedia of natural theology.

If Brewster's criticisms were colored by his own bitter experience of the mismanagement of scientific patronage in late Georgian Britain, he was not alone. Bell probably had in mind the recent vitriolic attacks made on the abuses of the medical establishment when he told Gilbert that he foresaw "great difficulty in reconciling people" to Lord Bridgewater's bequest.[51] True to form, the *Medico-Chirurgical Review*, a moderately reformist quarterly, greeted the Bridgewaters with the exclamation "When will the system of jobbery, and the misapplication of public funds and private bequests have an end in this country!" The reviewer claimed that "a worse plan could not have been devised for furthering the object of the testator, than that which has been adopted" and railed at some length against those responsible for mismanaging the bequest. This was not to detract

49. *Edinburgh Review* 58 (1833–34): 422–23, 426; Brewster to Babbage, 3 February 1833, in Morrell and Thackray, 1984, 158–60, on 159; Morrison-Low and Christie 1984; Morrell and Thackray 1981.

50. *Edinburgh Review* 58 (1833–34): 424–26.

51. CB to Gilbert, 14 September [1830], in Enys 1877, 15.

from the merits of the authors: they were "all men of talent and acquirements." But "had they written in expectation of competition, they would all have written better."

> If the prizes had been held out for the *best* essays, then there would have been ardent competition; and, instead of one languid dissertation on each subject, there would probably have been a dozen—to the most meritorious of which, the donation should have been awarded. The rejected essays would all have been published, and then the object of the dying donor would have been accomplished to an infinitely greater extent than it ever can be at present. In that way, talent would have been called forth, exercised, and rewarded on a large scale.[52]

Other medical journals had little to say on the matter, perhaps because rampant nepotism within the hospitals and abuses within the Royal Colleges provided ample scope for their attacks without reference being made to the Royal Society.

The suggestion of a meritocratic process of open competition found an echo in a number of religious monthlies. In Scotland the *Presbyterian Review* drew an interesting contrast between the manner in which the Bridgewater bequest had been administered and the pattern of state patronage in Paris. Referring to "the plan adopted by the French government in the construction of their extensive decimal tables," the reviewer sought to show that "in various branches of science it frequently happens, that investigations are most successfully made by the conjoined efforts of several inquirers." That the administration of English patronage was inferior to that across the channel was lamentable, especially to those like the reviewer who were "not ardent admirers of every thing that is French." However, criticism of the organization of the work was not restricted to those ideologically committed to the reform of patronage. The *Athenæum* gave a fair sense of the universal dissatisfaction in its review of Prout's treatise: "We were the first to condemn the distribution, according to which he and his fellow-labourers were obliged to work, and our condemnation has been repeated in one form or the other in every Journal, Magazine or Review of any note or character, which has had its attention directed to the matter." Even Whewell, answering Brewster's criticisms of his treatise in the High Church *British Magazine*, made no attempt to answer the Scotsman's attack on the administration of the bequest. He did, however, assert that "the problem which they had to solve was, undoubtedly, a dif-

52. *Medico-Chirurgical Review* 23 (1835): 400.

ficult one," observing "I have never yet seen any suggestion which would, in my opinion, have been likely to produce a better literary plan than the one which has been adopted."[53]

This did not stop one disappointed author from continuing to seek redress. Irish-born meteorological writer Patrick Murphy, whose eccentric physical theories had received unsympathetic treatment by London's scientific societies, wrote to one of the secretaries of the Royal Society in February 1834 to "set up a claim" to a share in Bridgewater's bequest. He had been encouraged to do so by reading Brewster's critical review of Whewell's Bridgewater, he reported, and now wished to point out that his *Rudiments of the Primary Forces of Gravity, Magnetism, and Electricity, in the Agency on the Heavenly Bodies* (1830) had done a much better job of "demonstrating the wonders of *Creative Wisdom*" than had "any preceding writer on the subject." Unsurprisingly, the secretary declined to respond to Murphy's letter, leading him to publicize his grievance in his next publication and to arraign the committee of the Royal Society "at the bar of public opinion." The bequest was not of the society's "*coinage*," he observed, and "*favouritism*" in its appropriation was thus "but another name for *injustice*."[54]

❋

In his presidential address to the Royal Society's anniversary meeting in November 1836, the Duke of Sussex announced that the Bridgewater trust had finally been fulfilled "by the appearance, which has long been anxiously expected, of the eighth Treatise of the series." Apparently sharing the common view that the management of the bequest had been far from ideal, he observed of his predecessor,

> It would ill become me to speak of the mode in which that important duty was discharged by him, or of the principles which guided himself and his distinguished assessors, in the selection either of subjects or of authors; but a list which is headed by the name of Whewell and closed by that of Buckland, can hardly be considered an unworthy representation of the science and literature of this country.[55]

53. Whewell 1834a, 267; *Athenæum*, 10 May 1834, 349; *Presbyterian Review* 5 (1834): 319. See also *Christian Teacher* 1 (1835): 507, and *Congregational Magazine* 13 (1837): 43.

54. Murphy 1834, 338, 348–49.

55. *Proceedings of the Royal Society* 3 (1837): 433.

Working in a piecemeal manner, Gilbert and his associates had certainly met Howley's stated objective of selecting "eminent persons." The average age of those appointed was fifty-one, and, with the exception of the Scottish cleric Chalmers, all were fellows of the Royal Society. Several were former or current members of its council, and six of those considered had been or were about to be awarded one of the society's medals. They were also experienced authors, most having at some time written for general audiences, and all except the rural parson-naturalist Kirby were involved in medical or university education. Yet Howley's view that the bequest was not intended to reward "ingenious" young men sat uncomfortably with the transformation that was under way in the administration of British science. Bridgewater's self-aggrandizing commitment to aristocratic patronage administered by fiat and viewed as evidence of the divine ordination of social inequality was under serious challenge.

# PART I

# *Writing*

# ✷ CHAPTER 1 ✷

# *Becoming a Bridgewater Author*

*We are presented with eight independent treatises,—written by authors who had no previous communication, who had never seen each other's productions, but were merely put in possession of the Cabalistic Titles of their respective Essays. The consequences of such an arrangement are obvious.*

Edinburgh Review, *January 1834*[1]

As the *Edinburgh Review*'s anonymous reviewer—the natural philosopher David Brewster—well knew, his characterization of the Bridgewater authors as strangers to each other was something of an overstatement. Most had at least met at the meetings of London's learned societies, and some were on very familiar terms. In Oxford, as a young man Buckland had been introduced to geology and mineralogy by his fellow author Kidd, now his colleague as professor of medicine, and from November 1831 he was also on the council of the Royal Society with Roget. Others of the authors were indeed unknown to each other. When Kidd's partner in the enterprise of writing on the "constitution of man," Thomas Chalmers, became concerned about the line of demarcation between them, he had to ask Roget to supply an address. In the event, both authors still included the human intellectual constitution, and Kidd wrote anxiously to seek Chalmers's reassurance that he had not encroached on the latter's territory.[2]

As the only one based in Scotland, Chalmers was the most obviously distant from his coauthors. Yet while the medical practitioners Bell, Prout,

1. *Edinburgh Review* 58 (1834): 425.

2. PMR to TC, 13 June and 13 October 1832, NC, CHA 4.189.11, 18–19. See also Blomfield to TC, 16 October 1830, and JK to TC, 23 April 1833, NC, CHA 4.132.45 and CHA 4.207.57–58.

and Roget were all based in London, Buckland, Kidd, and Whewell were located in England's ancient universities of Oxford and Cambridge, and William Kirby's rural rectory was in the farther reaches of Suffolk, a full day's stagecoach journey from London.[3] These far-flung collaborators maintained a degree of communication as they began to prepare their treatises. A number of them met on occasion, typically in London. More generally, communication between them occurred through letters, with Roget indefatigable as their de facto secretary. However, most correspondence related to the publication of the finished works and to payment rather than to the task of authorship.

In view of the limited extent of their interaction, it is not surprising that the Bridgewater authors produced works that were in many ways oddly matched. Moreover, while Gilbert and his colleagues had selected authors who shared a degree of scientific eminence, they also reflected some of the occupational and educational diversity of men engaged in scientific pursuits. Four were medical men; four were clergymen. Of the former, the surgeon Bell and two of the physicians, Prout and Roget, had been educated in Edinburgh's highly reputed halls but now practiced and taught in London, the hotbed of medical reform. By contrast, with the exception of his four years of clinical study at Guy's Hospital in London, the remaining physician, Kidd, had been based in Oxford since the age of seventeen. He thus shared with his former pupil, the clerical Buckland, an institutional milieu in which the sciences had to be justified in terms of their usefulness in relation to religious and humanistic education. The equally Anglican University of Cambridge had taken scientific studies, especially mixed mathematics, more to heart, but Whewell was nonetheless a keen advocate of the importance of keeping the sciences within the framework of a "liberal education."

As the newly appointed professor of divinity at Edinburgh, Chalmers might appear particularly distinct. However, his first love had been mathematics and natural philosophy. As a young parish clergyman in Fife, he had lectured on such subjects at his nearby alma mater of St. Andrews, claiming, when he applied in 1805 for the Edinburgh chair of mathematics, that his pastoral duties left five days a week free for scientific pursuits. His subsequent evangelical conversion had modified these views, but his new pastoral zeal had nevertheless prompted him to produce the acclaimed discourses on astronomy and Christianity and the important work of Christian political economy that had no doubt prompted his appointment. Kirby's life as a Cambridge-educated clerical naturalist was different again,

3. Daunton 1995, 308; *Morning Post*, 3 October 1831, 1d; Bates 1969, 53, 55–56.

FIGURE 1.1 William Kirby. Lithograph by W. B. Spence, 1848, after himself. Image: Wellcome Collection. Public Domain Mark.

although Kirby, too, had cherished hopes of securing a university chair before he became too old (fig. 1.1).

The religious contrast between Chalmers, as a leading light of the Evangelical party of the Church of Scotland, and Kirby, as an English country parson of a High Church cast, also deserves our notice. Theological orthodoxy was clearly a key consideration for the archbishop and his collaborators. The authors were all at least nominally Anglicans with the exception of Chalmers, who was a minister of Scotland's Presbyterian established church. Indeed, the High Church *Christian Remembrancer* considered it "an unnecessary stretch of liberality" to have selected a "Presbyterian divine, however distinguished."[4] Yet while Archbishop Howley and Bishop Blomfield were themselves both of the High Church party, there was no sense that a rigid party line was insisted on either theologically or politically, and the authors' views proved to be somewhat disparate.

With their different occupations and institutional settings, their different locations and religious views, how did these several authors come to write the treatises they did? This chapter follows them through the hard choices and practical demands involved in writing a Bridgewater Treatise. It begins by exploring the role of authorship in the working lives of such "men of science"—a topic of great concern in the age of reform—by considering what it meant for the eight Bridgewater authors, especially in relation to their varied strategies for securing the income necessary to do the scientific work that they wished to do. The chapter then focuses on the particular authorial challenge of the Bridgewaters. Historians have tended to think of the series as Paley's *Natural Theology* writ large, but, while the authors knew that their works would be seen in relation to that book, they did not emulate it. These were not theological treatises. Rather, the Bridgewaters in their different ways made the most of the fresh generic possibilities opened up by the new era of cheap print in order to reflect on the wider meaning of the sciences. Moreover, the authors were pragmatic about reusing and repurposing text that they had already prepared for other purposes. The chapter thus concludes by following the authors into the task of writing itself, discovering along the way that some of the inky fingers responsible for writing these works by the strikingly gendered "men of science" were actually female.

## *The Advantages of Authorship*

With the exception of Herschel, all those who were invited to become Bridgewater authors accepted the commission with alacrity. Such an invitation was difficult to decline. A thousand pounds was a small fortune, equal for most of the authors to more than a year's income. As critics including Herschel had been quick to point out, the financial support of the emerging cadre of men of science was one of the most pressing issues of the day. In his diatribe concerning the "decline of science in England," Babbage had neatly encapsulated the reformist complaint that the "pur-

4. *Christian Remembrancer* 15 (1833): 402.

suit of science" did not constitute a "distinct profession" as in "many other countries."[5] This situation needed to be rectified if the decline was to be stemmed, but Babbage was doubtful that the nation had the wit or the will to provide appropriate remuneration to men of science. In this context, prizes had the potential to provide useful encouragement for scientific work if only they were handled correctly, and to some the Bridgewater commissions seemed like so many prizes. What Babbage did not mention, however, was that authorship was itself becoming an important source of income for men of science.

The second quarter of the nineteenth century was a key moment in the emergence of the professional author in Britain. Authors were increasingly able to sustain a respectable living from writing in consequence of the enormous expansion of print. According to Harriet Martineau, the "vast spread of literature" had provided the support necessary for "a standing class of original writers." They also had a growing sense of collective identity as a literary profession, reflected in composite biographies found in monthly periodicals and other publications in the 1820s and 1830s. The identity that they forged involved a subtle negotiation of financial interests and more high-minded concerns with cultural leadership.[6] At a moment when questions of payment and status were being prominently canvassed by men of science, it is not surprising that similar issues also concerned them as they responded to the new possibilities for remunerative authorship. To what extent was it acceptable for them to make money from authorship? Could such paid work be combined with the aspiration to achieve scientific reputation? What were the potential pitfalls?

One man of science knew more than most about the perils of scientific authorship. Echoing Babbage's complaint about the lack of state sponsorship for science in the *Quarterly Review* for October 1830, Brewster reviewed the employment options open to those scientific men not engaged in the drudgery of university education.

> Some of them squeeze out a miserable sustenance as teachers of elementary mathematics in our military academies, where they submit to mortifications not easily borne by an enlightened mind. More waste their hours in the drudgery of private lecturing, while not a few are torn from the fascination of original research, and compelled to waste their strength in composition of treatises for periodical works and popular compilations.

5. Babbage 1830, 10–11.
6. Salmon 2013, 1–16; M[artineau] 1839, 264.

This reflected Brewster's own bitter experience. Over three decades his scientific research had been funded by his contributions to magazines and reviews; by his editorship of an encyclopedia, two scientific journals, and other authors' works; and by his contributions to such populist publishing ventures as John Murray's Family Library and Dionysius Lardner's Cabinet Cyclopædia, all alongside private tuition, invention, services to scientific societies, and prizes. But such work significantly reduced his capacity for original research. Just a few months earlier he had told his young protégé James David Forbes that, were he to become financially dependent on authorship for a portion of his income, he would become "a professional author, following the very worst of professions."[7]

Others, like geologist Charles Lyell, had altogether higher hopes of making scientific authorship a gentlemanly calling, and authorship became an increasingly important form of paid employment for many others, even when their main employment lay elsewhere. According to Babbage, the main professions that supported English men of science were the historic learned professions of medicine and the church together with a small number of university professorships. The eight Bridgewater authors were drawn from precisely these three groups, but while they were far from poor, they were mostly only moderately wealthy, despite their high scientific status. For each of them, professional authorship was also of financial importance, and the Bridgewater commission was a boon. By contrast, Herschel's scruples in declining the invitation were paid for by a sizeable patrimony of £25,000 in the funds and a quantity of real estate, which made him financially comfortable. As Babbage had so aptly observed, the possession of at least a "moderate fortune" was the only alternative to compromise. Even so, Herschel had not held back from the remunerative rewards of writing two volumes for Dionysius Lardner's Cabinet Cyclopædia.[8]

The extent to which the proffered money mattered to the eight Bridgewater authors was soon demonstrated by their actions. When Bishop Blomfield had assumed that the costs of publication would be met out of each author's £1,000, Whewell was quick to seek reassurance that this was not the case. Thereafter, the authors soon agreed among themselves that they would seek to publish the treatises at a rate of "not less than one half

7. Brewster to Forbes, 11 February 1830, in Shairp, Tait, and Adams-Reilly 1873, 59; [Brewster] 1830, 327. See also the chapters by Shapin and Brock in Morrison-Low and Christie 1984, especially page 19.

8. Secord 2014, 141–42; Babbage 1830, 11–13, 37–38; Will of William Herschel, NA, PROB 11/1662/76.

of the clear profits, the bookseller paying all the expenses at his own risk." Ultimately, therefore, they each received far more than £1,000.[9]

Nor was this the end of the matter. Once the first of the treatises appeared in March 1833, the trustees began to pay over sums of £1,000 to the completed authors. At this juncture, some of their number became uneasy about the fate of the interest that the earl's will had indicated that they should receive. Chalmers went so far as to seek legal advice, and on visiting Oxford in 1834, he mentioned his concerns to Buckland, who also consulted with Prout, Roget, and Kirby. Deferring the question until after the entire series had been published in the autumn of 1836, the authors then raised it with John Parkinson, the solicitor to Bridgewater's executors. Parkinson was not disposed to admit the claim, but Buckland obtained from the solicitor general the opposite opinion, and his fellow authors agreed to refer the two judgements to the attorney general "as an Umpire." The executors refused this plan, and at a meeting of four of the authors held at the Royal Society on 17 June 1837, it was considered "more advisable to abandon these claims" than to involve themselves in a law suit, "which would absorb all the profits in case of success, and assuredly expose the parties to public obloquy and consequently do dishonour to the subject on which they have been engaged." The authors settled for the interest that had actually accumulated on the stock at the time of its sale on 3 May 1833—approximately £26 each, after expenses.[10]

The executors had certainly been somewhat remiss, and, as the last author to publish, Buckland lost three and a half years' interest by this settlement. Moreover, it would be wrong to suggest that money was the authors' only or even necessarily their primary motivation. The satisfaction of the authors in being well paid was not incompatible with the long-standing sense of religious vocation that several had exhibited. The invitation also provided an opportunity to address a large audience in a way that appealed to several of the authors. Here was a chance to reflect on the subject of their researches on a grand scale with the potential of gaining a significant public reputation. Becoming a Bridgewater author was thus attractive on a number of levels. Nevertheless, the authors saw the financial rewards of authorship as an important part of their income and sought to maximize them. Just how important such payments could be to university profes-

9. Blomfield to WW, 4 September 1830, Gilbert to WW, [October?] 1830, and PMR to WW, 10 December 1830, TC, Add.Ms.a.201[44], 205[18], 211[114].

10. WB to TC, 20 February and 17 June 1837, NC, CHA 4.260.32–35; WB to Gilbert, 28 February 1837, in Enys 1877, 31–32; G Chalmers to TC, 5 July 1833, and WB to TC, 15 September 1836, NC, CHA 4.201.28–29, CHA 7.2.26.

sors, clerical naturalists, and medical practitioners becomes clear when we compare income from their different activities. Historians have rarely examined the lives of men of science from the perspective of their bank books, but in getting to grips with our eight authors and their attitudes to paid scientific work, the exercise proves especially informative.

From Brewster's position, it was galling when "well-remunerated professors" devoted themselves to "professional authorship" rather than original research. They were also, he observed, often kept from their research by pandering to the crowded lecture room in order to secure fees, becoming "commercial speculator[s]." Only where professorial chairs were entirely salaried, Brewster argued, or involved no teaching duties, did they promote scientific research. In consequence, no single discovery of note had been made in any British university in the preceding fifteen years. In fact, Brewster had an exaggerated idea of English professorial salaries, as Whewell eagerly pointed out in the January 1831 number of the High Church *British Critic*. The Scot had suggested that existing professorial salaries of £800 to £1,000 might be split two ways to provide incomes for "men of genius," who would advance science through original research and teach the most able students, and "popular" lecturers, who would carry out the more mundane teaching duties. But Whewell corrected Brewster's "happy ignorance" of English professorial stipends, reporting that the average income of the scientific professorships in Cambridge was under £200, with little to add from fees. Several, he stated from personal experience, yielded no more than £100.[11]

Such rates of pay were unquestionably low by middle-class standards. The central tranche of the middle classes—including "professional men, well-to-do clergy, the lesser gentry, superior tradesmen, and industrial managers"—had annual salaries in the range of £200 to £1,000. While the resurgence of scientific education and learning at England's two ancient universities in the late Georgian period led to the inception of a small number of new chairs and to the limited augmentation of some salaries from government funds, professorships were generally not by themselves such as to make scientific men wealthy. On the other hand, the increasingly exacting standards of mathematical education at Cambridge (and in a much more modest way at Oxford) had increased the demand for private and college tutors and for appropriate books for student use.[12]

11. [Whewell] 1831a, 72–73; [Brewster] 1830, 320, 326–28. Midcentury university commissioners estimated an average of a little over £300, including fees. See *Report into the University of Cambridge* 1852–53, 79–81.

12. Topham 2000b and 2013; Musgrove 1959.

Income from such sources had proved useful to the young Whewell as he entered on his career in the later 1810s. A master carpenter's son without money or connections to support or advance himself, he had supplemented his fellowship at Trinity College (initially worth around £90 annually) with private tutoring (bringing in perhaps £40) and with writing books for student use. His *Elementary Treatise on Mechanics*, published in 1819 by the university booksellers, immediately netted him over £70 and continued to be a steady earner over the next thirty years as a further six editions appeared. Several other student works followed, and only gradually was such income rendered less important by the rising value of his fellowship (worth over £225 by 1830) and by his modest stipends as a college tutor and, from 1828, as professor of mineralogy (worth £100 plus fees). Even then, his fellowship not only required him to take holy orders within seven years but also precluded his marrying until he had exchanged it for the lucrative college mastership in 1841. In such circumstances it was natural that the completion of his Bridgewater in December 1832 should leave him looking forward to being "affluent soon."[13]

Whewell's Oxford colleagues fared somewhat better financially. They thus had less cause, as they also had less opportunity, to write student works. As a physician Kidd had the advantage of being able to supplement his academic income from medical practice. Returning to Oxford to graduate as a bachelor of medicine in 1801, he took over an unestablished and unsalaried readership in chemistry. Shortly afterward, through the influence of his former headmaster at Westminster School, William Vincent, he was appointed to the newly endowed Aldrichian Professorship of Chemistry (worth £124 15s, plus fees). This enabled him to marry, and his university income continued to grow. First, the government granted an additional £97 to the Aldrichian salary. Then Christ Church appointed him to the Lee's Readership in Anatomy in 1816, which was worth £100 annually and £200 from 1827. Finally, in 1822 he replaced the chemistry chair with the Regius Professorship of Medicine, the pitiful stipend for which had been expanded by the addition of other roles to provide an annual salary of more than £450, independent of his practice. This was perhaps just as well, for the relatively marginal place of scientific and mathematical studies in Oxford's largely humanistic curriculum generated little

13. WW to R Jones, 21 December 1832, TC, Add.Ms.c.51[147]; *Cambridge University Calendar*, 1829, 41; Todhunter 1876, 1:9. The approximate value of his fellowship is calculated using Potts 1855, 356–57, and his seniority as recorded in the *Cambridge University Calendar*. For his textbook earnings, see the agreement dated 10 June 1820, in the Deighton, Bell, archive, CUL, Add. 9453, E1; see also TC, Add.Mss.83[3], fol. 37v.

demand for student texts. His Oxford-published *Outlines of Mineralogy* (1809) was an "elementary treatise" arising out of the mineralogy lectures he had elected to give as professor of chemistry, but the work did not go beyond a first edition.[14]

Buckland had a little more difficulty in establishing himself but ultimately did so in spectacular style. The son of a Devonshire clergyman of only moderate means, he was dependent from 1808 on his fellowship at Corpus Christi College (worth around £200 p.a.), supplemented by private tutoring. When in 1813 Kidd relinquished his unestablished readership in mineralogy to Buckland, his former pupil successfully petitioned the prince regent for a salary. The annual grant of £100 that he thus extracted from the government was almost doubled by student fees, and in 1818 Buckland obtained a further grant of £100 for a readership in geology, which yielded additional fees. While his lectures were not closely tied into the formal curriculum, they were extraordinarily popular, and he had thoughts of publishing them. In 1823 he expanded on his Royal Society memoirs outlining his extraordinary reconstruction of the fossil fauna of Kirkdale Cave in Yorkshire to produce a quarto volume concerning geological evidence of a recent flood. Published by the fashionable John Murray, *Reliquiæ diluvianæ* sold two editions in quick succession, earning him almost £370. He later boasted to his students that the cave findings had been important "as a mere matter of pocket," claiming the work had brought in £500. Not surprisingly, Buckland was soon planning a second volume and he later contemplated a smaller octavo edition.[15]

With an annual professorial income now well over £300, Buckland still felt unable to marry and forgo his fellowship until he secured a college living in 1825, which brought him a further £389. Shortly afterward, his influential patrons added a canonry of Christ Church that was worth well over £1,000. This sinecure put Buckland in command of an annual income in excess of £1,700 and left Lyell lamenting, "Surely such places ought to be made also for lay geologists." Lyell nevertheless later considered that had it not been for Bridgewater's £1,000, Buckland would never have produced another book after the *Reliquiæ*.[16]

Buckland's happy accumulation of posts as a clerical professor stands in

14. Kidd 1809, 1:[v]; *Report into the University of Oxford* 1852, 364–66; Gunther 1923–1967, 1:71, 3:118; *ODNB*, s.v. "Kidd, John."

15. WB to Murray, 10 October 1823 and 12 December 1831, NLS, MS.40165, fols. 19–20, 50–52; John Murray Copies Ledger B, NLS, MS.42725, 173–74; Boylan 1984, 59–60, 79, 640; Edmonds 1979. The original notes from Buckland's lectures transcribed in Boylan 1984 are in BGSA, IGS1/635.

16. Lyell 1881, 1:161, 386.

stark contrast to the scruples of Chalmers. Chalmers's father, a merchant of moderate means and large family, had striven to secure a parish living for his son, but while Chalmers initially saw no difficulty in using his rural Fifeshire living to support his scientific ambitions, his views about the duties of the parish clergy altered radically following his evangelical conversion between 1809 and 1811, and he campaigned against clergy holding multiple offices or "pluralities." His own livings were reasonably large (initially around £200 and later £400), but with a growing family his copious writings became a valuable source of income.[17]

Chalmers made a tidy sum from his contribution to Brewster's *Edinburgh Encyclopædia,* republished as *Evidence and Authority of the Christian Revelation* (1814), but his first great triumph as an author came in 1817. His Thursday afternoon lectures on astronomy and Christianity, delivered in 1815 and 1816 from his city center pulpit in Glasgow, created a great sensation among the cultured middle classes. When published, the *Astronomical Discourses* sold 20,000 copies within a year, earning him £1,800. The volume of sermons that followed in 1819 yielded a further £1,050. Chalmers, however, considered his publishers insufficiently attentive. Later that year he invested in the new book-selling business of his church elder, William Collins, establishing his brother as a partner in what was to become an enduring publishing house. Over the next quarter century, he produced numerous works with Collins, both original and edited, and as one of the most celebrated evangelical writers of the age, his sustained income from authorship exceeded his salary. One estimate suggests that by 1842 he had earned £14,000 from authorship. Such income was particularly desirable after he accepted the divinity chair in Edinburgh in 1828, since Chalmers declined to supplement the salary of £196 with a parish living, as previous professors had done.[18]

An increasingly familiar clerical figure in the decades that followed the publication of Gilbert White's *Natural History of Selborne* (1789) was the clergyman whose income was supplemented by writing on natural history. William Kirby's collaboration with entomologist William Spence between 1808 and 1826 to produce a popular *Introduction to Entomology* in four volumes made him one of the most reputable of such authors. It also earned him a handsome sum that supplemented his otherwise limited means. As a young curate in rural Suffolk, Kirby's income had been "slender," and he had taken in pupils. A small inheritance from his father, a solicitor, and his presentation in 1797 to the Rectory of Barham (worth £342 annually, in

17. Brown 1982, 20, 59, 60, 84–88, 93.

18. Brown 1982, 181; Keir 1952, 31–34, 148.

addition to the parsonage), eased his financial situation. With the exception, however, of unsuccessfully becoming a candidate for the Cambridge botany chair in 1815, he did not seek further preferment.[19]

Kirby claimed that his first extended publication in natural history, an authoritative monograph on the English bees, was prompted by a sense of religious vocation following a visit from his cousin, the educational writer and author of *An Easy Introduction to the Knowledge of Nature* (1780), Sarah Trimmer. The monograph was also successful in securing Kirby's scientific reputation, but it was published on commission and probably yielded little income. By contrast, his collaborator on the *Introduction to Entomology*, William Spence, had already "tasted the sweets of literary profit" and accordingly suggested that they prepare a "*popular*" introduction to the subject. The work's success was quite beyond its authors' expectations. The first volume had to be reissued straightaway, and the 1,500 copies netted Kirby £86. By the time the fourth volume was issued in 1824, the first had gone through four editions, and Kirby's total income now amounted to almost £1,200. Two further editions before his death brought in half as much again.[20]

Authorship was different again for medical men. Many of the thousands of medical practitioners in late Georgian Britain had little to do with professional authorship. However, it often played a significant part in the working lives of those who were ambitious, including those with scientific aspirations. The growing emphasis in the medical profession on keeping abreast of the progress of knowledge led to a rapid increase in demand for authoritative treatises emanating from acknowledged centers of excellence in the leading medical schools and hospitals. Booksellers seized this opportunity, and the publication of medical books was one of the earliest subject specialties within the book trade, proving significantly remunerative for publishers and authors alike. Those who were involved in medical teaching or who aspired to a lucrative practice at the top of their profession—such as Bell, Roget, and Prout—often found authorship to be a valuable source of both income and reputation.

That was certainly the case with Prout's *Inquiry into the Nature and Treatment of Gravel, Calculus, and Other Diseases Connected with a De-*

19. Freeman 1852, 160–61, 174, 345, 480; *Report into Ecclesiastical Revenues* 1835, 710–11.

20. Clark 2009, 28–30; William Spence to WK, 23 November 1808, in Freeman 1852, 286; Longman Divide Ledgers, *AHL*, A2–A6; Longman Commission Ledgers, *AHL*, B2–B3, B5–B9. The original manuscripts included in the *AHL* microfilm are preserved at the University of Reading Library, MS 1393. On clerical natural history writers, see Secord 2013, Allen 2010, and Lightman 2007.

*ranged Operation of the Urinary Organs* (1821), which was immediately translated into French and German and went through five editions over the space of twenty-five years. As the son of a small farmer in Gloucestershire, Prout had struggled to establish himself in practice in London. After graduating as an Edinburgh MD in 1811, he became a licentiate of London's Royal College of Physicians the following year, setting up in practice near the Strand, but he was discouraged by his lack of progress. His lectures in animal chemistry, delivered to a select audience at his house in 1814, as well as his articles contributed to scientific and medical journals helped to establish his scientific reputation, and he was elected to the Royal Society fellowship in 1819. However, it was Prout's treatise on urinary disease that was later considered to have established his reputation "as a chemist and practical physician," considerably augmenting his practice. It also added to his income, with later editions bringing in £105 each.[21]

Ironically, Prout's increased practice meant that he found much less time for his chemical researches. But the tensions went deeper. While the work contained many of Prout's experimental findings in urinary chemistry, he consciously restrained the scientific aspects of his writing for his medical audience. In particular, he stripped away a planned historical introduction that was to describe the original chemical experiments and instead sought to make the chemical details concise, practically oriented, and "intelligible to the general reader." Moreover, Prout was not alone in recognizing the potential danger for a medical author of straying away from a treatise of a "practical character." Thomas Young, one of the leading natural philosophers of the age, had struggled in practice as a London physician and relied heavily on authorial income. Yet as he told the editor of the *Encyclopædia Britannica*'s latest update in 1814, he found it "necessary to abstain as much as possible from appearing before the public as an author in any department of science not immediately medical." Later, William Benjamin Carpenter found that even his physiological writings hindered him in establishing a medical practice.[22]

For Bell, authorship was not only a valuable means of professional advancement but also a much more enduring source of income. As the son of a poorly paid Episcopalian clergyman who died when he was young, Bell's financial situation caused him great anxiety for many years. Even while attending lectures in Edinburgh in the 1790s, he assisted his brother John

21. Churchill Authors' Ledgers, *AHL*, M5–M7; Copy Letterbook, *AHL*, M3; Munk 1878, 3:109; Brock 1985, 52.

22. Carpenter 1888, 27–28; Young to M Napier, 9 August 1814, in Peacock 1855, 252; Prout 1821, [v].

in extramural lectures in anatomy and surgery, and the pair also earned income by producing a series of pedagogic works. Charles first used his considerable artistic skill to prepare a folio *System of Dissections* for student use (this went through three editions in a decade, the third earning him £100) and two quarto works containing engravings of the nervous system, one of which earned £105. He then contributed two volumes (and a supplementary volume of engravings) to the four-volume *Anatomy of the Human Body*, produced with his brother, a work that reached its seventh edition in 1829. He probably sold the copyright of this work—as he did of the engravings, for £150—but he subsequently received editorial payments totaling £275.[23]

These pedagogic works earned Bell credit as well as money, but he struggled to secure his longed-for "college life" of practical surgery, teaching, and research. Personal and party politics led him in 1804 to leave Edinburgh for London, where his publications had made a name for him. His London publisher, Longmans, became his de facto banker, and the dinners held by the firm provided important introductions to potential patrons. Moreover, Bell apparently hoped that his *Essays on the Anatomy of Expression in Painting*, not yet published, would sufficiently impress the president of the Royal Society, Joseph Banks, to secure him the professorship of anatomy at the Royal Academy. However, while the book again earned him praise and money—Longmans paid him more than £230 in profits on the first edition—it left him without a post. In 1805 Longmans offered Bell £300 for a *System of Operative Surgery* in two octavo volumes, which compared favorably with the £82 earned that year from pupils and the £25 from his anatomical lectures for painters. Bell hoped to profit by such works but he also told his brother, "I think my Surgery will [ . . . ] establish me in practice."[24]

Over succeeding years, Bell's private lecturing and his consulting practice became increasingly important sources of income. By 1810 his practice was worth £1,000 annually, and one of his former pupils later reported that his "professional income fluctuated between 1400*l.* and 2400*l.* a-year." This was a decent income for a leading London surgeon, but Bell contrasted it with those at the top of the profession, who made £9,000 per year. There

23. Longman Impression Books, *AHL*, H5, H7, H9–H12; Bell 1870; [Ferguson] 1843, 193.

24. Bell 1870, 20, 23, 29, 31, 38, 39, 66, 72; Longman Impression Books, *AHL*, H6 and H8; Longman Divide Ledgers, *AHL*, A1–A2; [Ferguson] 1843, 229. On Bell as a medical educator see Berkowitz 2015.

were also large expenses for such a fashionable London practitioner, including, in Bell's case, significant investments in his private medical school and museum in Great Windmill Street. In such conditions, the continuing income from old and new publications was invaluable. When an old debt was unexpectedly demanded of him in August 1821, his first recourse was to publish a new edition of his *Anatomy of Expression* with John Murray with a view to raising £100. New publications in the 1810s and 1820s—including his *Illustrations of the Great Operations of Surgery* (1821) and his *Exposition of the Natural System of the Nerves of the Human Body* (1824)—brought him further income of several hundred pounds. Moreover, Bell continued to believe that, along with his pupils, such publications were his "means of being known" and of securing patients.[25]

Roget was another medical man who lost his clerical father—the Geneva-born pastor of a Huguenot church in Soho—while still a young child, but there the resemblance to Bell ends. Following his graduation from Edinburgh in 1798, Roget's career was well supported by his uncle, the Whig lawyer and political activist Samuel Romilly. After a spell in Manchester, Romilly persuaded him back to London in 1808 with the offer of £1,500 to set up as a physician there, employing him as his cousin's tutor. Like Prout and Bell, Roget gave lectures, including at the latter's Great Windmill Street anatomy school, to secure a metropolitan reputation and practice as well as income. His uncle was also able to procure a hospital post for him by setting up the Northern Dispensary and later sought to secure his election as a physician at St. George's Hospital. Thus, by the time of Romilly's grief-stricken suicide in 1818, Roget (in whose arms he died) was reasonably well established in both reputation and practice without having published professional works like Bell's or Prout's.[26]

Roget had, however, written a number of anonymous contributions in the 1810s for Abraham Rees's *Cyclopædia* and Macvey Napier's supplement to the *Encyclopædia Britannica*, the latter alone earning him in the region of £200–£300. His later contributions to the cheap Library of Useful Knowledge—on electricity, galvanism, magnetism, and electromagnetism—were likewise insulated by anonymity from his professional reputation while earning a similar amount. Roget was nevertheless sufficiently wealthy to turn down Napier's invitation to write for the

25. Bell 1870, 153, 159; Murray Papers, *NLS*, MS.40078, fols. 87–90, MS.42725, 296; Impression Books, *AHL*, H8, H10; Divide Ledgers, *AHL*, A2, A3; Commission Ledgers, *AHL*, B3–B5, B7; [Ferguson] 1843, 208. On salaries, see also Loudon 1986.

26. Emblen 1970; Kendall 2008.

*Edinburgh Review* and Dionysius Lardner's invitation to earn £300 for a contribution to the Cabinet Cyclopædia.[27]

Despite being employed in medical, clerical, and academic professions, most of the Bridgewater authors found paid authorship valuable in making a comfortable living. Nor does David Brewster's view of such authorship as a degrading waste of scientific talent adequately characterize their experience. The increasing range of publications on the sciences in this period provided authors with substantial new opportunities to establish a scientific or medical reputation, often through reshaping knowledge on a larger canvas. The declinists had suggested that the publication of memoirs in learned transactions was the chief means of making an important contribution to the literature of science, but to restrict oneself to such publications was greatly to reduce one's capacity to take the intellectual lead. For the Bridgewater writers, authorship offered not only income but also a wealth of opportunity more generally.[28]

### *Following after Paley*

Gilbert and his episcopal advisers had considered it important to appoint authors who could write well. Those whom they selected were all experienced authors, and between them they had written in a wide range of different genres. But exactly what kind of books were they now expected to write? By whom were they meant to be read? To which genres were they to be referred? It was not obvious what the commission required, and Whewell was quick to raise the question. He asked Gilbert, How long were the works intended to be? To what extent were they "expected to be calculated for popular apprehension"? The reply was brief. "I think the work should be executed in a manner so as makes it instructive to all well educated persons, containing perhaps some more technical matter in notes, & certainly references to the best mathematical works."[29] This advice left much open to question, and in any case Whewell was alone in having the benefit of it.

There was one existing publication that all the Bridgewater authors

27. PMR to Lardner, 6 April 1829, WL, MS.7543/9; Committee Letter Book, March 1827–June 1830, UCL, SDUK8, pp. 69, 114; PMR to M Napier, [15 August 1815] and 8 August 1829, BL, Add. Ms.34611, fols. 248–49, Add. Ms.34614, fol. 148.

28. Babbage 1830, 155. On the changing role of publication in establishing scientific credit, see Secord 2009 and Csiszar 2018.

29. Gilbert to WW, n.d., TC, Add. Ms.a.205[18]; WW to Gilbert, 17 October 1830, in Enys 1877, 19. On writing well, see, for example, Blomfield to Gilbert, 1 October 1830, in Enys 1877, 16.

knew must come into their reckoning, if only as a point of contrast. Try as they might, they could not escape comparisons with a work that had become recognized as the standard discussion of the evidence of design in nature—Paley's *Natural Theology*. First published in 1802, the work had been a runaway success from the outset. Ten editions were produced in just four years, and it had continued to sell steadily. While its price remained well beyond the reach of most readers, almost forty thousand copies had been sold by 1830.[30] But what kind of a work was it? Like many successful publications, it drew on the conventions of multiple genres and was attractive to multiple audiences. Most obviously, it was a theological treatise that discussed religious doctrine in a manner suitable for use in teaching and learned debate. At the same time, it could be read as a work of Christian apologetics, answering the religious skepticism that worried many in the decades following the French Revolution; it could reassure the doubtful if not convert the disbelieving. Furthermore, Christians wishing to give substance to their appreciation of the wonders of God's creation could read it as a devotional work that prompted feelings of awe and adoration. By writing in a manner that combined these purposes, Paley had produced a work of wide and enduring appeal.

Several of the Bridgewater authors related their treatises more or less explicitly to Paley's work. Yet the book was not an easy one on which to build. Its great attraction was the extraordinary clarity and charm of its presentation. Paley reduced the complexity of eighteenth-century natural theology to a single argument, the argument from design. This he introduced using the familiar analogy between a human artefact—a watch—and objects in nature. The appearances of design in the first case, Paley observed, would lead us to infer the existence of a designer. Why then should they not in the latter? Once this simple argument was outlined, the bulk of the remainder of the work was occupied with providing a wide range of engaging and scientifically informed examples of design in nature drawn mostly from the living world. Five concluding chapters gave further attention to the kind of God that could thus be inferred and the wider bearings of natural theology. Yet while Paley made a limited attempt to forestall some of the objections that had been raised to natural theology by the Scottish philosopher David Hume in his posthumous *Dialogues Concerning Natural Religion* (1779), his work made comparatively few demands on the intellect. This mode of presentation led more than one reviewer to describe it as Paley's "Elements of Natural Theology."[31]

30. Fyfe 2002.

31. "Literary Review," *Monthly Visitor* 2 (1802): 200–201, on 200.

Paley would perhaps have been pleased with this description. In his dedication, he had explained that the work was intended to complete his system of theology by providing its foundation. It afforded "the evidences of natural religion" to complement "the evidences of revealed religion, and an account of the duties that result from both." These were all subjects that Paley had been responsible for teaching as a college tutor in Cambridge. Subsequently he had capitalized on the growing market for works intended for the use of university students to obtain sizeable payments from booksellers for his *Evidences of Christianity* (1794) and *Moral and Political Philosophy* (1785). These two parts of his theological system were widely used by university students, especially at Cambridge. Paley clearly hoped *Natural Theology* would appeal to the same market, and contemporaries agreed that it made a good "text-book" and that the three works together amounted to "one grand series of instruction." In the decades that followed it was used by students at a number of British universities, especially those attending courses in divinity.[32]

Considered as a theological treatise suitable for clerical education and use, Paley's *Natural Theology* had its detractors. From the outset there were evangelicals and High Anglicans who criticized what they considered its overemphasis on the role of reason in religion. Others, especially those of the more rationalist "Moderate" party in the Church of Scotland, reasonably concluded that Paley had only very partially developed the arguments of natural theology, and they sought to offer a more systematic account. Their works did not sell particularly well, however.[33] In any case, neither the terms of the Bridgewater bequest nor the invitations that went to the authors suggested that a theological treatise was what was wanted. Nevertheless, the Bridgewater authors might still hope to build on Paley's highly attractive work in indicating the religious value and safety of scientific knowledge, even if they did not emulate Paley's systematic theology. On its first publication reviewers had especially appreciated the way in which Paley's *Natural Theology* updated the scientific examples used in support of the design argument. A third of a century later, however, the sciences had moved on considerably. In addition, Paley had restricted his examples chiefly to the functional anatomy of animals, leaving many areas of the natural world unexplored. There seemed to be ample scope for an expanded and revised version.

32. "Literary Review," *Monthly Visitor* 2 (1802): 200–201, on 201; Fyfe 1997, 323; St Clair 2004, 626; *ODNB*, s.v. "Paley, William"; Paley 1802, vii.

33. On the evangelical side see, for example, Gisborne 1818; on the more rationalist side see, for example, Brown 1816 and Crombie 1829.

In fact, the Society for the Diffusion of Useful Knowledge's (SDUK's) Henry Brougham had conceived a plan for such a new edition in 1829, envisaging that "several men of science" could work together in its production. It proved difficult to arrange, and in the end Charles Bell was the only collaborator who signed up. The new edition was not advertised until the following December, after the Bridgewater authors had been announced and Brougham had been ennobled as lord chancellor in the new reform-minded Whig government. What it offered to add to Paley's original text were "Notes and Dissertations," numerous illustrations, and a "Preliminary Discourse on the Objects, Advantages, and Pleasures of the Study of Natural Theology." With Brougham and Bell both busy, the edition took some years to produce, running to five volumes. The first contained Brougham's *Discourse of Natural Theology* (1835), which set out to consider the "nature of the evidence and the advantages of the study." This amounted to a substantial (and controversial) reexamination of the philosophical basis of natural theology. The text of Paley's work was then annotated with notes that made it clear how far the sciences had developed since his time. Paley's famous opening contrast between a stone and a watch, for instance, was met with the observation that developments in geology now meant that even "common readers" would find a stone as full of evidence of design as a watch. An appendix to the main work included various "dissertations" by Bell, but Brougham's own dissertations, which ranged much more widely, filled a further two volumes published in 1839.[34]

Although the rationalism of Brougham's *Discourse* proved somewhat controversial, Brougham and Bell's annotated volumes of the *Natural Theology* were generally well received. The difficulty for anyone following in Paley's wake, however, was that the attempt to apply the argument from design to additional examples left the author open to the charge of making what had been beautifully expressed in the earlier work more wordy, tedious, or confusing. These, indeed, were all charges directed at the Bridgewaters. Nevertheless, several of the authors did seek to present their works as updating or supplementing Paley's. This was an especially obvious strategy for those of the authors who were medical men lacking the resources to develop sophisticated theological arguments. However, it remained problematic.

The predicament is especially well illustrated by the experience of Bell, who reported that he had been drawn into illustrating Paley's *Natural The-*

34. Brougham 1839; Brougham and Bell 1836, 1:1n–2n; Knight 1864–65, 2:157; *The Times*, 16 December 1830, 7c; Brougham 1835, 1–3.

*ology* by degrees. In 1810 he had told his future wife, "I see a God in everything, it is the *habit* of my mind." This became a characteristic of his lectures to medical students, taking on a polemical edge when he intervened in the controversy about materialism that shook the metropolitan medical world in the 1810s. The radical weekly *Lancet* satirically portrayed the Scot as one who never touched a "phalanx and its flexor tendon, without exclaiming, with uplifted eye, and most reverentially contracted mouth, 'Gintilmen, behold the winderful eevidence of *desin*!'" (fig. 1.2). In 1827, as a member of the newly formed SDUK, Bell prepared a treatise on *Animal Mechanics*, published in two sixpenny parts. With a minimum of theological argument the treatise cataloged a series of examples of design in the anatomy and physiology of animals. The publication was a great success, with its first part selling more than thirty thousand copies, and it was this that marked Bell out as Brougham's favored collaborator in the new edition of Paley. The project was agreed shortly before Bell received the invitation to write a Bridgewater, and over the course of the next year he prepared illustrative notes and appendixes and a set of "little drawings" for the new edition.[35]

By the time he was writing the preface to his Bridgewater, however, Bell was feeling overstretched. From "at first maintaining that design and benevolence were every where visible in the natural world," he explained, circumstances had gradually drawn him "to support these opinions more ostentatiously and elaborately than was his original wish." Inasmuch as he was writing on theology, Bell reflected, he was at the disadvantage of producing a work that was to some degree outside his professional competence. He could provide examples of design from the anatomy and comparative anatomy of the human hand, but he was not in a position to develop sophisticated theological arguments. Two years later Brougham was back onto the edition of Paley, causing Bell to complain to his brother, "It has always seemed to me that Paley's works are unfit to build upon—that their simplicity and almost childishness have been the sources of the popularity of that book, and that my illustrations would be liable to such criticism as is applicable to an artist who rears cumbrous heavy columns on a light ornamental frieze." Writing on such subjects had become a "real hindrance."[36] Yet while the attempt to illustrate Paley's *Natural Theology*

35. Bell 1870, 315–17; Brougham 1835, 2; Topham 1992, 414–15; "Mr Green's Sky-Rocket Lecture," *Lancet*, 27 October 1832, 151–55, on 154; CB to M Shaw, 2 November 1810, in Bell 1870, 180.

36. CB to GJ Bell, 29 March 1835, in Bell 1870, 339; Bell 1833a, xi.

FIGURE 1.2 Charles Bell. Lithograph from the *Lancet*, 7 September 1833, opposite p. 756. Image: Wellcome Collection. Public Domain Mark.

had become irksome and embarrassing to him, his Bridgewater was certainly intended to provide such illustrations, and he cross-referred to them in his notes to the edition of Paley.

The other medical authors also looked to the "unrivalled and immortal work of Paley" as the obvious point of reference in writing their treatises. Allusion to Paley's familiar statement of the logic of the design argument allowed the authors to move quickly to detailed scientific exposition of the evidence of design in nature within their respective domains. In addition, however, it allowed them to make claims to novelty. In his preface, Roget claimed that his emphasis on the "unity of design" among living creatures had enabled him to give a "scientific form" to his treatment of the subject that had been absent even in Paley's account. Similarly for Prout, the fact that Paley had only devoted a short chapter of eleven pages to chemistry left plenty of scope for expansion. He had, he confessed, "always thought that our excellent author has not made quite so much of his subject as he might have done."[37] For each, therefore, Paley became a convenient foil, while the reader was clearly to understand that these treatises were supplementary to that of the acknowledged master.

Responding to Paley's work was altogether more demanding for the three Bridgewater authors based in England's historic Anglican universities of Oxford and Cambridge. Both institutions were committed to educating all undergraduates in Christian theology, with well over half of the graduates subsequently going on to ordination, and Paley's *Natural Theology* was used in each. Yet its use was distinctly controversial. In Oxford, the prevailing High Church temper, with its emphasis on church tradition and the Christian scriptures, resulted in a growing resistance to a foundational natural theology. Cambridge had also leavened its focus on the role of reason in religion since the time of Paley, with an increasing stress on the Bible and Christian tradition reflecting the concerns of moderate evangelicals and High Anglicans. In both, the concentration was on scriptural education. Thus, instead of *Natural Theology*, Oxford turned especially to Joseph Butler's *The Analogy of Religion, Natural and Revealed, to the Constitution and Course of Nature* (1736), numerous editions of which issued from its press in the early nineteenth century. Butler's purpose of demonstrating that revealed theology was no more open to objection than natural theology, and that both had probability on their side, was much more in keeping with Oxford's High Church tradition. In Cambridge, significantly, it was Paley's *Evidences of Christi-*

37. Prout 1834a, 10–11; Roget 1834, 1:ix.

*anity* rather than his *Natural Theology* that was a set book for university examinations.[38]

Paley's *Natural Theology* was nevertheless still in use, especially by the many graduates who went on to ordination. Such individuals required a certificate of attendance on a course of divinity lectures to present to their ordaining bishop. In Oxford, the standard of the course rapidly improved under a series of High Church professors in the 1810s and 1820s who, despite having reservations about the value of natural theology, each recommended Paley's work. The emphasis here was not on natural theology as the foundation of Christianity but as part of the armory against religious skeptics. From 1823, for instance, Charles Lloyd began his lectures with the scriptures, taking issue with the usual maxim "Natural before revealed religion," but he later included a lecture on atheism that made due reference to Paley's book.[39] At the same period in Cambridge, Norrisian Professor of Divinity John Banks Hollingworth offered a system of theology that explained and justified Anglican doctrine rather than one that traced religious belief "from its origin," but he still found place for Paley's work in his discussion of "Objections of the Atheists." It was also used for more elementary education in some Cambridge colleges. At the start of the 1840s, for instance, one student found that he was examined on Paley's *Natural Theology* before being admitted to Trinity College (home to around a quarter of undergraduates) and again in college examinations at the end of the second year. University prize essays were also sometimes set on natural theological themes, albeit that they might invite a circumspect assessment.[40]

Although the role of natural theology in Anglican university education was thus limited and its theological value questioned, scientific professors in both universities also played on the ways that the sciences could assist in illustrating and confirming God's existence and attributes. In Oxford especially these claims were keyed to Paley's work. Much more than in Cambridge, the sciences were marginalized within Oxford's largely classical curriculum. Only a tiny proportion of students competed for honors in mathematical and physical sciences, and for the great majority atten-

38. On Oxford see Brock and Curthoys 1997, 10, 210, 482, 536, Hampden 1837, liii, Corsi 1988, and Rupke 1983; on Cambridge see Fyfe 1997, Searby 1997, and Thompson 2008.

39. [Parker] 1851; [Ffoulkes] 1892, 399–405, on 400; Brock and Curthoys 1997, 9; Baker 1981, chap. 4; Ince 1878, 24–26.

40. *Cambridge University Calendar*, 1833, vii; Bristed 1852, 1:15, 2:329; Searby 1997, 727–28; Hollingworth 1825, 1, 43; Fyfe 1997, 331–35.

dance at scientific lectures was thus entirely outside the curriculum. In such a context the case for scientific study needed to be vigorously made.

Kidd and Buckland had led the way as scientific professors in seeking to demonstrate the relevance of the sciences in providing a liberal education, claiming that they could usefully supplement Paley's work in protecting Christianity from skepticism. Appointed reader in geology in 1819, Buckland had argued in his inaugural lecture that the study of geology had a "tendency to confirm the evidences of natural religion" while also seeking to demonstrate that the findings of the science were "consistent with the accounts of the creation and deluge recorded in the mosaic writings." A few years later, when Kidd prepared his introductory lecture as professor of medicine, he also used the argument from design to connect his own teaching with the wider concerns of the university, stating more explicitly that his lecture was intended to be "illustrative of Paley's Natural Theology." Kidd claimed that this subject had been recommended to him by Charles Lloyd, the newly appointed professor of divinity, and that Paley's work was "so generally recommended and read" in the university that he could take the value of its arguments as read.[41]

This rhetoric was picked up by others. The following year, a local surgeon, James Paxton, issued the first revised edition of Paley's work, containing scientific illustrations and notes to help the "general reader" follow the details of Paley's examples. Sold by Oxford bookseller Joseph Vincent, the new work was a success, reaching a third edition within a decade. Alongside it Vincent marketed *Botano-Theology* (1825) by the new keeper of the university's Ashmolean Museum, John Shute Duncan, later issuing the two in a combined edition. Duncan, indeed, had set about reorganizing the museum's collections to illustrate Paley's work following his appointment in 1823. The catalog of the museum—edited by Duncan's brother and successor as keeper, Philip—later claimed that Paley's work, along with Kidd and Buckland's lectures, had generated a taste for natural history within the university.[42]

Not surprisingly, both Buckland and Kidd presented their Bridgewaters as supplements to Paley's work. Like his inaugural lecture, Buckland's treatise set out to demonstrate how (in words borrowed from one of Whewell's anonymous reviews) geology had "lighted a new lamp along the path of Natural Theology." A two-page preface outlined three new arguments in favor of divine design made possible by the findings of geol-

41. Kidd 1824, 1; Buckland 1820, [iii].

42. [Duncan] 1836, vi–vii; Duncan to WK, 6 August 1825, in Freeman 1852, 416–17, on 416. See also Rupke 1983, 238–39, and MacGregor and Headon 2010.

ogy and a brief conclusion elaborated further on how the new science had usefully supplemented Paley's account. As we shall see, however, the body of the text provided a connected account of the sciences of geology and mineralogy that was quite different from Paley's work in form. Kidd's Bridgewater similarly had much in common with his inaugural lecture, even to the extent that a significant proportion of it was taken verbatim from the earlier publication. Through a series of chapters, he elaborated on the design evident in the human species and its mutual adaptation with the world around it. Yet just as in his inaugural lecture, Kidd took Paley's claims for granted rather than making the case for them.[43]

The need to justify scientific studies was arguably less acute in Cambridge, where rigorous examination in mixed mathematics was central to the competition for academic honors. Nevertheless, professorial lectures in the sciences remained outside of the required curriculum for the bachelor of arts degree, and it is consequently not surprising that several professors pointed out the contribution that the natural sciences could make to knowledge of God's action in the universe. As in Oxford, however, they had to do so in a skeptical context. When the professor of geology Adam Sedgwick made the case for the sciences in a Trinity College sermon in December 1832, he knew that he faced opposition from those who objected to natural theology on the grounds that it misled people "on matters clearly put before them in the word of God." He had, he reported, heard preachers in the college assert that "knowledge of the attributes of God or even of his existence" came only from revelation. His view, however, was that undergraduates should make a habit of studying Paley's "delightful work."[44]

Recently appointed as professor of mineralogy, Whewell took seriously the concerns of those skeptical of the religious tendency of the sciences. While he was prepared to argue in his Bridgewater that knowledge of the sciences brought new insights into God's relationship with the creation, he pulled away from Paley's strong program for natural theology. Rather than a theological treatise, what Whewell had in mind was what has been called a "reflective" work, designed to examine the wider religious significance of modern "astronomy and general physics" in terms accessible to readers without prior knowledge of the subject.[45] More striking still, the Cambridge-

43. Kidd 1833, viii; Buckland 1836, 1:586. The quotation is from [Whewell] 1831c, 194.

44. Sedgwick 1833, 15, 87–88; [Otter] 1824, 574; Fyfe 1997, 329–31. See also Becher 1986 and Bellon 2012.

45. On "reflective" treatises, see Secord 2000, chap. 2, and Secord 2014.

educated High Churchman Kirby did not mention either Paley or natural theology at all. Far from providing an argument for the existence of God from design in nature, he used his treatise to show how the wonders of the living world could be better understood in the light of the truths of scripture.

The situation for Chalmers—in his third year as professor of divinity in Edinburgh—was different again. While most of Chalmers's students were studying to become ministers in the Church of Scotland, they were in a minority at the university, which had nothing like the narrowly religious ethos or rationale of Oxford and Cambridge. Moreover, students were not constrained in their studies in the same way. Few followed the official master of arts curriculum through to graduation, with its mixture of "Latin, Greek, Mathematics, Logic, Rhetoric, Moral and Natural Philosophy," and most were altogether more selective. Nevertheless, in Edinburgh as at Scotland's other four universities, those who studied what was called "moral philosophy" encountered natural theology as a significant part of the curriculum. Indeed, Chalmers's professorial career had begun in 1823 in the chair of moral philosophy in St. Andrews, and it was there that he had begun to develop the lecture course on natural theology that he later offered at Edinburgh.[46]

Chalmers had himself been taught natural theology as a student at St. Andrews by the Moderate George Hill, whose published *Lectures in Divinity* (1821) came to be widely used in Presbyterian theological education. Hill's *Lectures* referred readers to half a dozen works on natural theology, including Paley's, but while Chalmers used the work with his own students and likewise recommended Paley's *Natural Theology*, his preferred "text-book" was Butler's *Analogy of Religion*, with its strong emphasis on the necessity of revelation. His approach thus reflected evangelical priorities. Critical of the approach to systematic theology that started with "the constitution of the Godhead" and ended with "the consummation of all things," he preferred to proceed "in the natural order of human inquiry," beginning with the "darkness and the probabilities and the wants of Natural theology," moving swiftly to establish the scriptures as "a real communication from heaven to earth" before using them to deliver the answer to the human condition. In this way the study of theology could be made to fit with the "order of our practical Christianity."[47]

46. Francis A. S. Knox, "Dictations issued by Dr Chalmers in his Moral Philosophy Class [ . . . ] 1826," StAUL, MS.37483; *Report into the Universities of Scotland* 1831, 275; *Evidence on the Universities of Scotland* 1837, 37:103–4, 106; Hanna 1854, 2:43–56.

47. Chalmers 1847–49, 7:x–xi, xv, 9:xi, 130–32; Hill 1833, 1:13; *ODNB*, s.v. "Hill, George (1750–1819)"; "Notes from Dr Chalmer's Lectures on Theology," EUL, Dc.7.115; [Mason] 1864, 127.

As a theological professor, Chalmers was nevertheless the Bridgewater author who had the most cause to follow Paley's lead in preparing a theological treatise on natural theology that was suitable for use by students. He expressed strong views about the writing and use of "text-books"—a phrase at this time still used primarily for works where the "text" was to be discussed in class. There was no need, he felt, for each professor to prepare a comprehensive treatise on his subject. Rather, it was best to use standard textbooks as the basis for frequent class discussions and to put original thought into written lectures that might then be published and become textbooks in their turn.[48] In just such a way, Chalmers's original lecture notes furnished the materials for his Bridgewater. While he had to cut out aspects of his more systematic treatment to suit the wider audience and prescribed subject of his commissioned work, the general approach was similar, and the work had the argumentative structure typical of theological treatises. He even numbered the paragraphs within each chapter, emphasizing their sequential character and making them amenable to class use.

This did not make the treatise attractive to all readers. Indeed, some found Chalmers's treatment of his favorite arguments an unconscionable "splutter":

> He is [with an argument] like a child with a new wax doll, he hugs it, kisses it, holds it up to be admired, makes its eyes open and shut, puts it on a pink gown, puts it on a blue gown, ties it on a yellow sash; then pretends to take it to task, chatters at it, shakes it, and whips it; tells it not to be so proud of its fine false ringlets, which can all be cut off in a minute, then takes it into favour again; and at last, to the relief of all the company, puts it to bed.[49]

Yet within a couple of years, Chalmers had restored to his Bridgewater the excluded portions of his lectures, producing a more comprehensive work entitled *Natural Theology* that could be used with students. This occupied the first two volumes of a new edition of his collected works and provided a "general Treatise" in place of the "fragment of a more extended theme" that had appeared in his Bridgewater. Moreover, it was bound up by Collins with the next two volumes of Chalmers's works, on the evidences of Christianity, and offered for sale as "Chalmers's Theological Text-Book." The whole was used by his Edinburgh divinity students as a class book,

48. Chalmers 1847–49, 9:ix–xxi.
49. S Coleridge to A Brooke, 8 November 1836, in Coleridge 1873, 1:169.

with the professor also reading from it in lectures. Indeed, when Chalmers later prepared a mature work of systematic theology, boiled down from his lecture course, it began with a condensed account of the writings on natural theology and the evidences of Christianity that he had used to introduce his textbooks to his classes.[50]

It is surely no surprise that the Bridgewater author whose contribution most resembled a theological treatise was the one employed as a theology professor, but in this regard he was obviously unusual. While there were good reasons for most of the other authors to make reference to Paley's *Natural Theology* and even to present their own works as in some sense supplementary to his, the genre of Paley's work had little to offer the authors. Commissioned to write authoritatively about their several scientific specialties in relation to God's agency within the universe, the Bridgewater authors found altogether more suitable exemplars in the burgeoning market for print. In contrast to the situation in which Paley had written thirty years previously, there was now a rapidly expanding range of publications designed to interpret the meaning of modern science for the general reader. Such works offered much more attractive models for an author wishing to explore the religious significance of the sciences than did a treatise of natural theology, and the Bridgewater authors drew on them heavily for inspiration.

### *Interpreting the Sciences*

The authors were well aware that their treatises would be read in relation to the growing output of reflective treatises addressing a large and general audience about the objects, methods, history, and wider implications of the sciences.[51] However, negotiating a rapidly changing literary market is never easy. It involves imagining a new body of readers and imagining a literary form and style that such readers will find appealing. In addition, it has potentially momentous implications for one's reputation. The Bridgewater commission offered the authors an opportunity to impress influential readers but also an opportunity to damage their existing reputations. They inevitably found themselves learning on the job as they wrote their innovative treatises. Reviewing the varied and shifting array of genres man-

50. Watt 1943, 10; Hanna 1854, 2:706; Chalmers 1847–49, vols. 7, 8; [Mason] 1864, 127; "Prospectus of Dr Chalmers' Works," in Chalmers [1836]. See also TC to JW Cunningham, 20 April 1836, in Hanna 1854, 2:387.

51. Secord 2000, chap. 2; Topham 2007; Secord 2014.

ifested by existing publications, they combined and modified elements in ways that would allow them to accomplish multiple purposes. Perhaps most importantly, they drew on their previous experiences as authors, adapting both their conceptions and their techniques to the new task and the new market. The varied works they produced contributed to define and develop the generic possibilities of this new type of science writing.

The degree of sophistication that the authors brought to the task varied considerably. Kidd was at one extreme. The Oxford physician may not have received Gilbert's advice that they should write in a manner suitable for instructing all well-educated readers, but he had no doubt that a "popular rather than a scientific exposition of the facts" was called for. He would, he observed in his preface, "have shrunk from his present attempt, had he considered that any exact elucidation of the details of science was required in the execution of it." However, his notion of "popular" exposition had been carried over from the inaugural lecture to his Oxford anatomy course that he had plundered for materials. That lecture was addressed largely to classically educated gentlemen who viewed science as a "relaxation rather than a study," and Kidd's main object had thus been to manifest divine power and wisdom in creation "as a subject worthy of both philosophical and religious contemplation." His Bridgewater maintained much the same tone, complete with frequent quotations from classical authors (often in the original languages) and an appendix containing parallel passages from Aristotle and naturalist Georges Cuvier (in Greek and French) reprinted from a pamphlet prepared for use with his Oxford lectures.[52]

At the other extreme was Whewell, a man who was soon to become one of the age's foremost commentators on the sciences not least through the publication of his magisterial *History* and *Philosophy of the Inductive Sciences* in 1837 and 1840. Whewell's early publications had been written for undergraduates, but even then he had seen himself as having a vocation as an interpreter of the sciences. Writing to his close friend Herschel in 1818, he numbered himself among those who, while not involved in experiment, were employed "with twisting the results of other people into all possible speculations, mathematical, physical, and metaphysical." By the start of the 1830s he was ready to claim this interpretive role in the quarterly review journals, offering anonymous commentaries on several important reflective works of science, including Herschel's own. These high-profile journals encouraged reviewers to address the wider bearings of the sciences in a manner suitable for a general readership, and Whewell took

52. "Check Book, 1820–35," OUPA, OUP/PR/13/7/1/1; Kidd 1824, vii, 5; Kidd 1833, 347–75.

the opportunity to reflect on the emerging forms of scientific publication just as he began to plan his Bridgewater.[53]

Reviewing the first volume of Lyell's *Principles of Geology* in January 1831, Whewell began by considering what the work portended for geology's changing relationship to the public. For some years, he reflected, geology had been communicated privately among its cultivators through talk and action. This was quite at odds with the modern view that the press should be the "source of all information" and that "popular circulation" should be the "criterion of all progress." It had been almost impossible for "mere popular readers" to acquire clear or connected knowledge of progress in the science of geology. In part, this was a consequence of the geologists' reluctance to theorize, since "common readers" considered "general propositions" to be the only ones that had an "air of knowledge." Now, however, in offering a theoretical synthesis of the facts with all the skill and mastery of an "accomplished geologist," Lyell had finally broken the spell. Thus, while Whewell remained unconvinced by Lyell's theoretical position, he warmly welcomed the *Principles* as marking an epoch in the science and securing its reputation even among the "pious and religious."[54]

Another radical departure was Herschel's *Preliminary Discourse on the Study of Natural Philosophy*, published in 1831 as an introduction to the six-shilling scientific treatises of Dionysius Lardner's Cabinet Cyclopædia. Whewell reviewed his friend's book promptly, observing that its subject matter—the history and philosophy of science—was attractive "to many to whom science itself appears thorny and repulsive." Moreover, the world benefited enormously when eminent scientific men promulgated their "notions of the general character and bearing" of the sciences—ranging widely and providing a "more connected picture." It was no detriment to the scientific "vocation" to adopt this additional role, which could give the "common reader" a feeling for the otherwise "repulsive" realms of "professional" science. On the Continent, regular reviews of the recent history of science had been prepared, at the insistence of learned societies, by men of the eminence of Cuvier and Berzelius. Whewell's review suggested that the English deficiency in this regard would soon be addressed thanks to the "*Cabinet Cyclopædia*, and the *Family Library*." In fact, this was an editorial insertion and a puff for the review's publisher John Murray, who produced the latter. But Whewell was delighted that Herschel's

53. Yeo 1993, esp. 87–112; WW to Herschel, 1 November 1818, in Todhunter 1876, 2:29.

54. [Whewell] 1831c, 180–86, 205–6. On Lyell's *Principles* see also Secord 1997 and Secord 2014, chap. 5.

contribution to the former went well beyond a "mere survey" of the state of natural philosophy to offer a novel account of the "method of research" to which it owed its progress.[55]

Whewell's comments about the characteristics of such works culminated with his review of another John Murray production, Mary Somerville's *Connexion of the Physical Sciences,* just a year after his own Bridgewater was published. Somerville was, in Whewell's private estimation, "one of the best mathematicians in England," but her latest work was intended to introduce the physical sciences to a nonspecialist audience in a manner that explored and emphasized their interconnections. Whewell's review applauded and reflected on this achievement. Natural science, he observed, could be made "popularly intelligible and interesting" in one of two ways. Expositors could describe the subject matter and discoveries in detail, or they could choose instead to consider the science's history and bearings in more general terms. The former was something particularly well suited to the public lecturer, who could use models, machines, and diagrams. The latter, as Somerville had demonstrated, had the advantage of making the subject accessible to those with only the most rudimentary knowledge of science. It involved explaining the "progress and prospects" of a science, the relationship of new work to old, and the bearings of "new facts" in one subject on others. The increasing "dismemberment" of the sciences was a problem that had left men of science reaching out for a collective identity in the unpalatable neologism "scientist." Writing for the new, wider audience, however, Somerville had been able to show how the sciences had a tendency to be "united by the discovery of general principles."[56]

When he came to write his Bridgewater, Whewell notably sought to embody in it the approaches he admired in Lyell, Herschel, and Somerville. Following Gilbert's clarification about the intended audience, he set about writing a book that would introduce the physical sciences to uninitiated readers in such a way as to explain their wider implications. His particular focus, of course, was on their religious bearings, but from the outset he sought to introduce readers to the distinctive characteristics of natural philosophy and especially its emphasis on natural laws. As he explored the providential character of the laws revealed by natural philosophy—both on the earth and in the heavens—he provided a synthesis that focused on what could be learned about the divine administration, including its

55. [Whewell] 1831b, 374–77; WW to Jones, 15 July 1831, in Todhunter 1876, 2:123.

56. [Whewell] 1834b, 54–55, 58–60; WW to M Statter, [n.d], in Douglas 1881, 142. See also Yeo 1993, 92–102, 109–12, and Secord 2014, chap. 4.

progressive character. In the final section, containing religious reflections, he focused more exclusively on the humanistic aspects of natural philosophy, including how human minds worked and what could be learned from the history of science.

Whewell was by far the most sophisticated of the Bridgewater authors in reflecting on what was involved in addressing the new reading audiences for science, but others were also engaged with the question and had practical knowledge on which to draw. Two in particular had recent experience of writing for the useful knowledge society. Bell and Roget, who were both Whigs and moderate reformers, had joined the society upon its foundation in 1826 and thereafter served on its general committee for many years, becoming its "great authorities" on physiology. Roget—who was considered by Charles Knight, one of the society's leading publishers, to be the "most eminent" scientific man among its founders—was a diligent committee man and a "vigilant corrector of its proofs." Both he and Bell wrote treatises for the society's flagship series of sixpenny treatises for workers, the Library of Useful Knowledge.[57]

These useful knowledge treatises had done much to prompt the new fashion for scientific treatises directed to nonspecialist audiences, and there were obvious parallels with the Bridgewaters. Bell in particular felt so. Referring to Henry Brougham and Charles Knight, he told his brother: "From the Chancellor to his little bookseller (who writes better than any of us), the encyclopaedists are all writing the same stuff. And here are eight men more to wear the subject to the bone—all at the same work." This is perhaps not surprising, since Bell's contribution to the Library of Useful Knowledge, *Animal Mechanics* (1827–29), bore the subtitle "or, Proofs of Design in the Animal Frame." In general, the useful knowledge society shied away from potentially divisive references to theology, including natural theology, but Bell's treatise was one of the first to appear and was not unduly constrained by such qualms.[58] Thus, while Bell was noticeably reluctant to mention the Deity directly, his account of mechanical and vital instances of design in animal bodies bore striking parallels with his later Bridgewater, and portions of the text were recycled.

Bell's *Animal Mechanics* received particular praise from useful knowledge enthusiasts. Henry Brougham used the first part to publicize the Library of Useful Knowledge in the *Edinburgh Review*, observing that it was

57. Knight 1864–65, 2:123. On Roget and Bell's involvement in the society, see Topham 1992, esp. 413–19.

58. Brougham 1835, 2; CB to G Bell, 3 September 1831, in Bell 1870, 320. See also Topham 1992, esp. 405–6.

"perfectly scientific, and yet perfectly popular." When Cambridge mathematician George Peacock reported on the second part for the society's publication committee, he noted that it was "equally admirable for wise principles and correct and vigorous reasoning, and for clear and animated language." Years later, Knight acclaimed it a "model of popular writing upon subjects which demand high scientific knowledge." Its combination of authority and accessibility was exactly what the society sought in its elementary treatises. The useful knowledge society had been founded with the intention of making such works available to all classes, but it placed a particular emphasis on workers. When, however, Bell's brother referred to it as a "new society for mechanics," Bell begged to differ. In his conception, the society was "more intended for the rich than the hammermen." Thus, he was able to carry over a similar mode of address to his Bridgewater, aiming for a "slight and, if possible, an elegant *sketch*, a thing of easy digestion."[59]

While Bell expected that his *Animal Mechanics* would bring him a "vast accession of praise," he was aware that such writing would do little to bring him scientific or professional credit. In the midst of writing his Bridgewater he told his brother, "You look upon this as too great a matter. I think a lecture or a paper to the Royal Society of much greater moment." It would be good to appear in company with "fine writers," Bell felt, but he longed to be "at original matter and compositions—at science rather than writing." His "character" did not depend on such publications. Yet the derivative nature of such work could actually prove problematic for one's public character. Soon after the first part of *Animal Mechanics* appeared, Bell found himself charged by the *Lancet* with having plundered materials from Benthamite physician Neil Arnott's *Elements of Physics*, published just three months previously. In the second edition of his work, Arnott himself hinted at malfeasance on Bell's part and vented his spleen by pointing out what he took to be errors in Bell's vastly more successful work. Next, the *London Medical Gazette* (the *Lancet*'s new establishment rival) weighed in, arguing that both authors had drawn on Bell's earlier works. Arnott now found himself on the defensive, claiming that the originality of his essay was in the "selection, condensation, and arrangement" of facts.[60] The incident underlined that while popular or elementary works

59. CB to G Bell, 19 January 1827, 16 February 1829, and 3 September 1831, in Bell 1870, 295, 302, 320; Knight 1864–65, 2:123; [Brougham] 1827, 524. See also Brougham to CB, August 1827, in Bell 1870, 295n, and Arnott 1827, 220n.

60. Arnott 1828, 229n–31n; "Analyses and Notices of Books," *London Medical Gazette*, 26 January 1828, 217–20; Arnott 1827, 220n–23n; "Animal Mechanics," *Lan-*

might not make reputations, they might damage them, especially in such politically charged times.

The lightness of touch and breadth of reference of Bell's *Animal Mechanics* was not typical of the Library of Useful Knowledge. The works were more usually self-consciously pedagogic, providing exactly the kind of elementary introduction to the various branches of science that matched the society's rubric. Roget's four treatises on *Electricity* (1827), *Galvanism* (1829), *Magnetism*, and *Electro-Magnetism* (both 1831) were of this sort, offering a "succinct and connected account" of the many recent discoveries in this area, "collected from the various scientific journals and transactions" and "digested in a didactic order." In order to achieve this, he had sought to condense the materials as much as was consistent with "perspicuity." Roget's referee reports for the useful knowledge society reveal that he considered "order in the arrangement of the materials" as well as the "intelligibility of the descriptions" to be key in writing for a popular audience. In his own treatises he sought particularly to take the "student" through a "regular progression" from the simpler to more complex material, distinguishing established facts from hypotheses and illustrating Baconian principles by examples.[61]

To a significant extent, Roget carried over this pedagogic approach to his Bridgewater on animal and vegetable physiology, which in two large volumes amounted to over 1,250 pages. In his preface, he expressed the hope that by outlining the "general principles" on which the science was based, his "compendium" might prove a "useful introduction to the study of Natural History." As before, "careful selection" of materials was key, although here there was a particular need to select the best evidences of design. The facts were arranged in a "methodized order" and with "comprehensive generalizations" that would fix them in the memory. Only such topics were included as would be suitable to "every class of readers," and Roget excluded the history of the subject as an unnecessary addition. To assist the uninitiated reader, he carefully explained each of the "terms of science" as they were used, providing an index to help readers navigate back to such explanations, and there was also a basic taxonomic outline.[62]

In fact, however, Roget's composition of treatises for the useful knowledge society was only his most recent venture in addressing more general

---

*cet*, 27 October 1827, 151–55; CB to G Bell, 16 February 1829 and 3 September 1831, in Bell 1870, 302, 320.

61. Roget 1832, [iii]–iv; PMR to Coates, 18 October 1827 and 28 August 1829, UCL, SDUK24, SDUK26.

62. Roget 1834, 1:viii–xiii, xxxvi–xxxvii.

audiences, and his Bridgewater also reflected his years of public lecturing and writing for encyclopedias. He particularly had to consider how a "reflective" treatise should be composed when writing the article on physiology for the six-volume supplement to the *Encyclopædia Britannica* (1815–24). Conscious of the increasingly specialized nature of contemporary science, the encyclopedia's publisher, Archibald Constable, had paid unprecedented sums to have eminent men produce commanding accounts of the history, methods, and present condition of the several sciences. Roget's article on physiology was consequently designed to supplement the full account of the "principal facts relating to the *functions* of the animal economy" already published in the encyclopedia by offering an "outline of what may be considered the *philosophical* department of the science." His purpose was to analyze the "principal laws, or ultimate facts, to which vital phenomena are reducible," a concern later also exhibited in his Bridgewater.[63] Such an account enabled the reader to transcend the technical details of the subject and to engage with its implications at a higher level.

Roget reused material from this article repeatedly in the 1820s as he lectured on physiology to fashionable paying audiences at the Russell Institution, Royal Institution, and London Institution, and he later insisted that his Bridgewater built directly on those earlier lectures. Nor was his Bridgewater the first reflective treatise he had planned on that basis. As early as October 1823, following his first two Royal Institution courses, he had reported his intention to publish the substance of them "in a separate work." Then, at one of the first committee meetings of the SDUK in November 1826, he offered to write a popular treatise on physiology before agreeing in 1829 to contribute two volumes on physiology (including comparative anatomy) to Dionysius Lardner's Cabinet Cyclopædia. Both of these plans fell through in 1829, however, in the latter case because the publishers would not give him free rein to reuse the text in future. All along, it seems, he was hoping for the opportunity to write the physiology article in the next edition of the *Britannica*, which he finally did in 1838.[64]

While Whewell, Bell, and Roget were the most attuned to the new forms of scientific writing that were possible in the rapidly changing market for print, the remaining authors were similarly concerned to find an

63. [Roget] 1824, 182. On the transformation of the *Britannica*, see Yeo 2001, 251–76.

64. PMR to Napier, 19 February 1838, BL, Add.Ms.34618, fol.565; PMR to Lardner, 18 November 1828, 19 March, and 6 April 1829, WL, MS7543/4, 7, 9; PMR to Thomas Coates, 10 January and 8 June 1829, Coates to PMR, 6 June 1829, UCL, SDUK18, SDUK25, SDUK26; *Literary Gazette*, 25 October 1823, 686c; Roget 1846a, 483.

appropriate register to combine scientific exposition with religious interpretation. They did so with varying degrees of success. Arguably, one of the least successful was William Prout. Prout had delivered lectures on animal chemistry at his London house in 1814, using the opportunity to outline his distinctive chemical philosophy, and he returned to the theme as Gulstonian Lecturer at the Royal College of Physicians in 1831, but on both occasions the audience was relatively select. In his Bridgewater, by contrast, he offered an overview of the current state of knowledge in chemistry, meteorology, and "the chemistry of organization" that was intended to be accessible to the "general reader." However, no reference was made to design for long stretches, and religious reflections were often introduced rather abruptly in concluding sections. In part this reflected Prout's decision to use the opportunity to publish a more extended account of his controversial molecular theory, when doing so had otherwise seemed unachievable, and to recycle text from an unpublished medical treatise. As a result, he found himself repeatedly advising "general readers" that they might prefer to skip esoteric passages to reach the theological arguments at the end.[65]

The extent to which it was possible for a book to make an important or original scientific contribution while still being addressed to a general audience varied considerably across the sciences. Unlike with chemistry, for instance, scientific work in both natural history and geology depended on the production of expensive, heavily illustrated works of classification and description that could only be made financially viable by appealing to country gentlemen. These were also fields where reflective treatises written for a general readership were important in defining the research agenda. Lyell's three-volume *Principles of Geology* had been just such a book, and its publisher, John Murray, had been delighted to find in Lyell one of the few authors "who c$^{d}$ write profound science & make a book readable."[66] Buckland had achieved a similar feat in his *Reliquiæ diluvianæ*, and, from a rather different theoretical standpoint, he did so again in his Bridgewater.

According to one report, when Buckland's wife Mary (fig. 1.3) asked how he intended to set about his task, he replied, "Why, my dear, if I print my lectures with a sermon at the end it will be quite the thing." In the event, he took the work altogether more seriously than this throwaway comment suggests. Nevertheless, it conveys a good sense of his concep-

65. Brock 1985, 6–7, 9, 76–77, 86–89. For "general readers," see, for example, Prout 1834a, 22.

66. Lyell to M Horner, 17 February 1832, in Wilson 1972, 343–44, on 344; Secord 2009, 463.

FIGURE 1.3 "Professor and Mrs Buckland and Frank, c. 1830." From Gordon 1894, 103. Image: from the Biodiversity Heritage Library, www.biodiversitylibrary.org, contributed by Museums Victoria.

tion of the volume. Sandwiched between a lengthy preliminary chapter on the "Consistency of Geological Discoveries with Sacred History" and a rather shorter religious conclusion, the bulk of the text provided a systematic, authoritative, and up-to-date survey of the sciences of stratigraphy and paleontology that was "intelligible to those who know nothing of Geology." This was just what he had been doing in his Oxford lectures for over twenty years, aided by specimens and diagrams. Now he invested considerable labor in the preparation of an extraordinary volume of plates to illustrate the work. What Buckland hoped he had achieved was "a popular general view of the whole subject avoiding technical detail as much as possible."[67] In accomplishing this, however, he also put together a geological synthesis and paleontological survey that was an important point of reference for practicing geologists, complete with a comprehensive index.

67. WB to Anon, n.d., BO, Ms.Eng.Lett.d.5, fol. 251; C Lyell to G Mantell, 18 January 1832, in Lyell 1881, 1:367–68.

More literally than Buckland's, Kirby's Bridgewater began with a sermon, but its author also sought to make a useful contribution to the practice of science. Drawing on an abandoned volume of sermons concerning "the mystical language of scripture," his eighty-eight-page introduction outlined views about the use of symbolic interpretations of the Bible in the study of nature that most contemporaries found strange and archaic. Thereafter, however, Kirby's nine-hundred-page work provided a systematic account of the natural history of animals, especially invertebrates, with only the briefest of conclusions. Invited to write on the "Habits and Instincts of Animals," he had immediately expanded his territory to include their "History," informing the Bishop of London that there were "many points connected with the animal economy besides their habits and instincts" that evinced design, such as "their geographical distribution, relative numbers, adaptation to certain ends, affinities, analogies, &c." The form of the work he produced, complete with an extensive index, was intended to resemble an elementary treatise. As Kirby explained to the assistant conservator at the Royal College of Surgeons, Richard Owen, his object was to make his work "as generally *useful*" as he could, providing "short descriptions of each Class & Order" as he went through the animal kingdom "so as to make it a kind of Introduction to Zoology, mixing Religion with science."[68]

Such "mixing" was not a matter of paying lip service to the Bridgewater bequest. Concerned about his ability to rise to the challenge, the aging Suffolk rector confessed to his close friend and brother-in-law Charles Sutton that the decisive factor in his agreeing to take on the work had been that he would "incur blame justly" if he refused to apply his talents as a naturalist "to the service of the sanctuary." A similar concern had been evident in both Kirby's technical taxonomic study *Monographia apum Angliæ* (1802) and the *Introduction to Entomology* (1815–26). In the latter, Kirby and Spence declared that one of their "first and favourite objects" had been to direct the attention of readers, in Alexander Pope's often repeated phrase, "from nature up to nature's God." For Kirby, the Bridgewater was another opportunity for writing both comprehensibly and authoritatively about natural history in a way that was sanctified by constant reference to the divine author of nature. The account also resembled the *Introduction to Entomology* in the accessibility of its prose, although it was naturally not in the same epistolary form that had been adopted there with a view to

68. WK to Owen, 31 January 1835, RCS, MS0025/1/6/3; WK to CJ Blomfield, [November] 1830, in Freeman 1852, 437; see also Freeman 1852, 433, 440.

allowing the "digressions and allusions called for in a popular work" and the practical instructions required in an introductory work.[69]

In their different ways, each of the Bridgewater authors sought to reflect on and interpret the sciences for an audience extending beyond those actively engaged in research. In so doing, they drew on a growing repertoire of forms of popular writing about science. Several of the works, however, were intended to address multiple audiences, and some made important contributions to technical scientific debate. In part, the decisions taken by the authors concerning the audiences to be addressed and the appropriate genres for addressing them reflected differences between the sciences. More mundanely, though, the treatises took their diverse forms in part because of the pragmatic demands of authorship. Faced with an unexpected commission, the various authors had to find a way of putting together something appropriate in a timely fashion, and their strategies for doing so also shaped the resulting treatises to a surprising extent.

## *The Task of Writing*

While the practices of scientific authorship remain notably underexplored, it is easy to discover that practicalities were often as important as more high-minded concerns in determining the shape of the finished product. To a significant extent, for instance, the Bridgewater authors wrote by reference to their own previous publications and to their manuscript writings, including lecture notes and sermons. In some cases there was serious research to be done, which might involve reading, examining specimens, correspondence, and conversation. But always, in the end, there was the task of putting the text together—writing, editing, and revising—often with the input of friends and family.

Writing anonymously in 1839, the political economist Harriet Martineau advised the aspiring author that it was essential "that his social engagements should not interfere with his labours of the study," that he "keep his morning hours (and they must be many) not only free but bright," that his solitude be unprofaned by vanity, and that he also be "active in some common business of life." Earlier in the decade, as she was setting out on her own career as an author, Martineau had often discussed the practice of writing with her morning callers, finding that they differed significantly in their views about the best time for it, the degree of solitude required,

69. Kirby and Spence 1815–26, 1:ix–xi; Clark 2009, 28–30; Freeman 1852, 436.

the proper rate of progress, the appropriate amount of revision, and the value of comments from others. Somerville, for instance, wrote in the midst of her family, working from midmorning until 2 p.m. and oblivious to surrounding conversation. Lyell began late and produced only a "very few pages," since his work required research as he proceeded. Chalmers considered it a "heavy sin to write (for press) longer than two hours per day," since this was the "severest labour" that the brain could endure. It was, Martineau reflected, a mystery how he could produce so many books at that rate. The only point of agreement was that "the act of authorship is the most laborious effort that men have to make"[70]

Martineau thought that Bell must have written his Bridgewater "with the grave sincerity with which he did every thing." Fortunately, his wife Marion left a rather more detailed account of the process, and it was one in which she was intimately involved. Bell's attempt to address a nonprofessional audience was embodied in an ongoing conversation with Marion, and her involvement ensured that the resultant work was not only nontechnical and engaging but also suitable for a mixed readership. At the start of their marriage, when Bell had wished for his wife's empathy in his scientific work on the nervous system, he had shown himself to be "kindly sensitive" to her notion of feminine delicacy. For her, nerves became merely "connecting threads, traced by his pencil and showing 'design,'" all blood and gore washed away. In these early days, Bell promised his wife, "I'll write a book that may be on your drawing-room table." His *Animal Mechanics* and Bridgewater Treatise fulfilled that promise, but Marion had to help him to do the work.[71]

The initial prompt was bereavement. Following the death of Marion's sister and brother in 1827, she and Charles were both grief stricken. He hoped that involving his wife as an amanuensis on his design publications would occupy her thoughts, and so they began. Marion claimed that the works "came easily to him," because he was used to "thinking for his lectures." She recalled, "While dictating, he would stand, as if absorbed, then speak in a voice, I suppose, much the same as if he lectured, slow and deliberate, then come, put his finger on the paper, and almost breathless, finish his sentence or observation." For the lecturer, audience is key, but the audience in this case was different from the usual. The *Lancet* later claimed that the "entire matter" of Bell's Bridgewater had been included a couple of years earlier in his fifteen comparative anatomy lectures as professor at the Royal College of Surgeons but that it had now been utterly transformed

70. Martineau 1877, 1:429–35; M[artineau] 1839, 279.

71. Bell 1870, 405.

in form. "How Proteus-like is the character of modern literature!" the reviewer exclaimed, although this was an instance in which it was a pleasure to recognize the "old features beneath a new mask."[72]

Bell's difficulty in getting started with the new work, Marion recalled, focused on the question, What class of readers is it intended for? Addressing his wife directly helped Bell to answer that question. "After various beginnings and tearings up, one day, in driving from Sevenoaks, he said, 'Now, May, mark that down. It must do.'" His composition nevertheless came in fits and starts. Bell sometimes seemed "inspired." At other times he seemed "labouring" and then Marion would read to him while he drew, modeled, or etched. Being away from work in the countryside unfailingly revived his "sense of devotion and delight," especially when he was fishing. He later told his brother,

> If there be any best bits in the *Hand*, they were written after a day of comparative retirement and relaxation at Panshanger or Chenies [aristocratic estates at which he fished]. I have tasked myself while throwing a line how I should express myself on going to the little inn to tea. It is thus that one has the justest and fairest views of nature, which I believe would never rise into the mind of him who has the pressure of business upon him, at least such business as mine.

The only part of his working life that inspired him in this way, he claimed, was when he was surrounded by his class and felt their hunger for knowledge. On the whole, then, Bell's approach to writing for a nonprofessional audience depended on his finding the appropriate tone in domestic intercourse and rural relaxation. When his brother threatened to undo this accomplishment by offering advice about making the work more substantial, Bell was frustrated. It would, he felt, "make a large and consequently a tedious volume, which nobody would read, however good."[73]

Like Marion Bell, Mary Buckland acted as her husband's amanuensis, sitting up "night after night, for weeks and months consecutively" while her husband dictated his Bridgewater. It had to be at night, William reported, since it was a "physical impossibility" to work by day in Oxford.

72. *Lancet*, 26 October 1833, 166. The Hunterian lectures did not take place in 1831 due to reformist disruption at the college (Parr 1900, 6–7), and Bell's 1832 lectures were not printed, but see William Home Clift's notes in RCS, MS0239(6) and the following year's lectures in Bell 1833–34.

73. CB to G Bell, 5 December 1818, [1831], 3 September 1831, and 29 March 1835, in Bell 1870, 264, 318, 320, 340; Bell 1870, 405–7.

Once again, it was the wife who mediated between the lecture room and a larger public, and Mary's son later recalled that she gave to William's works an added "polish" as well as contributing illustrations. William was altogether more comfortable as a lecturer than as an author. In 1823, flushed with the success of his *Reliquiæ diluvianæ*, he had discussed the possible publication of his Oxford lectures with publisher John Murray, observing that it would be demanding and time-consuming work—a prospect he found unappealing. "I lecture only from short notes," he observed "& often without a single note at all—The only way w[oul]d be to employ a short hand writer to take Down what I say." Indeed, while Buckland's surviving lecture notes are typically on quires of papers that sometimes include a wide range of materials for consultation, it was the first page—outlining a series of heads to be covered—that clearly formed his prompt in the lecture room. Thereafter, his larger-than-life lecturing persona took over, as he played the part of a Regency showman, sometimes even imitating the behavior of extinct creatures. Thus, when Buckland set about building his Bridgewater on the framework of his lectures, he was faced with a not inconsiderable task. As one Oxford contemporary noted, he was a "good talker" but a "bad writer" and "afraid of proving himself so."[74]

Buckland typically delivered two lecture series each year, and he initially planned to begin his treatise with an account based on his mineralogy lectures before proceeding to one based on his geology lectures. He accordingly begged the loan of Whewell's recent report to the British Association on the "Recent Progress and Present State of Mineralogy," professing himself "anxious for the sake of avoiding errors in my Bridgewater to learn what is now the right faith." In the end he left the technicalities of mineralogy alone, adding a few juicy examples of mineralogical adaptations to the end of a treatise based broadly on his geological lectures, which nevertheless underwent revision in the process. In the 1820s the course had been advertised to cover in sixteen lectures "the Composition and Structure of the Earth, the Physical Revolutions that have affected its Surface, and the Changes in Animal and Vegetable Nature that have attended them." As Buckland began writing his Bridgewater in the early 1830s, however, his course reflected the work, becoming shorter and focused specifically on "the Organic Changes in the Structure of the Animal

74. HH Wilson to J Prinsep, 24 July 1834, private collection (courtesy of Andrew Grout); WB to Murray, 10 October 1823, NLS, MS.40165, fols. 19–20; WB to G Featherstonehaugh, 29 November 1835, CUL, Add.Ms.7652, II.LL.33; Buckland 1858, 1:xxxvi. On Buckland's lectures, see OUMNH–BP, Lecture Notes, and O'Connor 2007a, 80.

and Vegetable Kingdoms which have followed the Physical Revolutions that have taken place upon the Surface of our Planet."[75]

Student notes from 1832 make clear that the stated object of these lectures was to give "an account of the different states of our planet as a receptacle for animl. & vegb life." Beginning with the earth in a molten state, the lectures outlined the historical sequence represented by the strata. There then followed a detailed account of the history of life on earth during its different epochs. This broadly paralleled the distinctive structure of Buckland's Bridgewater, with its emphasis on the progressive history of the earth and its inhabitants as revealed by stratigraphic and paleontological research. Other aspects of the lectures foreshadowed the religious emphases of Buckland's treatise. His first lecture gave an account of how geology was to be reconciled with the scriptures that was similar to the one at the commencement of his published work. Elsewhere he emphasized (as in his Bridgewater) the value of geology in undermining the belief in species transmutation that he claimed was endorsed by a third of Continental philosophers. He also highlighted instances of anthropocentric design in the distribution of mineral veins that were later committed to print. Keen to show that he had been the first to correct Paley's disanalogy between a stone and a watch, Buckland reported that the relevant paragraphs of his Bridgewater came almost verbatim from the notes for one of his 1822 mineralogy lectures.[76]

While Buckland took the broad outline and many of the themes of his Bridgewater from his lectures, he entered into a major program of work to flesh them out. Indeed, the quantity of this "new matter" was one justifiable excuse for the endless delay in publication, although organizing the many illustrations also held things back, as he repeatedly pointed out. In May 1833 he claimed to have written as much as he "engaged to prepare by the time originally agreed upon" and that a requested extension would enable him "to extend the text as well as y[e] plates much beyond the amount originally contemplated."[77]

75. *Jackson's Oxford Journal*, 1 May 1830, 3a, and 19 May 1832, 3a; Registers of Attendance at Lectures, 1814–49, OUMNH-BP, Misc.mss./13; Edmonds and Douglas 1976, 148; WB to WW, 26 July 1832, TC, Add.Ms.a.66[28]. See the changing plans for Buckland's Bridgewater in OUMNH-BP, Lecture Notes 1/2.

76. Buckland 1836, 1:572–75 (cf. Boylan 1984, 597; OUMNH-BP, Lecture Notes 3/5–3/6); Boylan 1984, 619 (original notes in BGSA, IGS1/635). On the parallels, see Boylan 1984, 214–15, and the student notes at 582–653. Compare also Buckland 1820, 17, 19–21n, 31–32, with Buckland 1836, 1:70n, 541–47, 19.

77. WB to Children [draft], 8 May [1833], BO, Ms.Eng.Lett.b.35, fol. 44; WB to Featherstonehaugh, [25 April 1836], CUL, Add.Ms.7652, II.LL.32.

A series of folders in the Oxford University Museum of Natural History testify to this work. Some are keyed to individual chapters, others to various topics and themes, although the folders cannot be separated completely from the sets of notes kept for lectures and for other publications. Into these working folders, Buckland gathered a vast array of slips of paper, including references, notes, translations, printed pages, letters and illustrations from various informants, sets of queries, and drafts of text. Many of the materials relate specifically to sources published during the years in which the Bridgewater was written, as Buckland sought to bring his account right up to date, especially concerning paleontology. The paper trail often leads directly from Buckland's files to the text of his treatise, with confirmation appearing in the densely clustered footnotes. In addition, he examined a wide range of specimens first hand, including the newly acquired *Megatherium* at the Royal College of Surgeons. Buckland famously lectured on this specimen at the Oxford British Association meeting in 1832, but his most complete published account appeared in the Bridgewater.[78]

As Martineau had observed with Lyell, Buckland's combination of writing with research could only be expected to proceed slowly. However, it proved exceedingly protracted, not least because of the changes of plan. Extravagant rumors abounded about how he had destroyed the entire manuscript after completing it or had decided to rewrite large portions to accommodate new theories. Certainly some carefully drafted text on mineralogy became redundant, other text was excised because "too theoretical," and other text again was corrected and overwritten until it was almost illegible (fig. 1.4). Buckland consequently employed the underkeeper at the Ashmolean Museum to write out portions of the text neatly, but these drafts were subjected to further modifications, especially after friends had commented on them. One such friend recalled "nearly the whole of the MS" being in Mary's handwriting.[79] Altogether, Buckland's tortuous writing process became a byword among his peers.

78. Buckland 1836, 1:139–64 (cf. "Proceedings of the General Meeting, 1832," *Report of the* [. . .] *British Association for the Advancement of Science*, 1833, 95–110, on 104–7; Gordon 1894, 126–33; Boylan 1984, 650–51). For the folders, see OUMNH-BP, Notes by Subject and Lecture Notes 1/1–1/6.

79. [Broderip] 1859, 228; OUMNH-BP, Notes by Subject (Bridgewater Treatise) $1/4^{22}$, $2/7^{5}$, $2/9^{19-23}$, $4/6^{15}$; "The Late Rev. Dr. Buckland," *Literary Gazette*, 23 August 1856, 615–16, on 615; "Natural Theology," *Christian Remembrancer* 33 (1857): 70–117, on 80. On the assistance of friends, see, for example, Buckland 1836, 1:[ix], WJ Broderip to WB, 2 and 5 April 1834, Notes by Subject (Bridgewater Treatise) 3/6, WB to R Owen, 15 July 1835, RCS, MS0025/1/5/9, and WB to A Sedgwick, 28 October 1835, CUL, Add.7652, IB.44.

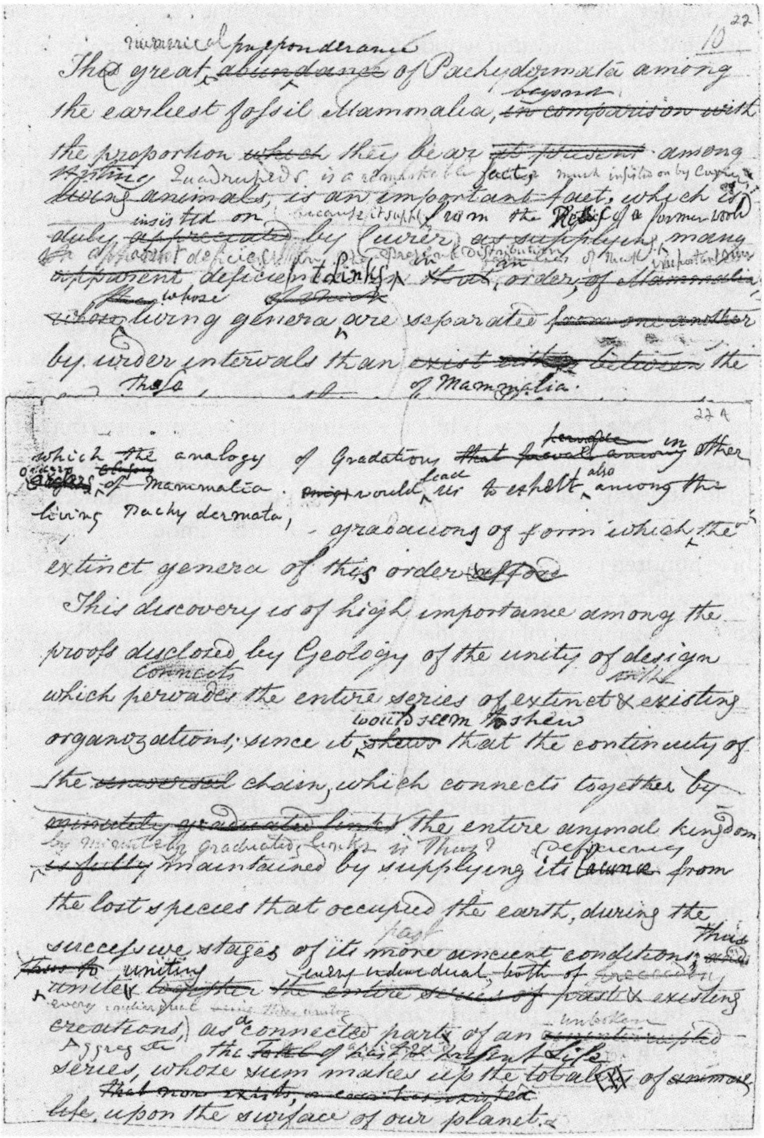

The great abundance of Pachydermata among
the earliest fossil Mammalia, in comparison with
the proportion which they bear at present among
living animals, is an important fact, which is
duly appreciated by Cuvier, as supplying many
apparent deficiencies in that order, of Mammalia
whose living genera are separated from one another
by wider intervals than exist between the
of Mammalia.

which the analogy of Gradations that prevail among other
orders of Mammalia would lead us to expect also among the
living Pachydermata, - gradations of form which the
extinct genera of this order supply

This discovery is of high importance among the
proofs disclosed by Geology of the unity of design
which pervades the entire series of extinct & existing
organizations; since it shews that the continuity of
the universal chain, which connects together by
minutely graduated links the entire animal kingdom
is fully maintained by supplying its lacunae from
the lost species that occupied the earth, during the
successive stages of its more ancient conditions;
uniting together the entire series of past & existing
creations, as connected parts of an uninterrupted
series, whose sum makes up the totality of animal
life upon the surface of our planet.

FIGURE 1.4 A portion of the draft manuscript for William Buckland's Bridgewater Treatise. Buckland Papers, Notes by Subject (Bridgewater Treatise), 1/4[22]. Image: © Oxford University Museum of Natural History.

Chalmers, by contrast, exhibited the iron discipline encapsulated in his comment to Martineau in working his lecture notes into a suitable form for publication. In fact, Chalmers was well practiced in making the transition between spoken and printed form, having done so not only with his sermons but also with his political economy lectures. He preached and lectured from manuscripts, so it is not surprising to find some of the sonorous phrases recorded in student notes of the 1820s repeated in his Bridgewater.[80] Clearly, however, adjustments of both style and content needed to be made.

For a start, Chalmers's lectures were intended to offer a relatively thorough grounding in natural theology, and they included an account of Samuel Clarke's a priori argument and a reply to David Hume's criticism of the argument for a first cause. While it was important in a divinity course to thus "clear away the injurious metaphysics" of the theology of "academic demonstration," this was not pertinent to the purposes of the Bridgewater. Indeed, even when he reinstated the excised matter (amounting to nearly three hundred printed pages) in his *Natural Theology*, he told readers that they could pass over much of it as merely prefatory. In the Bridgewater, however, Chalmers only provided a very brief overview of the philosophy of the subject before launching into his highly practical argument from the "supremacy of conscience." But there were also additions. Given his stated subject, Chalmers found it necessary to work up extra materials on the "intellectual constitution of man" and some on the "moral constitution of man" that were not retained for the *Natural Theology*.[81]

This recycling of materials is a strikingly recurrent theme among the Bridgewater authors and extended beyond the use of lecture notes to include the text of other books, whether already printed or in manuscript. In Prout's case, the commission gave an opportunity to rework the manuscript of the *Observations on the Functions of the Digestive Organs* that he had been close to publishing in November 1823. Prout's experimental program on the chemistry of living things had led to papers in the *Philosophical Transactions* that won him the Copley Medal in 1827, but the demands of his medical practice had made it difficult for him to complete his experiments and prepare them for publication in the form he wished. His planned book had been intended to combine his experimental findings concerning the chemistry of digestion and assimilation with practi-

80. [Mason] 1864, 127; Chalmers 1832, iv–v; *Evidence on the Universities of Scotland* 1837, 37:106.

81. Chalmers [1836], 1:x–xi; *Evidence on the Universities of Scotland* 1837, 37: 104, 106.

cal remarks, but for various reasons it had not appeared. Commissioned to write a third of his Bridgewater on the subject of digestion, he was now able to recycle the text wholesale. The manuscript does not survive, but one portion of it was quoted in another work by Prout's close friend, the physician John Elliotson, and from this it is clear that Prout incorporated some of the earlier work into his Bridgewater almost verbatim.[82]

In writing his work on digestion, Prout had no doubt followed the principle of his book on urinary diseases in making the chemistry accessible to the untutored medical reader. Nevertheless, he did not cut and paste the text into his Bridgewater without revision. The new work provided an opportunity for Prout to update his unpublished account using his later experimental findings on the presence of "muriatic" (hydrochloric) acid in the stomach and on the three classes of "alimentary matters" (saccharine, oleaginous, and albuminous). He also needed to incorporate reflections on the manifestations of God's attributes, although these typically appeared rather abruptly at the end of the scientific discussion. Nevertheless, the character of the exposition underwent relatively little modification. Indeed, when he subsequently published an expanded medical work covering both stomach and urinary diseases, he recycled a large body of text from the Bridgewater with little alteration. The medical work naturally had more to say about the chemistry of disease, but on other topics it was the Bridgewater that carried the more elaborate discussion, as footnotes pointed out. In his section on chemistry, Prout reworked and developed his contested matter theory, which to date had only seen the light of day in two initially anonymous articles in the monthly *Annals of Philosophy*. His medical labors, he noted, had made him despair of being otherwise able to submit his views "to experimental proof."[83] For Prout, the reworking of directly relevant material that might otherwise have languished unseen thus offered a double benefit.

By comparison, Kidd's reuse of approximately half the text of his 1824 *Introductory Lecture* on comparative anatomy appears distinctly cavalier. Some of the recycled passages certainly seem to have been included more for the author's convenience than for any more satisfactory reason. This applies, for instance, to the chapter on the powers of the human hand—properly the domain of Bell—which consisted almost entirely of a quotation from Galen used previously in the introductory lecture. Similarly, Kidd's recycled observations on "*lusus naturæ*" (abnormally developed fe-

82. Elliotson 1828, 310–12; compare 311–12 with Prout 1834a, 479–80.

83. Prout 1834a, xi. Compare Prout 1840, i–iii, ix–xxxii, with Prout 1834a, 415–16, 420–21, 473–81, 487–92, 494–505.

tuses) sat rather oddly in his chapter on the adaptation of external nature to the human intellect.[84] Most of the reused material, however, found its way reasonably naturally into the first five chapters on the physical condition of the human species, and the following four lengthy chapters on the anthropocentric adaptation of the atmosphere, minerals, vegetables, and animals were newly written, as was most of the chapter on adaptations to the human intellect.

Roget's expansion and reworking of his earlier publications and lectures operated in a more positive way. Building on the outline, some of the examples, and even some of the text of his public lectures, he was conscious that rapid developments in physiology meant that he nevertheless needed to update his materials.[85] When in 1829 he planned to write two volumes on animal physiology for Lardner's Cabinet Cyclopædia, he reported that, while he already had "a tolerable foundation for the work" in his "written Lectures," it would "require considerable labour to bring up the subject to the present improved state of the science." He was no less assiduous in preparing his Bridgewater. His preface outlined his extensive course of reading, notably in recent Continental comparative anatomy and physiology, but also acknowledged his debt to several lecture courses, including those of Robert Edmond Grant, the professor of zoology and comparative anatomy at the new London University.[86]

Roget's borrowings from Grant's lectures landed him in a bad-tempered dispute about alleged plagiarism. Grant was a political radical and materialist who had adopted Lamarckian views on the transmutation of species. He was also one of the best-informed British naturalists in regard to the "philosophical anatomy" of the prominent French naturalist Étienne Geoffroy Saint-Hilaire and his acolytes, and he incorporated their findings and theories into his London lectures. Roget was well aware of Continental developments, and Grant's lectures were an obvious means to acquire a more detailed and extensive knowledge. He consequently attended Grant's course of sixty-four lectures starting in April 1832, a further course of fifty-eight lectures starting the following October, and the end of the

84. Compare the parallel passages in Kidd 1824, 6, 7–9, 12–13, 25–29, 55–61, 61–62, 67–68, 69–72, and Kidd 1833, 17, 13–14, 14–15, 4–8, 53–59, 59–60, 67–69, 61–64. For the hand, compare Kidd 1824, 14–25, with Kidd 1833, 30–43. For *lusus naturæ*, compare Kidd 1824, 35–38, with Kidd 1833, 335–38.

85. Roget's Royal Institution lectures of 1822, 1823, and 1825 were all reported at length in the *Literary Gazette*. These were echoed by the series delivered at the London Institution in 1824, 1825, and 1826, which were in turn developed in Roget's Bridgewater. See [Thomson, Brayley, and Upcott] 1835–52, 1:xxxvi.

86. Roget 1834, 1:xi–xii; PMR to Lardner, 19 March 1829, WL, MS7543/7.

next course at the start of 1833. According to Grant, Roget took copious notes and received additional explanations from him after each lecture as well as "references to books, plates, and diagrams." Once his book began to be printed, Roget asked Grant to look over the proofs. He had made numerous references to Grant's original published work on invertebrate zoology, but Grant suggested that, considering how much of Roget's material had been derived from his lectures, he should also make an appropriate citation in such cases. Relations rapidly deteriorated, leading to an angry exchange of letters. Meanwhile, Grant sought to regain some degree of control by ensuring that his next course of sixty lectures appeared in the pages of the *Lancet*.[87]

Roget never denied having used materials from Grant's lectures in writing his Bridgewater. Rather, he robustly defended a position that the *Lancet's* editor Wakley had made his own—that public lectures (which Roget had paid £3 per session to attend) were public property. As Roget told Grant in November 1833, the "*Facts, and details of facts*" that had been communicated to him "as a regular pupil on the lectures of a public professor" were his to do with as he wished, especially when no claim to originality was made by the lecturer. Grant had accused Roget of giving a description and illustration of the elements of a vertebra and the parts developed from those elements that were "nearly identical" to those in his lectures. Roget replied that Grant's account expressed the same views of the subject taken by Geoffroy Saint-Hilaire and others and that his diagram was barely distinguishable from published ones. He nevertheless offered to give Grant the credit if he considered this or any other "views or modes of illustration" to be his own original work, and he claimed to have stopped the presses for over a week while awaiting a reply that never came.[88]

That Grant merely grumbled rather than taking further action tends to confirm that this was a matter more of etiquette than of strict legality. He was not surprisingly aggrieved that Roget capitalized on his expertise in a lucrative publication that interfered with his own plans. Perhaps more aggravating still was that Roget's treatise interpreted the work of Geoffroy Saint-Hilaire, of which Grant was the chief British exponent, in a Christian light.

Roget and Buckland were not the only authors to engage in extensive programs of research in the writing of their treatises. Kirby, whose treatise took him into realms of natural history far beyond his own areas of expertise, was planning as early as December 1830 to take lodgings in London for

87. Grant 1846; Roget 1846a. See also Roget 1846b and Desmond 1989b, 228–34.
88. Roget 1846a, 483; Roget 1846b, 420.

some months in order to be close to libraries and museums. He had spent heavily in building up his own library as he worked on the *Introduction to Entomology*, and it contained important works of reference that included Lamarck's *Histoire naturelle des animaux sans vertèbres* (1815–22) and the thirty-six-volume *Nouveau dictionnaire d'histoire naturelle* (1816–19) that his friend William MacLeay had helped him obtain from Paris. Not only did he carefully compare and evaluate such works against each other but he also examined specimens and sought authoritative advice from staff at the British Museum, the Zoological Society, and the Royal College of Surgeons. The surgeons' conservator William Clift and assistant conservator Richard Owen had been particularly helpful in repeatedly showing and explaining specimens to him, and Kirby had benefited greatly from Owen's "deep knowledge of comparative anatomy" and "familiar acquaintance with the classification of the animal kingdom." Even five months before publication, Owen was still receiving long letters from Suffolk urgently soliciting his taxonomic judgement.[89]

Like Buckland, Kirby found such work time consuming, and the two compared notes at the end of 1833. Believing that the agreed June publication deadline was binding, Kirby reported, he had "concocted" his first volume more hastily than he liked, getting it printed in the spring. Now he had learned that Gilbert was not unduly troubled by the prospect of a delay and that he had never intended hurrying the authors. Kirby thus took a further eighteen months to complete, and Buckland took almost three years. However, the delay meant that Kirby's treatise came out after Roget's, and he thus found himself with the extra labor of excising portions of the manuscript of his second volume to avoid repetition.[90]

Whewell also engaged in a significant research program. For this he needed the leisure of the long summer vacations, which he was obliged to spend in Cambridge since the libraries were "so indispensable" to what he was doing. By the summer of 1831 he was deep in reading on his new "pet" science, meteorology, finding the frustratingly unmathematical *Meteorological Essays* of John Frederick Daniell especially useful. This labor yielded "a most admirable set of Bridgewaterisms," which he was keen to commit to paper. Whewell was an inveterate notetaker and had maintained a read-

89. Kirby 1835, 1:civ; WK to Owen, 20 February 1834 and 31 January 1835, RCS, MS0025/1/5/9, MS0025/1/6/3; WK to C Sutton, 1 December 1830, in Freeman 1852, 438–39; Kirby and Spence 1815–26, 1:x–xi, 4:[573]–89; letters from WK to WS Macleay, 1817–21, LSL, MS237B; Corsi 1988, 239.

90. Kirby 1835, 1:civ–cv; Chiswick Press Cost Book A, BL, Add.Ms.41885, fol. 94; WK to WB, 21 November 1833, OUMNH–BP, Notes by Subject (Bridgewater Treatise), 3/2.

ing diary since 1817. By 1822 the folded sheets had needed indexing, and the whole thing now stood several inches thick. As he found his new subject growing on him, he was keen not to lose his passing "trains of thought," recording them in a "valuable copy book."[91] Moreover, even at this early stage, he was keen to have the comments of his close friend, the clergyman and political economist Richard Jones, on his musings. Given that he was being paid so handsomely, Whewell felt it inadequate merely to know that he was right; he needed to know that he could hold "people's attention and approbation." Thus, Jones found himself repeatedly importuned for advice as the manuscript took shape and Whewell routinely revised his drafts to achieve a "coherence and obviousness of object." In all of this, however, he had his female relatives in mind as ideal readers, and as publication approached he also sought advice from Elizabeth, Marchesa di Spineto—a popular Cambridge hostess who was married to a teacher of Italian and was one of his closest friends in the town.[92]

One of the most salutary lessons that comes from paying attention to the practicalities of scientific authorship is that far from its being an exclusively solitary activity, carried out in seclusion by these well-known "men of science," it proves often to have been a social activity that involved women in consequential and hitherto unsuspected ways. This was perhaps especially the case with the Bridgewaters, as the authors sought to find the appropriate register for a mixed audience, but in more fundamentally practical ways, some of the authors depended on female relatives to make it possible to get pen on paper. Writing depended on interlocutors and amanuenses; on informants, lecturers, and mentors; and on the repurposing of existing texts, all of which materially affected the outcome.

❋

Despite their best intentions, authors find that the practical demands of the task of writing obtrude on the written work produced. Given how the Bridgewater authors wrestled with the press of work, the frustrations of

91. Todhunter 1876, 1:147 (I have been unable to locate the reading notebook in the TC collection); Notes on Books Read, TC, R.18.9[1–13]; WW to Jones, 23 July 1831, in Todhunter 1876, 2:124–25; WW to A Whewell, 16 February 1830, in Douglas 1881, 138; WW to A Whewell, 22 July 1831, TC, Add.Ms.c.191[118].

92. WW to J C Hare, 17 February 1833, and WW to Jones, 23 July 1831 and 2 December 1832, in Todhunter 1876, 2:124–25, 149, 160 (for further discussion with Jones, see the letters in TC, Add.Ms.c.51); WW to A Whewell, 21 April and 5 June 1833, Add.Ms.c.191[129–30]. On Spineto see also Distad 1979, 53, and Clark 1900, 5.

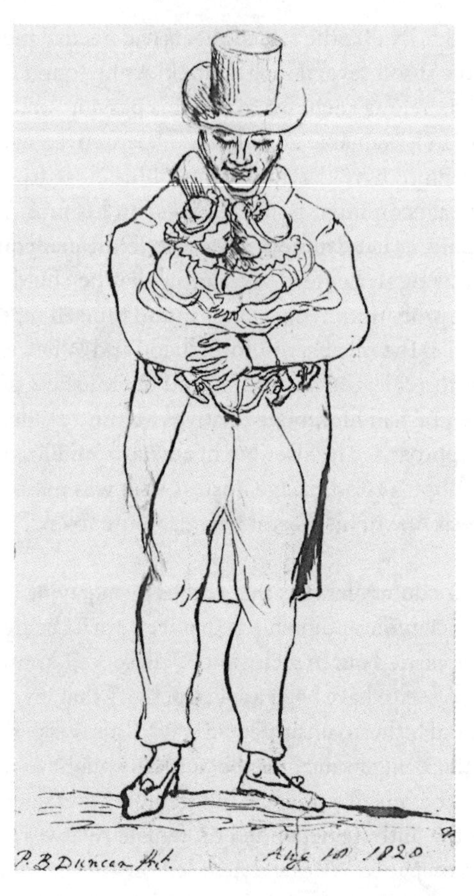


FIGURE 1.5 William Buckland holding his son Frank and "Master [A.] Loveday." Pen and ink drawing by P. B. Duncan, 18 August 1828. RCS Archives MS0035/1. Image: from the Archives of the Royal College of Surgeons of England.

unfinished manuscripts, the limitations of information sources, and the difficulties of getting appropriate words on the page, it would be naive to think that their treatises ever simply reflected an idealistic vision of the task in hand. Instead, they exhibited a degree of generic uncertainty and were shaped by an array of practical constraints that help to account for their sometimes striking idiosyncrasies.

Like most authors, those who wrote the Bridgewaters were harassed and strained by the demands of life. In February 1831 Whewell told his sister that the work would have to wait for the summer vacation, observing, "I can scarce find time for my common employments now, and shall want a little leisure for that." A year later it was still not written: he had a new edition of one of his textbooks to prepare. Peter Mark Roget had it harder. His wife, not yet forty, was dying of cancer and did not live to see his treatise published. In Oxford the Bucklands had plenty to distress them too. In 1833 Mary Buckland wrote to Whewell,

> By way of encouragement to my husband's labours, we have had the Bampton Lecturer holding forth in St. Mary's against all modern science, (of which it need scarcely be said he is profoundly ignorant) but more particularly enlarging on the heresies and infidelities of geologists, denouncing all who assert that the world was not made in six days as obstinate unbelievers etc. etc. [ . . . ] Alas! my poor husband—Could he be carried back a century, fire and faggot would have been his fate, and I daresay our Bampton Lecturer would have thought it his duty to assist at such an "auto da fé."

Then Mary was seriously ill following a miscarriage, and in the spring of 1835 their five children contracted the whooping cough, which proceeded to carry off the two infants (fig. 1.5).[93] It was against the backdrop of such personal, professional, and financial concerns that the authors wrote, and the choices they made in putting together their Bridgewaters reflected the daily exigencies of their lives. Yet they each also had a larger vision of how the sciences properly connected with the Christian religion, and the next chapter explores how the authors sought to embody that in their work.

93. *Gentleman's Magazine*, n.s., 3 (1835): 445; WB to S Compton, 10 May 1835, in Compton 2014, 12–13; Morrell and Thackray 1981, 230–31, 233–36; Rupke 1983, 215–18; M Buckland to WW, 12 May 1833, TC, Add.Ms.a.66[31]; WW to his sister, 16 February 1831, in Douglas 1881, 138.

✷ CHAPTER 2 ✷

# *Writing God into Nature*

*The subject proposed to me was limited: my prescribed object is to lead the friends of religion to look with confidence and pleasure on the progress of the physical sciences, by showing how admirably every advance in our knowledge of the universe harmonizes with the belief in a most wise and good God.*

WILLIAM WHEWELL, *Astronomy and General Physics*, 1833

As the 1833 Bampton Lectures at the University of Oxford made abundantly clear, the rapid advance of science in the age of reform seemed to some to be an unmistakable assault on Christianity. Speaking as the first of the Bridgewaters went through the press, the High Church lecturer Frederick Nolan felt sure that the publications of the Society for the Diffusion of Useful Knowledge, designed to diffuse science, possessed "a tendency hostile to Revelation" and were clear evidence "of an organized conspiracy in active operation for the subversion of all religion." While the scientific study of nature might provide evidence of God's attributes, scientific men typically sought to account for all natural phenomena in terms of physical laws. This was tantamount to supporting atheism, and, taking into account the damage science had done to the cause of revelation, one must be justly fearful of the consequences. A similar situation had, Nolan observed, not long before resulted in bloody revolution a few miles across the English Channel. These were "portentous" times, he warned, recalling the prophetic mood of his recent millenarian writings.[1]

Invited as leading men of science to write on "the Power, Wisdom, and Goodness of God, as manifested in the Creation," the Bridgewater authors had a perfect opportunity to answer such antagonists and to provide their own account of the religious bearings of the latest scientific findings. As the previous chapter showed, while four of their titles promised to exam-

1. Nolan 1833a, v–vi.

ine the sciences "with reference to natural theology," none of the authors felt called on to write a treatise of natural theology like Paley's. Rather, in their different ways, the authors sought to respond reassuringly to the wide range of anxieties manifested by their contemporaries concerning the tendency of scientific knowledge. This chapter examines each of the Bridgewaters in turn with a view to uncovering how each of their authors sought to demonstrate that modern science, far from undermining Christianity, confirmed and expanded its teachings.

Ranging across the several sciences and involving eight authors and twelve volumes, the Bridgewaters certainly encompassed the "infinite variety of arguments" that had been specified in the late earl's will. For the modern reader, as for readers in the 1830s, their bulk amounts to a significant challenge, and it is not surprising that few attempts have been made to offer an overview of them.[2] Taking the treatises in pairs, this chapter does just that, examining the purposes and strategies of the authors in making modern science speak of the "Power, Wisdom, and Goodness of God" and offering a sense of how each of the books was constructed. It is necessarily highly selective but draws to the front some of the most telling ways in which the authors sought to rebaptize the rapidly changing sciences at this pivotal juncture.

The first pair of treatises discussed—those by the theologically sophisticated clerical professors, Whewell and Chalmers—were among the first to be completed. Despite their rather different subject matter, they exhibited striking similarities in their claims concerning the significant limitations of natural theology. Both authors sought to show that scientific advances were no threat to Christianity and that the sciences could contribute meaningfully to the religious life, but they were also keen to show that Christian belief was rooted in appropriate habits of mind and an exercise of the will much more than in rational argument. Each was happy to allow that some knowledge of God was possible independently of divine revelation, but they nevertheless sought to make clear the very strict limits that they claimed surrounded natural theology.

In order of publication, Chalmers's Bridgewater was sandwiched between a pair of treatises by medical professors Kidd and Bell that were strikingly different. Neither of these authors felt comfortable in developing complex theological arguments, and Kidd was especially candid in declining to do so. Instead, what the pair offered, respectively, were accounts of

2. Notable exceptions include Gillispie 1959, 209–16, Robson 1990, Addinall 1991, 87–118, Topham 1993, 179–278, and Topham 2010. For older doctoral studies, see Thompson 1949 and Dahm 1969.

the human species and of the human hand that drew on their teaching of medical students, shoring up readers against perceived threats from materialism and transmutationism while offering examples of design in nature. Many of these examples bore strong resemblances to the kinds of examples of functional adaptation found in Paley's *Natural Theology*, but a generation on, the details were radically updated. Moreover, the debate about the relative importance of form and function in the study of anatomy that had been taking place in Paris between leading naturalists Georges Cuvier and Étienne Geoffroy Saint-Hilaire and had been echoed in London's medical schools left Bell in particular more aware of the need to make visible the "great plan of creation" that could be witnessed in animal form.

It was the following year before the next pair of treatises appeared, but they were also by medical men. While they similarly shied away from extended theological discussion, they were strikingly different from Kidd's and Bell's in that Roget and Prout both pitched their Bridgewaters as explorations into the religious significance of the expansion of lawlike explanation in physiology. For Roget, this meant considering the laws of animal form found in "philosophical" anatomy—notably, the "law of variety" and the law of "conformity to type." Far from threatening the Christian account of divine design, recent Continental work in this area offered, in Roget's view, an expanded sense of God's plan in creation. By contrast, Prout's focus lay primarily in the laws of chemistry, and he used his Bridgewater to outline much of his original work on matter theory and the chemistry of digestion that had been directed to establishing such laws. In doing so, however, he sought to demonstrate how the growing reach of natural laws, and more especially their limitations, served to reinforce the Christian's sense of divine action in the creation.

The final pair of Bridgewaters shared a concern to address the role and status of the Bible in relation to modern science, but Kirby and Buckland had strikingly different approaches to the issue. As a High Churchman interested in symbolic interpretations of the Bible, Kirby framed his Bridgewater on natural history in relation to those frankly idiosyncratic views, arguing that it was the Bible, not the arguments of natural theology, that held the key to the religious meaning of the natural world and that modern science went astray when it neglected to acknowledge as much. By contrast, Buckland swiftly made clear that he endorsed what was becoming a commonplace among scientific practitioners—namely, that, while scientific findings would always be found to be consistent with the Bible when that was properly interpreted, the natural world must be studied independently of it. Nevertheless, his Bridgewater offered a reassuring account of the consistence of modern geology with Genesis as well as a plethora of geological evidence in support of the claims of natural theology. In their distinctive ways, the eight authors thus strove to convince their readers that there was nothing in the rapidly developing sciences to give concern to the pious Christian.

## *Natural Theology and the Sciences*

Of all the Bridgewaters, the first to appear—Whewell's *Astronomy and General Physics*—was the most explicit in identifying its object as being "to lead the friends of religion to look with confidence and pleasure on the progress of the physical sciences."[3] Whewell knew only too well that there were sincere friends of religion who were uneasy about the tendency of the sciences. The point had been brought home to him forcibly by the strongly expressed opinions of his friend from undergraduate days, the Suffolk clergyman Hugh James Rose (see fig. 4.2). On 2 July 1826, the scarlet-clad dignitaries of the University of Cambridge had gathered in Great St. Mary's Church to hear Rose deliver the sermon for "Commencement Sunday." Taking as his text a passage from the book of Ecclesiastes—"No man can find out the work, which God maketh, from the beginning to the end"—he proceeded to deliver a strongly worded critique of the religious tendency of the natural sciences. The modern obsession with useful, wealth-generating knowledge was putting true religion in danger, Rose claimed. Where the study of literature fostered knowledge of moral and religious truths, the study of nature had little or nothing to contribute in this preeminent department of knowledge and only tended to encourage damagingly shallow habits of mind.[4]

Whewell was significantly troubled by his friend's comments. Responding later in the year to the published version of the sermon, he told Rose that he was convinced that there was nothing in the nature of experimental science "unfavourable to religious feelings." It might sometimes justify rebuke by "poaching" outside its own domain, but true inductive science must always "harmonize with the great truths of religion." It was not credible, Whewell asserted, that God had put "this tempting system of discoverable truths" in the path of humans while at the same time tainting it with the "poison of irreligion—a sort of tree of knowledge and of death, both in one, without the merciful prohibition attached to it." Rather, he claimed, the study of science could foster pious sentiments. Scientific ignorance left

3. Whewell 1833b, vi.

4. Rose 1826. On Rose, see Valone 2001, Thompson 2008, 56–61, and Bennett 2018.

a person "blind to many and wonderful views which, properly considered, it gives him of the relations of ourselves and the world to the Deity."[5]

Whewell had been ordained priest in May 1826. Invited to deliver afternoon sermons at Great St. Mary's the following February, he told Rose that he would take the opportunity to develop this response. His purpose, once again, was to offer reassurance that the natural sciences were supportive rather than undermining of religion. Now, however, he elaborated at greater length on the proper foundations of religion in God's self-revelation and the moral promptings of the heart. He had no interest, he insisted, in developing a theistic proof. Rather, he wished to manifest how modern science, properly understood, supported such religious feelings rather than undermining them, as Rose suggested.[6]

Many of the arguments developed in these sermons were reworked at greater length in Whewell's Bridgewater, a work that was characterized by the same modest tone of conciliation rather than by a strident claim for the religious value of a scientific natural theology. He had been commissioned—he reminded his patron, the High Church Bishop Blomfield, in his preface—to show "how admirably every advance in our knowledge of the universe harmonizes with the belief of a most wise and good God." This he was happy to do, but he was anxious not to be caught poaching, observing, "I feel most deeply, what I would take this occasion to express, that this, and all that the speculator concerning Natural Theology can do, is utterly insufficient for the great ends of Religion; namely, for the purpose of reforming men's lives, of purifying and elevating their characters, of preparing them for a more exalted state of being." The moral transformations of religion could only be achieved by the aid of revelation. This book had altogether more limited ambitions.[7]

Whewell's introduction laid out more explicitly the limitations that he believed to be applicable to natural theology. While asserting in his opening sentence that "most minds" found the study of the natural world to suggest "a creating and presiding Intelligence," Whewell remained noncommittal about the epistemological status of this connection, asserting that natural religion was "necessarily imperfect and scanty." His purpose, he claimed, was not to show natural theology to be "perfect and satisfactory" but rather to show how it related to modern natural philosophy. As scientific advances modified the understanding of natural phenomena, it

5. WW to Rose, 19 November and 12 December 1826, in Todhunter 1876 2:75–79.

6. These sermons and ancillary materials are in TC, R.6.17. On Whewell's sermonizing, see Brooke 1991a.

7. Whewell 1833b, vi–vii.

became necessary to "harmonize" such views with "belief in a Creator, Governor, and Preserver of the world." In particular, since "the peculiar point of view" of modern natural philosophy was that "nature, so far as it is an object of scientific research, is a collection of facts governed by *laws*," Whewell set himself the objective of showing that "this view of the universe falls in with our conception of the Divine Author."[8]

The view that a law-governed universe was consistent with divine creation was, of course, hardly new. Yet while eighteenth-century Newtonians had waxed lyrical on the subject, Paley's *Natural Theology* had more recently made isolated instances of organic mechanism the touchstone of design in nature. More importantly, the generation of outstanding French natural philosophers led by Laplace who had sought to extend and complete Newton's lawlike account of the universe had done so without reference to divine action. During Whewell's early career, Laplace's growing reputation as an atheist determined to exclude God from nature had played increasingly into the fears of science's critics. Several months before Rose's commencement sermon, Whewell's Sabbath reading—a passage from evangelical German pastor Christoph Christian Sturm's popular devotional aid *Reflections on the Works of God*—had set him reflecting on the topic. Sturm had observed that the migration of birds testified that the universe had been arranged with great wisdom. Whewell now questioned why, to some, the advance of lawlike explanation seemed to undermine a belief in providence. Laws were properly understood as the "rules of operation" of the creator, and knowledge of them provided the Christian with "a nearer and nearer knowledge of the thoughts of God as they regard the material world."[9]

Whewell developed this view in replying to Rose in his 1827 sermons, and he did so at greater length in his Bridgewater. Introducing his readers to natural laws, he promised to shed light on the character of the creator from the character of his administration. Since the sciences that had been most successfully reduced to laws were astronomy and meteorology, the analysis was to revolve around these two. Without further deliberation, Whewell proceeded to fill two-thirds of his treatise with a survey of adaptations in the laws of nature. The first of his three "books" concerned the "Terrestrial Adaptations" found in the relation between astronomical and meteorological laws and the properties of plants and animals. The second

8. Whewell 1833b, 1–3.

9. Todhunter 1876, 1:360–65. On Newtonian natural theology, see Gascoigne 1988 and Gillespie 1987. On growing fears concerning Laplacian physics, see Morrell 1971, Numbers 1977, Brooke 1979, and Schaffer 1989.

concerned the "Cosmical Arrangements" by which astronomical objects had been created and preserved. Many of the examples in the first were relatively conventional instances of adaptation, but it was in the second that the capacious reassurance of Whewell's emphasis on laws came into play.

In describing the laws by which the universe operated, Whewell's concern was to "trace indications of the Divine care." Thinly veiled behind this stood the specter of Laplace, whose sneering skepticism became Whewell's foil. Laplace had shown that the solar system was stable and consequently did not (as suggested by Newton) require God's intervening adjustments. Whewell asked, "Now is it probable that the occurrence of these conditions of stability in the disposition of the solar system is the work of chance? Such a supposition," he continued, "appears to be quite inadmissible." The point became most explicit when Whewell examined Laplace's own account of the "*primitive cause*" of this stability. In a chapter headed "The Nebular Hypothesis," he described Laplace's attempt to account for the origin of the solar system in a lawlike manner as a result of the condensation of nebulous matter under the influence of gravity. Whewell made it clear that he considered the scientific case for the nebular hypothesis to be unproven. However, his concern was to demonstrate that such a lawlike explanation, if established, would not undermine the Christian's belief in a creator. Rather, it would transfer "our view of the skill exercised, and the means employed to another part of the work"—to the laws and initial conditions that allowed such a development to occur.[10] Hence, as this extreme example made clear, the Christian had nothing to fear from the advance of science.

Having thus shown the harmony of modern science with Christianity, Whewell devoted the final book of his treatise ("Religious Views") to a somewhat miscellaneous series of important and wide-ranging reflections. Many of these themes had appeared in his sermons, and the short chapters were like a series of homilies. The first concerned the claim that the "Creator of the Physical World" was also the "Governor of the Moral World," reflecting the earnest religious intent manifested in his response to Rose. For religious purposes, it was knowing God as a "just and holy Governor" that mattered, and in this regard Whewell considered natural science to be of limited value. However, in terms similar to those used by the evangelical Chalmers, he argued that the mutual adaptation of physical laws to the moral constitution of humans—and most particularly the "principle of conscience"—clearly manifested God's moral character. Succeeding chapters elaborated on this point, as Whewell showed how mod-

10. Whewell 1833b, 150, 165, 184.

ern science enlarged the Christian's conception of God in the "Vastness of the Universe," while the conviction of a "moral governor" provided reassurance concerning "Man's Place in the Universe."[11]

Whewell next reflected further on the religious tendency of natural laws, again with the negative example of Laplace clearly in his sights. Where the first two books of the treatise had examined the adaptedness of nature's laws, he now reflected that, to most people, the very existence of natural laws in itself implied "a presiding intelligence." How did they come to this impression? Whewell was reluctant to commit himself. The "various trains of thought and reasoning" that led people from the natural world to God typically did not involve "any long or laboured deduction," he observed. The impression was "so widely diffused and deeply infixed" that some had questioned whether it was perhaps "universal and innate." Leaving the question unanswered, Whewell was content to rely on history. The impression that "law implies mind" had given rise to the "natural religious belief" of our species, he claimed. Moreover, the impression had continued to operate in the scientific age. Those who had been involved in scientific discovery, he argued at length in the next chapter, had been "peculiarly in the habit of considering the world as the work of God."[12]

The reassuring intent of this analysis became clear in Whewell's following chapter, which was the one on which he most wanted to hear his friends' comments. With his eye firmly on Rose, he observed,

> Complaints have been made, and especially of late years, that the growth of piety has not always been commensurate with the growth of knowledge, in the minds of those who make nature their study. [ . . . ] The opinion that this is the case, appears to be extensively diffused, and this persuasion has probably often produced inquietude and grief in the breasts of pious and benevolent men.[13]

Whewell considered such claims exaggerated but sought in any case to explain how the likes of Laplace could have fallen into irreligious views. The danger lay in the "deductive habits" of many eminent scientific men who, since they were engaged in unfolding the consequences of previously discovered laws, came to view those laws as necessary. By contrast, the "inductive habits" of scientific discoverers made them conscious that

11. Whewell 1833b, 292–93.
12. Whewell 1833b, 293–94, 296, 300, 308.
13. Whewell 1833b, 323.

they had deciphered a law that might have been otherwise but for the will of the legislator.

In a chapter on "Final Causes" Whewell now finally examined the logic of the design argument. Since the impression of design in nature was general, except in cases where the "exclusive pursuit" of deductive reasoning damaged the mind's capacity to apprehend other truths, it must have a "deep and stable foundation." Responding to the skeptical argument associated with David Hume that design could not be inferred from nature, Whewell asserted that design was something perceived immediately rather than being the product of a train of reasoning. Using the language of Immanuel Kant, he described it as a "regulative principle" by which humans make sense of the phenomena they experience. Only those like Laplace, who had indulged in a peculiar mental discipline, could throw off the impression of final causes or suggest that it merely amounted to an ignorance of real causes that would be banished by the extension of natural law. For those whose minds were properly constituted, scientific research merely transferred design "from the region of facts to that of laws."[14]

Whewell's idea that design was part of the basic mental apparatus that humans brought to nature was a long distance from the confident empiricism of eighteenth-century natural theology. His Bridgewater was consequently pervaded by a restrained mode of expression, often involving interrogatives or double negatives. "On any other supposition" than that of design, he observed of the adaptation of creatures to the diurnal cycle, "such a fact appears altogether incredible and inconceivable." He also suggested that knowledge of God that was supposed to have been derived from nature often owed much more than people realized to revelation. Finally, he concluded with two chapters that reflected on the very limited extent to which knowledge of God's administration of the natural world could be gained from science. Echoing the text of Rose's sermon, Whewell quoted Job's question, "Canst thou by searching find out God?" observing that it "must silence the boastings of science." As far as the material world was concerned, science had shown that God acted by general laws, but scientific ignorance of the higher laws governing life, mind, and morals was overwhelming. Science had shed light on the creator's plans, but "how incomparably the nature of God must be elevated above any conceptions which our natural reason enables us to form."[15]

Whewell has usually been seen as a "liberal Anglican." Like others to

14. Whewell 1833b, 342–43, 345–46, 349.

15. Whewell 1833b, 41, 252, 377.

whom that label has been applied, he believed that the religious natural philosopher should combine an "earnest piety ready to draw nutriment from the contemplation of established physical truths" with a "philosophical caution, which is not seduced by the anticipation of such contemplations, to pervert the strict course of physical enquiry."[16] Yet in the limitations that he placed on natural theology and the importance that he accorded to divine revelation and the need for moral regeneration, his Bridgewater reflected concerns more often associated with such High Anglicans as his friend Rose and patron Blomfield and with such evangelicals as Chalmers (fig. 2.1). Indeed, despite very obvious differences, Whewell and Chalmers's Bridgewaters bore striking resemblances. Both authors were professors concerned with developing a sophisticated understanding of the interconnections of scientific and religious truths, but both were also concerned with the practical, pastoral implications of such theological reflection. Far from being a complacent work of natural theology, Whewell's Bridgewater offered a nuanced vision of how modern science and its laws might be made to sustain a Christian faith.

❋

Thomas Chalmers's relationship with natural theology was an even more checkered one. As a young man growing up in late Enlightenment Scotland, he had been party to vigorous debates about the legitimacy and value of natural theology that had been pivotal to the reshaping of the Presbyterian national church. This began during his student days at St. Andrews in the 1790s, when he acquired the confidence in natural theology characteristic of the still-dominant "Moderate" party of the Church of Scotland only to have it brought crashing down by an encounter with the atheist materialism of Baron D'Holbach's *Système de la nature*. Subsequently, after attending the lectures of Edinburgh's John Robison, professor of natural philosophy, and reading the *Analogy of Religion* of eighteenth-century bishop and moral philosopher Joseph Butler, Chalmers rebuilt his Christian beliefs around the historical evidences of Christianity. From Robison's Baconianism and from Butler's emphasis on historical facts as the basis for Christian belief, Chalmers developed a distinctive science of theology. Setting natural theology aside, at least for practical purposes, he came to view the "external evidence" for the truth of the Christian revelation—that is, the miracles and prophecies that attested to its

16. Whewell 1833b, 355. On liberal Anglicanism and science, see Morrell and Thackray 1981, 225–29.

FIGURE 2.1 Thomas Chalmers. Mezzotint by William Ward after Andrew Geddes, 1822. Image: © The Trustees of the British Museum. All rights reserved.

veracity—as the factual basis on which a scientific inference of divine truths should be grounded.[17]

This change of view might seem to align Chalmers with the then re-

17. Topham 1999, 149–59.

surgent Evangelical party in the Church of Scotland, which was distinctly critical of the supposition that natural theology formed the foundation of Christianity. However, it was not until several years after his licensing as a Church of Scotland minister that Chalmers underwent a conversion to evangelicalism. At that period, he elaborated his critical views concerning natural theology and the supposed reasonableness of Christianity (one branch of Christianity's "internal evidence") in an article on "Christianity" that was published in 1813 in David Brewster's *Edinburgh Encyclopædia*. The account was soon republished in a successful book, but it saw him plunged into a veritable "theological war" with the Moderates, who endeavored to drag him through the church courts for his allegedly heterodox view.[18] In fact, even evangelicals considered that Chalmers had gone too far in his repudiation of large parts of the internal evidence of Christianity. For most, it was the inherent appeal of the gospel message to the human condition, rather than the examination of its historical evidences, that would bring belief.

In response to such criticisms, Chalmers began to modify his views. Even by the time his preaching first caused a sensation in London in May 1817, he was to be found urging the importance of the way the gospel message spoke to the human predicament. Invited in 1823 to leave his Glasgow pulpit for the chair of moral philosophy at St. Andrews, he developed his thinking further. Teaching the "celestial" branch of moral philosophy, Chalmers claimed, required that he "demonstrate the existence and character of a God" so far as the "light of Nature" would take him. Natural theology was very defective, he continued, but it helped to fix the unbeliever's attention on the evidences of Christianity, including the compelling internal evidence that came from its account of the human condition and its cure.[19] In developing this view, Chalmers continued to draw on Butler's writings, and as Edinburgh professor of divinity from 1827, he used them more fully to develop the distinctive approach to natural theology found in his Bridgewater on "the adaptation of external nature to the moral and intellectual constitution of man." In both he echoed Butler in emphasizing the moral obligation placed on the skeptic by the very possibility of God's existence, the presumptive evidence of such a possibility provided by the felt supremacy of conscience, and the need to search out and ex-

18. Topham 1999, 159–65.

19. *Evidence on the Universities of Scotland* 1837, 37:104, 106. See also Francis A. S. Knox, "Dictations issued by Dr Chalmers in his Moral Philosophy Class [ . . . ] 1826," StAUL, MS.37483.

amine any self-proclaimed revelation given the profound limitations of natural theology.

The dedication to Chalmers's Bridgewater, which emphasized that he had "derived greater aid from the views and reasonings of Bishop Butler" than from all else put together, highlights the continuities with his earlier lectures.[20] However, his Bridgewater's prescribed subject matter demanded some modification in the sequence and weighting of subjects, and Chalmers inserted copious amounts of new material. He was conscious that his approach to his subject, so heavily informed by Butler's apologetics and moral philosophy, might be considered idiosyncratic. There were, he explained in his preface, two possibly unexpected aspects to his analysis. First, since no author had been assigned to explicate "the mental constitution of man" independent of the external world, Chalmers considered that he must do so himself and demonstrate the design in that constitution too. In addition he argued that he would cut himself off from the most promising part of his subject if he were to define "external nature" to refer only to the material world. By defining it to include all that was external to the individual mind, including other people, Chalmers gave himself license to trace the many evidences of God's activity in human society, especially as revealed by political economy. In this way he was able to justify centering his Bridgewater on a Butlerian account of the evidence for God in the human conscience and the moral tendency of human society.

Thus, after an introductory chapter in which he discussed a number of preliminary matters relating to natural theology, Chalmers plunged into his first section on the "moral constitution of man" with three general arguments for God's existence and attributes, commencing with the argument from the "felt supremacy of conscience." He devoted the majority of the remaining seven chapters of the section to demonstrating how the social and economic aspects of external nature likewise provided evidence of God's moral governorship. The second section, on the "intellectual constitution of man," occupied only half as much space. Moreover, while it included some discussion of Humean skepticism about causation, the bulk of it was devoted to establishing—in keeping with Chalmers's earlier lectures—the ethical character of belief. This built up to the climax of the treatise, which reviewed the "defects and uses of natural theology" and emphasized how his earlier arguments had generated a moral obligation for the reader to examine the evidences of Christianity.

20. Chalmers 1833, 1:vi.

Chalmers's introduction began with a brief exploration of the logic of the design argument, especially as it applied to his own subject matter. The strength of evidence for the existence of a designing creator was proportional to the complexity of the contrivance, he argued. Thus, anatomy yielded more satisfactory evidence than astronomy. Indeed, this became ever more the case as astronomical phenomena were subsumed under natural laws by the likes of Laplace, although Chalmers was eager to point out (in a manner similar to Whewell) that design was manifested by the many "dispositions" of matter on which laws operated and that the "fiat and finger of a God" was required at points of origin, including in the formation of species. When it came to the human mind, the recognizable parts were few in number, and the workings were instantaneous and apparently simple, but there were compensations. That intelligence must itself be caused by intelligence might not be "demonstrable in the forms of logic," Chalmers observed, but it was "a thing of instant conviction, as if seen in the light of its own evidence," that could be rigorously defended using the intuitionist psychology that he had learned from the Scottish "commonsense" school of philosophy. In addition, Chalmers claimed, the human mind offered vastly superior evidence concerning God's moral attributes. The enigma of a suffering world could only be resolved by studying the constitution of the mind, which offered the strongest natural demonstrations of God's goodness and righteousness.[21]

These moral attributes of God quickly became evident in the three succeeding chapters on the moral constitution of man. Here, Chalmers avoided mental philosophy, except for occasional references to the important recent work of the Scottish moral philosopher Thomas Brown, concentrating instead on the bare "facts of the human constitution." The great fact of the human mind was the "felt supremacy of conscience," which had first been fully expounded in the first three of Butler's *Fifteen Sermons* (1726). This provided "nature's strongest argument, for the moral character of God" and was "the great prop of natural theology among men." Only a righteous creator would wish to make the ruling faculty of the mind such a strong voice in favor of righteousness. Chalmers's second and third general arguments also emphasized the manner in which the human mind had been constituted to favor morality. Drawing again on Butler's sermons, he claimed that virtuous action produced a pleasure that was quite independent of any sense of virtue and that vice was similarly inherently miser-

21. Chalmers 1833, 1:27, 35-37, 39. On Chalmers's debt to the commonsense school, see Rice 1971.

able.[22] Evidence for divine righteousness was also provided by the operation of habit on the human mind, since it served to reinforce virtuous action and afforded a natural punishment for vice.

Chalmers now moved on to the world external to the individual mind. This, he argued in strong echoes of Butler's *Analogy*, provided a moral "gymnasium" in which the mind was exercised and tested, again manifesting God's righteous character. The focus was especially on human society, and two long chapters were devoted to "those special affections which conduce to the civil and political well-being of society" and those "which conduce to the economic well-being of society." Following the argumentative thrust of his recently expanded lectures *On Political Economy, in Connexion with the Moral State and Moral Prospects of Society* (1832), Chalmers claimed that political economy was "but one grand exemplification of the alliance, which a God of righteousness hath established, between prudence and moral principle on the one hand, and physical comfort on the other." Life was a state of probation: the poor must either learn sexual restraint or face the divinely ordained punishments of the Malthusian law. Where political economy had long suffered from a reputation for radicalism and irreligion, Chalmers was now able to use it to illustrate God's attributes and enjoin public morality.[23]

From the start Chalmers had promised to correct certain errors in "jurisprudence and political economy" by demonstrating that they transgressed the divinely ordained laws of human nature. He thus expounded the God-given mechanism of familial affections as upholding "the patriarchal arrangement" of society and defended it from human meddling, especially the "cosmopolitism" of Owenites and utilitarians. In addition, he defended those "special affections" that God had designed to uphold property and the "well-being of society"— namely, the possessory principle and the principle of benevolence. This divine mechanism was, Chalmers claimed, wrecked by the English Poor Laws and by the system of tithes, against both of which he had long railed. God's wisdom was also to be seen in the operation of free trade, through which mechanism the selfishness of individual humans was brought to serve the common good. He was thus able to construct a natural theology grounded in laissez-faire politi-

22. Chalmers 1833, 1: 59, 72, 77, 108–9. On Brown and Chalmers's debt to him, see Dixon 2003, 109–33. On "Chalmers and the natural theology of conscience," see Hilton 1988, 183–88.

23. Chalmers 1833, 2:49. On the "rage of Christian economics," see Hilton 1988, 36–70; on the "'evangelicalization' of Parson Malthus," see 89–91. See also Young 1985, 31–39.

cal economy while at the same time taking many a sideswipe at the "desolating effect" of the ethics of utilitarianism.[24]

The shorter second part of Chalmers's treatise, on the intellectual constitution of man, began with a chapter on the "instinctive faith" in the uniformity of nature that formed, in combination with the principle of association, the Scottish "commonsense" answer to Hume's problem of induction. God's hand was to be seen in the manner in which this a priori conviction of uniformity "meets with its unexpected fulfilment, in the actual course and constancy of nature." He then devoted two chapters to the connection between the intellect, the emotions, and the will, drawing on his St. Andrews lectures and the novel analysis of Thomas Brown. Chalmers's purpose was to demonstrate the power of the will over both the emotions and the intellect and thus to establish the moral culpability of religious disbelief.[25]

This led naturally into the final statement of "the defects and the uses of natural theology," which was the only suitable climax for such a work, bearing the evangelical burden of his practical theology. For Chalmers, natural theology was neither more nor less than a preparation of the human heart for the message of that gospel that could only be found in the Christian Scriptures. Drawing a distinction between what he called "atheism" (skeptical unbelief) and "anti-theism" (dogmatic disbelief), Chalmers sought to demonstrate that the latter was a "glaring contravention" of scientific principles, since it would require omniscience to know the absence of God from all creation. Furthermore, he continued, the atheist was morally obliged to consider the possibility of God's existence, even in the absence of prima facie evidence, by the debt of gratitude that he would owe to any such being. Drawing on Brown's account of the "faculty of attention," Chalmers had established the general point that humans are responsible for their beliefs and emotions. He now sought to prove in a suitably Pauline manner the moral culpability of atheism: "Man is not to blame, if an atheist, because of the want of proof. But he is to blame, if an atheist, because he has shut his eyes."[26]

According to Chalmers, natural theology did not provide an irresistible demonstrative proof of God's existence and character but allowed inquirers "to conceive, or to conjecture, or to know so much of God" that, if they then encountered something such as a sacred text that purported to

24. Hilton 1988, 186; Chalmers 1833, 1:55, 225; 2:36, 51–68.

25. Chalmers 1833, 2:137, 257–59.

26. Chalmers 1833, 2:259, 272. Cf. Romans 1:18–20.

be a message from him, it was their "bounden duty to investigate." Thus, according to Chalmers, the chief *use* of natural theology was to fix the attention of skeptics on the claims of revelation. As seekers became more convinced of the existence and righteous character of God through natural theology, they would become likewise more convinced of their own sinfulness. Having taken the pilgrim thus far, the great *defect* of natural theology was that it could not reveal the path to salvation, which was the sole preserve of revelation. "Natural theology may see as much as shall draw forth the anxious interrogation, 'What shall I do to be saved?' The answer to this comes from a higher theology." Natural theology was not, then, the foundation of Christianity, since the latter had its own foundations and evidences. The purpose of natural theology, Chalmers explained, was to draw pilgrims to the foundations of Christianity and to cause them to examine the independent Christian evidences. It was not "the basis of Christianity" so much as "the basis of Christianization."[27]

Having for a time been profoundly skeptical about natural theology, the Chalmers of the Bridgewater had come to a point of giving it a useful role in the economy of Christian belief. In particular, his subject allowed him to emphasize the indications of a righteous creator in the individual's moral feelings and the moral tendency of society as revealed in laissez-faire political economy. At the same time, however, the profound limitations of natural theology, combined with Chalmers's analysis of the ethics of belief, allowed the evangelical preacher to end his treatise by urging his readers on to a distinctively Christian faith.

Whewell's and Chalmers's Bridgewaters thus presented natural theology as being of strikingly limited religious value. It had its place within Christianity, but it was the message of the Bible—especially the New Testament—that lay at the heart of true religion and brought about a transformation in human lives by moving the emotions and the will. Both authors nevertheless considered that the sciences—political economy as much as astronomy and physics—could contribute to such a limited natural theology. Moreover, as Whewell in particular urged, the advance of lawlike explanation did not undermine such evidence. Those men of science who fell into irreligious views as they reduced nature's phenomena to laws did so because they neglected to cultivate healthy habits of mind. The "religious views" offered in Whewell's treatise would assist readers to maintain altogether better mental hygiene.

27. Chalmers 1833, 2:281, 286, 291.

FIGURE 2.2 John Kidd. Pencil and chalk drawing. Image: reproduced courtesy of the Governing Body of Christ Church, Oxford, from a copy at the Oxford University Museum of Natural History.

## *The Dangers of Modern Medicine*

The intended partner of Chalmers's Bridgewater was that by Kidd on the "adaptation of external nature to the physical condition of man," but two more different books could hardly be imagined. Where the Edinburgh divinity professor had set out a carefully considered theological defense of a limited natural theology, the Oxford medical professor (fig. 2.2) assumed the truth of both natural and revealed theology and declined to give any theological defense of either. Having spent three decades seeking to justify the place of scientific education within Oxford's classical and theological curriculum, Kidd was more than happy to point to its religious value. But he was no theologian and in any case he knew that developing a system of natural theology in Oxford was fraught with difficulty, since High Anglicans had distinct reservations concerning its rationalist tendency. Instead, Kidd offered many examples illustrative of design in nature while seeking to show that a proper understanding of modern science could shore up both humanistic and medical scholars against atheism, just as he had in the inaugural lecture he raided for materials.

Kidd's preface set out this very restricted remit in blunt terms. It was not his object to "maintain a formal argument," he stated, but to "unfold a train of facts." He would not "attempt directly to convince the reader" that the adaptations he described were a proof either of God's existence or of his attributes, although he hoped that anyone who doubted these would nevertheless be convinced. However, his treatise was addressed "exclusively" to those who were already believers in both natural and revealed religion, a decision justified on the grounds that atheism involved an "intellectual absurdity." As he had earlier spelled out in his inaugural lecture, it was not necessary to "hold any argument" concerning God's existence. Such atheism as there was had its foundation not in reason but in human immorality—in moral depravity, intellectual pride, or vanity. Thus, after leading the reader through an edifying catalog of countless miscellaneous examples of anthropocentric adaptations drawn from the several kingdoms of nature, Kidd's grand conclusion was "that—whether from chance, (if any philosophical mind acknowledge the existence of such an agent as chance,) or from deliberate design—a mutual harmony does really exist between the corporeal powers and intellectual faculties of man, and the properties of the various forms of matter which surround him."[28]

For Kidd, habituated to Oxford's theological antirationalism, the design argument could not function as a source of Christian belief. Yet his book still had a serious intent. Kidd's long career as a university teacher had been built around the conviction that the natural sciences should be taught in relation to the main studies of a university—divinity, classics, and mathematics—acting like "the supernumerary war-horses of Homer's chariots; which were destined to assist, but not to regulate, the progress of their nobler fellow-coursers." His Bridgewater likewise used scientific exposition to support religious truths against contemporary attacks and sanctify sometimes sacrilegious classical learning. Addressing the university in the autumn of 1823 as the new Regius Professor of Physic, Kidd had justified the teaching of comparative anatomy to largely nonmedical

28. Kidd 1833, viii–ix, 339; Kidd 1824, 3–5.

students by arguing that, while Paley's design argument was not valuable as a means of convincing determined atheists, its value lay in "tending to counteract the influence of those who would inculcate atheistical opinions" and in assisting those with insufficient knowledge or acumen to protect themselves.[29] His inaugural lecture thus presented modern science as a means of shoring up religious belief against atheist materialism, both ancient and modern, and his Bridgewater followed suit.

In his lectures, Kidd claimed that the "infection" of atheism had reached only a very few of Oxford's thousands of students, but that even the occurrence of those few cases could effectively be prevented by appropriate scientific teaching. In the course of a classical education, he argued, students were brought into contact with atheist dogma, as Egerton had been sixty years before. If forearmed with scientific knowledge, they might readily see the folly of atheism. His lectures were thus calculated to assist those who were deficient in philosophical information. Some, for instance, were fearful of the damage that might be wrought by reading the writings of Lucretius, but Kidd boasted of how easily Epicurean errors could be counteracted by reference to modern science.[30] He also sought to demonstrate that a sound scientific education was invaluable in counteracting any danger from the materialist speculations emanating from modern medicine. Here Kidd had in mind the notorious *Lectures on Comparative Anatomy, Physiology, Zoology, and the Natural History of Man* (1819) of the surgeon William Lawrence and the widely debated phrenological theory of the Austrian physician Franz Gall, and he proceeded to discuss each in turn, calmly reassuring his auditors that neither made materialism plausible.

Expanding on his inaugural lecture, Kidd's Bridgewater adopted much the same approach. Reviewing the many anthropocentric adaptations in the kingdoms of nature in a manner suitable for what he had previously termed "religious contemplation," he sought to buttress his readers against errors that might undermine belief. His approach was not very confrontational or even very overt. As he had explained in relation to his earlier treatment of Lawrence's supposed materialism, harsh dealing was only liable to breed resentment and opposition.[31] Rather, he sought opportunities to counteract ancient and modern materialism obliquely. After the briefest of introductions, the treatise was divided into three main parts. The first four chapters considered the physical character of humans, paying particular attention to their distinctive features, most especially the

29. Kidd 1824, 13; Kidd 1818, 8.
30. Kidd 1824, 5.
31. Kidd 1824, 3, 47, 49.

brain. The next four provided a desultory examination of the adaptation to the human condition of the atmosphere, minerals, vegetables, and animals. Finally, Kidd devoted an idiosyncratic chapter to part of Chalmers's assigned topic, the "adaptation of the external world to the exercise of the intellectual faculties of man," adding an associated appendix. The middle four chapters apart, however, the body of the treatise built on Kidd's earlier remarks in correcting materialism and, more especially, transmutationism.

The shift in focus from materialism to transmutation clearly reflects the changing focus of concern in the decade since Kidd's inaugural lecture. The growing engagement of London's radical medical teachers with French transmutationist ideas in the 1820s made the subject highly topical. Moreover, Lyell's extensive critique of Lamarck's transmutation theory in the second volume of his *Principles of Geology*, published in January 1832, brought the topic to a wider audience. Kidd had acknowledged, in regard to Lawrence's materialism, that while such disturbing scientific discussions might be of limited concern within an Anglican university, they could subvert morality and religion when working-class radicals such as Carlile laid hands on them.[32] In his quiet and distinctly Oxonian way, he now set out to oppose such views.

Kidd's opening chapters on the physical character of the human species meandered through a range of topics, several of which were clearly suggested by the convenience of being able to lift around thirty pages of text from his inaugural lecture. The early chapters nevertheless had a thread running through them of opposition to any subversion of the unique status of the human species, especially from a transmutationist perspective. The treatise opened with a Shakespearean panegyric on the human species and continued in similar vein, extolling the dignity of the species as holding "the first rank among animals." Authors whose negative remarks might lead readers astray were corrected, including Lucretius and those who had "absurdly asked what claim man has, from his physical structure or powers, to be placed first in the scale of animal beings." The latter reference probably related to French naturalist Julien-Joseph Virey, who had denigrated the physical excellence of the human species. Kidd, by contrast, argued that man was "in every sense superior" to all other animals. Underpinning human supremacy, he explained, was the permanency of species, which could be seen not to have changed after thousands of years. And while some inadequate philosophers considered apes and humans

32. Kidd 1824, 52. On the growing concern regarding transmutation, see Corsi 1978, Yeo 1986, esp. 271–73, Corsi 1988, 236–40, and Desmond 1989b, esp. 228–30. On the early reception of Lamarck in Britain, see also Topham 2011, 328–32.

to be merely varieties of the same species, this was unthinkable if one considered all their structures and powers collectively. But the question was "puerile" since, whatever the physical resemblances might be, man's uniqueness was guaranteed by "his intellectual peculiarities and the moral and religious sense."[33]

Such relatively oblique references to transmutation were also to be found in Kidd's account of the nervous system. Describing the growing complexity of the nervous system as one ascended through the animal kingdom, Kidd was keen to shore up his reader against the transmutationist views of Geoffroy Saint-Hilaire. Geoffroy's leading disciple Étienne Serres had argued, in his widely read *Anatomie comparée du cerveau* (1824–26), that the nervous system of animals in higher classes passed through stages of development that resembled those in the adult forms of lower classes. On the basis of this recapitulation theory, Serres and Geoffroy claimed that "*lusus naturæ*" had failed to reach their proper stage of development. Geoffroy later suggested that such "monstrosities" might also be caused by the influence of the environment acting on developing embryos to make them pass farther up the scale of being, thus producing new species. Aware of these writings, Kidd argued that in the case of *lusus naturæ*, "nature never elevates the brain of an individual of a lower to that of a higher class."[34]

Kidd also argued that the human brain contained parts distinct from those of apes. Following an extended discussion of the localization of mental function, he suggested that it was highly probable that "the intellectual superiority of man, physically considered," depended on the "peculiarities of the human brain."[35] Repeating comments from his introductory lecture, he reassured readers that Gall's phrenological theory, in which mental function was considered to be localized within the brain, was neither materialist nor fatalist. Somewhat surprisingly, Kidd was happy to give the localization principle serious consideration despite offering criticisms of the detail and application of phrenology. Indeed, he even used physiognomy—the practice of determining character from facial appearances—to further emphasize the theory's plausibility. However, for Kidd the attraction of phrenological theory was that it was useful in urging the uniqueness and superiority of the human species in opposition to transmutation.

33. Kidd 1833, 1, 9, 18, 20–22. On Virey's prominence and the Bridgewater authors' reactions to him, see Corsi 1988, 237–42.

34. Kidd 1833, 51–52, 73–75. For a brief overview, see Desmond 1989b, 52–53.

35. Kidd 1833, 75.

Kidd returned to the subject of transmutation in his final chapter on "the adaptation of the external world to the exercise of the intellectual faculties of man." This was, of course, part of Chalmers's topic, but Kidd claimed that he also had originally been invited to handle the subject. His treatment was nevertheless idiosyncratic and patchy and particularly reflected the classical concerns of Oxford. He began, for instance, by desultorily considering "the rise and progress of human knowledge" before devoting a barely justified fifty pages to a comparison of "the progress which natural science had made in Europe, at a period shortly antecedent to the Christian era, with the state in which it now exists."[36] The advantage of this, however, was that it gave him an opportunity to discuss Lucretius's physics and to offer calm criticism of his atheism. More substantially, Kidd devoted a section to comparing Aristotle's *History of Animals* and Cuvier's *Le règne animal* (1817), which allowed him to elaborate at greater length on the failings of transmutationism.

Reviewing Aristotle's work, Kidd sought to correct several passages that had transmutationist connotations. First, he opposed Aristotle's doctrine of the "continuity of gradation" between different species, advocating instead the staunchly antitransmutationist Cuvier's nonhierarchical system for the classification of animals. While few English naturalists had been willing to endorse fully Cuvier's rejection of the linear ranking of animal forms inherited from the concept of a "chain of being," Kidd was not alone in rejecting such formulations at this juncture in the face of a transmutationist threat. He even went so far as to qualify the claims of progressionist geology, denying that only the simplest forms of animal life occur in the older strata with increasingly complicated forms appearing in more recent formations.[37]

Second, Kidd objected to Aristotle's "faint idea that the specific characters and dispositions of animals might be altered, from the effect of food and other circumstances." Cuvier, he pointed out, had opposed such transmutationist views on functionalist grounds, arguing that "in every animal the several parts have such a mutual relation, both in form and function, that if any part were to undergo an alteration, in even a slight degree, it would be rendered incompatible with the rest." Kidd also claimed Cuvier's authority for the view that, while environmental factors might affect individual development, there was no evidence that variation could exceed certain limits, leading through a "gradual alteration of structure" from a

36. Kidd 1833, 285; JK to TC, 23 April 1833, NC, CHA 4.207.57–58.

37. Kidd 1833, 309n, 310. See Yeo 1986, 271–73. On the British co-option of Cuvier, see Dawson 2016, esp. chap. 2.

lower to a higher species. Next, he followed Cuvier's lead in defending his antitransmutationist stance on historical grounds, referring not only to the evidence of species stability from Egyptian mummies but also to that from Aristotle's own descriptions, a point further evidenced by an appendix containing parallel passages from Aristotle and Cuvier. Finally, the antitransmutationist discussion led Kidd back briefly to the subject of monstrosities. In the face of Geoffroy's controversial views, he repeated his conviction "that in a *lusus naturæ* the character of the species, however obscured, is never lost."[38]

Overall, Kidd's treatise presented his readers with a broad and scientifically untaxing overview of anthropocentric adaptations in nature that was strongly marked by his experience as a medical professor in an Anglican and humanist university. Throughout, he took opportunities to gently defend believers from dangerous notions that included the latest transmutationist views as well as Epicureanism. However, he declined to present the argument from design as the answer for atheism, and the conclusion to his book was instead a homily well suited to his intended genteel Anglican readership. We should be grateful to God for his goodness in adapting external nature to our constitution, he wrote. But as believers in revelation, we should also know that external nature is adapted to our moral as well as our physical nature. In terms more reminiscent of the "powerful" Butler than of Paley, Kidd reminded his readers that God had provided a sphere in which humans could exercise their moral faculties in preparing for eternity. Ultimately, there was a choice to be faced: was it to be the humble cultivation of "pure pleasures" as exhibited by Newton or sensuous indulgence and "intellectual pride"? As for Chalmers, religious belief for Kidd was at least as much an ethical as an intellectual question.[39]

❋

If the professor of physic at the University of Oxford had cause to be concerned about the potentially materialist and transmutationist import of modern medical studies, the recently resigned professor of surgery, clinical surgery, and physiology at the secular London University had all the more cause to be so. Having taught medical students in London for almost thirty years, Bell knew their reputation for irreligion. As one of the reviewers of his treatise on *The Hand* later put it, the common view was that "*materialism* is constantly staring the medical student in the face"

38. Kidd 1833, 324, 328–30, 333–34, 336, 347–75.
39. Kidd 1833, 342; Kidd 1824, 51.

and that, without protection, "that Gorgon head may chance to petrify all the finer principles and feelings of his nature."[40] Moreover, some medical teachers, such as the notorious William Lawrence, seemed only too happy to promulgate materialism in their lectures.

Bell emphasized at the start of his preface that his religious reflections had arisen out of his role as an "anatomical teacher," a role in which he had taken every opportunity of impressing his reflections on the minds of his pupils. In evidence, he pointed to a worthy pedigree in the battle against medical materialism, particularly during the Lawrence affair. Twenty years previously, he reminded readers, the "Christian Advocate" at the University of Cambridge had attended his lectures and heard him demolishing "those French philosophers and physiologists, who represented life as the mere physical result of certain combinations and actions of parts, by them termed Organization."[41] And while he had subsequently found himself called on to elaborate his views to a point that stretched his theological competence to its limit, Bell hoped, like Kidd, that in introducing non-specialists to aspects of human and comparative anatomy he could shore them up against egregious errors. Moreover, again like Kidd, this ambition now extended to both Geoffroyan anatomy and Lamarckism as well as materialism.

After an insubstantial introductory chapter, Bell's treatise fell into two main parts concerning the "mechanism" and "vital endowments" of the hand mentioned in its title. The first of the five chapters on mechanism outlined Bell's functionalist, comparative approach before offering a lengthy comparison of the anatomy of the human hand with the forelimbs of other species. Next came briefer chapters on the muscles of the arm and hand and on "substitutes" for the hand in other species. A final chapter tackled transmutationist explanations of the comparative anatomy of the hand. The second part of the treatise, on the hand's vital properties, was briefer, with three chapters devoted to sensibility, touch, and the muscular sense. A concluding chapter considered the role of the hand in human ingenuity, returning to the question of species transmutation, and even an appendix of miscellaneous additional illustrations did not quite manage to fill up the required three hundred pages.

Like Kidd, Bell offered no developed theology. The restrained opening statement of his introductory chapter asserted that an informed examination of any living structure would as a matter of course lead to the grand conclusion "that there is design in the mechanical construction, benev-

40. *British Critic* 14 (1833): 73.

41. Bell 1833a, ix–x; Rennell 1819, 53. See Jacyna 1983.

olence shown in the living properties, and that good predominates." Bell promised to trace such features in the human hand and arm, contrasting them with the "corresponding parts" of other vertebrates. There was, he reassured his readers, no danger to any mind that was "free from vicious bias" in examining the mechanical structure of animal bodies. But humans had become rather habituated to the contrivances of their own bodies and needed to be roused to pay attention and feel appropriate gratitude. Bell hoped his account would assist in this, not least through the use of comparative anatomy.[42]

Right from the start, then, Bell's treatise examined the human hand in relation to the anatomy of other species. In part, this was to answer the ancient question about the grounds of human superiority, but much more important was the introduction of a modern framework for understanding anatomy. The first chapter thus outlined the principles of Cuvier's comparative anatomy, showing how they were opposed to more radical interpretations of animal form. An anatomical feature such as the human hand, he explained to readers, could not be understood independently of the human frame as a whole, since the parts were mutually adapted. At the same time, he observed, it could not be fully understood independently of the entire "system" according to which all vertebrate bodies were constructed. There was, Bell told his readers, "a greater design" to take in—a "more comprehensive view of nature"—in which the same bones in all vertebrates were adjusted to very different purposes by the slightest of changes. Yet in each case nothing could be "more curiously adjusted or appropriated"; it was, he suggested, as if the whole vertebrate system had been designed with each individual species in mind.[43]

The "greater design" exhibited by the body plans of animals offered a rich new vein of evidence of design for British naturalists to exploit in the 1830s. Bell certainly wished to make some religious capital from the notion, but he was far from imbuing it with the strongly idealist significance found in other writers. It is not clear whether, like Cuvier, he considered the uniformity of the vertebrate body plan to have been dictated, ultimately, by strictly functional requirements. However, Bell proceeded to outline how knowledge of this "great plan of creation," combined with the universality of functional adaptation, had made possible the reconstruction of fossil forms. Crucially, all animals had been "formed of the same elements," and the changes in them were "but variations in the great system,"

42. Bell 1833a, 1–2.
43. Bell 1833a, 19, 21–22.

always bearing "a certain relation to the original type as parts of the same great design." These views—including the choice of such words as *elements* and *type*—echoed the transcendental anatomy that was increasingly prevalent in London's medical schools. But while Bell was prepared to make use of such notions, the bulk of his treatise emphasized the functionalist perspective of Cuvier. Moreover, he also used them to attack the transcendentalist and transmutationist theories that had been Cuvier's target.[44]

Even Cuvier, Bell reported, had considered some species to be maladapted, but this was wrong. Throughout the history of the earth, the various creatures revealed by geology—including the apparently monstrous giant saurians—had all been well adapted to their circumstances, and their interactions at each epoch had provided a balance in nature. This amounted to a rebuttal of transmutationist theory, in which animals became successively more perfect. Understanding living creatures, Bell insisted, depended on the principle of adaptation between their "instincts, organization, and instruments" on the one hand and their situation within the natural order on the other. With this positive principle clearly articulated, Bell's first chapter closed by briefly dismissing three further dangerous "notions" concerning changes in the structure of animals in geological history—namely, that they could be accounted for environmentally, as a result of general laws, or by "the transposition and moulding" of elementary parts. Bell reassured his readers that he would return to this subject in more detail.[45]

First, however, Bell sought to outline the comparative anatomy of the hand, comparing each part of the forelimb with the similar parts in other vertebrates and exploring the functional adaptation in each. Such arguments had been used successfully by Paley but Bell now introduced his readers to the impressive way in which the "genius" of Cuvier had developed the science of comparative anatomy. As a result of his advances, the adaptation of animals to their conditions of life and the mutual adaptation of their body parts was more clearly visible than before. One striking consequence, as Bell amplified in the second edition, was that "one single bone or fragment of bone" was sufficient to recreate "a really accurate conception of the shape, motions, and habits of the animal" from which it came. Such principles and such a "mode of investigation" allowed naturalists to reconstruct not only fossil specimens but also the conditions

44. Bell 1833a, 23–24; Desmond 1989b. For a more strictly functionalist reading of Bell, see Bowler 1977 and Ospovat 1981.

45. Bell 1833a, 36–40.

under which they lived. A "new science" had been born out of an alliance between anatomy and mineralogy.[46]

Bell further fleshed out this picture in the next two short chapters before elaborating on "The Argument Pursued from the Comparative Anatomy." What his review had shown, he claimed, was that, independently of the marvelous way in which the parts of an animal's body were put together, there was a "more comprehensive system" embracing all animals and exhibiting a "certain uniformity in the functions of life." Deviations from this more general plan were not accidental; they were designed to achieve a "perfect accommodation" in each particular case.[47] But how were the varieties of animal form to be explained? Here, finally, Bell tackled Geoffroy's transcendental anatomy at greater length. He outlined to his readers the "very extraordinary opinion" that animals are all composed of the same elements recombined to achieve different ends. This was a theory that readers of the "more modern works on Natural History" might have encountered, and it had been advanced with the "highest pretensions" and as the "commencement of a new æra!" But Bell was clear that the contrary, functionalist principle was the only true guide in morphology: "parts are formed or withdrawn, with a never failing relation to the function which is to be performed." Reviewing one of Geoffroy's strongest examples, in the jaw of birds, Bell juxtaposed formalist and functionalist explanations, asking his readers to adjudicate whether the resulting anatomy was best explained as an "accidental result of the introduction of a bone" or as the consequence of an adaptation.[48]

This discussion of Geoffroyan anatomy led Bell to concede that not all "distinguished naturalists" possessed the mental qualities that one would expect—namely, that "combination of genius with sound sense, which distinguished Cuvier, and the great men of science." It was surprising with what "perverse ingenuity" people sometimes sought to "obscure the conception of a Divine Author," substituting the "greatest absurdities" or "interposing the cold and inanimate influence of the mere 'elements,' in a manner to extinguish all feeling of dependance in our minds, and all emotions of gratitude." Referring again to those who wished to make the variety of animals a consequence of changing circumstances or of "a desire and consequent effort of the animal to stretch and mould itself" (an obvious

46. Bell 1833a, 65, 97–99; Bell 1833b, 80, 82. On Cuvierian principles of fossil reconstruction, see Dawson 2016.

47. Bell 1833a, 132.

48. Bell 1833a, 39, 134–35, 137, 141.

allusion to Lamarckian transmutation), Bell reassured his readers that new characters were not introduced by hybridism and that species could be shown to have been stable for the last five thousand years. Changes in animal morphology all manifested "foreknowledge and a prospective plan," as both human fetal development and insect metamorphosis demonstrated. Bell's great conclusion from the comparative anatomy of the hand and arm was that their adaptedness could only be explained by the action of the creator and that species were thus separately created rather than arising by "gradual variation from some original type."[49]

Bell now turned his attention to the "vital endowments" of the hand. As an anatomist, his main achievements had been in relation to the nervous system (most particularly in distinguishing motor and sensory nerves), and he took the opportunity when discussing the sensibility of the hand to expound his views on the subject. His starting point was the conviction that "the sensibilities of the living frame" were designed: they were "appropriate endowments; not qualities necessarily arising from life; still less the consequences of delicacy of texture." Always conscious of the danger of mental materialism, he sought to establish "the mind's independence of the organ of sense." The way in which mental ideas consistently and precisely correlated with sensations could "neither be explained by anatomy nor by physiology nor by any mode of physical inquiry whatever." Bell's conviction was that sensibility was the result of an unconscious vital agency, but ignorance of the nature of vital action did not, he insisted, impair the perception of design. On the contrary, humanity's reliance on something so far beyond its conceptions should serve to stir up gratitude to the creator. In his earlier treatise on *Animal Mechanics* Bell had more explicitly championed the vitalist argument as especially "calculated to warm and exalt our sentiments." Elsewhere in that work he had used the vital force to explain not only sensibility but also animal development according to a "law of formation." Such a power was as great as that which moved the heavenly bodies, and if astronomy produced in the human soul a sense of cosmic insignificance, this could be corrected by studying the effects of vital agency in the human body.[50]

Bell's final chapter promised to prove that the hand was not the source of human "ingenuity" or consequently of "man's superiority"—a classical theme that was highly topical in the context of the ongoing debates

49. Bell 1833a, 142–43, 146–47.

50. Bell 1833a, 11–12, 165, 170–71, 278–79; Bell 1833b, 214–15; [Bell] 1827–29, 33, 47–48, 62–63.

about transmutation. Bell emphasized the mutual adaptation of the hand and other human mental and physical endowments. Where Buffon had pictured the first human as ill prepared for the world at his creation, Bell envisioned him with his desires and passions already formed in keeping with his conditions. The origin of such adaptation in particular acts of divine creation was, he claimed, made more manifest by the "great revolutions" found in the history of the earth and its inhabitants, which clearly necessitated the direct involvement of a creator. Moreover, the progressive changes in living forms showed that there was no eternal succession of causes, while the adaptation of a "general design" to different purposes clearly spoke of a creator.[51]

Throughout this somewhat meandering review of the adaptation of the human hand, Bell's combination of functionalist comparative anatomy and vitalist physiology was intended to provide readers with a reassuring sense that Christianity had nothing to fear from transmutationist and materialist doctrines that were increasingly prevalent but palpably unsupportable. His final few pages recapped the evidence that fossil and living creatures offered of "anticipating or prospective intelligence"—reiterating that this grand conclusion could even be established by the "examination of a part so small as the bones of the hand"—before ending rather abruptly with a brief reflection on the unique moral and religious qualities of humans.[52]

In later editions, Bell's main text concluded with a further half-dozen pages that in the first edition appeared in his "additional illustrations." Here, he frankly acknowledged that philosophical investigation did not necessarily have a beneficial effect on religious sentiments. Humans might come to feel cosmically insignificant, and there was real danger in the expanding vistas that philosophy brought. The danger was more substantial for the student of science than for the active researcher, who would not rest satisfied with the "discovery of secondary causes" but would have his mind enlarged and his thoughts elevated. Those who learned "at secondhand" had their natural religious sentiments degraded by philosophical explanations and by the pride of their "newly acquired knowledge."[53] The evidences of adaptation were important reminders of divine care, Bell insisted. But they were of no avail if one did not finally seek a closer relationship with God. Thus, like Kidd, Bell closed with a moralizing homily. And it was—somewhat ironically in a treatise on anatomical design—the

51. Bell 1833a, 216–17, 219; Bell 1833b, 220.

52. Bell 1833a, 221.

53. Bell 1833a, 277–78.

frailties of human bodies that Bell looked to in the end as administering divine discipline, cultivating virtue, and drawing people to God.

The Bridgewaters of these two medical professors had none of the theological sophistication of Whewell's and Chalmers's, although Kidd's especially conveyed a sense of how little weight many contemporary Christians felt was due to the arguments of natural theology. What they did offer, however, was a reassuring vision of the special status of humans within the natural order in explicit opposition to the materialism and transmutationism that were largely associated with recent French science. In the case of Kidd, this was in part tailored to the classical studies that dominated Oxford life, but his treatise nevertheless showed familiarity with developments in French science and their appropriation in Britain. Bell, however, was in the midst of the metropolitan fray, and he used his Bridgewater to outline and defend a functionalist and vitalist account of animal morphology against the encroachment of Geoffroyan idealism, transmutationism, and materialism. Yet for all its conservatism, Bell's treatise offered an exhilarating historical vision of life on earth quite different from Paley's apparently static creation combined with a new understanding of the patterns of God's "greater design."

## *Physiology and the Laws of Life*

Bell may have made new use of the "greater design" revealed by modern anatomy, but his response to the work of Geoffroy Saint-Hilaire was almost wholly negative. Peter Mark Roget—another prominent member of London's medical establishment—was prepared to go much further in using the new perspectives to reveal design in his treatise on *Animal and Vegetable Physiology*. Indeed, he had begun to incorporate philosophical anatomy into his lectures at a notably early stage. At the Royal Institution in 1823 and 1825, he emphasized the "uniformity of design" and "general plan" manifested by zoological "types," claiming that comparative anatomy gave evidence not only of functional design but also "of a most regular and studied plan of operations." By 1826 he was arguing that while some of the conclusions drawn from this by the "modern naturalists of the French school" might sometimes seem problematic, they were generally "satisfactory" and proved that "the author of nature" had employed "several distinct types" in creating animals.[54]

Roget did not visit Paris until 1828, but he found no difficulty in engag-

54. Roget 1826, 101n; "Royal Institution," *Literary Gazette*, 19 April 1823, 249–50, on 250; "Royal Institution," *Literary Gazette*, 12 March 1825, 170–71, on 170.

ing with transcendental approaches to anatomy at home in London. For example, he became an enthusiast for the idealist morphology elaborated in the *Horæ entomologicæ* (1819–21) of government official and entomologist William Sharp MacLeay, which drew on German *Naturphilosophie* in its emphasis on the underlying patterns of anatomical resemblance. He also attended the lectures on comparative anatomy given at the Royal College of Surgeons between 1824 and 1827 by the Coleridgean Joseph Henry Green, who again drew on *Naturphilosophie,* emphasizing the unity of creation in idealist terms. As a textbook, Green used Carl Gustav Carus's introduction to comparative anatomy, which was translated into English in 1827, and this became another of Roget's key sources. But Roget's reading program went far wider, including works by Johann Friedrich Meckel, Henri de Blainville, Pierre André Latreille, and Geoffroy himself, and he also appears to have made use of Julien-Joseph Virey's attempts to Christianize Geoffroy's theories. In addition, as we have seen, he was able to use the impressive London University lectures of radical Geoffroyan Robert Edmond Grant to flesh out his own developing perspective.[55]

Roget highlighted his novelty in his preface. Selecting from an overabundance of materials, he observed, he had naturally focused on examples that manifested design, but he had also sought to arrange and combine them methodically to give a larger sense of the "general plan of creation."[56] Roget nevertheless had no theological pretensions. His explanation of the theological underpinnings of his work was confined to two introductory chapters and a concluding one. In between were sandwiched over 1,100 pages providing a functionalist anatomy of animals and plants considered under four different headings relating to the mechanical, vital, sensorial, and reproductive functions.

Roget's first chapter, headed "Final Causes," briefly outlined the logic of the design argument in relation to the role of teleological reasoning in physiology, but it began by warning the reader about limitations. It was, he explained, the highest use of human reason to investigate our relations with the creator, but the knowledge of God thus acquired was "limited," "imperfect," and "faint." Indeed, the sheer scale of creation left humans feeling impotent to "form any adequate conception" of God. Roget's account of the reasoning processes by which such limited knowledge of the

55. Roget 1834, 1: xi–xii, 53–54; Kendall 2008, 224. On Roget's knowledge of "contemporary French natural sciences," see Corsi 1988, 239–40 and n27; on MacLeay, see Rehbock 1983, 26–30; on Green and the wider context, see Desmond 1989b, esp. 260–75.

56. Roget 1834, 1:ix.

creator was to be obtained built on the analysis provided in his article "Physiology," published in the supplement to the *Encyclopædia Britannica* ten years earlier, which itself bore evidence of his philosophical education in Edinburgh. As there, he now argued that natural knowledge was of two kinds, depending on whether the relations between the phenomena were of cause and effect or of means and end. While all nature was underpinned by causal laws, the phenomena of life were so complex as to be "utterly irreducible to the known laws which govern inanimate matter." The object of the science of physiology was to ascertain the distinct causal and teleological laws of life.[57]

Roget claimed that the study of final causes was "forced" on the attention by even a cursory examination of the natural world. Every detail of living creatures made it "impossible not to recognize the character of intention." To attribute such phenomena to chance was to say that they were uncaused, which was "contrary to the constitution of human thought," with its grounding in the principle of causation. Roget mounted a robust philosophical defense of the possibility of discovering such final causes, describing the inferential character of all causal reasoning and the analogical character of all inferences concerning other minds. As far as natural theology was concerned, this was "the foundation of our assurance" of the existence of a creator. The argument was felt with strongest force where the analogies between human and divine design were most palpable. But it was also, he repeated from Paley, a "cumulative" argument. Moreover, knowledge of the "grand comprehensive plan" of which contrivances were part helped to illuminate their purpose and to reveal "more general laws."[58]

On this final comment hinged the transition from Roget's functionalist, rather Paleyan first chapter to the new perspectives outlined in his second, on "The Functions of Life." According to Roget, God had four fundamental concerns in creating animals. The first three of these were functional—the welfare of individual animals, the continuation of species through reproduction, and the control of populations through death—and in outlining them Roget was able to introduce his readers to the subjects of the four parts of his treatise on the mechanical, vital, sensorial, and reproductive functions. But God's underpinning fourth concern was morphological—namely, the "systematic economy in the plans of organization."[59] The variety of animal forms was not accounted for by functional adaptation alone. Even where the purpose was the same, Roget observed,

57. Roget 1834, 1:1–2, 5, 10.
58. Roget 1834, 1:23–24, 27 and n, 33.
59. Roget 1834, 1:34–35.

the means adopted to achieve them were endlessly varied, "as if a design had existed of displaying to the astonished eyes of mortals the unbounded resources of creative power." This "law of variety" (which has some similarities to the "principle of plenitude" of earlier authors) was counterbalanced by another law, that of "conformity to a definite type." By this, Roget meant that individual species in the same class were formed as copies of a "certain ideal model," merely varying in the details with the same "elements of structure" endlessly recombined.[60]

Roget explained his law of conformity to a definite type in somewhat functionalist terms. Nature appeared to have "laid down certain great plans of functions to which she has adapted the structure of the organs"; the more subordinate functions had then been "accommodated to this general design." From this emerged what Roget called the "laws of the co-existence of organic forms." Like Cuvier's principle of the correlation of parts, these laws determined that changes in one part of an animal body necessitated changes in others. But elsewhere Roget made clear his belief in Geoffroy's "principle of compensation," according to which "the great expansion of one part is generally attended by a corresponding diminution of others." Morphological harmony was achieved, he reported, by a balance between the antagonistic principles of "*expansion* and *contraction*."[61]

In other ways, too, Roget steered a middle path between Cuvier and Geoffroy. Cuvier's extreme functionalism meant that the four body plans represented by his four *embranchements* (vertebrates, mollusks, articulates, and radiates) reflected the limited number of functional solutions to the problem of animal bodies between which there could be no intermediates. However, Roget had no such qualms, promising analogies and gradations between the different types, including among fossil forms, which would show them to be "parts of one general plan" that "emanated from the same Creator." The chain of being expounded by eighteenth-century Genevan naturalist Charles Bonnet thus proved to be more like a "complicated net-work, where several parallel series are joined by transverse and oblique lines of connection." Indeed, Roget considered the circular scheme of classification recently introduced by the idealist William MacLeay to be the most satisfactory idea.[62]

This graduated living world seemed to lend itself to a progressive history of creation that represented the working out of a preordained divine plan. Alluding perhaps to the Christianized interpretation of Geoffroy

60. Roget 1834, 1:48–49. On the principle of plenitude see Yeo 1986.

61. Roget 1834, 1:50–51, 304, 324.

62. Roget 1834, 1:51–54.

found in Julien-Joseph Virey's *Histoire des mœurs et de l'instinct des animaux* (1822), Roget announced,

> An hypothesis has been advanced that the original creation of species has been successive, and took place in the order of their relative complexity of structure; that the standard types have arisen the one from the other; that each succeeding form was an improvement upon the preceding, and followed in a certain order of development, according to a regular plan traced by the great Author of the universe for bestowing perfection on his works.

The fact that Roget offered no critical comment indicated that this was not a cause for concern: such a view was consistent with the separate creation of individual species, and many otherwise inexplicable "anomalies" were "easily reconcilable" with it. Indeed, in the body of his treatise Roget repeatedly referred readers to the way in which developments of structure were "prospective," with "rudimentary" organs appearing in lower forms in preparation for their use in higher ones as well as persisting in higher ones once they had ceased to be useful. The chapter concluded with further theory, outlining the law of the serial homology of parts, according to which there was "a tendency to the repetition of certain organs, or parts, and the regular arrangement of these similar portions either round a central axis, or in a longitudinal series."[63]

This set the blithe tone for the body of Roget's work. While he sounded a note of warning about transmutation in his conclusion, the bulk of the treatise showed little sign of alarm at the tendency of the work of Geoffroy, Meckel, and others on which it drew. The functionalist tone of his elaborate exposition of the mechanical, vital, sensorial, and reproductive functions of animals and plants was constantly laced with more morphological references. Within each section, Roget provided a comparative account rising from the least to the most complex forms, and much of Geoffroy's conceptual framework was introduced in passing. There was also surprisingly little direct reference to the creator. Roget preferred, he warned his readers, to follow "the common usage of employing the term *Nature* as a synonym, expressive of the same power, but veiling from our feeble sight the too dazzling splendour of its glory."[64]

63. Roget 1834, 1:54–57. On the debt to Virey, see Corsi 1988, 239–40. See also Rehbock 1983, 20.

64. Roget 1834, 1:14n. On Roget's emphasis on "the supportive connections in a feminized environment rather than in a personal relationship with a masculine, father God," see Gliserman 1975, 290.

Roget's final chapter summarized his findings under the heading "Unity of Design." The evidence for such unity provided by animal and vegetable physiology clearly pointed to "one Great and only Cause of all things." But now Roget explicitly associated his approach with the "*unity of composition*" doctrine "zealously pursued in all its consequences" by many eminent naturalists on the Continent. Eager to emphasize the hypothetical character of the doctrine, Roget nevertheless considered it well worthy of attention. It was, he explained, based on the supposed constancy of the "elements of structure" in related animals. From it had been derived a "law of *Gradation*" according to which living and fossil animals "arrange themselves" into "certain regular series." Each "fresh copy taken of the original type" was supposed to include an incremental advance, he reported, through "the graduated development of elements, which existed in a latent form in the primeval germ, and which are evolved, in succession, as nature advances in her course." Roget reported that this view received support from the theory that higher animals in any given "series" pass in sequence through embryological stages resembling the adult state of lower animals. Such phenomena were associated, he observed, with the existence of organs in "rudimental form" in lower orders, and a footnote explained that *lusus naturæ* were manifestations of these "established laws" being in some respect arrested.[65]

While considering this law of gradation to be supported by "numerous and striking" analogies, Roget warned that "great care" was needed not to stray beyond the facts. The evidential basis left it far short of a "strictly philosophical theory." And there were no grounds for adopting transmutationist views. Whatever their apparent similarities during their embryological development, species were separated by an "impassable barrier" and were "ever constant and immutable." Roget echoed Cuvier in asserting that, however fine the gradations between species, in no case was the series "strictly continuous" and "in many instances, the interval [was] considerable; as for example in the passage from the invertebrate to the vertebrated classes." By contrast, many Continental physiologists of the "transcendental school" had become carried away by "seductive speculations." Indeed, Étienne Serres had announced that the "simple laws" he had discovered had "explained the universe." Another "system-builder," Roget reported, had assumed an "inherent tendency to perfectibility" in living creatures and had in that way sought to "supersede the operations of Divine agency." It was left to a footnote to identify the miscreant as Lamarck, and the rather garbled account of the Frenchman's "celebrated the-

65. Roget 1834, 2:625, 627–35.

ory" concluded, "If this be philosophy, it is such as might have emanated from the college of Laputa."[66]

In the face of such French "speculation," Roget proffered the exemplary "humble spirit" of Newton. Human reason was limited, he noted, and while humans might be able to acquire some knowledge of nature and the creator by its light, they could not divine "the grand object" of creation. Only divine revelation could provide the basis of morality and the "nobler objects" of human existence.[67] Like Kidd and Bell, Roget thus ended with a rather cursory nod to Christianity, but the bulk of his large treatise was occupied with offering a synthetic overview of the new morphological perspectives rooted in the claim that they offered substantial new evidence for divine design. Its blithely authoritative prose suggested that there was nothing terribly alarming to the Christian in such views even if the inappropriately speculative character of transmutationist doctrine might require a brief warning note.

❋

Prout's strangely constituted treatise on *Chemistry, Meteorology, and the Function of Digestion* also addressed the question of the religious significance of natural laws in physiology (fig. 2.3). But while Prout was, like Roget, a London physician, his interests and approach were altogether more chemical than anatomical. Reaching for materials that were to hand—his theoretical work on the nature of matter and his physiological writings on animal chemical processes—he found himself pulled in two directions. On the one hand, his hypothesis concerning the unity of matter seemed to offer a clear indication of how chemical phenomena might be reduced to mathematical laws. On the other, his belief in the reality of a range of vital agents in the processes of organic chemistry seemed to put those processes beyond the scope of natural law for the foreseeable future and to bolster Christianity against materialism. In the somewhat disjointed parts of his Bridgewater, Prout sought to make religious capital out of both of these aspects of his work. As a result, the whole sat together somewhat uncomfortably.

Like the other medical authors, Prout clearly felt uncomfortable in writing theologically. His preliminary announcement "to the reader" was diffident in the extreme. Chemistry had not "hitherto been considered in detail with reference to Natural Theology," he announced, clearly think-

66. Roget 1834, 2:636–38.
67. Roget 1834, 2:638, 641.

FIGURE 2.3 "William Prout" (1902) by Henry Marriot Paget (after a portrait by John Hayes, 1830–39), detail. EU0060. Image: © University of Edinburgh Art Collection.

ing of Paley and discounting the writings of Boyle and Priestley. Since his was a "first attempt," there would be "numerous imperfections," he apologized. Moreover, the argument being "cumulative," Prout would illustrate the principles, often leaving the "application of facts to the argument" for the reader.[68] Sure enough, after a very brief introduction, he proceeded to

68. Prout 1834a, xi–xii. See also Brooke 1989, 45, and Brock 1985, 67.

divide his work into three parts—one devoted to chemistry, one to meteorology, and one to "the chemistry of organization"—making passing comments on adaptations as he outlined key aspects of the science but generally confining his religious reflections to separate concluding sections.

Prout's introduction was clearly an afterthought—perhaps added at the suggestion of a friend—and consisted of three unnumbered pages printed as part of the preliminaries. Its purpose was to outline what he considered the "leading argument" of natural theology, the design argument, using as an illustrative example the analogy between animal fur and the use of furs for warmth by humans. Prout claimed, however, that such examples, in which humans could identify divine designs, were "relatively few." The nature and the end of many things in the universe were "utterly beyond" human comprehension, leaving humans to infer that a design must exist without perceiving any to be present. Prout distinguished three types of cases in relation to the argument from design: those where the ends and means of divine design were evident, those where only the ends were evident, and those where neither was evident. The second class included "all the phenomena and operations of chemistry," from which he promised to derive his arguments for the being and attributes of the creator.[69]

In the second edition—produced just three months later—Prout expanded this desultory introduction by moving into it a discussion of religious skepticism that had appeared later in the text in the first edition. While arrangements in nature indicated design to "common understandings," he observed, some minds were "so obtuse, or so strangely constituted" as not to admit the argument. Prout was prepared to address briefly two such classes of objectors: those who attributed design to necessary laws of nature and those who denied that design could be proved to exist at all. To the first kind of skeptic, he answered that laws were almost all empirical and could not be proved to be necessary. Indeed, the progressive history of the earth suggested that there had formerly been "*different* 'laws of nature.'" It seemed unbelievable that natural laws had existed "from eternity," he observed, before concluding that there must therefore be a divine lawmaker. Prout also replied to the second, Humean kind of skeptic, but only briefly, arguing that while design might not have been proved, neither had it been disproved.[70]

After these brief introductory remarks, the treatise proceeded to Prout's

69. Prout 1834a, [xxv–xxviii].

70. Prout 1834a, 173–76. He elaborated a little further on the validity of analogical reasoning in regard to design in the second edition, probably borrowing from Chalmers and Roget in the process. Prout 1834b, 3–7.

book on chemistry, which began with some preliminary observations on "the rank of chemistry as a science" and its application to the design argument. This afforded further opportunity to reflect on the religious dangers posed by the laws of nature as Prout sought to reconcile his sense of the desirability of reducing chemistry to mathematical laws with his conviction that the mysterious agencies of chemistry offered distinctively powerful evidence of design. Drawing on John Herschel's recent *Preliminary Discourse on the Study of Natural Philosophy,* he located chemistry as one of the many branches of empirical science that were becoming more perfect and certain as they became more subject to "laws of quantity." However, he noted that this had been especially difficult to achieve in chemistry, where many of the phenomena were known primarily through the senses of taste and smell and where the causal processes remained obscure.[71] In fact, Prout knew better than most that new efforts were under way to establish such laws because he was at the forefront of them. Building on Dalton's atomic chemistry, he had made a significant stride toward a mathematically grounded account of the underlying unity of matter. By positing a fundamental form of matter, which he termed *protyle,* he had been able to explain the ratios in the atomic weights of chemical elements. However, while his Bridgewater offered a good introduction to this controversial theory, it was only in the conclusion that Prout had more positive things to say concerning the dawn of a mathematized chemistry. For now, instead, it was the difficulty of reducing chemical phenomena to natural laws that he wished to emphasize for theological purposes.

As Paley had shown, Prout observed, divine design was often to be observed in chemical phenomena even when the underlying operations remained unknown. While such evidence might be in some respects weaker than examples of designed mechanism, it also offered some advantages. When phenomena could be readily explained in terms of causal, mechanical laws, he argued, then God appeared "almost too obviously to limit his powers within the trammels of necessity." But in chemistry, where the laws were "less obvious" and often unknown, his operations seemed much more those of a "free agent." Like Bell, Prout considered that while "obvious mechanism" provided clear evidence of design, it could not arrest the attention nearly so well as the use of "means utterly above *our* comprehension," nor could it generate such "exalted notions" of divine wisdom and power. A concluding paragraph brought the message home forcefully. Should anyone claim that the order of nature was the result merely of laws rather than design, he would find himself answered by Prout's facts: pro-

71. Prout 1834a, 1, 4.

spective adaptations, beautiful adjustments, and "the subversion of even his favourite 'laws of nature' themselves, when a particular purpose requires it."[72]

The point was reiterated as Prout moved into the body of his first "book." A brief chapter on the interplay of "physical agents" and matter—and of the laws they obeyed—took a lengthy quotation from Paley as its "text" for exposition. According to the archdeacon, "God has been pleased to prescribe limits to his own power, and to work his ends within those limits" in order better to "exhibit demonstrations of his wisdom." It was as if one being had fixed the laws of matter and left another being to build a creation within those constraints, making contrivance necessary. Paley even went so far as to state that "there may be many such agents, and many ranks of these." Prout now took his remit to be providing a "summary view" of the laws of matter and of the "subordinate agents by which matter is capable of being influenced." In physical chemistry the main "subordinate agencies, or *forces*," were gravitational and "*molecular* or *polarizing* forces," but in the chemistry of living beings, there were many more such agents. The effect of this analysis was to emphasize the passivity of matter and to increase the sense of the dependency of the created order on divine activity.[73]

Prout's first substantive chapter expounded his speculative matter theory, which was offered as the fruit of "twenty years of close attention and no ordinary labour." Suggesting that the "general reader" might cut to the chase and read only the concluding summary and reflection, he proceeded to outline his view that molecules exhibited two kinds of polarizing forces, which were responsible for cohesive and chemical attraction and repulsion. These forces, he proposed, were analogous if not identical to electricity and magnetism, and they were also present in "the molecules of the imponderable principles, heat and light." Prout thus achieved a unification of forces, allowing him to "give a connected and popular sketch of molecular forces and operations" and, through his novel theory, to "display the beautiful simplicity and analogy that prevail throughout the whole." The ordered nature of matter, with its "*endless repetition of similar parts*," made clear its "manufactured" character and thus displayed the existence and intelligence of a deity.[74]

Prout next proceeded to the chemical elements, devoting a section

72. Prout 1834a, 11–12, 14.

73. Prout 1834a, 15–17.

74. Prout 1834a, 22, 83, 85, 87–88. On Prout's matter theory, see Brock 1985, 109–21.

to the "Laws of Chemical Combination" that drew on his earlier controversial argument that atomic weights were integers and that the supposed elements might thus all be composed of more fundamental building blocks—what in 1823 Jöns Berzelius had dubbed "Prout's hypothesis." He would not have advanced his "peculiar chemical opinions" in this form, he reported, had he not believed that they were "calculated, sooner or later, to bring chemical action under the dominion of the laws of quantity." Prout added a further note to the second edition, in response to criticism, making clear that his arguments for design did not depend on the truth of his molecular theory. But this was a little disingenuous. As his biographer relates, "the whole burden of Prout's *Chemistry* was to show how macrochemical phenomena could be reduced to the beautiful design of a molecular theory of matter." In successive editions, he revised, developed, and expanded this part of the text. Other molecular theories, he asserted in the third edition of 1845, were "utterly unworthy to be ascribed to the Deity," whose fundamental laws were "ever simple, general, and comprehensive."[75]

Alongside this grand argumentation from laws sat more conventional design arguments. Having described the various chemical elements and their compounds, Prout offered his theological reflections. He began by outlining what he saw as the adaptations within the hierarchy of elements and agents that resulted in a dynamic equilibrium in nature before turning to the adaptation of living creatures to that equilibrium. Within his hierarchy, living creatures could not have arisen of their own accord from "hydrogen, carbon, oxygen, and azote, with heat and light, &c." It was most likely that God had created new species directly, but Prout strikingly claimed that it would not affect the argument if God had "operated by delegated agencies and laws." Then there was the extraordinary way in which the elements appeared to have been created with a view to producing the compounds that would be needed, their own properties being "determined as the more general laws of matter might decide." According to Prout, this might explain the injurious properties of many elements.[76] Finally, he evoked the wondrous sense of divine omnipotence brought to mind by the regular behavior of the innumerable molecules in any chemical process.

Prout's account of chemistry thus reflected a somewhat uncomfortable

75. Prout 1845, 41; Prout 1834b, [xiii]; Prout 1834a, xi–xii, 121–40; Brock 1985, 65, 109, 176. Some of Prout's working notes relating to these revisions are to be found in WL, MS4018–19.

76. Prout 1834a, 150–70; see also Brock 1985, 86–87.

ambivalence about natural laws and their relation to divine design, and his second book, on meteorology, continued in similar vein. This book made fewer claims to scientific originality than the first, although it was here that Prout became the first person to apply the word *convection* to heat. In it, he offered an account of the effects of the "agents and elements of chemistry in the economy of nature," discussing especially the role of light, heat, water, and the materials of the surface of the earth in producing climatic phenomena. The climax of the book emphasized that these were all mutually adjusted to produce a system that was designed for the needs of plants, animals, and, above all, humans. God generally rigidly adhered to laws in nature, and, as a result, scientific discoveries only gave a stronger sense of the "simplicity and wisdom of design." At the same time, however, Prout offered high-profile examples of indispensable "exceptions" to natural laws in, for example, the anomalous expansion below a certain point of cooling water and the anomalous variability in the composition of air.[77]

The tension was particularly clear when Prout stood back to give the reader his theological reflections. Meteorological phenomena had long been the "objects of superstitions awe," but the scientific account he was now able to give reduced them to laws and arrangements. This need not "lessen our veneration" for God; rather, it should increase it. What little we can know of the incomprehensible Deity, Prout radically claimed, "we know nearly altogether from his works." It was students of nature alone who were qualified "to form an adequate conception" of God, and scientific pursuits were thus sainted. Anyone who substituted the laws of nature for divine design must have a "warped" mind, since these laws were clearly applied to designed ends. But meteorology offered some "exclusive" arguments for such skeptics in the instances Prout had given of laws being "infringed" just where such infringement was "indispensably necessary to organic existence."[78]

In his third book, on the function of digestion, Prout returned with vigor to God's mysterious "subordinate agents"—specifically the "*organic agent* or *agents*" responsible for the chemistry of organization. The book delivered an authoritative "summary view of those chemical properties, and laws of union, by which organized beings are distinguished from inorganic matter." Drawing largely on his earlier work, Prout outlined in turn the chemical composition of living bodies, their modes of nutrition, the process of digestion, and the subsequent use of digested food in the

77. Prout 1834a, 178, 190–93, 223, 233–34, 247–51. On "convection" see the *Oxford English Dictionary* (https://www.oed.com/view/Entry/40672) and Brock 1985.

78. Prout 1834a, 354–60.

body. The theological payoff, briefly elaborated in the concluding pages, lay chiefly in the functional adaptation of animals to the chemical operations occurring within them and to the chemical balance of nature. This was the more wonderful, he recapped, to the extent that it remained inexplicable.[79]

The point was that organic agents were, and would probably always remain, "altogether unknown to us." Prout nevertheless did his best to identify the limits of their power and the nature of their operation. With an "apparatus of extreme minuteness," an organic agent could bring together particular molecules and exclude others, allowing them to combine. Appropriating Paley's words, Prout sought to argue the case for there being "many such agents, and many ranks of them." This fitted in with his more general view of a hierarchy of agents in nature, each designed to achieve particular ends within the bounds defined by the laws of more general agents. Prout's hierarchy of agents corresponded with a succession of living creatures as found in progressionist geology, but it was "directly opposed to the notion of spontaneous development maintained by some distinguished French philosophers; as well as to the opinion that life is the *result* of organization." The strict hierarchy of inorganic and organic agents made it impossible that change could arise from below, and Prout claimed that "when a new and specific agent is required, a new and specific act of creation must be performed by the Great Architect of the universe." It was, consequently, "impossible that by any accidental concurrence of circumstances, a dog can, in the progress of time, be gradually converted into an ape, or an ape into a man."[80]

Prout's concluding remarks concerning the "future progress of chemistry" and the "tendency of physical knowledge in general" reiterated the consistent concern of his treatise to extend the scope of natural laws while guarding against their religious dangers. Picking up on his earlier discussion about the scientific status of chemistry, he noted that the progress of chemistry in recent years had been "truly astonishing," promising a golden age "when chemistry shall be brought more under the control of the laws of quantity." Chemists must not get ahead of themselves, however, especially in regard to physiology: they should confine themselves to examining the peculiar actions of the living principle rather than speculatively attempting to reduce that principle to an agent such as heat, electricity, or motion. Prout's target here was clearly such high-profile physiological materialists as William Lawrence, and he tartly asserted that the

79. Prout 1834a, 413, 537–43.
80. Prout 1834a, 429–37.

"imagined Pantheism" of the "complacent philosopher" who deified the laws of nature was no better than the superstition of the savage. The laws of nature were constantly being merged, and they were thus shown to be "mere delegated agencies." As a result, Prout was able to reassure his readers that the "tendency of knowledge" was "to fix the mind on the source of all knowledge and of all power—the GREAT FIRST CAUSE."[81] Yet over all, the treatise nevertheless conveyed a striking discomfort concerning the religious implications of natural laws.

Like the other medical authors, Prout and Roget offered their readers little theological sophistication. Instead, they used their Bridgewaters to give authoritative accounts of the cutting edge of comparative anatomy, physiology, and chemistry while emphasizing the extent to which these latest developments enhanced the Christian's sense of divine agency in nature. For Roget, it was the morphological laws being revealed by Continental naturalists that particularly enhanced the believer's perspective beyond Paley's emphasis on functional adaptation, although he warned his readers not to be seduced by their transmutationist speculations. For Prout, too, the advancing laws that were the central concern of science offered enhancements of the Christian's sense of divine perfection, as demonstrated by the simplicity of his own matter theory. Yet Prout remained ambivalent, wishing to stress the role of apparently anomalous adjustments in impressing the believer with a sense of design. For both, however, the progress of science could be seen to have a beneficial religious tendency.

### *The Bible and the History of Life*

Kirby was also concerned about naturalistic explanation and considered there to be divine "under-agents" involved in the phenomena of life, but his perspective was radically different from Prout's. Indeed, his hefty two-volume work on the *History, Habits, and Instincts of Animals* must be seen as something of an outlier among the Bridgewaters. Fully a generation older than most of the authors, the septuagenarian Kirby was also the only one working outside the context of universities or medical practice in the rural isolation of a country rectory. More importantly, he was alone among them in being a staunch High Churchman, associated with the Hackney Phalanx, who, as a group of young men in the 1790s, had played a large part in revitalizing the High Church party within the Church of England. It was these High Church commitments that accounted for Kirby's com-

81. Prout 1834a, 543–51, 557.

plete lack of interest in the project of a rationally constituted natural theology and for his distinctive view of the appropriate relationship between the Bible and natural science.

While Kirby shared the Hackney view that natural theology was both epistemologically weak and practically ineffectual against Christianity's opponents, he took up the challenge of his Bridgewater with a strong sense of religious duty. "Nothing, certainly," he wrote in response to the Bishop of London's invitation, "would be more gratifying to me than to employ my talents, such as they are, in the great cause of religion, especially in times like the present." Writing in the midst of the political ferment associated with the debate over the Reform Bill, Kirby regarded his Bridgewater as a suitable opportunity to serve God by stemming the tide of infidelity and sedition. This, indeed, was the acceptable service of one who considered his science to be an indulgence except insofar as it contributed to the paean of praise to God. Responding to his sister's congratulations on his appointment to the task, he asserted: "I am *far* behind you in what is better than all knowledge or scientific eminence,—unwearied and unceasing efforts to relieve the wants and instruct the ignorance of your fellow-creatures in humble life."[82]

Kirby's Tory paternalism pervaded his Bridgewater, but his was not a book designed to be read by the increasing numbers of unemployed peasants whom the work described as daily growing more immoral, lazy, intemperate, and insubordinate. Nor did it offer any rational justification for the arguments of natural theology even though Kirby promised Bishop Blomfield that he would "embody and concentrate a number of facts and observations that would strikingly demonstrate the being of God, and illustrate His adorable attributes." Rather, the great bulk of the work provided "a general sketch of the animal kingdom, its classes and larger groups, and so much of their history, habits, and instincts, as may be necessary to indicate their several *functions* and *offices* in the general plan of creation" with a view to illustrating God's attributes "more strikingly."[83] This lengthy account was prefaced by an introduction of almost a hundred pages that addressed more directly what Kirby considered to be the malaise of modern science, identifying its root cause and its cure.

In Kirby's view, science's ailment was the arrogant and delusional sub-

82. WK to Mrs. T Sutton, 1 December 1830, and WK to CJ Blomfield, [November] 1830, in Freeman 1852, 437–39.

83. Kirby 1835, 1:138, 398; WK to Blomfield, [November] 1830, in Freeman 1852, 43.

stitution of secondary causes, or natural laws, in place of God. This was to be found above all in the work of Laplace and Lamarck, but, unlike his coworkers, Kirby located the roots of their error in their failure to interpret nature by appropriate reference to the Bible. The distinctiveness of his diagnosis was clear from the French motto that appeared opposite his title page: "It is Bible in hand that we must enter into the august temple of nature, in order properly to understand the voice of the Creator." It had, Kirby began, always been his habit "to unite the study of the word of God with that of his works." Since Bridgewater's bequest had specifically directed that the divine attributes should be illustrated from "the whole extent of literature," he took this to justify his drawing on the Bible. And he was eager to point out that it was the failure of philosophers such as Laplace and Lamarck to do the same that had led to their "grievous errors." Their malaise was encapsulated in the saying of Jesus: "Ye do err not knowing the Scriptures."[84] Counteracting the errors of such philosophers, Kirby's introduction made clear, was the chief religious purpose of his treatise.

Kirby briefly described Laplace's nebular hypothesis, berating its author for refusing to refer cosmic arrangements to the Deity. The Frenchman had spoken of the "Author of Nature," he reported, "as perpetually receding, according as the boundaries of our knowledge are extended; thus expelling, as it were, the Deity from all care or concern about his own world." There then followed a twenty-page critique of Lamarck, one of a very few extended attacks on the naturalist's views to appear in English at this period. Kirby had been among the earliest British readers of Lamarck's *Philosophie zoologique*, in the summer of 1811, declaring it "one of the most infamous & at the same time absurd attempts at all but Atheism" while considering that "A child that has its senses, may answer it." A decade later, he was fearful that his friend the naturalist William MacLeay's "Eulogy" of Lamarck in his *Horæ entomologicæ*, together with his denial of separate species, might lead "young minds" to be perverted by the Frenchman's views on species change, for which Kirby now coined the term *Lamarckism*. Having thus concluded that it was his duty to expose Lamarck's "atheistic system to merited contempt," it was not until his Bridgewater that he finally did so, and by then he could claim that it was

84. "C'est, la Bible a la main, que nous devons entrer dans le temple auguste de la nature, pour bien comprendre la voix du Créatueur." Kirby 1835, 1:[ii], xviii–xix, translation mine. The motto is from Gaëde 1828; on Gaëde, see Morren 1865, esp. ix and xv. The biblical quotation is from Matt. 22:29.

unnecessary to provide a detailed critique of the transmutation hypothesis since Lyell had already fully considered it and satisfactorily proved the stability of species in the 1832 second volume of his *Principles of Geology*.[85]

Kirby had told MacLeay that he was inclined to "give the D[evi]l his duty" in acknowledging Lamarck's classificatory ability, and throughout the body of his Bridgewater he made deferential references to his taxonomic work. His introduction likewise described the naturalist as "distinguished by the variety of his talents and attainments, by the acuteness of his intellect, by the clearness of his conceptions, and remarkable for his intimate acquaintance with his subject." But he pulled no punches in ridiculing the suggested mechanism by which "in the lapse of ages a monad becomes a man!!!" Kirby briefly outlined how inconsistent Lamarck's theory was with the mutual adaptation exhibited by the parts of animal bodies before turning his attention to the Frenchman's materialism. This, he claimed, effectually excluded God from the government of his creation and made the human "an animalcule" rather than "a son of God." Lamarck had also effectively semideified a vacuous concept of "nature" in a way that robbed God of his powers, and Kirby took this to be his overarching purpose: to ascribe the works of creation to secondary causes and to leave no room for the "intervention of a *first*."[86]

Kirby's answer to such errors of modern philosophers was his distinctive biblical approach to the study of nature. With Bacon, Newton, and the Cambridge Platonists he endorsed the belief that there were active powers in nature—"*inter-agents*" between God and the material world through which he acted according to certain laws—but he considered that these needed to be understood in the light of biblical interpretation. Kirby was only too aware that introducing the Bible into scientific inquiry was out of step with contemporary mores, but he considered this attitude to be grounded in a misreading of Bacon. The latter's warning against using the Bible as a scientific "text book" was intended to proscribe biblical literalism and superstition, Kirby observed, but recently the edict had been interpreted so strictly that philosophers now paid no attention to the Bible, and a "clamour" was sometimes raised against those who did. The only exception to the rule were the "Hutchinsonians"—followers of the mys-

85. Kirby 1835, 1:xxii, xxxii; WK to A Macleay, 27 September 1811, WK to WS Macleay, 30 April 1821, LSL, MS237B (see also letters to WS Macleay of 9 April and 28 May 1821); Macleay 1819–21, 328–34. On Macleay, see Ospovat 1981, 101–13, and Rehbock 1983, 26–30. On the early reception of Lamarck's book, see Topham 2011.

86. Kirby 1835, 1:xxii, xxiv, xxviii; WK to WS Macleay, 30 April 1821, LSL, MS237B. See also Corsi 1988, 237.

tical system of natural philosophy and biblical interpretation devised by early eighteenth-century naturalist John Hutchinson—whom Kirby conceded had "perhaps gone too far in an opposite direction." Yet while he thus sought to distance himself from doctrinaire Hutchinsonianism, his approach nevertheless manifested an obvious debt to the school.[87]

Hutchinsonianism had always been especially popular among High Anglicans, and Kirby was no doubt introduced to it through the Hackney Phalanx—especially his Suffolk neighbor and friend William Jones of Nayland. According to Hutchinsonians, the Bible contained a definitive account of natural phenomena that had been lost as a result of corruption of the Hebrew text and could only be recovered through Hutchinson's own complex philological apparatus. Kirby's approach was somewhat more modest, but he nevertheless argued that diligent study of the Scriptures would yield information about the powers by which God ruled the universe that it would be impossible to acquire "by the usual means of investigation." Such analysis was arduous, he observed, because God had concealed "spiritual and physical truths under a veil of figures and allegory" in recognition that the "prejudices, ignorance, and grossness of the bulk of the people could not bear them." Thus, Kirby considered the symbolism of Moses's tabernacle and Solomon's temple to be especially informative about physical truth, since these had been constructed according to divine instructions and formed part of his plan to reveal his supremacy over the powers in nature. In particular, he was interested in the cherubim depicted above the mercy seat, which represented "the powers, whether physical or metaphysical, that *rule* under God."[88]

In all its mystical complexity, this diversion—drawn from Kirby's planned volume of sermons on the "mystical language of scripture"—filled fifty pages of his introduction. At its end, he had to remind the reader that, in addition to demonstrating his favored method of biblical interpretation, his purpose had been to counteract the tendency of some scientific men to lose sight of God in secondary causation. Kirby's desire was to make God evident everywhere "as the main-spring of the universe, which animates, as it were, and upholds it in all its parts and motions." The creator was always to be found "Maintaining his own laws by his own universal action upon and by his cherubim of glory." To Kirby's staunchly Tory mind, this vision of God's transcendent power had obvious political overtones, as did any Lamarckian attempt to undermine it. The cherubim not only

87. Kirby 1835, 1:xviii–l.

88. Kirby 1835, 1:l–li, lxxi. On Hutchinsonian interpretation of the Bible, see Cantor 1979, 12–17.

represented the powers of nature but also the powers in society. Thus, all government, heavenly and earthly, was "derived from Christ, as King of Kings and Lord of Lords, to whom *All power is given in heaven and earth.*" This emphasis continued throughout Kirby's treatise, where he stressed the role of a "kind," "watchful," and "fatherly Providence" that instructed organisms in their tasks. As in society, so in nature the creator fitted organisms for their "rank and station," their "office," and their "official duty."[89]

By the standards of the 1830s, Kirby's Hutchinsonian method was frankly bizarre, but it was largely confined to his introduction. In addition, however, the two preliminary chapters dealing with the creation and geographical distribution of animals also reflected a biblicism at odds with the scientific temper of the age. Like so many authors who have warned of the dangers of literal interpretations of Scripture, Kirby proceeded to contravene his own principles by giving a literal reading of Genesis, claiming that the earth was six thousand years old and that a miraculous deluge accounted for most geological phenomena. Reluctantly admitting the possibility that some species had become extinct after the Fall, Kirby nevertheless considered that many fossil forms, including Gideon Mantell's "giant Saurians," still existed, surviving in subterranean caverns (a view he considered was rendered plausible by the details of the Genesis flood narrative).[90] The biblicism continued in Kirby's second historical chapter, in which he sought to account for the postdiluvial distribution of animals using a combination of natural and supernatural means.

The descriptions of animals that filled the many remaining chapters of Kirby's book were much more conventional in making frequent reference to the wonderful functional design that animals exhibited by way of manifesting the power, wisdom, and goodness of the creator. Kirby's favorite example of this, which he considered a powerful answer to Lamarck, was animal instinct. References to remarkable instincts abound in Kirby's Bridgewater, and he devoted an entire chapter to a more systematic consideration of them, arguing that their disanalogy with human art made them appear "almost miraculous." Yet Kirby considered that they had a physical cause and cited with approval Julien-Joseph Virey's analogy of "a hand organ, in which, on a cylinder that can be made to revolve, several tunes are noted." Instincts were probably programmed into animal orga-

89. Kirby 1835, 1:c, ciii, 289, 324; 2:23, 25, 30, 51, 407, 440; Freeman 1852, 433, 440. See also Jacyna 1983, 325–26.

90. Kirby 1835, 1:21, 376–89, citing William Jones's *Physiological Disquisitions* (1781).

nization like the tunes on the cylinder. Running alongside Kirby's familiar discourse of design, however, were numerous rather more homiletic references drawing social, moral, and religious lessons from the animal kingdom. One remarkable passage, for instance, described how the system of predation confirmed the Christian doctrine of "vicarious suffering." Kirby continued: "Thus does the animal kingdom, in some sort, PREACH THE GOSPEL OF CHRIST."[91]

Kirby also went beyond Paleyan examples of functional adaptation in his numerous references to the underlying divine plan, which, like Roget, he gave a distinctly idealist cast. Over the preceding decade, Kirby had championed the "quinary" taxonomic system of MacLeay, whose *Horæ entomologicæ* had suggested that the plant and animal kingdoms were naturally divided into five basic divisions, each of which fell into five classes, and so on, in turn, through the taxa. According to MacLeay, each group of five was naturally arranged in a circular pattern, with those groups most closely connected having the greatest morphological similarity. In addition to these real anatomical "*affinities*," superficial resemblances could be traced between organisms that were otherwise dissimilar and taxonomically far removed. These "*analogies*" extended even so far as to apply between plant and animal kingdoms. MacLeay's system fitted well with Kirby's somewhat mystical notion of the divine plan, and many aspects of it appeared in his Bridgewater. In particular, Kirby used MacLeay's "analogies" to ward off Lamarckian dangers. The fact that there existed a "system of representation"—such as between zoophytes and plants—as well as a "chain of affinities" confirmed that there was an underlying divine plan.[92]

In its disregard of natural theology and its reliance on the symbolic interpretation of the Bible, Kirby's treatise stood in stark contrast to the rationalism of Paley's *Natural Theology*. Moreover, while Kirby found reminders of the paternal care of a patriarchal deity in the many examples of functional adaptedness to be found in the "history, habits, and instincts of animals," he was quite as much interested in the pervasive patterns of resemblance that spoke of a divine plan. As far as Kirby was concerned, the student of nature whose mind was imbued with Biblical truth would find nature replete with spiritual meaning. The sciences only became dan-

91. Kirby 1835, 1:xxv–xxvii, 291–94; 2:62, 220–21, 243; cf. 2:526. For other homilies, see 1:399, 2:91–92, 197, 387, 473–74, and 507–9. Regarding Kirby's views on instinct, see Richards 1981, 214–18, Grinnell 1985, 60–63, Clark 2006, 50–54, and Clark 2009, 44–53.

92. Kirby 1835, 1:191, 2: 109–10.

gerous when pursued in ignorant defiance of the Bible, as the examples of Lamarck and Laplace made abundantly clear.

❋

In their treatment of the connection of the Bible with science, and of the history of life on earth, the contrast between the treatises of Kirby and Buckland was acute. Where Kirby's work sought to restore what had only ever been a marginal approach to biblical interpretation, Buckland's *Geology and Mineralogy* authoritatively restated what had become the standard view among scientific men over recent decades. Moreover, this was what had been expected of him. Buckland was a clergyman and canon of Oxford's cathedral who had for two decades been responsible for teaching geology and mineralogy within the most Anglican of universities, and he had been appointed to write by the Bishop of London and the Archbishop of Canterbury. As a result, he was perfectly placed to claim the church's authority for a definitive account of the "consistency of geological discoveries with sacred history." In addition, his celebrated claim of a decade earlier, that geological findings furnished evidence of the biblical flood, gave a certain piquancy to his treatise. Many people shared Lyell's curiosity to know what he would "do about his diluvial theory" in the light of the ongoing failure to discover human fossils in diluvial deposits. Nor was Lyell the only one to have heard that he had "more than once changed his plan." In the event, Buckland used a footnote to quietly withdraw his earlier assertion.[93]

The situation was in some ways hardly propitious for such liberalizing moves. Nolan was by no means the only conservative Christian to be found denouncing on biblical grounds any account of the creation that involved more than six days or took place longer ago than six thousand years. Moreover, in Oxford in particular the rise in the 1830s of the High Church Tractarians polarized theological opinion within the university and made Buckland an obvious target. He nevertheless addressed the key question of biblical interpretation at the outset. Having introduced the subject of geology, he devoted a chapter to its proper relation to the Bible. He firmly separated the moral and natural realms, arguing for freedom of investigation within each according to the appropriate canons of evidence and reasoning. It was unreasonable, he argued, to expect to find in the Mosaic

93. Buckland 1836, 1:95n; Lyell to John Fleming, 7 January 1835, in Lyell 1881, 1:446.

narrative "the history of geological phenomena," since such details were "foreign to the objects of a volume intended only to be a guide of religious belief and moral conduct." The purpose of the Mosaic account had been "not to state *in what manner*, but by *whom*, the world was made."[94]

Like other liberal Anglicans, however, Buckland was still concerned to demonstrate the ultimate congruence of historical geology and Scripture. The reconciliation scheme that he offered was the hypothesis that the word *beginning* used in the opening verse of Genesis was intended "to express an undefined period of time, which was antecedent to the last great change that affected the surface of the earth, and to the creation of its present animal and vegetable inhabitants; during which period a long series of operations and revolutions may have been going on; which as they are wholly unconnected with the history of the human race, are passed over in silence by the sacred historian." This was hardly new. It had been Buckland's favored option when reviewing the possibilities in his *Vindiciæ geologicæ* some seventeen years earlier, and even then he had identified it as the opinion of "some of the ablest divines and writers of the English Church." Now, however, Buckland astutely included as supporting evidence a lengthy footnote on the meaning of the early verses of Genesis by the Oxford professor of Hebrew, Edward Pusey, known by the time of publication as a leading Tractarian.[95]

For all that it was bound to be controversial, this reconciliation scheme occupied only one of the twenty-four chapters of Buckland's Bridgewater. Indeed, it was not even mentioned as one of his chief objects. As earlier with his inaugural lecture, Buckland wanted not only to reassure readers that geology was consistent with the Bible but also to demonstrate that any "concession" that it might require from the biblical literalist was amply compensated "by the large additions it has made to the evidences of natural religion." This was what Buckland emphasized in his brief preface. As before, he described "three important subjects of inquiry in Natural Theology" that geology particularly advanced. The first related to minerals and their location in the earth, which gave evidence of adaptation to the needs of living creatures and especially humans. The second concerned theories of the eternity of the earth and of the origin of species by transmutation, to which geological phenomena were "decidedly opposed."

94. Buckland 1836, 1:15, 33. On biblicist critics of geology, see Rupke 1983, 267–74, Morrell and Thackray 1981, 230–33, and O'Connor 2007a, 2007b. Mortenson 2004 is also useful, but polemical.

95. Buckland 1836, 1:19, 22n–25n; Buckland 1820, 31–33.

FIGURE 2.4 "Ideal Section of a Portion of the Earth's Crust, Intended to Shew the Order of Deposition of the Stratified Rocks, with Their Relations to the Unstratified Rocks." Colored engraving by John Fisher after Thomas Webster. Buckland (1836), plate 1. Image: Wellcome Collection. Public Domain Mark.

Third came the project of extending to fossil species a Paleyan analysis of the functional adaptation of living bodies and of the "Unity in the principles of their construction."[96]

Of all the Bridgewater authors, it was arguably Buckland who most distinctly presented his work as an amplification of the evidence for design in nature found in Paley's *Natural Theology*. As we saw in the previous chapter, this had been his strategy for justifying geological study at the University of Oxford for more than two decades. However, he offered no theological defense of the argument from design, notwithstanding High Church and evangelical skepticism. Only in his conclusion did he discuss the matter directly, elaborating on the indissoluble connection between natural and revealed theology. Reason, he claimed, could discover plentiful evidence of the existence and some of the attributes of a creator, but no more. As far as what really mattered to humans was concerned—"the will of God in his moral government, and the future prospects of the human race"—all that reason could do was to demonstrate the need for revelation.[97]

Framing his book in this way allowed Buckland to get on with providing a synthetic account of geology and mineralogy, justified by its contribution to natural theology in much the same way as his Oxford lectures routinely were. Indeed, as the previous chapter showed, Buckland's book

96. Buckland 1836, 1:vii–viii, 14.
97. Buckland 1836, 1:589.

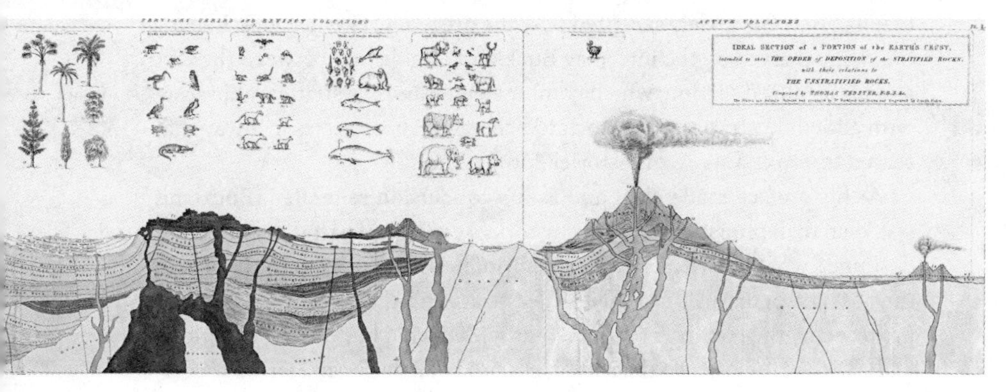


drew heavily on those lectures. It began with brief introductory chapters on the province and methods of the science of geology, including his discussion of its relation to the Bible. Next came a stratigraphic overview of the history of the earth and its inhabitants, illustrated by a breathtaking hand-colored idealized geological section that folded out to four feet (1.2 m) in length (fig. 2.4). However, the great bulk of the treatise—almost two-thirds of it—was taken up with five chapters outlining "Proofs of Design" in the structure of fossilized animals and plants. Finally, there were a few shorter chapters that examined proofs of design in the disposition of rocks and minerals.

At the outset, Buckland identified the "subject of the science of Geology" as the "history of the earth," comprising both its physical history and the history of life on it. Narratives of earth history had been largely avoided in British geological writing in the early decades of the nineteenth century for fear of appearing unduly speculative, and while Buckland's flamboyant evocations of former worlds in lectures had been important catalysts in shifting attitudes and expectations, the progressive history of the earth offered in his Bridgewater's opening chapters was still largely implicit. Running to fewer than sixty pages, it focused largely on the ascending strata and their characteristic fauna and flora—in combination with an account of "Geological Dynamics"—rather than on telling the kind of spellbinding historical story that was soon to be found in popular geological works. Nevertheless, readers often came away from Buckland's treatise with a narrative history vividly in mind, synthesizing the progressionist geology that had become widely accepted over the preceding two decades. In part, this was achieved by the interplay Buckland established between the text and his idealized section, which combined an extensive stratigraphic array with added illustrations of associated characteristic species in a way that served to evoke a sense of historical progression.[98]

As his preface made clear and as his conclusion reiterated, Buckland was keen to demonstrate that this historical geological narrative fatally undermined what he saw as significant anti-Christian theories. In particular, the absence of fossils from the earliest strata and the directional element provided by the cooling of the earth provided what he considered unique and conclusive evidence that there had not been either an eternal succession of the same species or "the formation of more recent from more ancient species, by successive developments, without the interposition of direct and repeated acts of creation." Buckland outlined the adaptedness of the different living forms to the earth's "progressive stages of advancement," summarizing the point in his conclusion: there was copious evidence of the beginning and end of "several successive systems of animal and vegetable life," he averred, each of which had its origin in the "direct agency of Creative Interference." This was a point that had previously been made by Whewell in an anonymous review of Lyell's *Principles of Geology*, and Buckland was now happy to quote Whewell's assertion that geology had thus made a distinctive new contribution to natural theology.[99]

This emphasis on the miraculous in the origination of new species was juxtaposed with an equally strong emphasis on the lawlike character of the changes in the physical globe. Thus, Buckland suggested that the earth had commenced in a fluid or nebular state and reported that the nebular hypothesis offered the "most probable theory" respecting the origin of the solar system, citing Whewell's claim that such a theory could only augment the sense that there must be a God. The "operations" involved in the historical development of the globe, including catastrophic upheavals, were the result of regular natural laws, a point that attested to divine involvement. Returning to this claim in his conclusion, Buckland reflected that the prospective design that had been involved in adapting the uniform properties of chemical elements to their future uses in rocks and minerals must elevate our conceptions of God's "consummate skill and power." The

98. Rudwick 2008, 428–31; Rupke 1983, 183, 200; O'Connor 2007a, 248–50; Buckland 1836, 1:36. On the use of images, see also Rudwick 1992.

99. Buckland 1836, 1:51–54, 76, 585–86. Like Kirby, Buckland was able to refer the reader to the more extended discussion of Lamarck's theory in the second volume (1832) of Lyell's *Principles of Geology*.

uniformity of law in the history of the globe provided additional proofs of God's ongoing activity.[100]

The series of chapters on fossil remains that followed Buckland's introductory stratigraphic survey provided further evidence against transmutation. Among the folders that Buckland put together in preparing his treatise was one marked "Species change of Lamarck," and the published work was littered with both implicit and explicit references to the transmutationism of both Lamarck and Geoffroy Saint-Hilaire. The latter's reptile ancestry, which loomed large in Buckland's notes, was hinted at in his printed repudiation of "every theory that would derive the race of Crocodiles from the Ichthyosauri and Plesiosauri, by any process of gradual transmutation or development." More fundamentally, while his account of life on earth showed a progression that matched the state of the globe, Buckland strove to qualify this in ways that rendered transmutation less likely. Thus, while fossil forms came increasingly to resemble modern forms through geological time, Cuvier's four fundamental taxonomic groups were "coeval with the commencement of organic life upon our globe." Likewise, Buckland cited many instances of apparent retrogression, which showed that it was "not always by a regular gradation from lower to higher degrees of organization, that the progress of life has advanced."[101]

Buckland was keen to demonstrate to his readers that this progress of life reflected the earth's "gradually increasing capabilities of sustaining more complex forms of organic life" rather than a natural tendency for living creatures to improve. In each geological epoch, the species present were perfectly adapted to the conditions and to their various places in the economy of nature.[102] The functional adaptation of living creatures in past epochs formed the main theme of Buckland's fossil chapters, allowing him to make good on his promise to extend Paley's analysis to geological phenomena. As he made clear, this was grounded in the rapid progress in the interpretation of fossil remains that had been made possible by Cuvier's comparative anatomy, and succeeding chapters worked their way through examples from Cuvier's four animal *embranchements* before concluding with fossil plants.

The pièce de résistance came early on, with an account of the fossil ground sloth *Megatherium* that even Cuvier had considered to be ill adapted. Buckland's widely applauded reconstruction of the habitat and

100. Buckland 1836, 1:40n, 46, 49, 580–82.

101. Buckland 1836, 1:61, 254, 312; OUMNH-BP, Notes by Subject (Bridgewater Treatise) 2/6. See also Rupke 1983, 175–76, and Desmond 1989b, 307.

102. Buckland 1836, 1:107n–108n. See also Rupke 1983, 158, 174, and Yeo 1986, 270.

habits of this huge, ungainly creature demonstrated that its structure was actually ideally suited to function. Yet while this emphasis on the functional adaptation of fossil species was central to his program of enhancing Paley's evidence, he also stressed the new confirmation it provided of an underlying divine plan in living creatures. Cuvier's triumphant fossil reconstructions from even a "single tooth or bone" were only possible because of the "Unity of Design and Harmony of Organizations" that "pervaded all animated nature." Fossil forms supplied the missing links in "the great chain whereby all animated beings are held together in a series of near and gradual connexions" and were "thus shewn to be an integral part of one grand original design."[103]

After this cornucopia of organic design, Buckland turned briefly to his remaining set of arguments concerning the arrangement of rocks and minerals in the earth's crust. This also provided evidence of design unknown to Paley, and Buckland became quite lyrical as he described the manner in which the earth's mineral wealth had been disposed so as to be best available for human use. An earlier chapter on the "Relations of the Earth and its Inhabitants to Man" had argued against a narrow anthropocentrism in the divine plan, but he now devoted five chapters to developing arguments from his *Vindiciæ geologicæ* concerning the human benefits resulting from basin-shaped formations and from the mutual proximity of iron and coal deposits. This kind of argument, with its implication that God had a particular concern for the prosperity of Britain, was of course well suited to the triumphalist mood of the first industrial nation.[104]

In his conclusion, Buckland briefly summarized the new and distinctive contribution of geology to natural theology. It had shown that the physical laws governing the earth's construction had been presciently designed to deliver a home suitable for a whole series of living populations. There was new evidence of the functional adaptation and unity of design in fossil organisms. Above all, geology had provided unique evidence of a series of miraculous beginnings in the history of life and of the untenableness of species transmutation. Whereas geology had excited "alarm" in the early days, it was now clear that the science, far from being at odds with true religion, was yielding new evidences in its favor. In a final rhetorical flourish, Buckland quoted extensively from authors as diverse as Bacon, Boyle, Locke, Herschel, Brougham, and Sedgwick concerning the distinct but complementary domains of physical science and revealed reli-

103. Buckland 1836, 1:109, 114; see also 583–85. For the *Megatherium*, see 1: 139–64.

104. Buckland 1836, 1:97–102, 524–79. See also Rupke 1983, 261–66.

gion. "Shall it any longer then be said," he inquired, "that a science, which unfolds such abundant evidence of the Being and Attributes of God, can reasonably be viewed in any other light than as the efficient Auxiliary and Handmaid of Religion?"[105] This was a rhetorical question that Oxford's geology professor had often asked, but his Bridgewater offered an answer on a new scale.

Buckland and Kirby were agreed that true science and the Bible would always be found in harmony, but their approaches could hardly have been more different, with the geologist's radical reappraisal of the text of Genesis standing in stark contrast to the entomologist's conviction that the text of the Bible had yet more to reveal about the natural world if only one used its symbolic key. The two were also far apart in their estimate of natural theology. For all that he was located at the heart of conservative Oxford, Buckland was among the more forthright of the authors about the theological value of the evidence of design, whereas the High Churchman Kirby was clear that scriptural knowledge had to be given primacy in order to understand the natural world. For Kirby it was the Bible that would set Lamarckian transmutationists aright, whereas Buckland was confident that modern geology would do so by itself, offering an independent testimony to the truth of the scriptures.

❋

As a clerical geologist based in Oxford, Buckland was especially attuned to the concerns that had been expressed about the irreligious tendency of geology. Yet for each of the authors, the religious jeopardy felt by contemporaries in relation to the current state of the natural sciences was a very real concern. Much more than developing a rationally constituted natural theology—which they all understood to be potentially problematic and concerning which the more High Church and evangelical of them had especially strong misgivings—the authors produced reflective treatises of a type characteristic of their era, offering ways in which to reconnect the modern sciences with Christianity. Whewell was the most explicit in identifying that the task in hand was to reassure the faithful that scientific developments were not at odds with orthodoxy, but in their strikingly varied ways, all of the authors sought to sanctify the developments in their several sciences.

On the whole, the message was a positive one. One pervasive concern

105. Buckland 1836, 1:593.

was the focus of the developing sciences on reducing phenomena to natural laws. Whether in chemical combinations, the operation of the solar system, the history of the rocks, animal morphology and physiology, or political economy, modern science was constantly extending the scope of natural laws. Yet the authors strove to show that such developments posed no threat to orthodox Christianity, and most felt able to claim that, on the contrary, they enhanced the Christian's appreciation of God's nature and actions. A second, related concern was with the increasingly historical perspectives that the sciences were offering of the process of divine creation. These might be felt to be at odds with divine revelation or to threaten the unique status of the human species, but, while the aged and conservative Kirby offered a minority report, others of the authors sought to reassure readers that historical and sometimes lawlike accounts of God's creative acts were not only consistent with the Bible but also offered new insights into the wonder of creation.

Set against these positive messages, however, were some more negative ones. The authors could hardly deny that there were high-profile philosophers and naturalists who had set themselves up in denial of orthodox Christianity or even of God's existence. Such examples, they insisted, did not show the sciences to be anti-Christian; rather, they showed that, in the sciences as more generally, individuals readily fell into bad mental and moral habits that led them away from the truth. Learning to view the natural world with Christian piety would ensure that the study was an asset to faith rather than the contrary. Such mental training would protect the believer from fallacious speculations regarding materialism and transmutation. These speculations, the authors urged, could be seen by the well-moderated mind to have no basis in fact. Indeed, they were doctrines at odds with scientific reasoning that had no place in modern science. It was, the authors intimated, only French speculators and disreputable infidels—enemies of Britain's Christian state—whose perversity had led them into such dangerous nonsense.

PART II

# *Publishing*

✷ CHAPTER 3 ✷

# *Distributing Design*

*Think of printing these treatises, designed for universal dissemination, in a style, and at an expense, that must limit their circulation to the narrowest compass. Between the lines of the work before us the Earl of Bridgewater might almost have driven his cab! Are scepticism and irreligion confined to a few of the upper classes?*

Medico-Chirurgical Review, *October 1835*[1]

Print had never been more politically charged than it was in the age of reform. To radical leaders, the printing press was the source and symbol of political freedom. Following the "Peterloo Massacre" of August 1819, when a peaceful Manchester reform meeting had been charged by the yeomanry, government repression had focused on restricting working-class access to printed news through punitive measures, and by the early 1830s a concerted campaign of defiance was under way to remove these "taxes on knowledge." The publication of hundreds of "unstamped" political periodicals, led by the *Poor Man's Guardian*, resulted in a steady stream of working-class agitators being consigned to prison, while radical journalists argued that the groundbreaking cheap publications of the Society for the Diffusion of Useful Knowledge (SDUK) were nothing better than a sop, designed to distract workers from their rights and enforce bourgeois ideology. Meanwhile, the flood of cheap secular publications provoked religious publishing organizations such as the Religious Tract Society and the Society for Promoting Christian Knowledge into redoubling their own efforts to rescue the soul of the nation through print.[2]

1. *Medico-Chirurgical Review* 23 (1835): 400.

2. On the "war of the unstamped" see Wiener 1969 and Hollis 1970. On the contest over cheap print, see, for example, Johnson 1979, Desmond 1987, Topham 1992, and Fyfe 2004.

It was inevitable that the decisions taken concerning the publication of the Bridgewater Treatises would be judged in the light of these wider developments. Were they to be books using all the strategies of cheap publishing developed over the preceding decade and consequently purchased by tens or perhaps hundreds of thousands? Or were they to be books fit for gentlemen, published on good paper, with wide margins and copious illustrations, and perhaps selling only the thousand copies required to be printed by Bridgewater's will? Although the *Medico-Chirurgical Review* expected such matters to have been settled by the earl's executors, it was the authors themselves who were left to search for a publisher, and it was in their negotiations with that publisher that both the form and the price of the Bridgewaters were fixed. The outcome—a set of large and expensive volumes—was far from giving universal satisfaction.

The authors' deliberations concerning their publisher and the appropriate form of publication betrayed multiple concerns. The series ought to reflect the dignity of the earl who commissioned it and the high-ranking officials who managed it, to whose patronage the authors were indebted. In addition, it ought to confirm and enhance their individual reputations, appearing in a form suitable for genteel readers and in some cases including strikingly complex illustrations. On the other hand, the authors had wished to make the series accessible to a more or less "popular" audience, which suggested that the costs should be kept down. To complicate matters still further, at least the majority of the authors were keen to maximize their returns despite having already been paid handsomely for their work.

In seeking to negotiate these several competing concerns, the authors found themselves engaged with a large assemblage of what we might think of as "technicians of print." The figure of the publisher looms large in the process, especially since the authors settled on a publisher who not only had quite exacting notions concerning book design but was also enmeshed in relationships with a printer and binders that significantly affected the final appearance of the volumes. However, their choice of publisher was far less carefully planned than one might expect, resulting to a significant extent from accidents of trade at a time of social upheaval. In addition, their publisher was not one used to producing heavily illustrated books, and most of those authors who wished to include illustrations had to spend a great deal of time either making or overseeing drawings and to involve themselves in negotiations concerning the rapidly developing printing technologies. Yet for all that the volumes that emerged from this complex process attracted considerable criticism concerning their size and expense, they nevertheless achieved a very surprising degree of commercial success, becoming one of the publishing sensations of the decade.

## *Choosing a Publisher*

The choice of a publisher has hugely important consequences for the appearance, price, market, and reputation of a book, but who was the appropriate publisher for a Bridgewater Treatise? The easiest decision was that it should be someone based in London, which was still the overwhelmingly dominant center of book production and distribution in Britain as well as for literary advertising and criticism. However, London had two main hubs for publishing, which differed markedly in their character. The historic heart of the book trade in the City of London had in recent years been challenged by the emergence of a new, more fashionable publishing center in the West End. The City was especially associated with publishers who combined their activity with wholesaling. Some of these houses, indeed, were almost exclusively wholesaling agents, not least for books produced in Edinburgh, Dublin, or increasingly in provincial towns. Others were heavily involved in publishing "Standard Works—Works on Education, Science, &c., and such as are in regular and constant demand." By contrast, West End publishers were "almost entirely devoted to what may be called the Literature of the Day—Works of Amusement and light reading, Travels, Memoirs, Novels, Tales, Poems, and other productions of a similar character."[3] It was not immediately clear where the Bridgewaters belonged on this rather crudely drawn map. While they were in some senses works of science, it was a reasonable question whether they might not also become part of the literature of the day.

Another consideration was the reputation of the publisher. While publishing continued at this period typically to be combined with other print trades, including retailing and wholesaling, a number of booksellers had acquired reputations as specialists in particular subjects. Publishing with one of these would create particular expectations. For instance, should the authors approach one of the well-known publishers of religious books, such as John Hatchard (in Piccadilly) or Rivingtons (in St. Paul's Churchyard)? But the authors were clear that theirs were not theological treatises so much as works to explicate the religious meaning of the sciences. Moreover, publishers with a large theological list were typically identified with religious parties, which would hardly have been appropriate with the Bridgewater Treatises.

What, then, about a publisher associated primarily with scientific and medical publications? In fact, there were no really prominent scientific specialists, but there were a number of medical publishers, including John

3. [Saunders] 1839, 55–57.

Murray's former partner Samuel Highley in Fleet Street or Elizabeth Cox near St. Thomas's Hospital, south of the river. A decade later, this was the approach of the anonymous author of the *Vestiges of the Natural History of Creation*, who, anxious to secure a scientific reputation for his work, approached the prominent medical publisher John Churchill.[4] However, the Bridgewater authors needed no such assistance. Their names and institutional affiliations, combined with their appointment by the president of the Royal Society, provided all the scientific credentials they needed. Moreover, they did not want their books to be viewed as narrowly scientific or even professional.

The authors also had a third kind of specialist publisher to consider. Committed, as they were, to addressing a general audience for science, they might have approached such pioneers of cheap "useful knowledge" publishing as Charles Knight (in Pall Mall East) or Baldwin and Cradock (in Paternoster Row)—both agents of the useful knowledge society well used to making scientific works available at low prices. The decision not to do so was at the heart of the ensuing controversy about the form and cost of the treatises, but it is clear that it was a decision made at an early stage. At their initial meeting, the authors had considered that the works should be in a gentlemanly octavo format, and Buckland later recalled that they had all felt that such a format, or a larger one, "was due to the dignity of the Thousand Pounds."[5] While the SDUK treatises had initially been in octavo, their cramped double columns of type were hardly such as to suggest dignity.

Much closer in character to the Bridgewaters than the useful knowledge treatises were such recent publications as Herschel's *Preliminary Discourse* and Lyell's *Principles of Geology*. These had been published by Longmans and John Murray respectively, publishing houses that were in many ways quite different from each other. The house of Longmans was already more than a century old and had long held a dominant role at the heart of the book trade in Paternoster Row. John Murray, by contrast, was a relative newcomer, but having moved his father's business to Albermarle Street in the West End he had become one of the most outstanding fashionable publishers of the age. What the two had in common, however, was that they were key players in what some contemporaries regarded as a "Golden age" of literary enterprise.[6]

4. Secord 2000, 111–16; Topham 2000a, 583–85.

5. WB to WW, 5 October 1832, TC, Add.Mss.a.66[29]; PMR to WW, 10 December 1830, TC, Add.Mss.a.211[114]; PMR to TC, 11 December 1830, NC, CHA 4.147.33.

6. ΠAN 1825, 483.

During the early decades of the century, older practices of publishing standard works collaboratively had begun to give way to a more competitive approach, involving literary entrepreneurs who were prepared to nurture able authors in the expectation of rich rewards. Murray, whose princely payments to Lord Byron are legendary, was one of the most notable of these. "He was," one contemporary noted, "among the very first booksellers of his day who treated authors, of character and connection, with the respect due to gentlemen," offering "prompt and spirited remuneration" as well as fine words.[7] On his move to the fashionable West End in 1812, he relinquished the conventional retail and wholesale trade of the bookseller, using his new premises to cultivate his gentlemanly persona as a patron of superior literature. This endeavor was supported by his *Quarterly Review*, founded in 1809, through which a coterie of able authors—now including Whewell and Roget—received payments and encouragement. In reinventing the business, Murray had relinquished his father's medical list. Moreover, the scientific books that could count as the "literature of the day" in which he specialized were relatively few in number. Yet the example of Lyell's *Principles*, Buckland's *Reliquiæ diluvianæ*, and the second edition of Bell's *Anatomy of Expression* demonstrated that those who could write about the sciences in a readable, accessible, and topical manner could find generous support from this prince of publishers.

In the same period Thomas Norton Longman had also steered his family's house through a remarkable transition. While the firm continued to publish trade books and to operate as a major wholesaler, it increasingly published books solely on its own account, and it separated its functions into different departments. The extra management and capital that this required came in part from an ever-increasing number of partners. Sir Walter Scott had joked that Longmans was the "long firm," but the model worked, and by 1831 the imprint read "Longman, Hurst, Rees, Orme, Brown, and Green." Output vastly increased, especially in medicine and science, leaving Longmans as the largest single publisher of such books. Moreover, while many of these counted as standard works, the firm's active entrepreneurship included an increasing number of books, scientific and otherwise, that were among the literature of the day. Like Murray, Longmans held regular literary dinners to cultivate relationships of mutual trust with rising and established authors as the basis of a long-term literary partnership. Bell had particularly benefited from this on his first arrival in London, and his relationship with the firm continued for many years to their

7. [Dibdin] 1832, 32n. On Murray, see Smiles 1891.

great mutual advantage.[8] Kirby, too, had found in Longmans a rewarding partnership for his *Introduction to Entomology*, and Roget had recently negotiated with them concerning a possible contribution to their epoch-making Cabinet Cyclopædia.

The established relationships that the Bridgewater authors had with these leading publishing firms provided good grounds for them to begin their search for a publisher there. Five of the authors, including Bell, Buckland, and Kirby, met in London on the evening of 8 December 1830 to determine a plan of action. Their proposal was that they should approach Longmans and Murray to ask about terms for a single edition of one thousand copies of each work, published at the bookseller's risk. Those absent from the meeting readily agreed. Even Chalmers, whose previous dealings with Longmans had left him dissatisfied, allowed that both were "respectable houses." Primed with the authors' minimum requirements, Roget was deputed to make the inquiry. The authors had agreed that they were not prepared to accept less than one-half of the profits of the publication. This was a standard form of agreement for established authors with publications that promised a modest but reasonably certain sale, and it was what Longman offered. Not surprisingly, the authors preferred John Murray's more generous offer of two-thirds of the profits.[9] Murray had long been noted for his flamboyant rewards for favored authors, and his offer in this case may perhaps reflect his friendly (and profitable) relationship with Bell and Buckland, or indeed his desire to outbid Longman to secure such a high-profile series.

Roget informed Murray of the authors' decision in the spring of 1831, but the publisher was very tardy about producing a written agreement. Murray was away from home when Roget made a trip to Albemarle Street with Prout in October. Roget soon returned with Buckland, and the three of them drew up an agreement. However, this was not circulated to the authors, and Murray later claimed that he had asked his solicitor to draw up contracts. In March 1832, prompted by a letter from Roget, he headed up his latest eight-page advertising insert with these "new works in the press." This produced further discontent, as the authors took issue with the titles assigned to them informally by Roget. Moreover, Murray failed to respond to the authors' increasingly anxious visits and letters through the summer of 1832, as the December publication deadline began to approach. By this time, Roget was beginning to doubt Murray's ability to complete

8. Bell 1870, 23, 53; Topham 2000a, 584–85; Briggs 2008, 4.

9. PMR to TC, 21 March 1831, NC, CHA 4.167.54; TC to PMR [copy], 4 April 1831, NC, CHA 3.13.17.

the commission. His suspicions were confirmed when he received a letter from Buckland informing him that the publisher wished either to be released from their verbal contract or to alter the terms. Murray had sent his foreman to Oxford to explain to Buckland that a slump in the book trade since the time when he had verbally agreed to the contract meant that he would incur "a certain loss" from the publication of the Bridgewaters.[10]

The depression in the book trade in 1831–32 was considered by many publishers to be more serious than the financial crisis of 1825–26 that had brought down the mighty Edinburgh publishing house of Murray's former business associate Archibald Constable. Murray had escaped relatively unscathed on that occasion. He had, however, suffered an immense loss of £26,000 with the failure in July 1826 of the *Representative*, a short-lived daily Tory newspaper he had planned with the young Benjamin Disraeli. This and other losses had made him more cautious than he had been in his prime. The latest stagnation in the book trade was precipitated by the crisis surrounding the attempts to pass the Reform Bill and by the arrival of a cholera epidemic claiming over thirty thousand British lives.

Thomas Frognall Dibdin, whose *Bibliophobia* (1832) documented the "present languid and depressed state of literature and the book trade," relayed Murray's view of the matter. According to Dibdin, Murray considered that, with the exception of his thriving *Quarterly Review*, "the taste for literature was ebbing." Above all, readers now expected their books to cost much less: "Men wished to get for *five*, what they knew they could not formerly obtain for *fifteen*, shillings. The love of quartos was well nigh extinct [...]. There was no resisting the tide of fashion, or the force of custom:—call it as you might. Clear it was to *him*, that the dwarf had vanquished the giant—and that Laputa was lording it over Brobdignag." Murray had embraced this inevitable development in 1829. Responding to the cheap publishing ventures of the useful knowledge society, his groundbreaking Family Library consisted of small octavo volumes of largely original nonfiction issued approximately monthly at 5s. Yet while the initial volumes promised commercial success, it had become clear by the end of 1831 that the series was more than £8,000 in the red and was failing. Not even cheap publishing could be made to pay.[11]

In this situation, Murray became anxious that some of the Bridgewaters

10. WB to PMR, 26 August 1832, BO, Ms.Eng.Lett.b.35, fols.30–31; PMR to Murray, 6 October 1831, NLS, MS.41032; PMR to TC 13 June and 20 August 1832, NC, CHA 4.189.11–13; CB to Murray, 14 March 1832, NLS, MS.49978, fols. 93–94.

11. Bennett 1982; Gettman 1960, 8, 10; Raven 2007, 304–5; Smiles 1891, 2:215; [Dibdin] 1832, 31.

might not sell as many as a thousand copies. In general, a thousand copies of a moderately priced book was considered to be a large edition. Not wanting to renege on his verbal promise of two-thirds profits, Murray suggested other changes. The works should be published serially, somewhat like the Family Library, at three or four month intervals in order to spread out costs and perhaps to generate demand. In addition, the account should be made up for the series as a whole, with Murray being at liberty to print more of the most successful treatises in order to make up any loss on the less successful ones. Buckland thought these proposals would cause delays and fail to guarantee at least the required thousand copies of each treatise and that they might also reduce the income of the more successful authors. His suggestion was that they should simply reduce their expectations to half profits, but Roget felt that Murray no longer had any heart for the undertaking. More seriously, he told Buckland that he had heard from "several different quarters" that the publisher's business was in an "embarrassed state." To both it seemed likely that Murray would become insolvent, and Whewell had heard "on all sides" that "glorious John" was ruined.[12]

Roget circulated a copy of Buckland's revised terms to the other authors but made clear that his own preference was to reopen negotiations with Longmans instead. Longmans, he discovered, continued to be happy to offer a half-profits agreement for an edition of a thousand copies, but they wished to indemnify themselves against losses. The draft agreement required the authors to publish all future editions with them on the same terms, since "some of the Treatises may not sell so well as others, & may remain upon their hands." It also gave the publishers the right to remainder unsold copies. Longmans also suggested, as Murray had done, that the treatises should appear "in monthly succession, & that two of the most popular subjects should take the lead," in order to generate demand. Roget relayed the details to his fellow authors but with a negative gloss. Having recently turned down an invitation to write on physiology for Longmans' Cabinet Cyclopædia on the grounds that they would not allow him to reuse his material in other forms, he was now extremely unhappy to find the same restriction in operation again. With the publisher still not finalized less than three months away from the promised publication date, Roget obtained his colleagues' permission to approach another publisher.[13]

12. WW to Richard Jones, 13 September 1832, WB to WW, 5 October 1832, TC, Add.Ms.c.51[141], Add.Ms.a.66[29]; WB to PMR, 26 August 1832, PMR to WB, 29 and 30 August 1832, BO, Ms.Eng.Lett.b.35, fols. 30–35; Grant 1837, 1:207–8.

13. PMR to Lardner, 6 April 1829, WL, MS7543/10; PMR to WB, 9 October 1832, BO, Eng.Lett.b.35, fols. 36–37; PMR to WW, 30 August 1832, 9 October 1832,

As these negotiations went on, it became increasingly clear that the initial publication deadline of 30 December 1832 was not practicable. In any case, Roget, whose wife was dying, knew that he could not be ready, and Buckland (among others) was also "sadly behind hand." With the authors' consent, Roget consequently sought to extend the deadline to the following June. To this the new president of the Royal Society (the Duke of Sussex), the bishop and archbishop, and Bridgewater's executor (Lord Farnborough) all reluctantly agreed. Meanwhile, weary of the prolonged negotiations, the Oxford and Cambridge authors decided to visit the metropolis to resolve the matter with Roget, Prout, and Bell. On 11 October, these six authors drew up a list of conditions for publication, the most important of which were that the copyright should rest with the authors, that the publisher should not discount or remainder any of the works until at least three years after publication, and that all the treatises should be published between 1 March and 1 July 1833. Within a day or two, they had agreed that on these terms (and on a half-profit basis) they should employ William Pickering of Chancery Lane (fig. 3.1).[14]

Pickering was hardly the most obvious choice of publisher. With premises amid the Inns of Court—within easy reach of both the British Museum and the Society of Antiquaries—he was an antiquarian bookseller whose new publications were aimed at a learned and wealthy clientele. In contrast to Longmans and Murray, he lacked the capital to become a patron of literature, and his business was on a vastly smaller scale. His standard publishing practice was to produce relatively small editions of books which, while they were unlikely to make his fortune, were equally unlikely to make a significant loss. In particular he found that reprinting classic and standard works, often in series, met with a relatively predictable demand. Such publishing also fitted well with his antiquarian interests and clientele and allowed him to capitalize on the editorial and production qualities that came to characterize his list. Many of those who published new books with him were also customers for his antiquarian stock, and his published output was thus characterized by traditional literary and religious values.[15]

While Pickering was not one of the leading or most fashionable pub-

---

TC, Add.Mss.a.211[116–17] (cf. PMR to TC, 30 August 1832, 8 October 1832, NC, CHA 4.189.14–17). For the terms, see also BO, Ms.Eng.Lett.b.35, fols. 65–67.

14. PMR to TC, 13 October 1832, NC, CHA 4.189.18–19; PMR to Gilbert, 13 October 1832, in Enys 1877, 17–18; Blomfield to WW, 20 October 1832, TC, Add. Mss.a.201[45]; PMR to WB, 29 August and 9 October 1832, BO, Eng.Lett.b.35, fols. 34–37; WB to WW, 5 October 1832, TC, Add.Mss.a.66[29].

15. On Pickering, see Keynes 1969, Warrington 1985, 1987, 1989, 1990, 1993, and McDonnell 1983.

FIGURE 3.1 William Pickering with Charles Whittingham in his summer house at Chiswick. Drawn by Frank Dodd from an oil painting by Charlotte Whittingham. From Warren (1896), 209.

lishers, he was thus eminently respectable. There was no question that, in his hands, the treatises would appear in a form that gave due distinction to the Earl of Bridgewater's purpose. Indeed, Pickering's reputation depended on the quality of his product. This much was satisfactory, but otherwise the decision seems to reflect the growing desperation of the authors. The suggestion arose with Roget, and we may speculate that it was Pickering's growing role as one of the booksellers to the British Museum that brought him to Roget's attention. Roget's fellow secretary at the Royal Society, John George Children, was on the curatorial staff at the museum, as was Antonio Panizzi, whom Roget was in the process of engaging to catalog the Royal Society's library. From these or other quarters the authors assured themselves of "the high reputation of Mr Pickering, & of his ability to execute the publication in the most satisfactory manner," and they were very pleased with his response to their proposed terms.[16] Within days, he had drawn up agreements, which the authors duly signed.

Given the concerns of both Murray and Longman concerning the likelihood that some of the treatises might not sell as many as a thousand copies, we might question why so conservative a publisher as Pickering should have agreed to take them on. Most of his books were published in small editions of five hundred copies, and his commitment to print eight thousand Bridgewater Treatises required no small investment. To a relatively small publisher like Pickering, however, the series promised the possibility of a significantly enhanced reputation. Here were several high-profile authors associated in a project patronized by an earl and administered by an archbishop and a learned president, and the books in question had shortly before been advertised by one of the leadings publishers of the age. Moreover, as we shall see, Pickering was in the midst of a trade dispute with London's leading booksellers, and picking up such a notable series proved to be a useful bargaining counter.

The incongruity of Pickering becoming the publisher of the Bridgewaters did not escape contemporaries. They were the kinds of works that were to be expected from the likes of Longmans or Murray. Only by commercial accident did the series come to be published instead by a small but eminently respectable publisher whose trade was mostly with scholarly gentlemen. As we shall see, however, that accident had significant implications for the shape of the books that were produced.

## *Making the Books*

Two days before Christmas 1832, Whewell told geologist Adam Sedgwick that he hoped to be soon "tormenting the men of ink" with his "Bridgewater hieroglyphics." Given the dreadful handwriting of several of the Bridgewater authors, it is no doubt just as well that their finished books were not—as we commonly say—"written" by them. Their pleasant appearance rather reflects the processes of manufacture selected and executed by an array of print technicians. The decisions of such technicians concerning how the Bridgewaters were put together had immediate consequences for their price and thus for their availability to the growing numbers of readers of all classes. Year by year, publishers were being encouraged by the increased demand for print to experiment with new mechanical technologies of book production in producing ever cheaper pub-

16. PMR to TC, 13 October 1832, NC, CHA 4.189.16–17.

lications. Even as the Bridgewater authors struggled to find a publisher, the useful knowledge society was engaged in a radical experiment with the launch of its weekly *Penny Magazine* on 31 March 1832. Printing on steam-powered presses from multiple stereotypes cast from the original set type, the society achieved new levels of cheapness resulting in unprecedented sales of as many as two hundred thousand copies per week. Britain was on the cusp of a communication revolution mediated through the centuries-old combination of paper and print.[17]

It was publishers who were primarily responsible for negotiating the rapidly changing technical and commercial complexities of book manufacture on behalf of their authors, engaging with an array of other technicians in the process, including paper manufacturers, printers, engravers, binders, and retailers. As the example of Murray illustrates, however, even the publishers felt themselves to be all at sea as both the market and the technologies of production shifted beneath their feet. Against this backdrop, the Bridgewater authors had selected a publisher noted for his commitment to producing books in keeping with his antiquarian book stock. The substantial octavo volumes that emerged were printed in a large and spacious font, and the wide margins surrounding the text contributed to the luxuriant feel. In an age when paper typically accounted for a third of the cost of manufacturing a book, such design choices came at a significant cost. Even before customers inquired about the prices of the finished treatises, their form clearly signaled the kind of market intended—the aristocracy, the gentry, and the more affluent middle classes—but the very substantial reality of the price tag certainly confirmed that impression. With the treatises costing between 9s 6d and £1 15s each and a total of £7 15s 6d, purchase of the series would have represented a whole month's income for many middle-class families.

The decisions taken and the reasons behind them become clearer when we reconstruct the process of manufacture and consider the technicians involved. However, one of the most important decisions in regard to the price of the series had been taken by the authors themselves as early as December 1830. This was the decision that each of the treatises should, when published, fill "one or more octavo volumes of not less than 300 pages." The term "octavo" here refers to the eight leaves (sixteen pages) into which each printed sheet was to be folded. Depending on the size of the original sheet, the octavo format could yield volumes of various sizes. However, as

17. Stoddard 1987, 4; WW to Sedgwick, 23 December 1832, in Todhunter 1876, 2:150–51. For the "industrial revolution in communication," see Secord 2000, 24–34; see also Topham 2000a and Fyfe 2012.

the *Author's Printing and Publishing Assistant* (1839) explained, historical or scientific works were usually produced in octavo on sheets of "demy" size, producing a substantial tome of around 8¾ × 5½ inches (22 × 14 cm).[18] It was this size of volume that the authors had in view from the outset.

Publishing in demy octavo gave a work gravitas. While novels, poetry, and children's books typically appeared in smaller sizes, works of learning were expected to be dignified by size. In the case of the Bridgewaters, the authors considered that the large extent of the Earl of Bridgewater's patronage demanded such recognition, as Buckland reminded Whewell in October 1832. With his manuscript well advanced, Whewell had begun to think that his populist style was better suited to a smaller duodecimo size. As far as Buckland was concerned, dignity and practical considerations concerning the inclusion of plates made this unthinkable, for the first edition at least.[19]

From the commitment to demy octavo, other design decisions followed naturally. Nonfiction works in demy octavo were usually printed using a relatively large (Pica) type. This was the type size suggested both by Murray and by Pickering in their negotiations with the authors, although the publishers communicated it obliquely by asking them to look at existing publications.[20] However, the significance of this conventional decision becomes strikingly apparent when alternatives are considered. The cheap treatises of the SDUK's Library of Useful Knowledge had been produced in demy octavo but, like the cheap weekly miscellanies that became common in the 1820s, they used a much smaller type, printed in two columns. This was how Bell's *Animal Mechanics* had been printed, selling in excess of thirty thousand copies. Had his Bridgewater been printed in the same way, the text would have filled around 110 pages, which at SDUK prices would have retailed for 1s 9d. As it was, it filled three times as many pages, and sold for six times the price.

The point was not lost on reviewers. The *Magazine of Natural History* complained of Kirby's treatise: "The type is large, and the page not so, and the quantity of words in the two volumes, we guess, about equal to the quantity in a few ninepenny parts of the [SDUK's] *Penny Cyclopaedia*; yet 30s. are asked for this quantity and the plates! Who can profit by the teaching of a Kirby while the bookseller precludes access to his lessons?"

18. [Saunders] 1839, 31; Gaskell 1974, 224; PMR to WW, 10 December 1830, Add. Mss.a.211[114].

19. WB to WW, 5 October 1832, TC, Add.Mss.a.66[29].

20. PMR to TC, 30 August and 13 October 1832, NC, CHA 4.189.14–15, 18–19; [Saunders] 1839, 35.

By adopting the traditional design values of learned books, rather than the cramped layout of useful knowledge publications, the authors certainly confirmed the dignity of their project. However, the *Medico-Chirurgical Review*'s quip concerning the Earl of Bridgewater being able to drive his cab between the lines of type reflected more particularly their choice of publisher. Pickering was noted for his typographical taste and discrimination, producing works characterized by the simplicity and elegance of their design. He was assisted in this by his close association with the printer Charles Whittingham. Based nearby in one of the side streets off the top end of Chancery Lane, Whittingham had been introduced to Pickering in June 1829 as "the most accomplished printer in the town" and soon became his chief printer.[21] The two became close friends, lunching together daily at the Crown Coffee House in Holborn, opposite Chancery Lane. Their businesses also became intimately intertwined, with Pickering by far Whittingham's largest client and Whittingham one of Pickering's financial backers.

Working together, printer and publisher were a formidable force in the production of fine books. Their friend the American bibliographer Henry Stevens later recalled, "It was their custom, as both used repeatedly to tell us, to each first sit upon every new book, and painfully hammer out in his own mind its ideal form and proportions. Then two Sundays at least were required to compare notes in the little summer-house in Mr. Whittingham's garden at Chiswick, or in the after-dinner sanctuary to settle the shape and dress of their forthcoming 'friend of man.'" One characteristic feature of Pickering's books was his title pages, which were notable for their stylishness, and the Bridgewaters were no exception. Each treatise bore a title page on which its individual title and author were laid out clearly in capitals. The other half of the page gave his own name together with his dolphin and anchor device, signifying that he was the English disciple of the Renaissance humanist printer Aldus Manutius. Other information—such as the series of which the treatise was a part, its ordinal place in that series, the edition number, and any authorial epigraph—was carefully removed to the half-title page, which preceded it. This not only maintained an elegant air but also allowed for the possibility that purchasers might wish to treat the volumes as separate rather than as part of a set.[22]

For the text, Pickering and Whittingham selected modern Roman typefaces of the sort that had become ubiquitous in recent decades, with a no-

21. Warren 1896, 135; *Magazine of Natural History* 8 (1835): 471.

22. Pickering to TC, 7 February 1833, NC, CHA 4.212.12; McDonnell 1983, 67–86; Stevens 1884, [22–23].

table stress on the vertical strokes and with horizontal serifs. These did not have quite the clarity and regularity of the Caslon Old Face types that the partners were famously responsible for reintroducing a decade later. Nevertheless, the types were spaciously appointed, with either the body of the type notably large in proportion to the face or the lines of type "leaded" with spacers, so that the printed lines felt open rather than cramped. The effect was enhanced by the fine presswork for which Whittingham was renowned. Throughout his career, Whittingham's small staff of pressmen used only hand presses. The steam-powered cylinder presses that in the years after their introduction in 1814 permitted the rapid production of newspapers, periodicals, and large editions of books, were uneconomical on the small editions that Whittingham was responsible for printing. More than that, however, they were at odds with the printer's commitment to produce a clear and delicate impression from both type and wood engravings. Impressive as were the steam-printed wood engravings of the useful knowledge society's *Penny Magazine*, they could not compete with those of the Bridgewaters (fig. 3.2).[23]

While the books were thus printed to an impressive standard, they were not significantly more expensive to print than comparable works, but they did use a large supply of the fine machine-made paper that had only recently supplanted handmade wove paper. Whittingham made up his accounts largely according to the standard trade scales, although not all printers adhered to these.[24] Moreover, while he and Pickering were particularly well placed to produce books of quality and beauty, they were by no means incapable of doing so on a smaller, cheaper scale. In partnership with his uncle in the mid-1820s, Whittingham had produced the dainty little 32mo. volumes of Whittingham's Cabinet Library, sold by the reprint and remainder bookseller Thomas Tegg. Produced in small type, but to a high standard and with fine wood engravings, these volumes tapped into a growing middle-class market for reprints priced at as little as a couple of shillings, although part of the saving of course came from not having to pay the authors of out-of-copyright works. Had the text of Kirby's Bridgewater been produced in this form, alongside Whittingham's *Natural History* compilation cheekily marketed as being by "Mrs. Mary Trimmer"

23. Warren 1896, 338–40; [Saunders] 1839, 11; Gaskell 1974; McDonnell 1983, 92–94. On the strategies used to print wood engravings on steam presses, see [Chatto] 1839, 696–707, "The Commercial History of a Penny Magazine," *Penny Magazine* 2 (1833): 377–84, 417–24, 465–72, 505–11, on 420–21, and Knight 1864–65, 2:115–16.

24. McDonnell 1983, app. 2; Hansard 1825, 778–800; Savage 1841, 736–50; Babbage 1832a, 166–71.

320 THE MECHANICAL FUNCTIONS.

they possess in common with all articulated animals, the typical form of which consists, as we have seen, of a series of rings, or segments, joined endwise in the direction of a longitudinal axis. The principal portions into which the body is divided are the *head*, the *trunk*, and the *abdomen*: each of which is composed of several segments. I have here given in illustration, the annexed figures, showing the successive portions into which the solid framework, or skeleton, of one of the beetle tribe, the *Calosoma sycophanta*,* may be separated. The entire insect, which presents the most perfect specimen of a complete skeleton in this class of animals, is represented in Fig. 149; and the several detached segments, on an enlarged scale, in Fig. 150. The head (C), as seen in the latter figure, may be regarded as being composed of three segments; the trunk (X, Y, Z), of three; and the abdomen (B), of nine. Fig. 151, is a view of the head separated from the trunk, and seen from behind, in order to show that its form is essen-

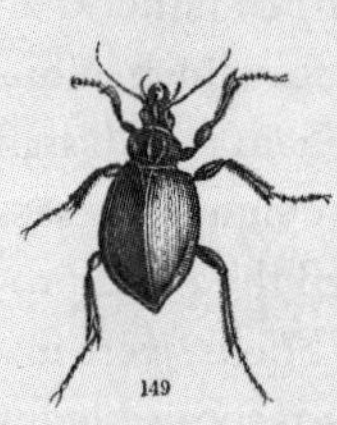

149

* *Carabus sycophanta.* Linn.

STRUCTURE OF INSECTS. 321

tially annular, and that it resembles in this respect the rings of which the thorax consists, and to which it forms a natural sequel.

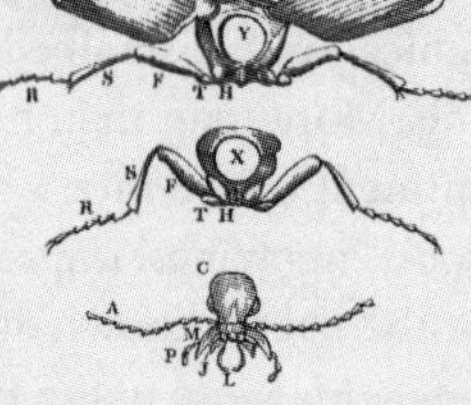

The *head* contains the brain, or principal enlargement of the nervous system, and the organs of sensation and of mastication. Its size, as compared with the rest of the body, varies much in

VOL. I. Y

FIGURE 3.2 Delicate anatomical detail on display in entomological wood engravings from Roget's Bridgewater Treatise (1834).

(doubtless to call to mind Kirby's cousin Sarah's successful publications), it could have fitted into three small volumes and cost around a third of its actual retail price of £1 10s (fig. 3.3).

Pickering, too, knew how to make cheaper books. Indeed, looking back over his first ten years as a publisher in 1832, he reflected that his example had not only improved the appearance of books but also contributed to reducing prices. Pickering's reprint series had, like Whittingham's, combined small size with high quality to achieve cheaper editions of classic works that would be attractive to middle-class readers. The 5s volumes of his Aldine Poets, for instance, were advertised as combining "cheapness, convenience, and beauty." Printed in octavo, but on the smaller foolscap paper ("to range with The Family Library, the Waverley Novels, and the Cabinet Cyclopædia"), they used less than two-thirds of the paper of larger

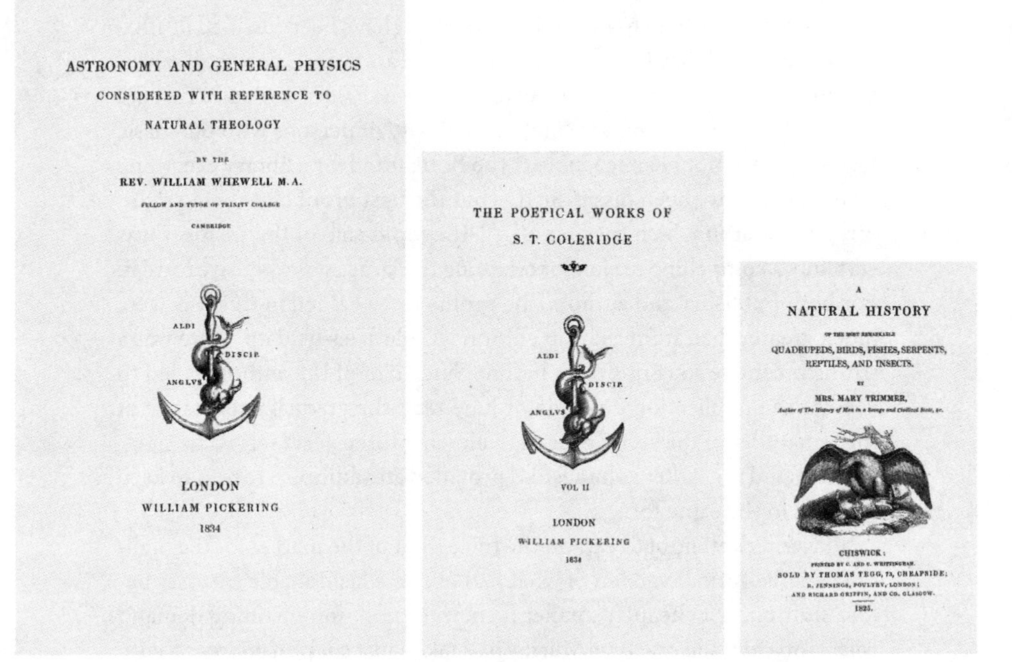

ASTRONOMY AND GENERAL PHYSICS
CONSIDERED WITH REFERENCE TO
NATURAL THEOLOGY
BY THE
REV. WILLIAM WHEWELL M.A.
FELLOW AND TUTOR OF TRINITY COLLEGE
CAMBRIDGE
ALDI DISCIP. ANGLVS
LONDON
WILLIAM PICKERING
1834

THE POETICAL WORKS OF
S. T. COLERIDGE
ALDI DISCIP. ANGLVS
VOL II
LONDON
WILLIAM PICKERING
1834

A
NATURAL HISTORY
QUADRUPEDS, BIRDS, FISHES, SERPENTS,
REPTILES, AND INSECTS.
BY
MRS. MARY TRIMMER,
CHISWICK:
SOLD BY THOMAS TEGG, 73, CHEAPSIDE;
R. JENNINGS, POULTRY, LONDON;
AND RICHARD GRIFFIN, AND CO. GLASGOW.
1825.

FIGURE 3.3 Title pages of Whewell's Bridgewater Treatise (1834), Pickering's edition of *The Poetical Works of S. T. Coleridge* (1834), and Whittingham's *A Natural History of the Most Remarkable Quadrupeds, Birds, Fishes, Serpents, Reptiles, and Insects* (1825) by Mary Trimmer, showing their relative size.

works like the Bridgewaters.[25] However, Pickering's activities in producing such cheaper editions of standard works had significantly angered other members of the trade who were keen to maintain higher prices even for books without copyright protection.

Whittingham and Pickering could thus clearly maintain their high book-production standards in smaller and cheaper volumes. However, even after the first editions of the Bridgewaters had been produced in dignified demy octavo, Pickering was keen for future editions to appear in the same form. He explained his reasoning in November 1833 to a distressed Chalmers, who was smarting at comments in the previous month's *Quarterly Review*. The anonymous reviewer had argued that eight such expensive treatises were unlikely to achieve wide circulation in the new age

25. *Quarterly Literary Advertiser*, May 1830 (stitched into copies of the *Quarterly Review*); Pollard 1978, 43–47.

of cheap literature, lamenting that they were "offered for sale at exactly the same sort of price which the booksellers might have been justified in affixing to them" had they been required to pay the authors £8,000 themselves. But Pickering was unabashed: "As to the size & form I am quite certain, that no other would have satisfied the majority of persons who purchase. If the type had not been readable & the Book fitted for a library every one would justly have been dissatisfied—but the best proof of its being right, is the manner in which they go off."[26] The rapid sale of the treatises was certainly a compelling reason for retaining the large and expensive format. For both publisher and authors, the profits to be gained in this way were much greater than from a cheap edition. While sales held up, there was a strong incentive to continue as before. Since half of the authors failed to keep to the publication deadline of June 1833, the gradual appearance of the remainder of the series over the following three years served to maintain demand for earlier volumes and provided an additional reason to keep them all in the same format.

Pickering's attempt to exploit the full extent of the market for the high-priced large-format version of what had become fashionable treatises before shifting to a cheaper, smaller format to mop up remaining demand was standard trade practice. Murray had taken just such an approach with Lyell's *Principles*, waiting until the third edition before finally moving away from expensive demy octavo to reach new readers. With the Bridgewaters, it was not until May 1837, after the last of the series had been out for nine months, that the first smaller format edition appeared. Despite his reservations concerning the octavo format, Whewell had been pleased with Pickering's design, telling a friend that his treatise made "a very pretty-looking book."[27] By 1837, a stupendous 7,500 copies had been produced at the higher price of 9s 6d but Whewell still clearly harbored ambitions of reaching a larger readership. The sixth edition was thus produced in a smaller and cheaper foolscap octavo volume, priced at 6s, just like the volumes of Longmans' Cabinet Cyclopædia.

Moving to the smaller format immediately reduced the amount of paper used by more than three-fifths. With Whewell's text not needing further emendation, Pickering also took advantage of stereotyping to reduce costs. This technology had gradually been making ground since the start

26. Pickering to TC, 18 November 1833, NC, CHA 4.212.23; *Quarterly Review* 50 (1833): 2–3.

27. WW to Richard Jones, 24 March 1833, in Todhunter 1876, 2:161; Secord 2014, 161. On the financial and presentational issues pertaining to cheap editions, see also Secord 2000, chap. 4.

of the century, and by the 1830s it had become an indispensable part of the apparatus of cheap publishing. By taking a cast of the hand-set moveable type, a stereotype foundry could provide permanent plates for use in printing while the expensive type was redistributed to its cases for use in further books. In this way, the publisher avoided having to pay for the text to be repeatedly made up from scratch by the well-paid compositors or having to accept the risks and costs of tying up a large bulk of expensive paper in a single printing. Having used stereotyping as a means of reducing costs on his Aldine Poets, Pickering was familiar with its start-up costs. Stereotyping Whewell's sixth edition meant that the printing costs were actually almost half as much again as for the comparable third edition. Only when a further impression was taken off the same plates in 1846 did stereotyping repay the investment. Having plates ready in the warehouse meant that the total printing cost per sheet was now just over a quarter of the average for previous editions.[28]

The lower retail price of Whewell's sixth edition was also partly achieved by both the author and publisher receiving less profit. Pickering's accounts unfortunately do not survive, but a reasonable estimate of costs suggests that profits on the sixth edition were less than half of what they were on the third edition, indicating the considerable financial incentive to exhaust the market for more expensive editions before producing cheap ones. Chalmers, who was the author most concerned to maximize his audience, managed to turn even this principled concern to his financial advantage, however. The rewritten version of his Bridgewater that appeared as the first two volumes of his collected works in January 1836 was packaged to take advantage of the growing market for small 6s volumes. Printed on demy paper but folded to make a duodecimo (so two-thirds of the size of his Bridgewater), Chalmers was thus able to sell his much enlarged work, *Natural Theology*, for three-quarters of the price of the original. Meanwhile, those who wished to complete sets of the Bridgewater Treatises continued to purchase editions in the original, more expensive form.

Many of the sets that were completed in this way are now to be found bound in more or less sumptuous leather bindings, suitable to the library of a gentleman or an institution. Initially, however, they appeared in a binding consisting of blue cloth over stiff boards (see fig. I.5). Such cloth and board bindings, so familiar to the modern reader, were a striking novelty in the 1830s. At the turn of the nineteenth century, books were typ-

28. This is based on McDonnell 1983, 2xi–2xxi (esp. 2xii), which offers a detailed analysis of the evidence in the Chiswick Press ledgers (BL, Add.Mss.41885–92, 41914–15, 41927–28) concerning the financial history of Whewell's Bridgewater.

ically supplied to retail booksellers in unbound sheets for them to have bound up locally in leather (described as "bound") or in paper-covered boards ("in boards"). As publishers embraced a more specialized role in the new century, some chose to take this aspect of book manufacture under their own control, sending batches to the binders as needed. The use of cloth developed out of this, and William Pickering had a significant role in that development. As a publisher, Pickering cared about the appearance of the books that bore his name more than most. Seeking to offer a temporary binding more robust than paper-covered boards, he pioneered the use of cloth in the early 1820s, working closely with Archibald Leighton, a representative of a large London bookbinding dynasty. The pair soon transformed ordinary calico into a stiffened and glazed cloth that provided a reasonably robust covering, and from the mid-1820s this became a typical feature of Pickering's books.[29]

Pickering did not have entire editions put into cloth-covered boards at once. Nor did he represent it as a permanent binding. The innovation nevertheless lent itself to a permanent edition binding that was produced using machinery and a division of labor, making it significantly cheaper than bespoke binding. The folded sheets of paper were still stitched together by hand (by women), but instead of the boards being attached to the sewn pages and covered in paper or leather, a standard, prefabricated "case" of cloth and boards could now be manufactured separately and attached when the volume had been sewn. With large editions the economies to be achieved by such a process were readily apparent. Following the introduction of a new arming press in 1832, Leighton experimented with lettering and decorating such standard case bindings using blind or gold blocking. Cloth was also increasingly embossed to give it a "grain" or pattern, so that books in cloth boards could now have an elaborate, decorative appearance. By 1833, the SDUK and other publishers of cheap editions were familiarizing book purchasers with the notion that an attractive cloth binding might provide an acceptable alternative to leather.

By contrast, Pickering was a publisher of quality who had no interest in suggesting that his cloth bindings were a means of economizing. They were, rather, a superior form of temporary binding. Hence, the Bridgewaters appeared in a plain blue cloth—initially unembossed and later with a simple grain effect. They had no decorative blocking, and their authors and titles were identified using paper labels on the spine. Except when demand dictated, Pickering sent the volumes in small batches to be bound

29. Warrington 1993; Gaskell 1974, 245; McDonnell 1983, 96–99. On the development of cloth and board binding, see Sadleir 1930 and Carter 1932, 1935.

by one of several regular trade binders in the city. There was no attempt to make every possible saving from edition binding, although the fixed price of 7d per volume was reasonably modest. Sometimes sheets were delivered direct to him from the printer, suggesting that some customers wished to have them bound in leather from the outset, either by their own binder or through Pickering by one of the craft binders in the West End. Nevertheless, many of the surviving copies of the Bridgewaters are still to be found in the original cloth binding. Reasonably robust and soberly presentable, some readers—even wealthy ones such as the Whig politician Constantine Henry Phipps, 2nd Earl of Mulgrave—found no need to have them rebound in leather.[30]

The 1830s were a pivotal moment in the history of the book trade. Emerging technologies of steam printing, stereotyping, and edition binding combined with rapid changes in the market for print such that publishers were presented with radical new possibilities in book manufacture. In many ways a traditionalist, William Pickering had nevertheless made important contributions to the development of cheaper publishing, working closely with his printer and binders in the process. Yet both he and the authors of the Bridgewater Treatises were motivated in their design and manufacture by a desire to produce dignified books that would look at home in a gentleman's library and would maximize their earnings. The consequence was that their prices were such as only gentlemen could afford. As we shall see, moreover, this was exacerbated by the desire of several of the authors to have their works adorned with illustrations that only added to the costs, although the use of recent technological innovations ensured that this was done more cheaply and more lavishly than could hitherto have been anticipated.

## *Figuring It Out*[31]

As on so many matters, the authors were offered no advice as to whether their works should be illustrated, and there seems to have been little discussion between them on the matter. Not surprisingly, therefore, the treatises varied considerably in this regard. Three had no illustrations at all, whereas Buckland's treatise included eighty-seven plates, one of which was his spectacular foldout section. Given the very significant costs and

30. Thanks to the generosity of his relation Sophie Forgan, Mulgrave's cloth-bound copies of Bell and Whewell's Bridgewaters are in my possession.

31. I owe this apt phrase to Shteir and Lightman 2006.

labor involved, decisions on such matters were far from trivial. To some extent, of course, they were governed by expectations in relation to subject matter. The natural historical sciences, for instance, had come to be associated with forms of publication that often included sumptuous illustrations. But such expectations were in flux. As the audiences for the sciences changed, so did conceptions of the role of illustrations in both instruction and amusement. Debate on the issue was further stirred up by the transformation of the technologies and economics of illustration. In addressing the question of appropriate illustration, the Bridgewater authors thus found themselves having to consider new audiences, to experiment with a wide range of new technologies, and to engage with an ever-more diverse workforce of artists and technicians.

For centuries, publishers, authors, and illustrators had been obliged to choose between cheap but generally crude woodcuts, which were resilient and could be printed alongside the text, and much more expensive copper plates, which were printed separately using high-pressure rolling presses and consequently lost definition after only a few hundred impressions. But as with other aspects of book production, illustration underwent revolutionary change in the decades around 1800. The revival of fine wood engraving using the hard end grain of boxwood—initiated in the 1770s by Thomas Bewick—had made possible the publication of high-quality images alongside the letterpress. Since this was also cheaper, the technology was appropriated in the early nineteenth century by publishers eager to gain an advantage, not least in the growing market for cheap publications. At the same time, the introduction of lithography, invented in 1796, offered a cheaper and more convenient alternative to copper plates. The process depended on the selective affinity of printer's ink for the greasy marks by which the image was applied to a smooth stone in advance of its being wetted. Finally, the introduction of more resilient steel plates in place of copper in the 1820s also contributed to reduce the costs of such fine printing where a longer print run was required, since the lines engraved or etched into the surface, from which the ink was printed, were slower to degrade.[32] Such copper and steel plates were sometimes collectively referred to as "intaglios," from the Italian word for carving or engraving, *intagliare*.

Taken altogether, these developments underpinned a radical transformation in the visual appearance of printed matter during the second quar-

32. For an overview of the illustration revolution, see Twyman 2009. On the illustration revolution and the sciences, see Rudwick 1974, Secord 2002, and Topham 2020. For a more socially oriented analysis, see Anderson 1991.

ter of the nineteenth century, and this was nowhere more evident than in relation to the sciences. While the most triumphant productions of Bewick's pioneering wood engraving had been works of natural history—the *General History of Quadrupeds* (1790) and the *History of British Birds* (1797–1804)—it was some time before such illustrations began to become more common in scientific books. Meanwhile, the use of copperplate illustrations abounded not only in scientific treatises and learned transactions but also in spectacular plate works in which sumptuous natural history illustrations were accompanied by limited amounts of descriptive text. The cost of such plates was challenging, however, and from the 1820s lithography began to compete. When, at the start of the second series of its *Transactions* in 1822, the Geological Society decided to introduce some lithographic plates, the printer's estimate suggested these would cost £5 each, around a third of the cost of the copper plates, and since illustrations still accounted for three-fifths of the total budget of production, this was a hugely significant saving. Lithography could not yield the fine lines required for some natural history illustration, but, given the economies it offered in time and money, it is not surprising that it was soon also being adopted by the Linnean Society.[33]

It was the new cheap publications of the 1820s that blazed the trail in the extensive use of wood engravings, offering what were surprisingly sophisticated images to the tens or even hundreds of thousands of readers of twopenny periodicals such as the *Mirror of Literature, Amusement and Instruction*. The useful knowledge society was especially prominent in these developments. Emulating the *Mirror*, the society included wood engravings in its sixpenny useful knowledge tracts from 1826 before moving on to produce breathtaking images in its flagship *Penny Magazine* from 1832 and in an array of other illustrated publications. The *Penny Magazine* set new standards and created new expectations, spawning the illustrated press that is so emblematic of the Victorian era. At the same time, wood engravings also soon became a familiar aspect of book publishing, coming to be incorporated in the growing range of scientific and medical works.

By 1830 illustrations were rapidly becoming ubiquitous, appearing in a wide variety of publications and in a much more diverse range of forms than ever before, often alongside text. Yet the role and value of such illustrations in scientific books remained a matter for debate. Sometimes, they seemed critical in recording information that could not be conveyed adequately by words. At other times, they were viewed as necessary in training both the eye and the hand of the medical student or naturalist. But at

33. Topham 2020.

other times again, the images were viewed as being important for the pleasure they conveyed, whether in making a book saleable, in advocating the attractions of scientific learning, or in securing recruits to the practice of natural history. However, as the audiences for print expanded to include workers whose supposed sensuality was a cause of widespread concern, the affective power of illustrations was sometimes a cause for concern. Sensational images that indulged the senses ran counter to the supposed rationalizing tendency of scientific education.[34]

In such circumstances, it was by no means obvious that works like the Bridgewaters should be illustrated. Chalmers's treatise, for instance, with its more theological approach and its subject matter of moral philosophy and political economy, had little need for illustration. Perhaps more surprisingly, neither Kidd's nor Whewell's treatise was illustrated. In Kidd's case, various anatomical and other subjects would have lent themselves to illustration, but his text was avowedly nontechnical, giving no justification for explicatory diagrams, and he had no experience of such matters. Whewell did have experience, having used engraved plates of geometrical diagrams in his university textbooks. The work in hand, however, was different; it made no pretense to explain the technical details of science, so the models, machines, and diagrams of the public lecturer were not required.[35] As with Herschel's *Preliminary Discourse*—which was also unillustrated except for its engraved title page—the reader was not to be distracted from the argument about the character and tendency of the sciences by superfluous illustrations. The remaining five Bridgewaters, however, all included illustrations. For several, illustrations seemed critical for reasons of scientific utility, educative or otherwise, but concern with visual appeal also clearly played a part.

In many ways, the most conventional of the authors was Kirby, who included twenty plates illustrating the appearance and anatomy of a wide range of species described in the text, much as he had done in his *Introduction to Entomology*. Like that production, the Bridgewater was not an original work of natural history, describing specimens new to science. Rather, its images were generally intended to familiarize the reader with unfamiliar animal forms or more specifically to make more intelligible complex descriptions of structure or behavior. They were thus an eclectic mix. Some were drawn from specimens, while others were redrawn from existing publications, often with little or no indication of the source. Many, such as the trapdoor spiders in Plate XI B, were visually striking, while others,

34. Secord 2002.
35. [Whewell] 1834b, 54.

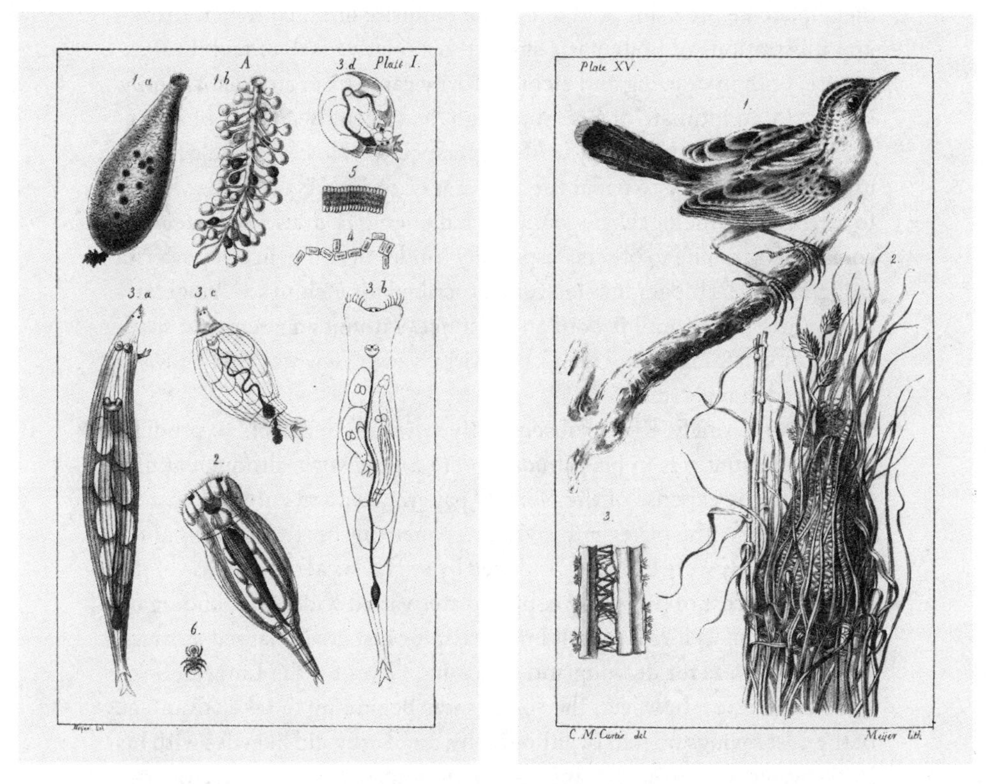


FIGURE 3.4 Plates I ("Infusories," lithograph by Meijer) and XV ("Birds," lithograph by Meijer after C. M. Curtis) from Kirby's Bridgewater Treatise (1835).

such as the tailor bird in Plate XV, had obvious aesthetic appeal (fig. 3.4). At the same time, however, they reflected Kirby's aspiration to provide a useful introduction to natural history. Thus, many of the images and the associated texts were sufficiently technical to function in training the eye of the budding naturalist to observe in nature what the adept considered to be of importance.

Kirby had many years of experience producing such illustrations, which meant that he was familiar with both the high costs incurred and the practical difficulties of managing the work of the artists and technicians employed, especially when located in the further reaches of rural Suffolk. So expensive was the commissioning of copperplate illustrations that, as a young man engaged in producing a privately published monograph on the English bees, he had learned how to etch and had undertaken fourteen of the eighteen plates himself. Etching was a relatively accessible process in which the draftsman made incisions in a wax coating applied to the metal plate before using acid to cut the requisite lines. However, Kirby's 404 illustrations of body parts showed his relative lack of technical expertise, both in drawing and etching. Kirby came from an artistic family. His uncle, an intimate of Gainsborough, had been the author of a standard mid-eighteenth-century work on perspective. However, while he was used to "committing to paper the appearance of objects which he wished to retain in his memory," he found it challenging to draw adequately "to convey his meaning to others," especially "under the influence of powerful magnifiers." His biographer tactfully described his etchings as "singularly characteristic, the outline hard and distinct, without an attempt at shadowing or softening of any kind," but Kirby's own view was that he was a "wretched draughtsman."[36]

Not surprisingly, Kirby subsequently relied on engravers to produce the fine illustrations in his *Introduction to Entomology,* although at first this was at the expense of the Norfolk papermaker and entomologist Simon Wilkin.[37] The plates in his various papers in the *Transactions of the Linnean Society* were likewise produced by engravers at the expense of the society. The cost of preparing copper plates varied widely, depending on the artist. The well-regarded John Curtis, for instance, charged as much as seven guineas for drawing and engraving plates for the Linnean Society. By the 1830s, however, the society was beginning to take advantage of the cost savings offered by lithography, and Kirby did likewise with his Bridgewater. Moreover, the large production run of Kirby's Bridgewater—many times the number of fine impressions that might be expected from a copper plate—made the greater resilience of the lithographic stone an important factor.[38]

Another advantage of lithography was that it had the potential to increase artistic control over the image, since draftsmen untrained in the technicalities of engraving could in principle draw directly on stone. However, Kirby had the drawings done by an experienced natural history draftsman and engraver, Charles Morgan Curtis, before a lithographic artist transferred the images to stone. They were then printed by Britain's leading lithographic printer, Charles Hullmandel. The value to Kirby of this arrangement was that it allowed him to employ an artist whose

36. WK to WB, 21 November 1833, OUMNH–BP, Notes by Subject (Bridgewater Treatise), 3/2; Freeman 1852, 193–94.

37. Kirby and Spence 1815–26, 1:xx. On Wilkin's troubled involvement in the publication, see Kirby's correspondence with A and WS Macleay, LSL, MS237B, and WK and Spence to Bacon and Wilkin, 21 May 1814, NHM, KIR E 1:1.

38. Dyson 1984, 59–60, 160; Topham 2020, 79. Cf. WK to A MacLeay, 26 December 1817, LSL, MS237B.

eye he could trust—something that had always been important to him. First, there had been Curtis's elder brother John, a young local naturalist whose patron Simon Wilkin had sent him to learn copper engraving. Based just fifty miles away, near Norwich, Curtis had repeatedly worked under Kirby's scrutiny in preparing the plates for the *Introduction to Entomology*, so that the naturalist had come to rely on him implicitly. However, having secured work for Curtis with the Linnean Society and spurred him to produce his vast plate work, *British Entomology* (1823–44), Kirby had been obliged to train up another Norwich youth, Henry Denny, to help him with illustrations, sending him also to learn copper engraving before employing him on the final two volumes of the *Introduction to Entomology*.[39] By the 1830s Denny had also moved on, and Kirby settled on Curtis's younger brother to illustrate the Bridgewater.

Charles Curtis was by now also an experienced entomological engraver who inspired the trust between author and illustrator that Kirby valued. It is not clear whether Curtis was based in London or Suffolk, but he was dispatched to draw specimens belonging to the Royal College of Surgeons, the Zoological Society, and its "bird stuffer" John Gould, as well as specimens from Kirby's own collection, including a fish jaw found on the compost heap of a nearby clerical colleague that had been identified for him by the Surgeons' assistant conservator Richard Owen. What Curtis could offer in preparing such drawings was a shared sensibility concerning the appearances that made the specimens "characteristic." Kirby had come to trust the older Curtis's entomological judgement to such an extent that when figures did not correspond with the "character" he had given specimens in the text, he began to fear errors on his own part. Achieving correspondence between text and image was a significant concern, and it is in the appropriation of images from other publications—a common practice little explored by historians—that the processes involved are most easily recovered. When, for instance, Kirby had illustrations of the highly changeable *Rotifer vulgaris* redrawn from one of Christian Gottfried Ehrenberg's memoirs on infusoria for plate I (3a–3d), the eyespots and limb-like appendage were rendered more vividly anthropomorphic (see fig. 3.4). This perfectly reflected Kirby's use of Ehrenberg's findings concerning the complexity of infusoria to gainsay the spontaneous generation that underpinned Lamarckian transmutation.[40]

39. See Kirby's correspondence with A and WS Macleay, LSL, MS237B, and Freeman 1852.

40. Kirby 1835, 1:xxii–xxiii, 145–46, 349–52; WK to A MacLeay, 10 May 1819, LSL, MS237B. Compare the plates in Ehrenberg 1832, although note that Kirby does

The illustrations of Kirby's Bridgewater were thus in keeping with the established practices of natural history publication, offering authoritative, characteristic representations of specimens that were intended to train the reader's eye to see nature aright while simultaneously offering aesthetic pleasure. The investment required to achieve these effects was considerable, both in time and money, but Kirby did what he could to mitigate these costs, using a biddable artist, redrawing existing images, making the most of lithography, and choosing not to have the plates colored. At the same time, he made a point of working with artists whom he could trust unreservedly to represent nature consistently with the truths that he sought to convey in his text.

By comparison, Prout's Bridgewater needed far fewer illustrations. In relation to meteorology, however, it did also include a lithographic plate—a map of the world illustrative of Alexander von Humboldt's isothermal lines reproduced from an intaglio plate in the article on meteorology in the *Encyclopædia Metropolitana*. The work was done by a little-known lithographer, but once again, there was transformational intent.[41] Prout had a visually complex folded section inserted into the center of the map that was intended to illustrate the diminution of light and heat from the equator to the poles, the correlation of this with the fauna and flora of different zones, and the parallel between such geographical zones and the zones caused by elevation. By thus combining mapping, visual scales, diagrammatic representation, and description all into the one plate of 14½ × 8½ inches (37 × 22 cm), Prout hoped to clarify large parts of his conceptual discussion at minimal cost. Color was crucial to achieve the effect, and extensive hand coloring was used to distinguish the several continents, to fill in the detail of the scales of light and color, and to indicate the different vegetation of the zones.

In addition to his plate, Prout also had a small number of simple wood engravings made to illustrate diagrammatically various aspects of his matter theory as well as electromagnetic and optical phenomena. These certainly contributed to the clarity of the exposition. Moreover, while not long before such illustrations would have necessitated the expense and inconvenience of separately printed copperplate engravings—as, for instance, in John Dalton's *New System of Chemical Philosophy* (1808–10)—Prout now had the advantage of being able to incorporate much cheaper wood en-

---

not identify this as his immediate source. On "characteristic" images, see Daston and Galison 2010, 55–113; on Curtis, see Williamson 1903–5, 1:362, and *ODNB*, s.v. "Curtis, John (1791–1862)." Curtis died in Suffolk in 1839.

41. On the lithographer, Bradbury, see Twyman 1976, 26.

gravings into his text. For a chemist, however, illustration was altogether less significant than for a naturalist, especially in a treatise that was not technical and did not need to represent the use of apparatus.

Prout's diagrammatic wood engravings resembled the rather lackluster wood-engraved diagrams recently made familiar by the scientific treatises of the SDUK's Library of Useful Knowledge, which were intended to elucidate the sometimes rather abstruse meaning of the text. Roget's useful knowledge treatises on electricity and magnetism had used simple diagrams in just this way, and when it came to his Bridgewater he likewise scattered an extraordinary array of 463 wood-engraved figures across its pages, explaining that "mere verbal description" could never convey "distinct ideas of the form and structure of parts" without the aid of figures. However, while the great bulk of the images were thus unspectacular, the overall effect was striking (see fig. 3.2). Almost one in six of Roget's pages carried an illustration, making this a visually groundbreaking book that looked vastly different from the more sedately illustrated treatise of Kirby. Accustomed as we now are to seeing books in which images are liberally scattered across the pages, the effect of this innovation is easy to underestimate. Even by the 1850s, when Henry Gray's *Anatomy Descriptive and Surgical* exploited the form so triumphantly, the shock factor was long gone. In the 1830s, however, such illustrations were only beginning to be used in medical publications, and Roget's work stood out. Indeed, when physician John Conolly later urged the useful knowledge society to improve its effectiveness in the provinces by preparing large display diagrams for lectures, Roget's was the first on his list of works in anatomy and physiology that might be plundered for such purposes.[42]

The cost and labor involved in illustrating a book in this way was very considerable. Most of the illustrations were redrawn from other publications, but a number were drawn from original specimens in institutions such as the Hunterian Museum, and others were idealized diagrams of structures, including one that Robert Edmond Grant claimed had come improperly from his blackboard.[43] Most of Roget's illustrations were drawn by Agnes Catlow, who had joined the family as a governess in January 1832 as Roget's wife became progressively more ill. It seems likely, indeed, that his choice of governess in part reflected his illustrative needs.

42. Conolly to T. Coates, 18 January 1840, UCL, SDUK 37; Richardson 2008; Roget 1834, 1:xii.

43. Grant to PMR, 15 November 1833, in Roget 1846a, 483. Compare Grant 1834, 539, and Roget 1834, 1:393. Sources are given minimally in Roget 1834, 1:xxii–xxxv, and often more fully in the text and footnotes.

Catlow's father, Samuel, had been a schoolmaster and Unitarian minister in the Midlands town of Mansfield, and Agnes was a well-educated naturalist who later went on to author a number of favorably regarded books of popular natural history, becoming a reasonably adept artist. Catlow's labors on Roget's behalf earned her £25, but the bulk of the cost of illustration was due to the engraver John Byfield, who had a considerable reputation and was responsible for many of the engravings that graced Pickering's books. For his finest work, such as vignettes, Byfield charged as much as two to three pounds, but this less taxing work totaled £200 for 463 figures. The expense was thus considerable, probably making Roget's treatise second only to Buckland's in the cost of its illustrations, but the robust wood blocks were suitable for repeated reuse, being employed to produce many thousands of impressions as well as being stereotyped for use in the United States.[44]

Roget's decision to invest so much in wood engravings was especially well rewarded, since Whittingham was renowned for his "very superior printing" of wood blocks, producing altogether more impressive results than most printers.[45] The engravings thus made for a visually very attractive book, but they also made for one that was highly effective in conveying an authoritative overview of the increasingly technical sciences of physiology and comparative anatomy. Roget had written a book that was intended to be scientific in form and suitable as an introduction to the subject. By exploiting the new availability of wood engraving, he had managed to illuminate technical discussions to striking effect with a range of depictions and diagrams that had rendered them more intelligible to general readers and more informative to knowledgeable ones.

Bell's Bridgewater also included numerous wood engravings, but Bell took a rather different approach to anatomical illustration. He was himself an accomplished artist, having been taught as a boy by the painter David Allen, and he had from the start of his professional life made the most of his talent in a range of publications. Moreover, these ventures were underpinned by a deep-seated commitment to the use of drawings and of the

44. Engen 1985, 39; copy of account between Roget and Pickering, BO, Ms.Eng. Lett.b.35, fols. 69–70; see also John Murray Copies Day Book 7 1830–33, NLS, MS 42892, p. 298. On the stereotyped images, see Barnes 1976, 376–77, Kaser 1963, and Chiswick Press Debit and Credit Ledger A, BL, Add.Ms.41927, fol. 136. On Catlow, see Roget 1834, 1:xii, Kramer 1996, 80–85, "Science Gossip," *Athenæum*, 25 May 1889, 667, [Watkins and Shoberl] 1816, s.v. "Catlow, Samuel," and Census Returns of England and Wales, 1851, NA, HO107/1717, fol. 629.

45. [Britton] 1834, 260n; but see also "Our Wood Engravers," *Monthly Magazine* 15 (1833): 496–506.

act of drawing itself in the practical education of the surgeon's hand and eye. Thus, Bell introduced drawings into his lectures at his Great Windmill Street medical school, frequently drew on the blackboard, and produced oil paintings and wax models for student use in his museum. When he prepared his two treatises on anatomy for the useful knowledge society, he took it upon himself to draw thirty-five figures of animal anatomy and mechanical equivalents with the primary purpose of helping the reader to understand the mechanical arguments being made in the text. Indeed, the diagrams often included annotated letters for purposes of reference. In among the more mundane images, however, were to be found flying buttresses on a cathedral, a classical male nude, and two jolly peasants on an overloaded cart. Such artistic or playful illustrations reflected Bell's sense that these treatises were at least as much for genteel as for plebeian readers.[46]

Of course, Bell was not constrained to use wood engravings in his Bridgewater. However, when he had previously written for a nonmedical audience—in his *Anatomy of Expression*—he had contrived to integrate images and text by printing text around copperplate engravings, and he now achieved the same effect through the altogether cheaper expedient of wood engraving. Like Roget he scattered images over almost one-sixth of his pages, but Bell's range was altogether richer, reflecting his more general sense of his audience. While the illustrations usually related to the text to some extent, the connection was sometimes tenuous. Indeed, Bell's first illustration was a vignette of a dancing Pan that was a mere jeu d'esprit. Many of the images were of animals, but these were often in naturalistic settings that were designed to stimulate the imagination. The standing position of a bear was illustrated by the depiction of a chained dancing bear in a bear pit watched by smiling children. Although some images exhibited anatomical details, these were typically in a purified skeletal form suitable for a mixed audience. Very occasionally strap-like muscle illustrated the mechanics described in the text. Only in three, nonhuman cases was the gore of the dissecting room alluded to: two were illustrations of Bell's 1809 dissection of a lion's paw, while the third pathetically depicted a dissected bird's wing (fig. 3.5). As this suggests, although the drawings were typically idealized, they sometimes betrayed more particular origins. The bones of the forelimb of the Plesiosaurus and Ichthyosaurus, for in-

46. CB to G Bell, 29 January 1827, in Bell 1870, 295; CB to T Coates, 11 July 1827 and 21 January 1829, UCL, SDUK 24 and 26. In addition, Carin Berkowitz emphasizes the extent to which Bell considered beauty to enhance the "pedagogical efficacy of his drawings"; see Berkowitz 2011, 273.

58 ANATOMY OF THE

If we observe the bones of the anterior extremity of the horse, we shall see that the scapula is oblique to the chest; the humerus oblique to the scapula; and the bones of the fore arm at an angle with the humerus. Were these bones connected together in a straight line, end to end, the shock of alighting would be conveyed as through a solid column, and the bones of the foot, or the joints, would suffer from the concussion. When the rider is thrown forwards on his hands, and more certainly when he is pitched on his shoulder, the collar bone is broken, because in man, this bone forms a link of connection between the shoulder and the trunk, so

80 STRUCTURE OF BIRDS.

observe how the whole skeleton is adapted to this one object, the power of the wings.

Whilst the ostrich has no keel in its breastbone, birds of passage are recognisable, on dissection, by the depth of this ridge of the sternum. The reason is that the angle, formed by this process and the body of the bone, affords lodgement for the pectoral muscle, the powerful muscle of the wing. In this sketch of the dissection of the swallow, there is a curious resemblance to the human arm; and we cannot fail to observe that the pectoral muscle constitutes

FIGURE 3.5 Artistically composed wood engravings from Bell's Bridgewater Treatise (3rd ed., 1834), depicting the anterior limbs of humans and horses, and displaying the dissected wing of a swallow.

stance, were drawn in simplified form from specimens at the Royal College of Surgeons.[47]

The overall effect of Bell's images was certainly to familiarize readers in a general way with aspects of anatomy, but this was accompanied by a conscious effort to entertain a nontechnical audience. Constructing drawings of this sort was a form of recreation for Bell as he wrote his treatise, and the playfulness of the images no doubt in part reflects his resolve to create an appropriate mood. While Bell could etch, however, he was dependent in this case on wood engravers to redraw his images. Only a few of the engravings were signed, but those that were came from the prominent engravers Ebenezer Landells, Charles Gray, and John Jackson, all trained in Newcastle-upon-Tyne under Bewick or his disciples. Moreover, Bell had

47. Bell 1833a, 101–2; Bell 1870, 145.

exacting standards, and even such able engravers could expect to have to make alterations. With forty-six engravings in total, some of which were sizeable or complex, the cost must have amounted to tens of pounds, and as late as 1851 John Murray paid Pickering £30 for the old blocks.[48] But the effect in setting the accessible tone of Bell's work was palpable.

❋

The forgoing accounts reveal the expense, labor, and complex choices involved in scientific book illustration at this transitional moment, but Buckland's Bridgewater—which was illustrated on an altogether grander scale—is remarkable for bringing all the new technologies into juxtaposition. Like Kirby, Buckland was used to the visual demands of natural history, having labored hard to organize appropriate plates for his many papers in the transactions of the Geological and Royal Societies. Reprinting some of these, his *Reliquiæ diluvianæ* included no fewer than twenty-seven plates. In addition, he had, like Bell, found illustrations to be indispensable aids in the lecture room, expending considerable sums in developing his collection of charts that were intended to engage the interest and imagination of students as well as to illustrate technical details (fig. 3.6).[49] Not surprisingly, therefore, Buckland ended up devoting one entire volume of his Bridgewater to his eighty-seven plates, featuring 705 figures together with over 110 pages of explanatory text. He thus had strong incentives to keep abreast of the latest developments in illustrative technology, and despite its conventional appearance, his volume was trailblazing in this regard, with costs kept far below what they would otherwise have been.

The importance that Buckland accorded to the plates is evident from the labor he invested in them. From the outset there was "endless trouble about them," but as Mary explained to Whewell in May 1833, the geology was "utterly unintelligible without the plates," and they became Buckland's standard explanation for his delay. In explaining geological phenomena to a general audience, Buckland was accustomed to being able to refer to visual images. A key object in gathering together a large set of plates in a separate volume—rather than inserting some as wood engravings in the

48. John Murray Copies Ledger E, 1846–76, NLS, MS42730, p. 201 (see also John Murray Copies Day Book 7 1830–33, NLS, MS 42892, p. 218); CB to T Coates, 21 January 1829, UCL, SDUK 26.

49. Boylan 1984, 234, 331, 562–63, 641; O'Connor 2007a, 75–80. A large collection of teaching illustrations survives in the OUMNH–BP and is currently the subject of doctoral research at the University of Leeds by Susan Newell.

FIGURE 3.6 William Buckland lecturing in the Old Ashmolean Museum, 15 February 1823. Colored lithograph printed by C. Hullmandel after N. Whittock. Image: Wellcome Collection. Public Domain Mark.

text, as Buckland had initially contemplated—was to allow his readers something like the combination of visual and verbal experiences encountered in his lecture room, with the two volumes open side by side. The sometimes extensive descriptions of the plates were intended primarily for the specialist reader and were hidden in notes "which the Country Gentlemen may skip."[50] For the majority of readers, however, the images were intended to render the main text more intelligible by making exotic geological phenomena visually familiar.

Right at the outset, Buckland referred readers to his hand-colored fold-

50. WB to Featherstonehaugh, [25 April 1836] and 29 November 1835, CUL, Add. Ms.7652, II.LL.32, 33; WB to Sowerby, 21 September 1831, NHM, DF LIB/601/45, Let. 216; WB to G Harvey, 16 April 1834, BO, Ms.Eng.Lett.d.5, fols. 204–5; M Buckland to WW, 12 May 1833, TC, Add.Ms.a.66$^{31}$.

out plate containing an idealized geological section, which juxtaposed a representation of the stratigraphic column with a temporal scale and with reconstructions of 120 forms of life characteristic of the different geological epochs (see fig. 2.4). Next followed eighty-one plates to illustrate the great body of the text, depicting paleontological specimens, often many to a page, and a handful of skeletal or schematic reconstructions. Finally, five plates of geological sections illustrated the closing chapters on the dispositions of rocks and minerals. Throughout, Buckland shied away from the more highly imaginative representation of "scenes from deep time" that were soon to dominate more popular geological works but concerning which geologists were notably cautious at a period when they were anxious about being accused of unwarranted speculation. He included only one such image: a small imaginary coastal scene depicting pterodactyls (plate 22). Yet, while Buckland's plates thus enhanced the scientific status of the work, their number, range, and size offered the same kind of spectacular visual experience that captivated those who attended his lectures.[51]

In preparing his illustrations, Buckland depended for source material on those in a wide array of existing publications, which were roughly identified in the notes. However, the act of redrawing was once again transformational. The redrawing of Thomas Hawkins's *Plesiosaurus triatarsostinus* in plate 17, for instance, not only involved the stripping away of the specimen's stone ground (which had emphasized Hawkins's expertise as a collector and restorer) but also involved a reclassification (fig. 3.7). Moreover, Buckland supplemented the plate with another containing further original drawings of Hawkins's specimens at the British Museum (plate 18). Similarly, having had an Oxford artist redraw from the *Philosophical Transactions* a representation of the head of an ichthyosaurus, Buckland dispatched a London artist to the British Museum with the woodblock to add "certain oblique lines of structure" and other features.[52]

Even more demanding was organizing the many original drawings of specimens. Not surprisingly, Buckland sourced a good number of these in Oxford, where he had greatly enlarged the Ashmolean collection, while others came from the collections of the Geological Society, the British Museum, and the Royal College of Surgeons. At the last of these, Buckland had arranged for Richard Owen to prepare and draw the pearly nautilus—of the living parts of which Owen had recently given the first account—in order to illustrate Buckland's theory concerning the function

51. O'Connor 2007a, 91–95, 330–33, 357–61; Rudwick 1992, esp. 233–35.

52. WB to Sowerby, 12 December 1833, NHM, DF LIB/601/45, Let. 235; Carroll 2008, 108–11; Storrs and Taylor 1996, 404.

of the siphuncle (a tube running through the chambers of the spiral shell). Many other specimens were borrowed from private individuals, and in other cases Buckland depended on informants to supply drawings of specimens at a distance. Especially remarkable was Miss S. C. Burgon, who not only drew fossil sharks' teeth from the cabinet of a relation but in one case prepared the intaglio plate herself (plates $27^{e-f}$).[53] The assembly of such a rich array of materials was a major organizational endeavor, and it is no wonder that the process delayed Buckland's volume, especially given that he kept finding additional and improved subjects for illustration.

On first inspection, Buckland's volume of plates appears technologically conservative. Printed on one side of heavy-gauge paper of the sort usually used for copper plates, many superficially look like they might have been produced using such plates, which indeed had been Buckland's expectation. In the event, they were "chiefly steel & wood or transfer lithography"; indeed, more than half were wood engravings. These technologies enabled him to combine relative cheapness with resilience, a quality that became ever-more necessary as the likely number of copies required grew. While some accounts report that Buckland spent the entirety of his thousand pound honorarium on the illustrations, the costs of drawing and engraving were apparently charged to the publishing account, and Buckland reported that they amounted to £500. This was still a very substantial figure, working out at an average of £5 15s per plate. However, the large print run of the first edition—ten times that of many scientific memoirs—meant that the cost of engraving contributed only two shillings (rather than a pound) to the final cost.[54] While printing was on top of this, the new technologies also meant that the illustrations did not need to be re-engraved three or four times for the first two editions, as would have been necessary with copper plates.

There were, however, several factors to take into account in choosing between the different technologies. In addition to cost, Buckland was concerned both with the visual effect produced and with the ability of particular artists to render the details knowledgeably and accurately. In all

53. Owen to WB, 28 July 1832, 14 and 17 December 1833, Temple University, Owen Correspondence, 1:17, 23, 25. Miss Burgon was almost certainly Sarah Caroline Burgon, whose father was a merchant and archaeologist. See *ODNB*, s.v. "Burgon, John William."

54. *Quarterly Review* 56 (1836): 62n; *Monthly Review*, n.s., 3 (1836): 351; *Edinburgh Review* 65 (1837): 15; Allibone 1870–71, 1:277; *British Critic* 20 (1836): 307n; Chiswick Press Cost Book B, BL, Add.Ms.41886, fol. 110; WB to Featherstonehaugh, [25 April 1836] and 29 November 1835, CUL, Add.Ms.7652, II.LL.32, 33; WB to Sowerby, 21 September 1831, NHM, DF LIB/601/45, Let. 216.

cases he maintained the same close scrutiny of the work being done and insisted on revisions and corrections. Buckland's friend William Daniel Conybeare reported that the geologist had found it necessary to "instruct the draughtsmen employed, in the necessity of accurately representing minutiæ of detail, of the importance of which they could not be themselves aware; and again, from the same motives, to repeat the same lessons to the engravers to secure fidelity in their copies."[55]

The issues are well illustrated by Buckland's iconic foldout geological section, which involved considerable negotiation with a range of individuals. First, he asked London geological lecturer Thomas Webster whether he would be willing to provide a suitably scaled-down drawing of the ideal section he used in his public lectures. An accomplished artist, Webster had been employed at the Geological Society until 1827 as librarian, curator, and draftsman, drawing the society's geological map in 1820 and many illustrations for the *Geological Transactions*. Now earning his living independently, he planned to use the section in a work of his own but agreed to draw a version for Buckland. This incorporated additions devised between them, in part in emulation of various published geological sections, such as Henry De la Beche's recent *Sections and Views Illustrative of Geological Phenomena* (1830). Buckland also wished to incorporate reconstructions of characteristic fossil forms, and for a spell he planned to include illustrations of minerals. At first, the intention was that Webster would draw the fossil reconstructions using various published sources, but ultimately Buckland turned to a local artist, Joseph Fisher, on whose services he came to rely heavily.[56]

Printing an illustration of this size was a specialist business, and Buckland sought the assistance of the Ordnance Survey's official map seller, the cartographer and engraver James Gardner. Gardner first arranged the engraving of Webster's drawing on copper, and there was a three-way exchange about the details (were Gardner's engravers capable of adding smoke to the volcano, Buckland wished to know, or would there be an extra cost?). Then Fisher engraved the fossil reconstructions in Oxford, perhaps on a separate plate. In any event, Gardner organized for the section to be printed by transfer lithography using the "large press" of the promi-

55. *British Critic* 20 (1836): 307n.

56. Buckland 1836, 2:1–2; WB to Webster 3 December 1831, 14 January [1832], and 13 November 1836, FMC, Perceval Papers, Q28–30; Challinor 1961–63, 19:296. Webster employed GJ Scharf to draw a twenty-foot version of the ideal section in January 1834 for use at the Royal Institution; see Scharf's Journal, 1833–35, NPG, NPG7/3/7/2/1.

nent lithographic printer William Day. By transferring the engraved image directly onto lithographic stone (quite possibly repeatedly), many more copies could be produced without the loss of definition characteristic of copperplate engravings; printing from stone was in any case cheaper. At the same time, the precision of the copper engraving was largely retained, as Gardner was eager to demonstrate to Buckland by sending a direct impression from the copper plate for comparison. Finally came the coloring, carried out by hand in watercolor paint, probably by women, according to an agreed color scheme that Buckland was again involved in correcting.[57]

In regard to the other plates, Buckland's views about the appropriate technologies to use gradually developed as he learned more about their potential. At the outset, he had been very dubious about the radical idea of using wood engraving for plates. While it was the "best & only method" for inserting images into text, he was not sure whether it was adequate to produce "delicate lines," and he also wanted to know whether blocks could be obtained the size of his page and how much it would cost. Wood engraving readily lent itself to simple line drawings, such as the geological sections in plates 65 to 69 and the reconstructions of pachyderms from the Paris basin in plate 3, redrawn from the intaglio plate in the second edition of Cuvier's *Recherches sur les ossemens fossiles* (1821–24). However, the extended possibilities of the technology for scientific illustration were just beginning to become more apparent, and many of the forty-eight wood engravings in Buckland's volume depicted much more complex subjects, such as the skull of *Ichthyosaurus platyodon* in plate 10, copied from an intaglio plate in the *Philosophical Transactions*. Many of these were drawn by local artist Joseph Fisher, and they were all engraved either by him or by Pickering's engraver, John Byfield. Buckland clearly valued having Fisher close at hand, where he could discuss desiderata, showing him his own rough drawings and commenting on work in progress.[58]

57. Gardner to WB, 9 May and 20 October 1834, BO, Ms.Eng.Lett.b.35, fols. 49–50; Pickering to WB, 1 March 1835, OUMNH–BP, Notes by Subject (Bridgewater Treatise) 3/6. On transfer lithography, see Hullmandel [1824], 80, Twyman 1970, 130, and Hullmandel 1821, 128–35. The same copper plate was apparently reused as late as 1858; see the correspondence in Frank Buckland's Scrapbook, RCS, MS0035.

58. OUMNH–BP, Notes by Subject (Bridgewater Treatise) 1/1, 27 and 29v; Buckland 1836, 1:[ix], 2:2; WB to Sowerby, 21 September 1831, NHM, DF LIB/601/45, Let. 216. Buckland had hardwood supplier John Jaques send Fisher woodblocks in December 1831 (John Murray Copies Day Book 7 1830–33, NLS, MS.42892, p. 188). On Buckland's dealings with Byfield, see WB to Sowerby, 20 August 1833 and [November 1834?], NHM, DF LIB/601/45, Lets. 233, 208. The wood-engraved plates are 2–4, 6–14, 16–25′, 26a, 27–27d, 46′–46″, 51, 54–62, 64–69.

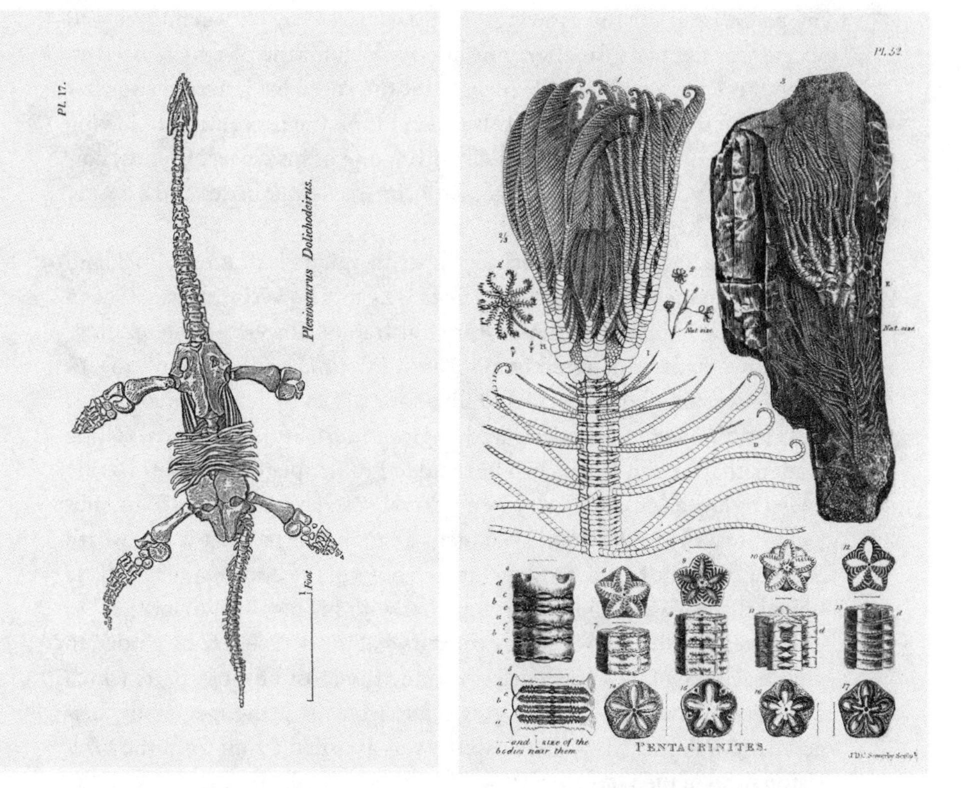


FIGURE 3.7 Plates 17 ("Plesiosaurus Dolichoderus," wood engraving after Thomas Hawkins) and 52 ("Pentacrinites," engraving by James Sowerby), from Buckland's Bridgewater Treatise (1836).

In fact, Buckland's decision to have all his illustrations as separate "plates" worked to his advantage, since printing wood engravings separately could assist in achieving a higher quality impression. Still, where fine detail was required, there was no question that intaglio plates had the advantage over wood engraving. For instance, Buckland had Fisher draw and engrave a collection of fossilized feces on metal rather than wood (plate 15). The greatest concentration of intaglio plates were those of cephalopods (plates 28–36, 38–39, 42–44″), trilobites (plates 45–46), and crinoids (47–53) (see fig. 3.7). Fisher drew many of the simpler cephalopods and the trilobites, but these were then put onto steel plates by London painter and engraver John Christian Zeitter. Buckland was still feeling his way with the new technology and wanted to know whether the steel plates would allow for "alterations" in the way copper did. Moreover, while the new technology might be more resilient, it was not cheap. Having paid Fisher for the drawings, he now had to pay between three and six guineas per plate for engraving and 15s each for the plates themselves. Then there were additional expenses for the specialist lettering engraver to add the captions. Moreover, Buckland took the precaution of having the prominent sculptor Francis Chantrey, one of his closest friends, correct Zeitter's proofs against original specimens—four boxes and a basket cluttering Chantrey's house.[59]

With the finer and more technically demanding illustrations, Buckland drew on the specialist services of James De Carle Sowerby, the scion of an outstanding family of natural history illustrators. Sowerby's paleontological prowess, demonstrated by his *Mineral Conchology* (1812–46), meant that he was much in demand to illustrate works by leading geologists. Buckland now had him draw and engrave most of the crinoids, telling readers that "much value" had been added to the plates "by their having been engraved (except Pl. 48) by a Naturalist so conversant with the subjects." Sowerby understood specimens as an insider, so that it was worth Buckland's while to risk sending him portions of the book manuscript. He hoped that forwarding this text together with his own "rude outlines" of encrinites would allow Sowerby to "see what is my object & be guided in your tracing by this knowledge—making the most of those parts which I have to speak about—." Such insight into what made the specimens characteristic was critical. When Sowerby was awarded a sum from the Wollaston fund of the Geological Society in 1840, Buckland observed that the "drawings and engravings of fossil shells and plants" with which he had graced learned publications expressed their characters "with a degree of accuracy and truth, which no pencil or burine but those of a scientific artist could possibly accomplish."[60]

In total Sowerby engraved a dozen plates for Buckland's Bridgewater, and he also supplied drawings for wood engravings. Since Buckland only had opportunity to see the artist on trips to society meetings in London, he sent endless letters discussing the details. A surviving bundle of more than fifty of Buckland's letters, together with numerous proof sheets, provides a fascinating insight into the process. Sowerby's good nature was proverbial, but Buckland must have come close to testing it, with constant requests over the space of four years to modify details, add new illustra-

59. Chantrey to WB, 7 May 1833, BO, Ms.Eng.Lett.b.35, fols. 10–11; Gordon 1894, 128n; Zeitter to WB, 7 January 1833, NHM, Owen Collection 62, 27/291–92; Zeitter to WB, 9 April 1834, OUMNH-BP, Notes by Subject (Bridgewater Treatise), 1/1, 76; Twyman 2009, 125–27.

60. Buckland 1840, 7; WB to Sowerby, 17 Dec 1832 and 6 Apr 1834, NHM, DF LIB/601/45, Lets. 219, 237; Buckland 1836, 2:79.

tions, undertake new plates at short notice, and try out new technical possibilities. A variety of objects passed back and forth between them. Besides boxes of specimens, Buckland would sometimes begin by sending his own rough sketches of what he wanted, and in the case of two of the plates of the squid genus *Loligo*, there were drawings for engraving that had been produced by Mary, who was altogether a more talented artist than William. Back the other way came Sowerby's outlines of plates—"tracings preparatory to setting them on ye steel plate"—followed by repeatedly corrected proofs. On some occasions, Buckland relied on others to check the adequacy of the image. And then there were wood blocks, with drawings by Sowerby and others, and instructions on one occasion to have Byfield replace a section of the wood so that Sowerby could make a last-minute addition.[61]

The correspondence is especially revealing about the rapidly developing technologies of reproduction. Having established the desirability of intaglio plates rather than wood engraving for the crinoids, Sowerby at first etched on copper, but a year later the talk was of engraving on steel. By the following spring, Buckland was asking Sowerby to let Hullmandel experiment with transferring the initial copperplate image to lithographic stone so that he could compare the two. How did the costs compare? he then wanted to know. The steel plates, at least, were more robust than the copper, although they were vulnerable to rust, as Buckland was only too aware. Now, though, Buckland wanted to try out transfer lithography with the steel plates, too, with a view to reducing printing costs. With his wife's illustrations of *Loligo*, he had followed Sowerby's advice to use mezzotint on the steel plate, with tonal effects produced using a toothed rocker, since "lines w[oul]d interfere with the lines of Structure." But while he feared that this would make transfer lithography impossible, he found it worked "perfectly well." Generally the experiments seem to have gone well, but Hullmandel was not always available to undertake such work, and Buckland found his leading competitor Day less satisfactory, so it is not clear just which plates were finally printed by transfer, although Buckland's comments to a friend indicate that there were a number.[62]

Apart from the images transferred from intaglio plates, the number of

61. WB to Sowerby, 17 December 1832, NHM, DF LIB/601/45, Let. 219. The letters from WB are in NHM, DF LIB/601/45, with some of Sowerby's replies in OUMNH-BP, Notes by Subject (Bridgewater Treatise), 1/1, 1/2, and 3/6. Sowerby engraved plates 28–30, 44′–44″, 47, 49–50, 52–53, and 63; he also drew plate 51 on wood, as well as some smaller illustrations.

62. WB to Featherstonehaugh, 29 November 1835, CUL, Add.Ms.7652, II.LL.33; WB to Sowerby, 18 August and 8 December 1834, NHM, DF LIB/601/45, Lets. 242, 249.

lithographs used was relatively small. While the technology's obvious advantages had led to its widespread adoption in natural history illustration, Buckland considered that it was not well suited to many of his purposes. At an early stage he reported his dissatisfaction with some lithographed plates, perhaps of ammonites, observing that they were "not so good" as he had imagined they would be and that "many of the outlines are not decided enough & rather scratchy." He continued: "I see lithography will not do for the very fine lines I want to express in a plate I am preparing & I find that as the subjects are light I can get them engraved on steel nearly as cheap as on stone." In addition, he was under the impression that lithography would not give the large number of impressions needed without redrawing. Steel engraving, however, may have proved more expensive than expected, for he soon afterward shifted his images of corals (plate 54) from lithographic stone to wood engraving.[63] The ten lithographic drawings that made it into the final volume were ones in which shading, more than fine lines, was called for: petrified fish intestines standing proud from a slab of slate, several sets of fossil footprints, and a couple of the ammonite plates. The pièce de résistance was Buckland's reconstruction of the skeleton of the ground sloth *Megatherium*, which loomed out of the page in a striking manner (fig. 3.8).

Such textured effects had previously been achieved using intaglio plates. Not only was lithography cheaper, however, but it offered Buckland another opportunity to draw on the services of an artist whom he could trust. George Johann Scharf had learned lithography at its source in Munich, and in the years following his arrival in London in 1816 he became a significant figure in the lithographic revolution there. Taken under the wing of leading lithographic printer Charles Hullmandel, he rapidly gained a reputation with the Geological, Zoological, and Linnean Societies as they adopted lithography in the production of their transactions. Buckland had early employed Scharf to draw original specimens for his *Reliquiæ diluvianæ* and to put Thomas Webster's cave illustrations on stone, and he continued to employ him regularly, including in the preparation of diagrams for teaching.[64]

63. WB to unknown, 11 [January] 1833, BO, Ms.Eng.Lett.d.5, fols. 202–3; WB to Charles Stokes, 7 April 1833, OUMNH–BP, Notes by Subject (Bridgewater Treatise) 3/2. Buckland considered lithography "very cheap, and artists very good" in Germany, but when the geologist Louis Agassiz used his own lithographic press to prepare an edition in Switzerland, it was a financial failure. WB to Featherstonehaugh, [25 April 1836], CUL, Add.Ms.7652, II.LL.32; Agassiz 1885, 281–87; Marcou 1896, 1:117–22.

64. See OUMNH–BP, esp. Large Drawings and Prints and D Series Drawings. That intaglio and lithography could produce very similar results is illustrated by the

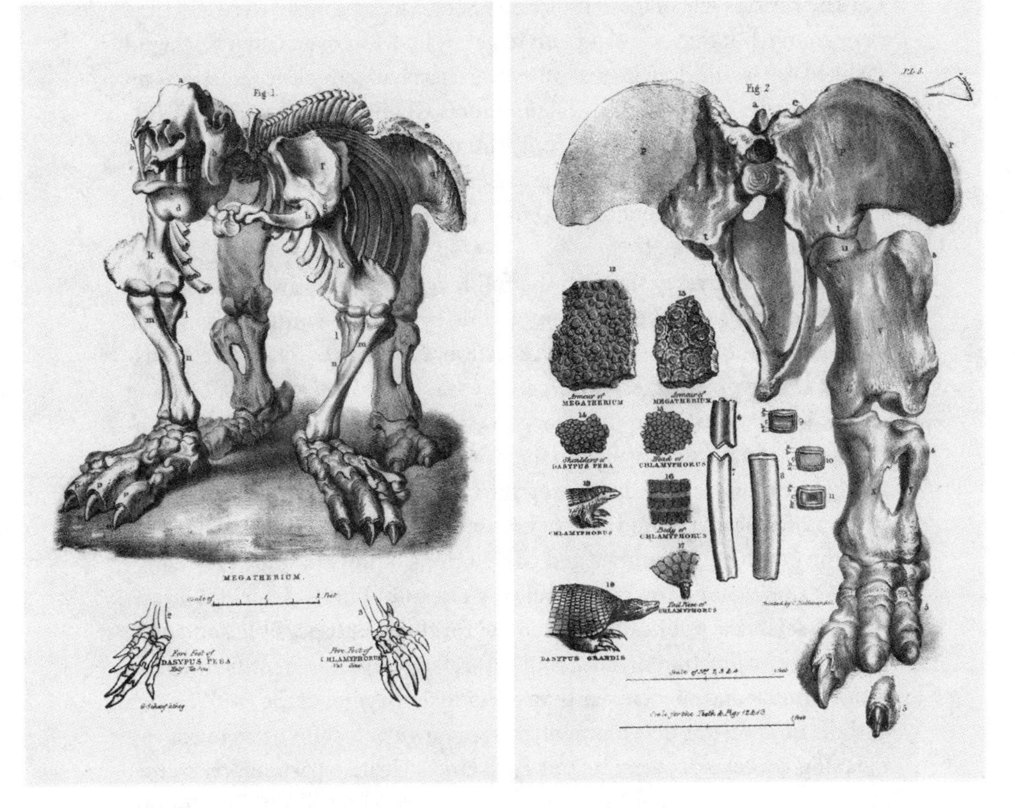


FIGURE 3.8 Plate 5 ("Megatherium"), from Buckland's Bridgewater Treatise (1836). Lithograph by George Scharf.

Scharf's surviving diary dramatizes the work involved. In the case of the iconic *Megatherium*, Buckland initially had Scharf redraw an evocative reconstruction of the famous Madrid specimen from Christian Pander and Eduard D'Alton's *Die vergleichende Osteologie* (1821–38) in February 1834. This took sixty-four hours and earned Scharf £5, although he calculated it should have been £6 8s at his hourly rate of 2s. Next, however, Buckland arranged for Scharf to stay with him in Oxford for two nights in November 1834 in order to draw the leg and pelvis from the plaster cast of Woodbine Parish's new specimen, which Buckland had commissioned for the Ashmolean. The initial drawing took fourteen hours, but back in London it took another twenty-five to draw it on stone, earning Scharf just short of

replacement between the first and second editions of a chalk lithograph for Plate 15′ (fish intestines) by a stipple engraving.

£4. The various sets of fossil footprints Scharf was assigned were less time consuming, but they were difficult to get right. Moreover, when Buckland disliked the initial drawings of one set, Scharf had to make corrections on his own time. Later, Buckland had Francis Chantrey oversee Scharf's attempt to accurately render the footprint of *Ornithnites diversus* for one of the foldout plates, although Chantrey reported that it would have been better done as a wood engraving, since its effect depended chiefly on "the true direction of *lines*."[65]

Buckland's odyssey through the full range of illustrative technologies clearly reveals both the new opportunities and the complex decisions faced by men of science at this transitional moment. They were accustomed to the practice, especially in the life and earth sciences, of using costly illustrations when necessary to enable readers to visualize the appearances of nature according to their own perceptions. But the new technologies that were rapidly coming into use meant not only that illustrations, now cheaper, could be used more widely than previously but also that the graphic possibilities and conventions of illustration were changing. As the useful knowledge society's cheap publications had demonstrated, scientific publications intended for the widest possible audiences could now be illustrated, rendering natural objects and scientific theories more familiar and spectacular than previously. Several of the Bridgewater authors built on this development, using a range of technologies to convey meaning and evoke pleasure, but Buckland's treatise particularly stands out. That a work so laden with apparently costly plates could be sold to many thousands of nonspecialists was a wonder of the age. Yet being innovative meant that Buckland had weighty challenges to face in navigating between the practical, graphic, and financial advantages and disadvantages of the different technologies.

In the face of all the challenges it is no surprise that the process of illustrating Buckland's Bridgewater significantly delayed its publication, so that Francis Chantrey feared the work would "be blasted by *mismanagement alone*." Buckland was not, moreover, the only one delayed by illustrations. Of those to include illustrations, only Bell—who was his own artist—was on time in delivering his manuscript. By the beginning of March 1833 five treatises were in press, including the first volume of Kirby's, but Pickering

65. Chantrey to WB, 15 June 1836, BO, Ms.Eng.Lett.b.35, fols. 18–19, 24; Scharf's Journal, 1833–35, NPG, NPG7/3/7/2/1; WB to Scharf, 12 November 1834, NPG, NPG7/3/7/2/2/3; D'Alton 1821, plate II; Gordon 1894, 128n–129n; Rudwick 2008, 433.

told Chalmers that he expected Prout, Roget, and Buckland to be "rather late." Soon afterward, Buckland wrote to Roget explaining that his treatise could not be published by the end of June, since more time was necessary in order to complete the engravings. Roget reported that, although his text was ready, he, too, was delayed by the need to have the engravings "properly executed," as well as by the "rapidly declining health" of his wife, now within days of her death. Kirby, whose second volume was far from ready, had, Roget reported, obtained an extension from Lord Farnborough, and Buckland proceeded to obtain an extension for Roget and himself through the president of the Royal Society. Despite these difficulties, all the Bridgewaters except Buckland's were either published or in press by the end of July 1833. Prout's and Roget's appeared early in the following year, and Kirby's appeared in June 1835.[66] All that remained was to sell them.

## *Selling Science*

In his *Reminiscences of a Literary Life* (1836), Thomas Dibdin pictured himself on an imaginary tour of literary London. Arriving in Chancery Lane, he exclaimed, "How does Mr. Pickering do this morning? And where are his *Caxtons* and *Wynkyns*, and *Pynsons*—his *Alduses*, *Elzevirs*, and *Michel Le Noirs*? But Mr. Pickering has a note of louder triumph to sound, in being publisher of the BRIDGEWATER TREATISES [ . . . ] which bid fair to traverse the whole civilized portion of the globe." Published in a form and at a price suitable only for the libraries of the wealthy and by a bookseller noted for his antiquarian stock and clientele, the Bridgewaters did not seem to be marked out for extensive sales. Until very recently, this had been the norm: most new books by reputable authors had been expensive and consequently limited in circulation. According to pioneering cheap publisher Charles Knight, the average price of a complete book in 1828 had been 16s, with an average of 12s 1d per volume. Such prices were prohibitive for the vast majority of readers. Even a "respectable" £10 householder, enfranchised by the 1832 Reform Bill, might easily have had a weekly income of as little as 48s, so that Buckland's Bridgewater costing 35s represented a vast sum and even the cheapest (9s 6d) more than a day's income. Yet the Bridgewaters nevertheless became one of the literary sensations of the

66. PMR to WB, 5 April and 11 May 1833, WB to Sussex [Draft], [April 1833], Children to WB, 2 and 16 May 1833, WB to Children [draft], 8 May [1833], BO, Ms.Eng.Lett.b.35, fols. 39–40, 42–48; Pickering to TC, 2 March and 22 July 1833, NC, CHA 4.212.15, 20; Chantry to WB, 17 June 1836, BO, Ms.Eng.Lett.b.35, fols. 20–21. For the progress of publication, see appendix B and Topham 1993, table 5.1.

decade and established for Pickering a rather more eminent reputation as a publisher than he would otherwise have had.[67]

Looking back in 1843, one of John Murray's obituarists considered that turning the Bridgewater series away had been one of only two lapses in judgement in the publisher's entire career. How did the series come to achieve such success against the expectations of one of the most astute literary judges of the age? In part, as we shall see, the answer lies in the variety of ways in which readers found them to be valuable. But the mechanisms by which they were distributed and sold within the book trade also had their part to play. As modern authors, publishers, and booksellers can testify, decisions about marketing and distribution, as well as the contingencies of passing events, can make the difference between literary triumph and disaster. Pickering had a strong incentive to make the Bridgewaters sell. Normally, he limited his financial risk as a publisher by restricting production. Two-thirds of the editions printed for him by Whittingham comprised fewer than a thousand copies.[68] In undertaking to print a thousand copies each of the treatises, Pickering exposed himself to significant jeopardy. Such an investment required a matching commitment to marketing through advertising and the presentation of copies to review journals.

In Pickering's favor was the fact that the Bridgewater bequest had been a matter of ongoing discussion in the public journals over the preceding four years. In March 1832, Murray had even given the series top billing in his latest eight-page catalog of "New works in the press, or preparing for publication." It was standard practice for such catalogs to be stitched into a publisher's existing publications, and their contents were rapidly picked up in the literary intelligence sections of several monthly magazines. A year later, Pickering also gave over the cover of his latest eight-page catalog to what he now for the first time called "The Bridgewater Treatises." He invested around £40 in having twenty thousand copies of the catalog printed and paid more to have many of these stitched into the quarterly numbers of the leading review journals, carrying it to some of the most regular book buyers in the country. The March catalog was apparently stitched into Murray's *Quarterly Review*, and the July catalog, also advertising the Bridgewaters, was stitched into the competing *Edinburgh Review*.[69]

67. Knight 1854, 261; Dibdin 1836, 904–5. See the sample domestic budgets in Peel 1934, 1:104–8, 126–34; see also Altick 1957, 276.

68. McDonnell 1983, 123; "Mr John Murray," *Athenæum*, 1 July 1843, 610–11.

69. Chiswick Press Debit and Credit Ledger A, BL, Add.Ms.41927, fol. 104; copy of account between Pickering and Roget, n.d., BO, Ms.Eng.Lett.b.35, fols. 69–70.

The series featured prominently in Pickering's list over succeeding years, but his advertisements also peppered the newspapers, weekly literary journals, and monthly magazines. The first advertisements for Whewell's Bridgewater appeared in the newspapers on 15 March 1833, promising that the remaining works in the series were in "great forwardness" and would soon appear. Over coming weeks and months, Pickering had frequent recourse to such advertisements. When the publisher made up the bill for the first edition of Roget's treatise, he charged £61 13s for advertising and for Roget's share in having 29,500 catalogs printed and inserted in reviews. This was a notably large sum. For instance, when Charles Knight had published Babbage's *On the Economy of Manufactures* the previous year, he reportedly spent £40 on advertising. A decade later, medical publisher John Churchill spent less than £25 in advertising the first edition of *Vestiges of the Natural History of Creation,* although its anonymous author was especially keen to invest in the work's promotion. Pickering, it seems, was taking no chances. Moreover, further advertisements were placed in local newspapers by booksellers seeking to cash in on the trade.[70] Then there were review copies. If Roget's treatise is anything to go by, Pickering was more restrained here, giving out only a dozen copies for review in the first instance. However, as reviews started to appear, this brought additional publicity, since the contents of the leading magazines and reviews were themselves advertised and reviewed.

Importantly, the individual Bridgewaters benefited from the effect of being part of a series, even if not one formally scheduled for periodic publication. Pickering had both Whewell and Kidd's treatises printed by the start of March 1833, but he decided to lead off with Whewell's, following it with Kidd's in April and with Chalmers and Bell's treatises as they became ready in May and June.[71] From the moment that Whewell's *Astronomy and General Physics* began to be advertised, all of the volumes were kept regularly in the public eye. It took more than three years for the last of the treatises to appear. Pickering advertised the appearance of Prout's Bridgewater in March 1834 and Roget's in May. In June 1835 came Kirby's and finally the much-hyped Buckland's in September 1836. In this way, the series made endless demands on public attention over the course of four successive London seasons, each contribution advertising all the others.

Pickering's engagement with the wholesale and retail trade produced rather less visible effects, but these complex commercial mechanisms were also critical to the success of the Bridgewaters. Within a month of

70. *Hull Packet,* 23 August 1833, 1c; Secord 2000, 128; Babbage 1832a, 167.
71. Pickering to TC, 2 March 1833, NC, CHA 4.212.15.

Whewell's treatise being advertised for sale, Pickering told the Edinburgh bookseller John Anderson that it had "very nearly sold off." Moreover, a "great proportion" of Kidd's treatise, which had not yet been advertised for sale, had also gone.[72] What lay behind these statements was that wholesalers and retailers in London and beyond had decided to purchase a significant number of copies from Pickering and offer them to the public. They now also had an interest in promoting the series, and their decision to offer copies for sale contributed to its success.

Securing the interest of the trade in a new work was of great importance to the entrepreneurial publishers of the early nineteenth century. The traditional booksellers around Paternoster Row often continued to take shares in editions of "trade" publications, dividing the copies between them. By contrast, fashionable West End publishers such as John Murray took on themselves the entire responsibility for selling their editions. Such publishers could not afford to wait for orders to come in from other booksellers in response to advertisements and reviews. Rather, before new works were published, they canvassed the leading London booksellers to "subscribe" for copies at a discounted price. Sometimes this took place at trade dinners—convivial occasions when publishers entertained an invited group of leading booksellers at a coffee house or tavern before seeking to enroll them as subscribers to their forthcoming works. On such occasions, the larger publishers were able to sell several thousand pounds' worth of stock, but even smaller publishers held such sales, sometimes in combination with each other. It was through these mechanisms that the increasingly large and specialized wholesaling booksellers, such as Simpkin and Marshall, exerted significant sway over the success of new works. As Pickering explained to Chalmers, "when the larger purchasers take a considerable number" of a new edition being offered for subscription, "all the others who have had numerous enquiries for it, follow like a flock of sheep & subscribe liberally." It was reported in 1819 that one desperate publisher had secured "ten times as many subscriptions among other booksellers" as he would otherwise have done by the devious act of changing Longmans subscription of "50" copies to read "500."[73]

It is unclear whether William Pickering offered the Bridgewaters for

72. J Anderson to TC, 12 April 1833, NC, CHA 4.198.31. The earliest advertisements for Kidd's treatise appeared in the *Standard*, 22 April 1833, 2a.

73. "Tales of My Landlord," *Literary Gazette*, 18 December 1819, 802–5, on 805; Jerdan 1852–53, 3:129–31; Pickering to TC, 3 January 1834, NC, CHA 4.227.52–53. On trade sales, see Rees 1896, 64–65, Shaylor 1912, 78, 145, 160, 247–67, Plant 1939, 404–5, Babbage 1832a, 262, and Babbage 1832b, xi.

subscription at a trade dinner or more informally. For one thing, the publishing season was almost upon him before he acquired the series. More seriously, however, he was in the midst of a major dispute with the leading London booksellers. Early in 1832, in a four-page document headed *Booksellers' Monopoly*, Pickering announced that he was being refused new books at the standard trade discount and that some booksellers were claiming that his own publications were "not out," "out of print," or "discontinued." The situation had arisen on suspicion that he had contravened a set of regulations drawn up in December 1829 by a committee of the most powerful London booksellers and publishers. Determined to protect the high price of standard "trade" works, the committee had required booksellers to respect advertised prices on pain of exclusion from the wholesale trade. Pickering was suspected of having supplied books to two London retailers who had been blacklisted for underselling, and his name was consequently circulated around the trade on a placard. To Pickering it seemed clear that it was his own perfectly legitimate publishing of cheap reprint editions in competition with standard trade editions that had brought him to the committee's notice. In retaliation he named the ten-man committee responsible (which included both Murray and Longmans) and threatened legal action, apparently without success. Only after he became publisher of the Bridgewaters did the committee relent, seeking to restore normal trading relations.[74]

Why was the trade so enthusiastic about the Bridgewaters before even a single copy had been sold? Murray and Longmans, of course, had detailed knowledge of what was planned. Perhaps, also, with all the advance publicity, the series was being asked for in bookshops. More fundamentally, however, commentators on all sides acknowledged that the publishing season had been shockingly thin. While the Reform Bill had finally passed the previous summer, it had been followed by the advent of the first Whig administration in a generation, and politics continued to dominate the public attention to the detriment of literature. In February a lead article in the weekly *Literary Gazette* complained that in such a climate, "capitalists" would not "speculate upon the higher classes of mental labour." Literature suffered as a result, and instead of teeming with "the best productions of talent, science, and genius," the journal now found its weekly list "generally stinted." Quickly reviewing the books received in the preceding eight

74. Barnes 1964, 1–18; Pickering's *Booksellers' Monopoly: Address to the Trade and the Public* (London, 1832) (reproduced in Pollard 1978, 43–47); Babbage 1832a, 264–65; *Retail Booksellers' and Bookbuyers' Advocate*, January 1837, 23 (quoted in Warrington 1987, 621).

days, the writer concluded, "it will, we think, be acknowledged that the publishing concerns of Great Britain are at a very low ebb."[75]

Other commentators agreed that the new books of the season amounted to a "ragged regiment." In such a market, the Bridgewaters had much to commend them. There was something of novelty and note about their being prompted by an earl's bequest, which could be expected to generate sales. Moreover, several of the authors were well known to the fashionable world, and the works of a Bell, a Buckland, or a Chalmers could be expected to attract notice. Whewell was especially pleased to see Chalmers's treatise out early, as he was "a man of great literary name" with a reputation that would "command readers." Buckland, however, considered the excellence of Whewell's Bridgewater—which he considered the best of the series—to have been important in giving the "whole series a good name." It was, he considered, lucky for some of the rest that it "came forth first into the world," a view that seems justified by the positive reviews that Whewell's treatise received in the early weeks.[76]

The judgement of the trade in favor of the Bridgewaters helped to foster a rapid retail sale, and Pickering soon had a runaway commercial success on his hands. New editions were rapidly required, and the publisher found himself ordering uncharacteristically large print runs. From the outset he had felt certain that Chalmers's reputation as a popular author meant that his treatise would "sell in much greater proportion than some of the others." He consequently printed a larger first edition of 1,500 copies. However, before Chalmers's treatise was published in May, the demand for Whewell's treatise had led him to put a second edition of 1,500 copies in the press. Since Bell's treatise had begun to go to press at the start of the year, it was again in an edition of a thousand copies, but by the time it was published in June, demand was such that it was oversubscribed by three hundred copies. Pickering now prepared a second edition of two thousand copies. Bell was delighted, telling Lord Brougham, "I am happy to say that before a copy is in the shops another edn is called for." Having the editions oversubscribed helped to generate demand. As Pickering explained to Chalmers, the "trick" was to "have the works '*well asked for*'" before committing to a reprint, so that wholesalers and retailers would

75. "Review of New Books: English Literature," *Literary Gazette*, 16 February 1833, 97–98.

76. WB to A Irvine, 25 February 1837, CC, Ms.531; WW to A Whewell, 21 April 1833, TC, Add.Ms.c.191[129]; "Our Weekly Gossip on Literature and Art," *Athenæum*, 23 February 1833, 122.

subscribe handsomely.[77] By September 1833 larger second editions of all four of the published treatises had appeared, but these also sold rapidly. At the start of 1834 Pickering published third editions of between 1,500 and three thousand copies.

Such figures were modest compared to the tens or even hundreds of thousands of copies sold of some of the cheap publications published since the advent of the "March of Mind" a few years earlier. Priced as the treatises were, however, their sale was spectacular. A couple of years previously, political economist John Ramsay McCulloch had reported that the average edition size for books stood at 750 copies. Even so, he continued, fifty out of 130 works recently published by "an extensive publishing concern in the metropolis" had taken a loss. Only thirteen had subsequently been reprinted, and most of the second editions had proved unprofitable. More generally, McCulloch estimated that a quarter of books took a loss, and only one in eight or ten reached a second edition. In bucking this trend, the Bridgewaters bear comparison with the "sensational" evolutionary work, *Vestiges of the Natural History of Creation*, a decade later. Priced at a more modest 7s 6d, *Vestiges* also seized the attention of wealthy and institutional book purchasers, becoming one of the books of the season. Comparably with the Bridgewaters, the four editions printed in the first critical year yielded 5,250 copies.[78]

For Pickering these increasingly large editions were unfamiliar territory. Whittingham only ever printed twelve editions of more than two thousand copies for Pickering, and seven of these were Bridgewaters. When Chalmers questioned his move to larger editions in November 1833, Pickering explained that he was "generally averse to printing large editions, unless the merits & sale of the works" warranted his doing so, but that it was "a waste to pay for composition twice over." He could, he pointed out, use variant title pages to suggest multiple editions from the same printing, but he did not think the public were concerned whether a book was in its third or its fourth edition. In any case, with all the volumes still to come, he remained confident about continued demand.[79]

77. Pickering to TC, 2 March and 22 June 1833, 3 January 1834, NC, CHA 4.227.15, 18–19, 52–53; CB to Brougham, [June 1833], UCL, Brougham Papers, 45606.

78. Secord 2000, 131; [McCulloch] 1831, 430, 432. Grant 1837, 1:134, reported that only one in fifteen books paid its expenses, one in two hundred reached a second edition, one in five hundred a third, and one in a thousand a fourth, although this estimate apparently included books published on commission, including many unsalable pamphlets.

79. Pickering to TC, 26 November 1833, NC, CHA 4.212.24; McDonnell 1983, 122.

His confidence was well placed. When Prout's and Roget's treatises were finally ready in the spring of 1834, Pickering printed editions of two thousand only to have to print a further three thousand immediately. A year later, with Kirby's treatise finally ready, he printed five thousand copies with variant first and second edition title pages. Kirby, however, denied having given Pickering permission to increase the edition from two thousand to five thousand copies and obtained an injunction to stop him from selling the work. His complaint was that "he had been deprived of the opportunity of correcting the new edition," which might consequently harm his reputation as a "man of science." In the end, the matter was resolved through arbitration, with Pickering agreeing to reissue the second edition with two pages of errata and six additional pages of notes. By the time the long-awaited geological Bridgewater was published in September 1837, Pickering could be sure of a vast sale. "From coming at the fag end," Buckland quipped, "mine had the advantage of making up every bodies set." This time, Pickering sold the first printing of five thousand copies as a first edition and, recognizing the public notoriety it had attained, immediately set about printing a further five thousand. A shocked Harriet Martineau noted in her diary that this was a "much greater sale than novels!" and that Buckland had probably "united the religious and scientific world" in his readership.[80]

For four successive publishing seasons, the Bridgewaters were thus prized commodities that attracted extensive sales among wealthy and institutional book purchasers. Indeed, other publishers eagerly used their celebrity as a means of selling their own publications. As early as April 1833, Edinburgh publishers Oliver and Boyd advertised *The Testimony of Nature and Revelation to the Being, Perfections, and Government of God* with a puff from the weekly *Spectator* claiming that it was "an attempt to do in one volume what the Bridgewater Treatises are to do in eight." Written by Henry Fergus, a minister in the Scottish Relief Church and director of the Dunfermline Mechanics Institute, its advantage was that it sold for "somewhere about half the price of one Bridgewater octavo." Other publishers sought to sell more substantial volumes as "uniform" with the series. Early in 1834 John Murray advertised *An Argument to Prove the Truth of the Christian Revelation* by the Earl of Rosse in this way. Longmans followed suit in January 1834, announcing that their *Introduction to the Study of Nature, Illustrative of the Attributes of the Almighty, as Displayed in the Creation* (written by physician and writer John Stevenson Bushnan)

80. Martineau 1877, 3:215; WB to A Irvine, 25 February 1837, CC, Ms.531; "Kirby v. Pickering," *Morning Post*, 19 December 1835, 4a, 13 January 1836, 4a.

would be published "uniform with the *Bridgewater Treatises*." Smith, Elder later did the same with William Sidney Gibson's *The Certainties of Geology* (1840), as did Pickering with George Crabbe's *Outline of a System of Natural Theology* (1840).[81]

Other publications appropriated the brand in the title. One of the first to do so was an anonymous *Of the Power, Wisdom, and Goodness of God, as Shewn in the Works of the Creation*, published in Worcester in July 1835. Similar phraseology had appeared in book titles over many years, but one reviewer felt that it now had more specific resonances, describing the work as "a 'Bridgewater Treatise' for the young and thoughtful inquirer." Pickering protested that the title interfered with his copyright and threatened legal action, and the publisher promptly withdrew the work from sale, reissuing it under the title *Popular Illustrations of Natural History, Exemplifying the Wisdom of the Deity, and His Beneficent Designs*. This did not prevent later appropriations, of which Babbage's *Ninth Bridgewater Treatise*, published in 1837 by John Murray, was only the most shameless. The following year, surgeon Charles Mountford Burnett's *The Power, Wisdom, and Goodness of God, as Displayed in the Animal Creation* borrowed the Bridgewater title and was also largely "made up of quotations from the Bridgewater Treatises and other similar productions."[82] In 1840 Longmans described the second edition of Frederick Collier Bakewell's *Natural Evidence of a Future Life* as "a contribution of natural theology designed as a sequel to the Bridgewater Treatises," and popular lecturer Daniel Mackintosh branded his *The Highest Generalizations in Geology and Astronomy, Viewed as Illustrating the Greatness of the Creator* (1843) as a "Supplement to the Bridgewater Treatises." Others were more diffident. It is poignant that, with the manuscript of his three-volume *Proofs and Illustrations of the Attributes of God from the Facts and Laws of the Physical Universe* in hand by the spring of 1830, geologist John MacCulloch felt he had to hold back its publication until after the Bridgewaters appeared, so that it consequently only appeared posthumously, in 1837.

Having for several years captured the public imagination, the initial market for the Bridgewaters was largely exhausted by the early 1840s.

81. *Literary Gazette*, 26 December 1840, 843; *Morning Chronicle*, 8 May 1840, 1g; "Literary Notices," *Magazine of Natural History* 7 (1834): 96.; *Caledonian Mercury*, 27 April 1833, 1d, quoting from "Books on the Table," *Spectator*, 20 April 1833, 360. Murray's advertisement is in my copy of Coleridge's *Aids to Reflection* (1843).

82. "Mr Burnett's *Power, &c. of God*," *British and Foreign Medical Review* 7 (1839): 227–28, on 227; "Critical Notices of New Publications," *Analyst* 2 (1835): 427–42, on 441–42; *Berrow's Worcester Journal*, 2 July 1835, 1d; 23 July 1835, 2c; and 6 August 1835, 1d.

At that point, Pickering might have been expected to have launched the cheaper editions previously discussed with the authors. However, the 1837 foolscap edition of Whewell's *Astronomy* had taken six years to sell, suggesting that the cheaper market was limited, perhaps because so many copies were already available in libraries. Only in the early 1850s did other publishers begin to issue the Bridgewaters at lower prices, and by that period they had come to be seen as "standard" works, as signified by the inclusion of many of them in reprint publisher Henry G. Bohn's Scientific Library.[83]

The spectacular sales of the Bridgewaters over the course of a decade earned a small fortune for their authors and publisher. Although full accounts have not survived, these early editions yielded many hundreds of pounds for each of their authors. The account for Roget's first edition showed an income to him of £337, that for Chalmers's first two amounted to £318. Geologist William Henry Fitton had heard that Buckland had earned £2,000 on each of his first two editions, although this seems more credible as a combined sum. Since Pickering was publishing on a half-profits contract, his profit on the series clearly ran to many thousands in total. Nevertheless, while the authors were well pleased with their publisher's effectiveness in producing and marketing an imposing product, they were less happy when it came to his settling accounts. Those who had published in 1833 were due to receive their first payments at the end of the year, but none were made. When Chalmers's first statement of account finally arrived in June 1834, he was shocked to find that it included a commission charge of 10 percent on the amount sold. Bell, who agreed with Chalmers that the charge was unprecedented, visited Pickering to remonstrate. The publisher was unable to substantiate his claim that it was in their agreement but insisted that he needed it to cover such incidental expenses as postage, package, bad debts, "Commission paid to traveller [ . . . ] Expenses of trade sales & many other incidentals."[84]

The matter remained unresolved for a further eighteen months. Chalmers voted with his feet by taking the next edition of his treatise to be printed and published by Collins, although they allowed Pickering to attach his own title page to the number of copies he required for customers wishing to make up sets. As other authors received their accounts, they

83. Appendix B; Chiswick Press Stock Book, BL, MS 41914, fol. 63.

84. Pickering to TC, 18 June 1834, CB to TC, 14 June [1834], G Chalmers to TC, [June 1834], NC, CHA 4.227.50–51, 4.218.43, 4.219.69; TC to WW, 4 March 1834, TC, Add.Mss.a.202[25]; Lyell 1881, 1:473; copy of account between Pickering and PMR, n.d., BO, Ms.Eng.Lett.b.35, fols. 69–70.

joined in the complaint, and Bell and Prout were reportedly "very angry" about Pickering's "unusual charge." Accordingly, several of the authors met at the library of the Royal Society in the first week of November 1835 with the purpose of "resisting Mr. Pickering's claim."[85] They determined to avail themselves of a clause in their agreement that allowed for any points of dispute to be referred to "two independent persons, one chosen by each party, with power to nominate a third, whose decision shall be binding." The authors nominated John Murray, who privately told Buckland that he was surprised that Pickering did not know "his own interest better than to have entered on, or persisted in such injudicious litigation." Pickering, who had previously heard Murray's unfavorable view of such charges, prevaricated before finally nominating the prominent wholesaler William Sherwood. These two publishers adjudicated that 5 and not 10 percent was a reasonable charge for "incidental expenses," and the matter was finally resolved.[86]

Pickering also had further sources of profit to exploit. The Bridgewaters had considerable success both in the United States and on the European continent. With the exception of Roget's, which took slightly longer, the series was republished within a year by the leading American publisher of British books, the Philadelphia publishing company of Carey, Lea and Blanchard. Yet as Buckland had explained to George Featherstonehaugh, republication of some of the illustrated treatises was likely to be ruinously expensive if the images had to be remade. Thus, in the cases of Roget's and Buckland's treatises, Pickering secured further income by arranging, through the publisher's agent John Miller, for casts of the engravings to be made and shipped out to Philadelphia. With no international copyright agreement, however, the authors could make no similar profits out of such activity. Not surprisingly, when the campaign for an international copyright agreement reached a peak in 1837, the Bridgewater authors were no-

85. WB to WW, 29 October 1835, TC, Add.Mss.a.66[34]; JK to WB, 3 November 1835, BO, Ms.Eng.Lett.b.35 fol.51; Collins to TC, 5 March, 13 April, 9 May, 20 June, 28 August, 27 October, 27 November 1835, NC, CHA 4.233.43–44, 47–48, 50–55, 58, 61–62; Chiswick Press Debit and Credit Ledger A, BL, Add.Ms.41927, fols. 137, 231, 268, 382, 439.

86. Murray to WB, 21 November 1835, John Murray Jr to WB, 18 January 1836, BO, Ms.Eng.Lett.b.35, fols. 55–56, 59–60; Memorandum of Agreement, TC, Add. Ms.a.210[194b]. See also PMR to WB, 17 November 1835, 4 January 1836, WB to Murray [copy], 17 November 1835, and Murray to WB, 15 February 1836, BO, Ms.Eng. Lett.b.35, fols. 52–54, 57–58, 61; Murray to Pickering [copy], 11 August 1834, NLS, MS.41910, fol. 152; and WB to Murray, 6 January 1836 and 16 February 1836, NLS, MS.40165, fols. 57–60.

table for their willingness to join in a petition to the US Congress, numbering seven of the fifty-seven signatories.[87]

❋

The initial distribution of the Bridgewaters was to a significant extent governed by the authors' determination both to produce substantial volumes that would dignify the noble patronage that had spawned them and to secure credit and a handsome profit for themselves. Finding themselves contracted to a publisher who was used to producing beautiful works for gentlemanly connoisseurs, they settled on a form of publication that confirmed the status of the series as something worthy of serious discussion in reviews and in fashionable society. However, while these were certainly books for the wealthy, Pickering's use of an innovative publisher's binding and the authors' use of innovative illustrative technologies meant that they were in some ways transitional at this moment of rapid change. In addition, two of the authors sought to extend the readership of their treatises by producing smaller and cheaper editions. Thus, in physical form, as well as in their contents, the Bridgewaters combined significant conservative elements with innovations characteristic of the industrial age. Moreover, the state of the trade, the public profile of the bequest, and the growing public demand for accessible, authoritative, and religiously orthodox works on the sciences meant that sales far outstripped what the authors or publisher might have expected of them.

To commentators of many different political and religious persuasions, however, the authors' decision to publish the series in such an expensive form seemed out of keeping with the spirit of the age, especially since many believed that poorer readers were precisely those who most needed such works of substantial and orthodox instruction. Some considered that a portion of the original bequest should have been used to subsidize the series, as had happened with the second edition of George Combe's phrenological treatise, the *Constitution of Man* (1828). Others considered that they should have been manufactured using the new techniques of mass production to reach a much larger readership, as also later happened with the *Constitution of Man*. Issued in a "People's Edition" costing just 1s 6d, Combe's work became one of the most widely read scientific works of the

87. "International Copyright Law," *Metropolitan Magazine* 18 (1837): 412–18; WB to Featherstonehaugh, [25 April 1836], CUL, Add.Mss.7652, II.LL.32; Barnes 1976, 376–77. On foreign editions of the Bridgewaters, see Topham 1993, 293–96; on the internationalization of copyright, see Seville 2006.

century, with 101,000 copies in print by 1865.[88] But such a scheme of radical cheapening was far from the minds of the Bridgewater authors or their publisher, preoccupied as they were with profits and prestige.

One consequence of the decisions of authors and publisher, however, was that other mechanisms of distribution became especially important. As demand for print ran ahead of the cheapening of books, new means were developed by which readers gained access to expensive books, and many of those who bemoaned the difficulties created for readers by the high price of the Bridgewater series set about rectifying the situation in various ways. Thus, the Bridgewaters were bought extensively for many of the proliferating libraries that were to be found in the towns and villages of Britain during the age of reform, ranging from church and chapel libraries through gentlemanly subscription libraries and reading clubs to the libraries of the newly widespread mechanics' institutes, catering to the petit bourgeoisie as well as aspirational workers. In such settings the high price of the Bridgewaters was a less significant barrier, and many poorer readers gained access to them and to other gentlemanly productions through such means. Probably even more significantly, however, the series was also widely discussed and excerpted in a wide range of periodicals and other publications, undergoing what James Secord has called a process of "literary replication." In this way the Bridgewater Treatises became household names far beyond the few thousand aristocrats, gentry, and professionals who were able to purchase them directly. In the process, however, they underwent significant reinterpretation and repackaging that was largely beyond the control of their authors or publisher.[89]

88. Combe 1866, viii; Gibbon 1878, 1:255–64; *Congregational Magazine* 13 (1837): 43n. See also Secord 2000, 69–73, and Van Wyhe 2004, 133–34.

89. Secord 2000, 126. A survey of the acquisition of the Bridgewaters by such libraries is given in Topham 1993, 297–312. See also Topham 1992.

✵ CHAPTER 4 ✵

# *Science Serialized*

*Among the numerous advantages for the acquisition of knowledge which the present age possesses, none appears more conspicuous than that derived from the great diffusion of periodical literature. The press, daily, weekly, monthly, and yearly, pours out a flood.*

Mirror of Literature, *January 1834*[1]

The physical form of books like the Bridgewater Treatises had significant consequences both for the meanings that they bore and the ways in which they became available to and were used by different classes of readers. Yet one of the most important consequences of the vast expansion of periodical literature that took place in the age of reform was the extent to which it made information about and materials from such expensive books available to a far wider range of readers. Often, periodical reviews, notices, excerpts, and advertisements prompted their readers to obtain copies of the Bridgewaters by purchase or loan. More often still, they served as proxies for the books, interpreting and transforming the originals. In an age dominated by the periodical press, the meaning of the celebrated Bridgewaters was significantly shaped by the experiences readers had of their representations in newspapers, magazines, and reviews.

The extent to which periodicals were coming to dominate literary production was a frequent subject of comment in the 1830s. When in February 1834 Charles Knight offered an estimate of the financial returns of the book and periodical trades for the preceding year, he noted that books only accounted for just over a quarter of the total of £2.4 million. A similar proportion came from periodicals and serials, including twenty-one respectable weeklies (selling six hundred thousand copies, including in monthly parts), 208 monthlies, thirty-five quarterlies, and ten annuals,

1. I. J., "Periodical Literature," *Mirror of Literature*, 18 January 1834, 35–36, on 35.

while most of the remainder was accounted for by the 577,000 copies of English newspapers sold each week. Knight acknowledged that there had been a notable increase in the number of periodical titles in the preceding couple of years, but he saw this as part of a general pattern, pointing to a trebling since the start of the century. In that time, moreover, periodicals had diversified almost beyond recognition, now representing a bewildering array of formats, prices, and intended audiences.[2]

Many of these new periodical forms depended heavily for their content on other publications and especially on the new books of each passing season. Some offered lengthy and commanding essay-reviews that were connected to the original publication only in a tangential way. Others provided synoptic reviews that consisted of excerpted passages strung together with a minimum of critical comment. Still others turned extracts into stand-alone articles. Sometimes books became a subject for news commentary or a matter of controversy among a periodical's correspondents. And on other occasions the periodicals drew on each other so that a writer might not even have seen the original publication before committing some account of it to print. In taking hold of the Bridgewaters, the myriad periodicals of the 1830s thus reflected a wide range of purposes, priorities, and practices. Across this array of periodicals, the Bridgewaters were re-presented as through a kaleidoscope, coming to the attention of many readers who would never open their pages but who would nonetheless claim familiarity with them (table 4.1).

The series was reviewed extensively in the literary press over the first few months, starting with the weeklies and soon extending to the monthlies and quarterlies. The chapter begins by exploring how, for all their differences, these periodicals exhibited a striking degree of agreement in suggesting that the primary attraction of the Bridgewaters was not that they contributed to natural theology but, rather, that they offered a vision of the sciences that sat comfortably with Christianity. In the following two sections, the chapter examines the extensive coverage of the series in the religious and the scientific press. Here again, in a huge range of periodicals, readers were advised that natural theology was potentially problematic but that the religious safety of the Bridgewaters' authoritative science could be made useful to a wide variety of readers, whether in sustaining religious feelings and habits, in rendering scientific education religiously safe, or in encouraging scientific readers to reflect on religious questions. Not that there was an entire absence of dissenting voices. As the last part

2. [Knight] 1834, 5. On the expanding periodical market, see also Cantor et al. 2004, 7–10.

of the chapter shows, these were especially to be heard in the storm of press coverage generated by the publication of Buckland's much-delayed Bridgewater. Yet the overall effect of the controversy was again to convey an impression that Buckland's conciliatory view should be read as the new orthodoxy.

## *Safe Science as Literary News*

In a season short on literary news, the refreshing novelty of the first Bridgewaters represented an attractive opportunity for Britain's literary periodicals over the weeks and months that followed their publication in the spring and summer of 1833. Commissioned from some of the nation's best-regarded men of science at the behest of the highest dignitaries in the church and in science, they were nevertheless offered as accessible reflections on the wider implications of the sciences. As such, they were fit meat for most of the literary weeklies, monthlies, and quarterlies even if they were a little too soberly scientific for the most fashionable of the monthlies. Moreover, the impression conveyed by the bulk of these early reviews was generally positive. The overwhelming message to those keeping up with the books of the season was that these were works that authoritatively but appealingly confirmed the religious safety of the natural sciences.

Of course the relationship between periodical reviews and readers is a complex one. For the habitués of London's fashionable and literary circles, perceptions of new publications were to a significant extent shaped by polite conversation. Even for such readers, however, published accounts in periodicals played an important role. Like the "literature of conduct"—conduct manuals and etiquette guides—they helped men and women find appropriate ways of incorporating new publications into their social intercourse. With the publication of the first four of the Bridgewaters falling in the spring and early summer, many of the reviews that promptly appeared in the weeklies and monthlies did so in good time for the very height of the London social season.[3] For those who were geographically or socially excluded from London's fashionable society, periodicals were arguably all the more important, offering a sense of connection with Britain's cultural elite that was pertinent in drawing rooms and public spaces the length and breadth of the country.

It helped the cause of the Bridgewaters that the first reviews appeared in the weekly literary journals that had, over the course of fifteen years,

3. Secord 2000, 164.

TABLE 4.1 Timeline of periodical reviews (excluding newspapers)

*1833*

**Bridgewaters Published:** W (March), K (April), C (May), Be (June)

**Other Works Published:** George Fairholme's *General View of the Geology of Scripture* (March), Henry Fergus's *Testimony of Nature and Revelation* (April), Frederick Nolan's *Analogy of Revelation and Science* (November)

**March:** *Athenæum* (W)

**April:** *Athenæum* (Kd), *Spectator* (W, Kd)

**May:** *Brit. Mag.* (W), *Christ. Rememb.* (W), *Lit. Gaz.* (W)

**June:** *Athenæum* (C), *Christ. Rememb.* (Kd), *Examiner* (W), *Gent's Mag.* (W), *Lit. Gaz.* (Kd, C), *Metropolitan Mag.* (W), *Mon. Mag.*

**July:** *Athenæum* (Be), *Brit. Critic* (W), *Christ. Rememb.* (C), *Fraser's Mag.* (C/Kd/W), *Gent's Mag.* (Kd), *Ladies' Cab.* (C), *Mon. Rev.* (C)

**August:** *Brit. Mag.* (C/Be), *Christ. Rememb.* (Be), *Gent's Mag.* (C), *Lit. Gaz.* (Be), *Mon. Rev.* (Kd, W)

**September:** *Fraser's Mag.* (C/Be), *Ladies' Cab.* (Be), *Lond. Med. & Surg. J.* (Be), *Spectator* (Be)

**October:** *Brit. Critic* (C), *Christ. Exam.* (C), *Edinb. New Phil. J.*, *Lancet* (Be), *Quart. Rev.* (W/Kd/C/Be)

**November:** *Christ. Exam.* (C), *Edinb. Christ. Instr.* (C), *Lond. Med. Gaz.* (Be), *Mon. Rev.* (Be)

**December:** *Phrenological J.* (C)

*1834*

**Bridgewaters Published:** P (March), R (May)

**January:** *Brit. Mag.* (Kd), *Edinb. Rev.* (W), *Med. Quart. Rev.* (Be), *Westminster Rev.* (C)

**February:** *Christ. Reformer* (C), *Gent's Mag.* (Be), *Lancet* (Kd)

**March:** *Christ. Reformer* (C/Kd), *Lancet* (Kd), *Lond. Med. Gaz.* (P), *Presb. Rev.* (C), *Lit. Gaz.* (P)

**April:** *Mo. Rev.* (P)

**May:** *Athenæum* (P), *Christ. Reformer* (W)

**June:** *Spectator* (R)

**July:** *Christ. Rem.* (P), *Presb. Rev.* (Kd), *Med. Quart. Rev.* (P), *Athenæum* (R), *Lit. Gaz.* (R), *Lond. and Edinb. Phil. Mag.* (P)

**August:** *Metropolitan Mag.* (Kd), *Christ. Reformer* (Be), *Mechanics' Mag.* (Be)

**September:** *Brit. Mag.* (P), *Dublin Univ. Rev.* (P), *Mechanics' Mag.* (series), *Presb. Rev.* (W)

**October:** *Metropolitan Mag.* (R), *Mon. Rev.* (R), *Edinb. Rev.* (R)

**November:** *Presb. Rev.* (P), *Mag. of Bot. and Gard.* (R), *Veterinarian* (R)

**December:** *Brit. Mag.* (R), *Christ. Reformer* (P), *Christ. Rememb.* (R), *Gent's Mag.* (P)

*1835*

**Bridgewaters Published:** Ky (June)

**Other Works Published:** Henry Brougham's *Discourse of Natural Theology* (May)

**January:** *Edinb. Med. & Surg. J.* (R)

**March:** *Presb. Rev.* (Be)

**April:** *Veterinarian* (P)

**May:** *Veterinarian* (R)

**July:** Lit. Gaz. (Ky), *Spectator* (Ky), *Veterinarian* (Be)

**August:** *Athenæum* (Ky), *Christ. Teacher* (Ky), *Mag. of Nat. Hist.* (Ky), *Veterinarian* (Be)

**September:** *Brit. Mag.* (Ky), *Gent's Mag.* (Ky), *Mon. Rev.* (Ky), *Sunday Sch. Teacher's Mag.* (C)

**October:** *Edinb. J. of Nat. Hist.* (Ky), *Ent. Mag.* (Ky), *Fraser's Mag.* (Ky/R), *Lancet* (Ky), *Medico-Chir. Rev.* (Ky), *Sunday Sch. Teacher's Mag.* (C), *Veterinarian* (Ky)

**November:** *Lond. Med. Gaz.* (Ky)

**December:** *Christ. Rememb.* (Ky)

*1836*

**Bridgewaters Published:** Bu (September)

**Other Works Published:** Henry Duncan's *Sacred Philosophy of the Seasons* (October)

**January:** *Medico-Chir. Rev.* (Ky), *Presb. Rev.* (Ky)

**April:** *Medico-Chir. Rev.* (Ky, P), *Quart. Rev.* (Bu), *Sunday Sch. Teachers' Mag.* (W)

**May:** *Presb. Rev.* (R), *Sunday Sch. Teachers' Mag.* (W)

**June:** *Sunday Sch. Teachers' Mag.* (W)

**July:** *Medico-Chir. Rev.* (P)

**October:** *Spectator* (Bu), *Brit. Critic* (Bu)

**November:** *Scottish Mon. Mag.* (Bu), *Christ. Teacher* (Bu), *Brit. Mag.* (Bu), *Mag. of Pop. Sci.* (Bu), *Mon. Rev.* (Bu), *Intellectual Repository* (Bu), *Lit. Gaz.* (Bu)

**December:** *Brit. Mag.* (Bu), *Dublin Univ. Mag.* (Bu)

*1837*

**Other Works Published:** John Macculloch's *Proofs and Illustrations of the Attributes of God* (April), Charles Babbage's *Ninth Bridgewater Treatise* (May)

**January:** *Christ. Rememb.* (Bu), *Cong. Mag.* (Bu), *Eclectic Rev.* (Bu), *Lit. Gaz.* (Bu), *Presb. Rev.* (Bu), *Sunday Sch. Teachers' Mag.* (Ky)

**February:** *Athenæum* (Bu), *Christ. Rememb.* (Bu), *Gent's Mag.* (Bu), *Mining Rev.* (Bu), *Sunday Sch. Teachers' Mag.* (Ky), *Veterinarian* (Bu)

**March:** *Christ. Rememb.* (Bu), *Sunday Sch. Teachers' Mag.* (Ky)

**April:** *Christ. Rememb.* (Bu), *Edinb. Rev.* (Bu), *Mon. Repository* (Bu)

**May:** *Christ. Rememb.* (Bu)

**October:** *Christ. Reformer* (Bu), *C. of E. Quart. Rev.* (Bu)

**December:** *Fraser's Mag.* (Bu)

*Note:* W = Whewell, Kd = Kidd, C = Chalmers, Be = Bell, P = Prout, R = Roget, Ky = Kirby, Bu = Buckland. A comma between abbreviations indicates separate reviews; slashes between abbreviations indicate all in the same review.

# THE ATHENÆUM

Journal of English and Foreign Literature, Science, and the Fine Arts.

No. 282. LONDON, SATURDAY, MARCH 23, 1833. PRICE FOURPENCE.

☞ This Journal is published every Saturday Morning, and is received, by the early Coaches, at Birmingham, Manchester, Liverpool, Dublin, Glasgow, Edinburgh, and all other large Towns; but for the convenience of persons residing in remote places, or abroad, the weekly numbers are issued in Monthly Parts, stitched in a wrapper, and forwarded with the Magazines to all parts of the World.

## REVIEWS

*The Inferno of Dante.* Translated by Ichabod Charles Wright, M.A. London: Longman & Co.

The business of translation demands, in different cases, a very different order of mind, inasmuch as to transfuse the spirit of an author, is a more noble task than to convey his meaning; though with some writers the letter is everything, and the spirit nothing. Homer and Virgil afford instances in point, from the highest class of poets, in their respective styles. The true translator of Homer must be altogether a different intellectual being from the most accomplished translator of the Roman bard; while Horace, whose light, but sweet and tender spirit, seems always impatient of the weight of words, demands a translator whose mind can make its way through the intricacies of poetry, with a clue of phrases, thin and silken as his own. It is the same with the poets of modern times. Redolent of beauty as are Tasso and Ariosto, —pure, acute, and elegant, as the minds must be who can transplant their graces, and give them a new life in a foreign soil, the translator of Dante requires a bolder mental constitution, a stronger vision—a more intense and deeper perception of elemental poetry. We have, accordingly, looked with much interest, not unmixed with wonder, at the work before us; and this feeling was not a little increased by the circumstances under which it appeared. The intellectual habits of the day are not calculated to foster minds either capable of producing, or fitted to enjoy such productions. Dante, of all poets, is the most opposite to the conventional tastes of the nineteenth century. He lived, it is true, in an age of strong excitement, and so do we: he nourished and expressed the bitterest hate that could be generated in the strife of parties, and there is no want of such feeling in this our day: he drew with a broad, strong pencil, gave huge and distinct forms to his groups, and made heaven and hell speak the language of earth; and all this is conformable with the wishes and appetite of our own time;—but the excitement of Dante's age was the fierceness of a giant in his youth; its politics were the direct calculations of power and freedom; its utilitarianism, the abstractions of poets and patriots. When Dante described in colours that flash and glare upon the imagination, and gave a tongue to the darkness of the abyss, he did not sacrifice the mystery to make the truth palpable, for he doubted not the ability of the muse to make visible, things invisible. He spoke, therefore, with the voice of one having authority; and the moral obedience then accorded to the poet, directed him to employ the power of his vast genius on themes which might influence the destinies of his race. But this reverence for the inspirations of poetry no longer exists; and great, consequently, as the appetite may be for whatever rouses and excites, it is not for what the poet may utter in his enthusiasm;—that which, at one time, would have bowed the hearts of thousands, can now scarcely command the attention of a few solitary readers; because the poet has no longer the office of either a prophet or a teacher.

This was the first thing which made us look with surprise at the new translation of Dante: then came that derived from the nature of the task itself. It should be observed, in estimating the difficulties which a translator of Dante has to encounter, that there is one peculiar in his author—the perfect originality of his poetical diction and phraseology. He lived in the very dawn of Italian poetry, and was himself almost its father. There are, consequently, in his verses, none of the common-place, universal sentences of other writers—none of those phrases which have since become the property of poets in every country, and for which there are well known and acknowledged equivalents in every language. In Dante the metal is sculptured by the poet's own hand, and each line is sharp and distinct. The running into moulds was unknown in his day. It is evident from hence that the translator of Dante has none of those helps which the conventional language of poetry offers to others. He has to grapple with a stern and almost primitive phraseology; to exercise the highest powers he possesses in simple interpretation; and to be forcible by being literally enact.

In the next place, there are difficulties peculiar to the matter of the poem. There is no real or immediate connexion between the different parts of it: they hang together solely by the continuous exercise of an intense creative power, the counterpart of which must be possessed by the translator, or the Divina Commedia will present little more than an unintelligible series of loose exaggerations.

When we look at the versification, another difficulty appears, and one which, on the simplest consideration of the subject, is seen to be of no ordinary kind. Italian verse cannot be translated into English verse, so to speak—that is, be rendered into similar musical cadences: it can only be imitated, and the imitation, as such, must in many cases be a very faint one. So far as versification, therefore, is concerned, the translator is left almost to his own will, having nothing to guide him but the bare form of the metre, or stanza. To this, except in some few instances, English verse can scarcely be made to submit; and he has, consequently, to choose out of the various stanzas and measures of his native language, the one which his ear alone tells him affords the nearest approximation to the mellifluous flow of the Italian. In the case of Dante, this is a most important consideration. His versification is singularly adapted to his style and subject. After centuries of study and experiment, the Tuscan muse still found it the best medium for the conveyance of her truest and loftiest inspirations. Monti escaped the enervating trammels of his age, immortalized his name, and recalled to memory the brightest period of his art, by imbuing himself with the spirit of Dante, and speaking as Dante spoke. It should not be lost sight of, that the language of the Commedia is essentially dramatic, and the terza rima affords by its construction the utmost facility for dramatic dialogue. Its frequent and complete pauses, together with the provision it makes for recurring to, and keeping up, the sense of successive verses, are its most remarkable properties in this respect; and it is easily seen that a translator must find something to answer them in the structure of his own stanza, or fail in an important part of his object.

These are some of the chief difficulties with which Mr. Wright had to contend, when he boldly undertook to give another version of the 'Inferno.' Of the manner in which he has executed his task, we shall now endeavour to give our readers some idea. A close adherence to the original, we have stated to be the first quality requisite in a translator of Dante; but an adherence to the letter and spirit of the original is not to be accomplished by the substitution of words which answer, but of words which are equivalent, to those of the poet. This the English translator, if he possess the proper qualities of his office, need not fear of accomplishing. His language will, with scarcely any exception, furnish him with equivalents to the Italian; and Mr. Wright merits the no ordinary praise of having availed himself of this noble advantage presented by our native language, the richness, power, and flexibility of which are, on the other hand, proved by the excellence of his translation. We have followed him through several passages line by line, and find him as exact as the most scrupulous admirers of the great Florentine could desire. The following extract will confirm our praise to those who can compare it with the original, while the rest of our readers can scarcely fail of being struck with the fine and earnest spirit of the English. The first passage we select is from the eighth canto, in which the poet describes his passage with Virgil over the infernal lake, and their arrival at the gloomy city of Dis.

Then smite mine ear a loud and shrill lament,
Whereat I stretch'd mine eye to whence it came.
"Behold, my son," to me the master cried,
"We now draw near the city named of Dis,
Where crowds of wretched citizens reside."
"Master," quoth I, "already I discern
Its bright vermillion mosques in the abyss,
Which, as in furnace heated, seem to burn."
"The fire," he said, "that glows eternally
Within the walls, that ruddy hue supplies,
Which in the infernal valley thou may'st see."
Then we arrived within the trench profound
That compasseth this wretched land of sighs;
And framed of iron seem'd the walls around.
A tedious circuit made, at last we came
Where, "Lo the entrance—quit ye now the boat,"—
We heard the pilot's thundering voice exclaim.
More than a thousand on the gates I spied,
Rain'd down from heaven;—disdainfully they shout:
"Say who is this, that (death's dread power untried)

FIGURE 4.1 *Athenæum*, 23 March 1833. Image: reproduced with the permission of Special Collections, Leeds University Library.

become the primary source of literary news. These journals sold just a few thousand copies in total, but they were widely purchased by libraries and reading rooms and were extensively read. Moreover, while differing in politics, they shared a scientific ethos. The Tory *Literary Gazette* and Whig *Athenæum* (fig. 4.1) both billed themselves as journals of "literature, science, and the fine arts," providing extensive coverage of the meetings of scientific societies and often relying on men of science for their reports.[4] In addition, the reformist *Spectator*, which was rather more like a weekly newspaper, nevertheless billed itself as "a weekly journal of news, politics, literature and science," devoting several pages each week to book reviews. The general outlook of these journals was thus comfortably in keeping with new books designed to establish the religious safety of the sciences before a pious public, and they all promptly welcomed the individual Bridgewaters in turn, offering brief characterizations and sometimes extracts, albeit with some criticisms of the management of the bequest.

The review of Whewell's *Astronomy and General Physics* that appeared in the *Athenæum* just a week after its publication in March 1833 was penned anonymously by the Irish writer and journalist William Cooke Taylor, who was a keen statistician and supporter of the British Association. Taylor was one of the journal's regular reviewers, and it was only to be expected that in a Whig journal during the age of reform he would begin by criticizing the Bridgewater bequest and its handling. Having got that off his chest, however, he moved on to praise Whewell's contribution in high terms as "learned, eloquent, and convincing; the matter well arranged, the arguments logically disposed, the inferences fully justified by the premises, the style at once simple and elegant." Whewell might have had his "conclusions supplied by authority," but the *Athenæum* would have "confidently recommended" the treatise to the young "as a safe guide to science, and as worthily inculcating [ . . . ] sublime truths" if only it had been published in a suitable format and at a suitable price.[5]

This was a reasonable start, and the *Athenæum*'s comments were soon echoed elsewhere. Reviewing Whewell's treatise over three columns in mid-April, the *Spectator* also began by describing and questioning the handling of the bequest while pointing out that the authors were of unimpeachable ability. Whewell probably had the "most universal mind" that could have been found, and he had produced a masterful work that was "deeply interesting" to ordinary readers. In regard to the "grand object" of the treatise, the reviewer took his cue from Whewell. The evidence of design was obvious to everyone, and "in this view the Bridgewater Treatises may be held as unnecessary." However, it was "important to be shown that the extensive modern discoveries of science—the nearer view our philosophers had got into its arcana—did not militate against the conclusions

4. Holland and Miller 1997; James 2004. On the *Literary Gazette*, see Jerdan 1852–53; on the *Athenæum*, see Marchand 1941.

5. *Athenæum*, 23 March 1833, 184. The early reviews are listed with full bibliographical details in appendix C; see also table 4.1.

obtained in former times." The "still more important view" that might be taken of the series, the reviewer continued, was not "the light they cast upon Natural Theology—which, truly, wants none—but the light they reflect on natural science." Whewell's Bridgewater conveyed a vision of science that would cause the student to "abjure the grovelling intrigues, the selfish passions, the degrading influences of the world, and dedicate himself [ . . . ] to the ennobling pursuit of knowledge."[6]

The Bridgewaters were not, these early reviewers of Whewell informed their readers, primarily works of theology. Rather, they were scientific works written by leading scientific men in a manner that was accessible to ordinary readers and in such a way as to show that the findings of modern science were not at odds with accepted religious truths. They provided, in the *Athenæum*'s terms, an authoritative and "safe guide to science."

From this point of view, it was as well that Whewell's Bridgewater appeared first. Before April was out, both the *Spectator* and *Athenæum* had reviewed Kidd's treatise and found it wanting. It was "vague and prolix," reported the *Spectator* in a brief notice, and "but a moderate thousand pounds worth." The *Athenæum*'s reviewer—a young Irish surgeon, Perceval Lord, whose *Popular Physiology* was published by the Society for Promoting Christian Knowledge the following year—had much more to say about the "feelings of unpleasant surprise" Kidd's work elicited. Kidd's name had led him to expect "solid information delivered in a clear and agreeable form," but instead the treatise meandered away from its subject, introducing incorrect facts and inconclusive arguments. Over four columns, he criticized dubious claims that might have been forgivable in a "merely speculative and popular writer" but were not in a reputable "man of science." The point was well made by contrasting Kidd's inaccurate generalizations with the magisterial account of the ground sloth given by Buckland at the climax of the previous year's British Association meeting in Oxford. Despite his standing, Kidd's work failed to meet the high standards set by leading men of science.[7]

The rapid succession of reviews did not, however, allow this negative impression to stay long in the mind. In May, the *Literary Gazette* echoed its competitors' high praise of Whewell's treatise and, after strongly criticizing Kidd's Bridgewater at the start of June, it greeted Chalmers's work with superlative praise. This was a work of originality and profundity, the reader was informed, worthy of lengthy extracts. The *Athenæum* agreed.

6. *Spectator*, 13 April 1833, 331–32.
7. *Athenæum*, 20 April 1833, 248; *Spectator*, 20 April 1833, 360.

No single work since Joseph Butler's *Analogy of Religion* had displayed "more profound philosophy, clearer and more cogent reasoning, or a larger share of the pure 'religion of the heart.'" It was the work of a "master mind," Taylor reported, before providing a full synopsis of its arguments. Bell's treatise attracted more moderate praise from Lord: the Scot's logic and organization were weak but his detailed expositions were generally good and worthy of extensive quotation. By the end of the summer, the *Literary Gazette* could summarize the situation half way through the "Bridgewater octave of authors." Each had a different tone but they were "on the whole harmonious" even if, "like the notes of the octave when not properly tuned," they ran into each other.[8]

These first reviews were none of them very long, but they served to build up a picture of the Bridgewaters as a reassuring manifestation of the religious tendency of the latest scientific findings. Having made scientific news a core part of their journalistic format, both the *Athenæum* and the *Literary Gazette* were eager to support this conciliatory view. It was in keeping with their extensive coverage of the British Association meetings, and when the *Literary Gazette*'s report on Whewell's address at the start of the Cambridge meeting in June referred to his "exalted reputation," a footnote referred the reader to the journal's review of his Bridgewater in confirmation.[9]

This connection was made more apparent following the publication of Frederick Nolan's Bampton Lectures at the start of November 1833. Nolan's intemperate attack on what he considered the irreligious tendency of modern science and the British Association provoked a leading review in the *Literary Gazette* that ran over two issues. While "controversial divinity" was a "proscribed topic" in the journal, it could not stand aside while scientific studies and institutions were attacked "under an ostentations and mistaken zeal for religion." Noting that Nolan had himself been one of the members of the British Association the previous year in Oxford, the reviewer allowed himself to speculate about his apparent change of heart. Since the only works relating to science and religion published by members of the association in the interim were the first four Bridgewaters, these—and especially that by the incoming secretary, Whewell—must be the cause of Nolan's ire. However, "ordinary minds" had only found in Whewell's book "reason to admire that union of manly piety and

8. *Literary Gazette*, 17 August 1833, 519; *Athenæum*, 22 June 1833, 396.

9. "The British Association: Journal of a Week at Cambridge," *Literary Gazette*, 6 July 1833, 417 and 417n.

of comprehensive views of science, so well calculated [ . . . ] to conciliate the interests of philosophy and religion." The Bridgewaters were perfectly calculated to reassure those, perhaps alarmed by Nolan, who feared the religious consequences of the British Association.[10]

❋

Long before this article appeared in November, other periodicals had also begun to review the Bridgewaters in earnest. Most literary monthlies no longer viewed themselves as magazines (literally "storehouses") of literary and scientific information. Instead, they offered altogether more spirited fare in the shape of essays, sketches, jeux d'esprit, and poems, generally penned by well-paid authors of literary reputation or even notoriety. As the *New Monthly Magazine* observed in 1820, these new-style literary magazines were "full of wit, satire, and pungent remark—touching familiarly on the profoundest questions of philosophy as on the lightest varieties of manners—sometimes overthrowing a system with a joke, and destroying a reputation in the best humour in the world."[11] Only the venerable patriarchs of the monthly press in Britain—the centenarian *Gentleman's Magazine* and the octogenarian *Monthly Review*—continued to be produced on the older model, and these two soon began to issue synoptic reviews of the Bridgewaters. Over the space of four years, their fifteen reviews filled more than 150 pages of text, welcoming the Bridgewaters as generally authoritative works that amply manifested the religious safety of modern science.

Again, Whewell smoothed the way. At the start of June the *Gentleman's Magazine*, which was particularly associated with the antiquarian interests of the country gentry and clergy, gave publicity to Whewell's stated object of showing the harmony of philosophical and religious truths. The politically liberal *Monthly Review* was slower off the mark and reviewed the more celebrated Chalmers first, but, while it was critical of the handling of the bequest and of Chalmers's redefinition of his subject, it warmly commended the volumes as containing a lifelong "fund of materials for solemn and valuable meditation." The following month, its reviews contrasted the "exalted rank" of Whewell's treatise with the "comparative failure" of the "eccentric professor" Kidd. Despite criticisms, however, the general effect of the *Monthly*'s growing digest of the series, with its lengthy specimen passages, was to suggest that it offered an account of natural philos-

10. [Daubeny] 1833, 769. See also Nolan 1833b, 1834, and [Daubeny] 1834a.
11. [Talfourd] 1820, 309; reprinted in Talfourd 1842.

ophy "in her proper colour and shape, as the hand-maiden of truth, and akin to revealed religion."[12]

These two old gentlemen of the magazine market had loyal if modest readerships, and their accounts of the Bridgewaters as confirming the religious character of the sciences boded well with more old-fashioned readers. The Bridgewater authors had engaged a publisher who could produce books suitable for such readers, and it is no coincidence that Pickering purchased a share in the *Gentleman's Magazine* in the course of 1833, becoming copublisher. The magazine's anonymous reviews of the series were written by Pickering's collaborator and new editor, the Suffolk clergyman John Mitford, and, while they were far from being shameless puffs, they were well directed to securing the reputation of the Bridgewaters.[13]

Systematic reviewing of this sort was alien to the more popular literary monthlies. Where the *Gent* sold around 1,200 copies, the most successful magazines achieved sales as high as seven thousand copies, and their avid readers awaited the latest number with bated breath, hoping for some new literary sensation. Sometimes, this would be a dashing essay in response to a new book, but otherwise the two leading magazines—*Blackwood's Edinburgh Magazine* and *Fraser's Magazine*—rarely troubled themselves with reviews. Some of the other magazines did retain a section devoted to shorter reviews and literary notices, not infrequently extending their interest to scientific and medical publications intended for a general audience.[14] However, most readers could be assumed already to have heard of the Bridgewaters, and only the reformist *Metropolitan Magazine*—begun by the poet Thomas Campbell and then under the editorship of the naval officer and novelist Frederick Marryat—bothered to notice the series in this way.

The *Metropolitan*'s emphasis was again on the manner in which authoritative authors had shown that the sciences did not need to be considered dangerous to religion. In June 1833 a prominent and unusually long five-page review of Whewell's treatise eulogized the author's high talents and elaborated on his account of the potential dangers of scientific practice. "Nothing," the reviewer reflected, could be "more unfavourable to the liveliness of faith, and the holy confidence so warmly commended in Scrip-

12. *Monthly Review*, n.s., 2 (1833): 387, 500, 568; n.s., 3 (1834): 237. Until January 1834, issues of the *Gentleman's Magazine* were published at the start of the month following the stated date.

13. de Montluzin 2003; McDonnell 1983, 37–38, 72, 100, 109.

14. [Johnstone] 1834, 493; Grant 1836, 2:297. A reviews and notices section continued to be a regular part of the *Monthly Repository* and of the *Monthly*, *New Monthly*, *Tait's Edinburgh*, and *Dublin University* magazines.

ture," than the mental habit, fostered by the experimental method, of "suspending belief up to the actual demonstration of the subject." Likewise, a growing awareness of material causation readily "estranged" inquirers from "spiritual considerations." In such a context, the "devotional spirit" of Whewell's work would assist in introducing an appropriately religious habit of mind into the study of science despite the fact that elaborate natural theologies were ineffectual in counteracting the lunacy of atheism.[15]

In other magazines, the Bridgewaters only featured where they could be turned into the subject of a lively essay, as in the June 1833 issue of the liberal *Monthly Magazine*. The *Monthly* had originally had a particularly strong scientific bent, and even in its more literary form scientific topics and books continued to be of interest. The Bridgewaters thus became the occasion of a short essay that contrasted their authors with an altogether less satisfactory class of "Amateur Naturalists." Responding to the *Spectator's* suggestion that a Bridgewater polymath might have written a master work on all the sciences, the writer begged to differ. In the modern age, each branch of science required "long research, and patient industry" to master, and few could attain eminence in even two or three such branches. "What does Dr. Buckland know of entomology?" the reviewer asked. "But, in geology, Buckland is a giant—and it is fit that he should 'stick to his wax.'" The Bridgewaters allowed their authors to exhibit scientific authority in their respective domains. By contrast, the first of the competing works to appear, Henry Fergus's *Testimony of Nature*, attempted to cover the whole ground, which was an impossible task. Acting from honorable intentions and with obvious ability, he was nevertheless in danger of "becoming ridiculous, by being an Amateur Naturalist."[16]

At the start of July 1833, *Fraser's Magazine for Town and Country* also turned its attention to the Bridgewaters in the first of what became a series of four rumbustious essays. In the three and a half years since its founding, *Fraser's* had become the most talked about of the London magazines, rivalling the dominant *Blackwood's* with its similarly Tory politics and its combination of controversial articles and more intellectually demanding matter. Like its Scottish archetype, few of *Fraser's* articles were devoted to scientific subjects. David Brewster's early series of "Portraits of Eminent Philosophers," for instance, had ended after only two instalments. Subsequent articles relating to scientific topics were typically oriented toward

15. *Metropolitan Magazine* 7 (1833): 34–35, 37.

16. *Monthly Magazine*, 2nd ser., 15 (1833): 616–18. Such pejorative uses of *amateur* are usually associated with the later nineteenth century. See, for example, Alberti 2001.

the magazine's more usual interests in politics and religion, and this was the case with the essays on the Bridgewaters. They were written anonymously by the magazine's assistant editor John Abraham Heraud with the assistance of the editor William Maginn, and they turned the series into a pretext for offering a peppery critique of natural theology from the magazine's conservative Coleridgean outlook.

A disciple of the German idealist philosopher Friedrich Schelling, Heraud was not an acute philosopher—Thomas Carlyle considered him a "dud"—but his essays had all the dashing, partisan bluster necessary to attract significant attention. The first began by criticizing the neglect of Bridgewater's suggested theme of "*discoveries, ancient and modern, in arts, sciences, and the whole extent of literature.*" A treatise on this, Heraud later clarified, would have included "the Book of Books, with all that has been produced concerning it." It would thus have provided what was lacking from the other treatises—namely, an emphasis on the ascendancy of divine revelation. "If revealed religion be a pleonasm," *Fraser's* asserted at the outset, "natural religion is a contradiction [ . . . ]. All our knowledge of nature will not serve us to comprehend either the Divine Essence or Existence." However, God's self-revelation was not confined to the scriptures; it was also to be found within the human mind. It was true, the review claimed, that one could trace the laws of nature and the law of duty back to one and the same creator, as Whewell wished to do. However, the laws of nature were products of the human mind. They were, as Kant had shown, "the forms of the understanding" that regulated perceptions of the physical universe, and they echoed the "presence of God in the soul of man."[17] Compared with such transcendentalism, the Bridgewaters lacked "philosophical depth," and *Fraser's* especially regretted that Chalmers's subject had not been assigned to Coleridge, a portrait of whom had adorned the page preceding the review.

This first essay attracted sufficiently favorable attention (one newspaper hailed it as a "beautifully written philosophical paper") that *Fraser's* continued the subject in the lead article of the September issue. Discussing Chalmers's and Bell's treatises, the reviewers were now more conciliatory and appreciative—especially of Bell's work—but elaborated further on their earlier general criticisms about the management of the bequest and the inadequacies of natural theology. This article became the subject of widespread comment. Newspapers not uncommonly reported on the contents of the fashionable monthlies, but as Scottish magazine editor Chris-

17. *Fraser's Magazine* 8 (1833): 65, 73, 77, 80, 259–60; T to J Carlyle, 30 May 1834, in Sanders et al., 1970–2021, 7:202.

tian Johnstone observed, *Fraser's* "enjoyed the favour and regular notice of the newspaper press" and was thus "constantly kept before the public eye." Papers ranging from the metropolitan *Morning Post* to a large number of local weeklies now reported favorably on the magazine's treatment of the Bridgewaters. They agreed that the essay was a little on the sober side for the magazine but considered it a "high intellectual treat—a precious gem of concentrated moral, natural and divine philosophy" that was leavened by more lively matter in the issue.[18]

When *Fraser's* added a third article in October 1835, discussing Kirby's and Roget's treatises, it congratulated itself on the praise it had received for its handling of the Bridgewaters, observing, "No other periodical, great or small, weekly, monthly, or quarterly, has taken up this great argument in its integrity." *Fraser's* had "driven the ploughshare of the transcendental philosophy through the statements of our natural theologians, heaping up on either hand of the furrow the ridges of error."[19] It was not that *Fraser's* was generally critical of the view that the sciences could be prosecuted in accord with Protestant orthodoxy. Rather, the magazine proclaimed the inappropriateness of natural theology as a means of achieving such a consilience, which must be rooted instead in transcendental philosophy and the primacy of scripture.

The response of *Fraser's* to the Bridgewaters is strikingly at odds with the well-known claim of Robert M. Young that the "common intellectual context" found in the more fashionable monthlies and quarterlies of early nineteenth-century Britain was rooted in a "relatively homogeneous and satisfactory natural theology." Moreover, other fashionable monthlies also considered natural theology to be a controversial subject, both theologically and politically, although they did not follow *Fraser's* lead in offering reviews of the first few Bridgewaters. A much more obvious occasion for a spirited essay on natural theology was offered a couple of years later by the *Discourse of Natural Theology* of newly ousted Whig lord chancellor, Henry Brougham, which was perfectly suited to point up the subject's controversial character. On the Tory side, for instance, the recently founded *Dublin University Magazine* claimed that natural theology had been "the almost uniform weapon of infidel philosophy" and that Brougham's attempt to reformulate it as a science was liable to be made an excuse for deism. On the radical side, by contrast, *Tait's Edinburgh Mag-*

18. "Literature," *Exeter and Plymouth Gazette*, 14 September 1833, 4b; [Johnstone] 1834, 494; "Literature—Frazer's Magazine for This Month," *Freeman's Journal*, 10 July 1833, 2d.

19. *Fraser's Magazine* 12 (1835): 415–16.

*azine* argued that Brougham's work stood on the side of the poor in the battle against "priestcraft" and that the more natural theology was studied as an inductive science, the more the "reign of dogmatic tyranny" would be undermined.[20]

For the fashionable monthlies, then, natural theology was a contentious subject, and the Tory magazines in particular were far from convinced of its safety. Nevertheless, while the Bridgewaters were to a significant extent rather too soberly scientific for the more self-consciously literary magazines to notice, there was broad agreement that the dangers and limitations of natural theology did not undermine the contribution of the Bridgewaters in offering a safe guide to science. When a later reviewer in the *Dublin University Magazine* went off script in praising Buckland's contribution to natural theology, the editor felt it necessary to offer a corrective emphasizing that the Bridgewaters had only been successful insofar as they had illustrated the "true theology of nature" rooted in the Bible. But they were nonetheless useful for all that.[21]

❋

As fashionable society prepared to gather in London again in the autumn of 1833, the Tory *Quarterly Review* added its weighty testimony to the view that the Bridgewaters would do good. The rise of the intellectually demanding quarterly reviews, beginning with the *Edinburgh Review* in 1802, had been one of the marvels of the age. Their learned and often opinionated review essays—written by paid writers who were typically prominent in literature, politics, or learning—had attracted unheard of sales of as high as fourteen thousand copies. While this had fallen back below ten thousand by 1833, sales of the *Edinburgh* and *Quarterly* still far exceeded all the fashionable magazines except *Blackwood's Edinburgh Magazine*, and they were widely commented on in newspapers. They became important vehicles of public debate among leading literary, political, and scientific figures for whom the formal anonymity of the articles was often an open secret. But other readers also found in them an insight into the debate among the nation's cultural leaders. While their judgements might be slow

20. [Horne] 1835b, 807; "Natural Theology," *Dublin University Magazine* 6 (1835): 448–66, on 448; Young 1985, 127–28; Cantor et al. 2004, 13–14. For other reviews of Brougham's work see Topham 1993, 553–615; Horne also contributed a review of it to the *Westminster Review* ([Horne] 1835a).

21. *Dublin University Magazine* 8 (1836): 700; "The True Theology of Nature," *Dublin University Magazine* 9 (1837): 190–95. The attribution of the second review to John Scouler in Houghton et al. 1966–89 is highly doubtful.

in coming, the quarterlies provided the preeminent public forum in which the wider bearings of the sciences were discussed.[22]

The proportion of scientific essays in these journals remained relatively modest. The challenge for the editors, in an age of growing scientific specialization, was to match suitable scientific books with authoritative and talented reviewers in such a way as to yield essays that were appropriately accessible and engaging for the quarterlies' wide readership. What was needed was reflective writing concerning the nature, tendency, and implications of scientific knowledge rather than any unduly detailed engagement with scientific arcana. However, many scientific contributors struggled to produce the kind of lively prose that the quarterlies relied on to maintain their sales. In 1839 the *Quarterly*'s editor, John Gibson Lockhart, complained, "Our Whewells, Brewsters, Lyells, &c., are all heavy, clumsy performers; all mere professors, hot about little detached controversies, but incapable of carrying the world with them in large comprehensive *resumés* of the actual progress achieved by the combined efforts of themselves and all their rivals." In addition, it was a challenge to identify scientific publications that were well suited to prompt attractive essays.[23]

In this context, the Bridgewaters were perfect fare. While being largely nontechnical and addressing the wider bearings of the sciences, they were also written by leading men of science. In addition, their origin in an unparalleled bequest gave them a welcome degree of notoriety. Not surprisingly, the three leading reviews all published essays on the series by the start of 1834, and the *Edinburgh* and *Quarterly* both subsequently returned to the subject.

It was fortunate for the Bridgewaters that the *Quarterly* was first in the field, opening its October 1833 issue with an essay on the four so far published. On the whole the anonymous reviewer (fashionable physician Henry Holland) considered the treatises "creditable to the literature of the age," manifesting the support offered by the sciences to Christian belief. For centuries millions had depended on revelation for their religious convictions, but now the time had nearly arrived "when science and conviction ought to walk hand in hand with faith." As the sciences had been cultivated more successfully, the signs and demonstrations of a divine creator had become brighter and more numerous. The Bridgewaters showed that the tendency of modern science was unfailingly to support Christian

22. Shattock 1989, 97–99; [Johnstone] 1834, 492–94; Yeo 1993, 82. See also Butler 1993.

23. Smiles 1891, 2:454; Yeo 1993, chap. 4.

belief, and the reviewer offered an attractive synopsis of how the progress of astronomy yielded a growing feeling that the universe was a divine creation, enforced by references to his own scientific observations. Gradually, Holland incorporated details from Whewell's, Bell's, and Kidd's treatises into his essay, interspersing discussion of recent developments in the sciences with rhetorical flourishes concerning the "innumerable and powerful" arguments they provided for God's existence.[24]

This discussion gave credit to the arguments of natural theology, but, notwithstanding the reviewer's Unitarian roots, his concern was not to examine the formal contribution that the series made to such arguments. The point became clearer the following March, when the *Quarterly* dusted off Alexander Crombie's *Natural Theology* (1829) for review, noting that it made up for one deficiency of the Bridgewaters by providing "the most comprehensive view of the whole science of natural theology that has hitherto appeared." A much more serious deficiency of the Bridgewaters, however, was that the bequest had been handled in such a way as to lose an opportunity to forward the "diffusion of really useful knowledge." Was it likely, Holland asked pointedly, that so "numerous and expensive" a series would "attain any wide circulation in these days of cheap literature"?[25] As far as the *Quarterly* was concerned, the Bridgewaters were no great addition to the literature of natural theology. Rather, they were books that usefully made clear the religious tendency of the several sciences in a way that would benefit even poorer readers.

In January 1834 the *Edinburgh Review* was rather more forthright in proclaiming the inadequacies of Whewell's Bridgewater as a work of theology. The fundamental complaint was with Whewell's claim that conceptions of the creator required revision in the light of new scientific discoveries and his consequent ambition to "bring up our Natural Theology to the point of view in which it may be contemplated by the aid of our Natural Philosophy." According to the reviewer—the evangelical David Brewster—natural theology was not imperfect in this way. God's power, wisdom, and goodness were plainly "*manifested* in the CREATION," so that people were without excuse for unbelief, as the epistle to the Romans made clear. While scientific advances might enlarge our conceptions of God, they could never render them more perfect. Whewell had introduced scientific reasonings that were unnecessary for the purposes of natural theology,

24. *Quarterly Review* 50 (1833–34): 4–7, 34.

25. *Quarterly Review* 50 (1833–34): 1–3; "Crombie's Natural Theology," *Quarterly Review* 51 (1834): 213–28, on 213.

insecurely hypothetical (and thus harmful to natural theology if proved false), and in some cases rather demeaning of God's power.[26]

The *Edinburgh*'s criticisms in part reflected the tendency noted by Lockhart for men of science to engage in hotheaded quarrels of little wider interest. Brewster and Whewell had only recently crossed swords in public over Brewster's article in the *Quarterly* concerning the decline of science in England (which Whewell privately claimed had exhibited "great folly and impudence"), and Brewster now reechoed this theme in his opening attack on the management of the Bridgewater patronage. But there were also principled divergences between these natural philosophical heavyweights concerning the character of natural theology, its proper connection with the sciences, and matters of physical theory. Whewell arranged to insert a reply to Brewster in the *British Magazine*, the High Church monthly edited by his friend Hugh James Rose. He first defended his use of detailed examples to evoke religious sentiments before urging that, as the sciences progressed, it was inevitable that people would seek to connect their scientific with their religious views. Whewell claimed, moreover, that no scientific truth was "different *in kind*" from a good theory and that "the really essential and valuable part" of any good theory would survive its fall. Brewster's suggestion that natural theology should operate independently of scientific theory would leave the subject bereft of content, he concluded.[27]

Whewell's riposte thus served to restate what the hostility of Brewster's review had partially obscured—namely, the religious value and safety of modern science—and it was widely publicized. The *British Magazine*'s publishers boasted about Whewell's letter in advertisements, newspapers reported on it, and the *Cambridge Chronicle* even reprinted it in full.[28] In general, moreover, this was a view to which the *Edinburgh* coterie was committed, and when Brewster twice revisited the Bridgewater series in the review, his statements were altogether more positive.

A contrasting view was to be expected from the radical *Westminster Review*, and the journal obliged in its leading article on Chalmers's Bridgewater in January 1834. This Benthamite review was much more focused on political matter than either of its contemporaries, and partly in conse-

26. *Edinburgh Review* 58 (1833–34): 427–30, 435. Cf. Romans 1:18–20.

27. Whewell 1834a, 266; *Edinburgh Review* 58 (1833–34): 426; WW to H. Wilkinson [copy], 16 February 1831, TC, O.15.46[64]; Morrell and Thackray 1981, 51–52, 66. On Whewell and Brewster's deeper theological disagreements, see Brooke 1977, esp. 248–51, and Desmond 1989b, 63–64.

28. See, for example, *Standard*, 17 February 1834, 2b, 21 February 1834, 3d, *Leamington Spa Courier*, 1 March 1834, 4b, and *Cambridge Chronicle*, 7 March 1834, 3c, 4a–b.

quence it gave much less attention to scientific publications. In any case the religious purposes of the Bridgewaters were hardly likely to have wide appeal among the *Westminster*'s readership. However, the publication of Chalmers's work offered the journal an excellent opportunity to follow up on its recent criticism of that author's *Political Economy*, and the reviewer's stated object was to answer Chalmers's remarks against utilitarianism.

The great bulk of the essay, written anonymously by the editor, Thomas Perronet Thompson, was devoted to ridiculing Chalmers's statements in his first two chapters concerning the arguments for God's existence and character derived from the human conscience and the pleasure of virtue. Had these been stated in utilitarian terms, the review conceded, they would have been acceptable, but while Thompson suggested that natural theology was not inconsistent with "Christian Utilitarianism," he nevertheless took the opportunity to take a few sideswipes at the whole Bridgewater project. Criticizing Chalmers's attack on Laplace's views concerning the necessity of gravitation, the review suggested that this reflected the clergy's long-standing reluctance to accept new scientific views as exemplified by the cases of Galileo and modern geologists. Since the church was made of "stretching materials," it could always give "up to a point" when forced to do so, but in this radical journal the Bridgewaters hardly seemed to bear out the vision of Christianity and science in mutual support.[29]

Contemporaries knew how to read such anticlerical sentiment in the *Westminster*, and criticism of the Bridgewaters from this quarter was hardly likely to cause their authors to lose sleep. In the *Quarterly* and *Edinburgh*, however, men of science commented authoritatively on each other's publications before the largest possible audience of leading politicians, churchmen, lawyers, medical men, and literary figures. While this allowed them to give vent to personal animosities or to argue about principled disagreements, reviews in the quarterlies were an important means of making a case for the cultural value of the sciences and of presenting a united front to the world. Certainly, while fault lines were clearly visible in the case of the Bridgewaters, the reviews generally emphasized the ways in which the sciences were supportive of Christianity. Moreover, while substantive theological issues undoubtedly arose, the importance of the series was not seen primarily in theological terms.

Taken as a whole, the Bridgewater authors had good cause to be pleased with the reports of their treatises in the main literary journals during the

29. *Westminster Review* 20 (1834): 3, 21. See also Pickering to TC, 22 July 1833, NC, CHA 4.212.15, 20, and TC to WW, 4 March 1834, TC, Add.Ms.a.202[25].

first few months. A wide range of respectable weeklies, monthlies, and quarterlies from across the political spectrum had welcomed the series for its rebaptism of the sciences. Many, it is true, had criticized the handling of the bequest, but that was hardly the authors' concern. Many, also, had expressed reservations about the contribution of the Bridgewaters to the literature of natural theology, but reviewers did not consider their chief value to lie in that direction so much as in their religiously safe vision of the latest scientific views. Moreover, while such reservations about natural theology were if anything rather more forcibly expressed in the religious press, some reviewers there had also been swift to welcome the series for the religious safety of its presentation of science.

## *Natural Theology and True Religion*

Religious periodicals were vocal about the Bridgewater Treatises from the outset. The very earliest monthlies to notice the series (in May 1833) were two High Church magazines, the *Christian Remembrancer* and the *British Magazine*, and the earliest quarterly to do so (in July) was the High Church *British Critic*. These periodicals returned to the topic repeatedly over the following months, and others of an evangelical cast soon joined them. Across these reviews there was, at best, circumspection concerning natural theology, and many were frankly as hostile on the subject as *Fraser's Magazine*. Only Unitarian magazines were more generally positive concerning natural theology, and even here the mood was changing. Nevertheless, many religious reviewers were appreciative of the Bridgewaters for reconnecting the several sciences with Christianity in an accessible fashion.

No other class of periodicals in the age of reform could compete with the religious press in number and diversity. One survey lists more than ninety religious periodicals published in Britain during the four years in which the Bridgewaters were first published. These included some of the cheapest periodical titles, for which sales were often considerable. Surveying the "cheap and dear periodicals" in 1834, *Tait's Edinburgh Magazine* reported on magazine sales from one Scottish town. While the *Edinburgh Review* sold eight copies per issue and *Blackwood's* fourteen, the sixpenny *Evangelical Magazine*, *Scottish Missionary Register*, and *Edinburgh Christian Instructor* were selling seventeen, twenty, and twenty-five copies respectively, and the twopenny weekly *Scottish Pulpit* was selling sixty.[30]

30. [Johnstone] 1834, 495. Altholz 1989 provides an invaluable overview of the religious press in nineteenth-century Britain.

Moreover, unlike the readers of more expensive publications, the readers of these cheap religious magazines often had little else to read, devouring their sectarian publications with particular avidity.

Even discounting the missionary and sermon periodicals, many of the most widely distributed religious monthlies—typically priced at sixpence per month—were also very narrowly religious in their focus and gave little quarter to publications that were not avowedly so. This applies, for instance, to the nonsectarian *Evangelical Magazine* and *Gospel Magazine* as well as to many of the denominational magazines, including the Anglican evangelical *Christian Guardian*, the evangelical *Church of Scotland Magazine*, the *Baptist Magazine*, the *Wesleyan-Methodist Magazine*, the *Primitive Methodist Magazine*, the Irish *Orthodox Presbyterian*, the *Unitarian Magazine*, and the *Catholic Magazine*. Not only were the Bridgewaters far too expensive for most readers of such periodicals to consider purchasing but also their character was too scientific and insufficiently theological or devotional to warrant discussion. There were, however, a number of more expensive religious magazines and reviews that set themselves to review a somewhat wider range of publications—sometimes with a high degree of theological acuity—even occasionally extending to works of science. It was in these more expensive and intellectually demanding journals, aimed at the clergy and other wealthier readers, that the value of the series was debated, often at considerable length.

It is perhaps no surprise that the first of the religious journals to notice the Bridgewaters was the recently founded *British Magazine*, the brainchild of Whewell's friend and sparring partner Hugh James Rose (fig. 4.2), which celebrated the publication of Whewell's treatise in the number for April 1833 and offered a laudatory synoptic review the following month. The *British Magazine* was intended to be a general magazine, but it was aimed especially at the clergy and expressed a theology that had at least as much in common with the nascent Anglo-Catholicism of the Tractarians as with the more traditional High Church. In general, it reflected its editor's views about the dangerous tendencies of science. Indeed, Rose had recently reprinted his earlier sermon on the subject, albeit with additions reflecting his interchange with Whewell.[31] Unsurprisingly, therefore, while the magazine reviewed some scientific books, they were usually either works of natural history or ones that discussed the religious bearings of science.

The *British Magazine*'s review of Whewell's Bridgewater reflected his relationship with Rose in its eloquent appraisal of the author's personal

31. Rose 1831, 183–209; [Rose] 1832, 1; Thompson 2008, 59–60.

FIGURE 4.2 Hugh James Rose. Lithograph after unknown artist. LIB11. Image: reproduced with the permission of Lambeth Palace Library.

merits. Whewell had, the reviewer declared, extended the design argument as found in Paley to include the natural laws that characterized modern science—a task of considerable difficulty involving a rare combination of skills. But by far the most "original, interesting, and powerful" part of the book was the final section, which gave a rare insight into the "views and feelings generated in a mind of the highest order" by modern science. Whewell was a scientific exemplar whose behavior stood in marked contrast to the prevailing "evil fashion" of reducing phenomena to laws without deriving any spiritual instruction from the process. The author had a unique understanding of how scientific habits affected the human mind and had given an "impressive" demonstration of the "imperfection of all the views of the Deity which the most exalted human wisdom can form." Drawing on these insights, the book was above all effective in appealing to the heart as well as the reason and in producing reverence as well as conviction.[32]

This was a good start for the Bridgewaters in a potentially hostile corner, and a second High Church monthly also greeted Whewell's treatise warmly in its lead review for May. This was the well-established and now narrowly theological *Christian Remembrancer*, through which the Hackney Phalanx had endeavored to defend the Church of England from its dissenting and Unitarian critics. The magazine considered Whewell's Bridgewater "admirable," following his lead in asserting that, while revelation alone could "effect the great end of religion," the mind could be "led to acknowledge the perfections of the Divine Legislator" by "looking through nature up to nature's God." Thus, while Whewell was interested in leading readers to a higher estimate of divine perfection, the *Remembrancer* was especially pleased with his determination to root such reflections in the "moral government in the world" as experienced in the human conscience and to point readers to revelation.[33]

The *British Magazine* and *Christian Remembrancer* followed on rapidly with reviews of Kidd's and Chalmers's treatises, and in July they were joined by another High Church journal, the quarterly *British Critic*. This was another publication friendly to Whewell and, although it was now more theological in focus than formerly, it took the earliest opportunity to notice his Bridgewater. The forty-page review was again very positive, but the High Church analysis of the limitations of natural theology now came much more to the fore.

Written anonymously by one of the journal's leading reviewers—Charles Le Bas, the mathematical professor at the East India Company's college at Haileybury in Hertfordshire—the review discussed Whewell's book in company with Scottish physician John Abercrombie's very successful *Philosophy of the Moral Feelings* (1833), a book that presented a studiously pious exposition of the moral philosophy of the Scottish "commonsense" school. Starting from Abercrombie's assertion of the existence of certain "*first truths*" of morality, "which admit of no demonstration and which need none," the reviewer turned the same principles to religion, suggesting that belief in the being and attributes of God was similarly "*instinctive*." Moreover, he contended that argument on the subject would never convince those who disputed these conclusions. To reach such a skeptical state, the reviewer asserted, their minds must have "gone

32. *British Magazine* 3 (1833): 589.

33. *Christian Remembrancer* 15 (1833): 258, 261.

through a course of unnatural and artificial discipline," the result of which was the "banishing of moral sentiment and emotion from their philosophy" in favor of "mere speculative reason." One might as well try to convince Charles Babbage's famous calculating engine of God's existence as one of these skeptics.

Here the analysis took on openly Kantian terms, as the reviewer ascribed the instinctive belief in divine existence and divine attributes to the "Practical" as opposed to the "Speculative" reason. Like Kant, he argued that the belief in a morally perfect God arose from the natural tendency to "hunger and thirst after what is benevolent and good," a tendency founded on the proper development of the instinctive "moral powers and perceptions" of humans. And like Kant, too, he suggested that, to the speculative reason, the phenomena of the universe could never demonstrate the existence of a morally perfect God.

This analysis was close to Whewell's own, and the *British Critic* welcomed his Bridgewater on its own terms. Here was a "distinguished master of physical science engaged in the glorious office of showing—to use his own words—'how admirably every advance in our knowledge of the Universe harmonizes with the belief of a most wise and gracious God'." This was the more valuable, Le Bas noted, since "physical researches" had not always been "signally favourable to the development of moral and religious sensibilities." The efforts of those who, like Whewell, sought to manifest God's character through his contrivances were to be applauded because of the emotional rather than the intellectual effect of their writings. The effectiveness of such illustrations was to be judged on the basis of their capacity to "take captive the affections." Not surprisingly, Le Bas warmly commended Whewell's analysis of the effects of different scientific habits of mind on devotion as being perhaps "the two most powerful and original chapters of the book."[34]

It was natural that the *Critic* should select Chalmers's treatise for review in the following issue. Again, the review echoed the author's own clear statement of the limitations of natural theology. The unknown reviewer frankly admitted that some had objected to the whole scheme of the Bridgewater Treatises on the grounds that natural theology was "well nigh an obsolete thing" in the age of revelation and might detract from the latter. Yet in responding to the fear that confronting atheists with natural theology might lead to deism rather than Christianity, the reviewer followed Chalmers's analysis, answering that natural theology was "a bridge

34. *British Critic* 14 (1833): 76–78, 81, 90, 93, 95, 107. On the *British Critic*, see Houghton and Altholz 1991.

constructed, not by way of a foundation whereon men are to erect their dwelling-places, but merely as a pathway along which they may travel in safety to the realms of a higher theology." Even then, however, its use was compromised. Asserting that the conviction that design demands a designer is a matter of "intuition," the reviewer admitted that argument could never persuade those who refused to admit the intuition. This being so, what was the value of works such as the Bridgewaters, which amassed evidence of design? The reviewer was happy that they did have a value. First, they might confirm the faith of believers. But second, they might also impress skeptics with a "salutary horror" at setting themselves up against such a vast "fortress of testimony" and recall them from their "outrageous folly" before they became "irreclaimable adept[s] in the mysteries of impiety."[35]

These meaty theological engagements with the Bridgewaters nicely reflect High Church ambivalence toward natural theology, an ambivalence that the reviewers found to resonate well with the treatises of Whewell and Chalmers. A rationally constituted natural theology might be limited and potentially dangerous, but in the hands of such authors its arguments could function to protect and foster religious sensibilities in an increasingly scientific age. Thus, while all three High Church journals were very religious in focus and rarely contained reviews of more secular scientific books, they gave the Bridgewaters extensive attention. With their religious frame of reference and their opposition to materialism and transmutation, the series represented the acceptable face of modern science.

❋

The more intellectual magazines and reviews of evangelical Christianity—both in the established churches of England, Scotland, and Ireland and in the various denominations of Protestant dissent—were markedly less interested in the Bridgewaters than those of the Anglican High Church. Once again, these were periodicals that were largely religious in matter, and they only occasionally noticed scientific books—chiefly ones that were of devotional value or that had important theological bearings. Most of them expressed more or less serious reservations concerning natural theology, often in approving reviews of Chalmers's *Natural Theology* or in critical ones of Brougham's *Discourse of Natural Theology*. Nevertheless, several of them found something in the Bridgewaters worthy of bringing to their readers' attention.

35. *British Critic* 14 (1833): 240–41, 252–53.

First in the field, in October 1833, was the monthly *Christian Examiner and Church of Ireland Magazine*, which embodied the combination of High Church and evangelical attitudes characteristic of Irish Anglicanism at this period. The magazine occasionally included brief notices of popular scientific works, but the only one of the "valuable series" of Bridgewaters it reviewed was Chalmers's, which was treated as a theological work in a two-part leading review. The pseudonymous reviewer described Chalmers's contribution as "masterly," but what he especially valued was the claim that the proof of God's existence and attributes derived more from feelings, prompted by conscience, than from reasoning. With the man who had closed his ear to the voice of conscience, the review echoed Chalmers, the "demonstrations of the schools" were ineffectual. God had never "intended that His existence should be made out by the demonstrations of Natural Theology," nor were such demonstrations effective on "bold and intelligent" atheists. The reviewer ended by reminding his readers, as Chalmers had, that "the everlasting Gospel alone is the source of true philosophy."[36]

The following month, the *Edinburgh Christian Instructor*—which had for a number of years been the main organ of the increasingly dominant Evangelical party in the Church of Scotland—reviewed Chalmers's treatise in a similar tone. As one of the most prominent leaders of that party, Chalmers commanded considerable respect, and the lengthy synoptic review was suitably favorable. As the reviewer acknowledged, however, there was no agreed position within the party regarding the soundness or usefulness of natural theology. Indeed, on few subjects had a "wider variety of opinion" existed: while some held it up as "all-sufficient," others "denied its existence, or pronounced it to be pernicious." The *Instructor*'s reviewer took a negative view, observing that natural theology could not discover "either the existence or the character of God" and that it was only after the existence of God had been "announced to us" by revelation that what was called "natural theology" could furnish us with "solid arguments in proof of his existence, and [ . . . ] his character." As Chalmers's chapter concerning the "defects and uses of natural theology" had emphasized, the knowledge critical to any religious inquirer could only be obtained by examining the Bible. It was also precisely this aspect of Chalmers's treatise that was emphasized in the lengthy extract that appeared in the Belfast *Orthodox Presbyterian* for September 1833.[37]

36. *Christian Examiner*, n.s., 2 (1833): 667, 674–75, 743, 746.

37. "Defects and Uses of Natural Theology," *Orthodox Presbyterian* 4 (1832–33): 423–25; *Edinburgh Christian Instructor*, 2nd ser., 2 (1833): 767. See also "Chalmers

As with the *Christian Examiner*, Chalmers's Bridgewater was the only one of the series to be reviewed by the *Edinburgh Christian Instructor*. Increasingly occupied with church politics, the magazine had become markedly less inclined to review nonreligious works in recent years and, while it took notice of a number of works about the natural world, they were typically such as exhibited "scriptural piety," relating the book of God's works to that of his word and stirring the religious feelings of the evangelical faithful. Among them were the popular scientific works of Thomas Dick, a former minister in the Scottish Secession church, who considered science a form of religious devotion. Reviewing Dick's *On the Improvement of Society by the Diffusion of Knowledge* in October 1833, the *Instructor* observed, "There are two ways of connecting religion and science; the one is by making religion scientific, and the other is by making science religious." Dick had chosen the correct path in insisting on science "being made to assume the garb and speak the language of religion." Much more than the Bridgewaters, such works connected the latest findings of the sciences with the claims of the gospel and with feelings of religious devotion. The Bridgewaters might be useful in strengthening the arguments of natural theology, the reviewer suggested, but Dick's works and their ilk were altogether more valuable to evangelical believers.[38]

Of the evangelical periodicals, only the bimonthly *Presbyterian Review* reviewed all eight of the Bridgewaters. Begun in July 1831, shortly after the death of the *Instructor*'s editor, by a talented group of Chalmers's divinity students in Edinburgh, the review provided the Evangelical party in the Church of Scotland with a more intellectual forum, containing lengthy and demanding reviews that were much read and discussed in Scotland and Ulster. Starting in March 1834 with an anonymous review by Patrick C. MacDougall (later Edinburgh's professor of moral philosophy), the journal devoted much space to praising the series. However, little attention was devoted to the legitimacy or value of the claims of natural theology.

One exception was when the *Presbyterian* reviewer of Prout's Bridgewater began by noting that he had heard the Bridgewater bequest "by some ridiculed, by others blamed" in consequence of its having been "misrepresented as an attempt to teach natural religion to the exclusion of

---

on Natural Theology," *Edinburgh Christian Instructor*, 3rd ser., 1 (1836): 359–70, and Baxter 1985, esp. 108–11. The Belfast journal later reviewed Chalmers's *Natural Theology* positively, referring back to his Bridgewater. "Review," *Orthodox Presbyterian* 7 (1836): 280–88.

38. "Dr Dick on the Diffusion of Knowledge," *Edinburgh Christian Instructor*, 2nd ser., 2 (1833): 708–10, on 709. For an excellent account of the *Instructor* and competing periodicals, see Currie 1990, 10–138.

revealed,—to demonstrate through the works of creation, that there is a God, and there to leave men to seek, as they best may, for a Saviour." Noting that belief in the being of a God was too deeply rooted in an innate principle to "require the aid of argument for its support," the review argued that the Earl of Bridgewater had wished his authors to provide proofs from nature of the attributes rather than the existence of a God. More fundamentally, he had "intended to teach men, not the science of natural religion, but the religion of natural science." The "religious illustration of natural science" had the "great advantage" of "making Christians more intelligent, of increasing their capacities for enjoyment, and rendering them better prepared, when called upon, to give a reason for the hope that is in them." In this way the Earl of Bridgewater had been "made an instrument in the hand of God for good, by enlarging men's ideas respecting the Divine attributes and ways, and thereby leading them more fervently to love, more humbly to adore."[39]

South of the border the leading evangelical magazine of the Church of England, the *Christian Observer*, hardly mentioned the Bridgewaters. As the organ of the "Clapham Sect" grouping of Anglican evangelicals, the magazine had for some time been an important vehicle for the cultured evangelicalism of William Wilberforce. By the 1830s, however, the magazine was—like the *Edinburgh Christian Instructor*—more narrowly religious in outlook than it had been previously, and its editor, clergyman Samuel Charles Wilks, generally felt it necessary to apologize when introducing any scientific topic into its pages. Moreover, compared with the magazine's early days, when a reviewer had warmly welcomed Paley's *Natural Theology*, the *Observer* now took a much less favorable view of natural theology. In the spring of 1833, when it might have been noticing the Bridgewaters, it carried articles about Hutchinsonianism that claimed natural theology was neither valid nor safe.[40]

Two years later the "celebrity" of Henry Brougham meant that Wilks felt obliged to offer a very lengthy and rather critical review of his *Discourse on Natural Theology*, which made clear why the *Observer* had given no attention to the Bridgewaters. According to Wilks, the "spirit of the age"—as exemplified by Brougham's engagement in secular and scientific education—was manifest in a "spurious liberality, a passion for generalizing in religious sentiment, a desire to amalgamate all creeds in simpler

39. *Presbyterian Review* 6 (1834–35): 1–3. Cf. 1 Peter 3:15.

40. Beth, "On the Hutchinsonian Physico-Theology," *Christian Observer* 33 (1833): 154–58; A Hutchinsonian, "Defence of Hutchinsonianism," *Christian Observer* 33 (1833): 219–21; Topham 1999, 148.

elements, the deification of Knowledge as man's chief good, and a disposition to concede the most important points of religious doctrine and sentiment for the fancied interests of science." Such "generalized religion," or practical deism, suited the mercenary interests of booksellers, who could thus maximize the market for their books, and it also suited the "lukewarm readers" who wanted amusing or interesting books without what they saw as the "bigotry of Saintship." The religion of such works—no doubt including most of the Bridgewaters—was an "unacknowledged plagiarism on Christianity" that left the heart unmoved and the soul unredeemed.[41]

Similar circumspection regarding natural theology was also evident in periodicals of evangelical dissent. Neither the *Congregational Magazine* nor the *Eclectic Review* (an intellectually demanding periodical read especially by Congregationalists and Baptists) noticed any of the Bridgewaters until moved to do so by the controversy prompted by the publication of Buckland's. Moreover, when reviewing Brougham's *Discourse of Natural Theology* the *Eclectic*'s proprietor-editor Josiah Conder made clear that natural theology, strictly defined as knowledge of God acquired by reason "in the absence of revealed knowledge" was a "science falsely so called." According to Conder, a work like Brougham's was only useful in protecting religious feelings from the chilling effects of secular scientific study by offering a reminder that God revealed himself to those who sought him "in His works, as well as in His word." The *Congregational Magazine* was distinctly more positive about Brougham's work, but, while it allowed that the Bridgewaters possessed "great merit," it complained that mismanagement had resulted in their being prohibitively expensive.[42]

The *Wesleyan-Methodist Magazine* kept its readers apprised of the progress of the Bridgewaters, announcing the appointment of the authors, listing the treatises in its carefully restricted "Select List of Books Recently Published, Chiefly Religious," and publishing extracts from two of them. Yet the clear party line, which was again made especially explicit by its review of Brougham's *Discourse of Natural Theology*, was that what was thought of as natural theology was ultimately dependent on revelation. Moreover, while the Bridgewaters might be useful in giving a Christian gloss to the sciences, their usefulness was tempered by their lack of ref-

41. Review of Lord Brougham's *Discourse of Natural Theology*, *Christian Observer* 35 (1835): 687–99, 738–50, 804–19, on 691–92. See also Topham 2004b, 46–47, 54–55.

42. *Congregational Magazine* 13 (1837): 43; "Works on Natural Theology," *Congregational Magazine* 13 (1837): 242–51; [Conder] 1835, 165, 185 (for the attribution of authorship see Conder 1857, 339). The *Eclectic*'s review of Buckland's Bridgewater was somewhat more positive regarding natural theology.

erence to specifically Christian doctrines. This seemed to be clearly in view when the magazine noted that Henry Duncan's *Sacred Philosophy of the Seasons* (1836–37) was, in its "occasional references to the peculiar doctrines of holy Scripture" and its "popular" quality, "better adapted to general use than are several modern works of much higher pretensions." There was no question, however, that the Bridgewaters could be made suitable sources for extracts in the right hands. Indeed, the Methodists' fourpenny *Youth's Instructer*, which was edited by the same editor as the adult magazine, was happy to reprint extracts once the cheaper edition of Whewell's treatise appeared in 1837, recommending it as a standard part of the juvenile library.[43]

Similar attitudes found expression in the publications associated with the pan-evangelical but largely dissenting Sunday School Union. In 1813 the union's secretary, William Freeman Lloyd, had begun a *Sunday School Teachers' Magazine* containing essays and correspondence as well as reviews of books interesting to Christian educators. The magazine did not at first notice the Bridgewaters, and in 1833 and 1834 it carried a vigorous and polarized debate among its contributors concerning the value of natural theology in religious instruction. Yet while some considered that natural theology carried children away from true religion, even they were willing to concede that the study of nature had its place alongside the study of the scriptures, not least as a preservative against skeptical attacks. When the magazine began to notice the Bridgewaters in September 1835, the reviewer was one of those who had been on the positive side of the argument. Nevertheless, the lengthy reviews he devoted to Whewell's and Kirby's contributions suggested that what was most useful about them was the religiously safe account they gave of the sciences.[44]

It is striking how few reviews of the Bridgewaters appeared in the extensive evangelical press. The series did not generally commend itself as the most appealingly accessible or scriptural set of devotional resources concerning the natural world. Moreover, misgivings about natural theology were as widespread in the evangelical periodicals as in the High Church ones, and reviewers were not inclined to welcome the Bridgewaters for their contribution to the subject. However, they could be useful to the

43. *Youth's Instructer*, n.s., 2 (1838): 61–63; "Select List of Books," *Wesleyan-Methodist Magazine*, 3rd ser., 16 (1837): 515–18, on 518; "Select List of Books," *Wesleyan-Methodist Magazine*, 3rd ser., 14 (1835): 544–48, on 547–48. On the *Wesleyan-Methodist* see also Topham 2004d.

44. Topham 1992, 423–29. On the Sunday School Union, see also Topham 2004a; the Union's fourpenny *Youth's Magazine* also contained numerous extracts (see chap. 5).

extent that they allowed evangelical readers to cultivate their feelings of religious devotion while reading about the sciences. In addition, some evangelical reviewers found that they offered a valuable means of sanctifying scientific education.

In their encounters with the Bridgewaters, readers of the great bulk of religious periodicals—High Anglican and evangelical alike—were left in no doubt of the limitations and dangers of natural theology. Moreover, even among Unitarians the rationalist emphasis on natural theology was beginning to be questioned by the 1830s. While the long-established *Christian Reformer* obediently endorsed the rationalist outlook of Priestley in its lengthy reviews of the Bridgewaters, the newly founded *Christian Teacher* reflected a new mood in English Unitarianism—partly inspired by the American Unitarian leader William Ellery Channing—which modified and ultimately supplanted such rationalism with an emphasis on religious sentiment. Thus, when the *Teacher* caught up with the Bridgewaters in a review of Brougham's *Discourse of Natural Theology* in 1835, it pronounced them to be "on the whole" a failure. A subsequent review of Buckland's treatise described it as a valuable contribution to the literature of natural theology, but other articles in the *Teacher* argued clearly that while the advance of science would add to the strength of the arguments of natural theology, such arguments were of no avail if the heart were not engaged. Belief was less a matter of "logic" than of "feeling."[45]

The religious periodicals, then, largely confirmed and elaborated on the impression given by the literary periodicals. Welcome as the Bridgewaters might be for their religiously safe vision of the sciences, any pretensions they might have to contribute to a formally constituted natural theology were at best of limited value and at worst a serious danger to true religion. In an increasingly scientific age, however, in which every soul potentially had access to scientific publications, the safe science of the Bridgewaters was not something to disdain.

## *Making "Useful Knowledge" Safe*

Religious reviewers often expressed concern about the rapid expansion in secular publications on the sciences, but that expansion was nowhere

45. *Christian Teacher* 2 (1836): 664–71; H. G., "On the Study and the Spirit of Natural Theology," *Christian Teacher* 3 (1837): 384–92, on 391; [Thom?] 1835 (attribution in Yule 1976, 255). On natural theology in Priestleyan Unitarianism, see Webb 1990, 126–30, and Yule 1976, 255–58. On the periodicals, see Altholz 1989, 73–74.

more evident than in the periodical press itself. Most recent were the cheap weekly miscellanies intended for mass consumption, but there was also a rapidly growing array of scientific and technical magazines suited to a range of pockets—from the learned *Philosophical Magazine* to the threepenny *Mechanics' Magazine*—not to mention an ever-more prominent medical press.[46] Many of these periodicals found cause to comment on or draw from the Bridgewaters as they appeared, and, despite their widely differing purposes, they commonly agreed that the treatises were valuable in offering an up-to-date account of scientific knowledge that was at once useful and religiously safe.

The various journals encompassed a great variety of ideas about how scientific knowledge could be rendered innocuous to religion, and this was nowhere more evident than in the cheap miscellanies. In the spring of 1832 the unprecedented success of *Chambers's Edinburgh Journal* and the Society for the Diffusion of Useful Knowledge's (SDUK's) *Penny Magazine* quickly brought religious anxiety about secular scientific publications to fever pitch. Both magazines aimed at religious "neutrality," but to their critics that was tantamount to irreligion. Moreover, within weeks, sales of *Chambers's Journal* (which was priced at 1½d.) had reached fifty thousand, while sales of the *Penny Magazine* averaged over two hundred thousand during its first quarter.[47] The wide distribution of magazines of such an "Anti-Christian tendency" prompted the Society for Promoting Christian Knowledge (SPCK) to begin a *Saturday Magazine* in July 1832 (fig. 4.3), identical in format and price to the *Penny Magazine*. This was a radical departure for a society whose stock in trade had hitherto been religious tracts. Now, however, it formed a committee to produce cheap and educational works with a "Christian character and tendency" that might swamp nonreligious or materialist fare. Working with printer and publisher John William Parker, the society rapidly turned out a range of publications of which the *Saturday Magazine* was by far the most successful, with regular sales standing at around eighty thousand after a year.[48]

The Bridgewaters readily lent themselves to the SPCK's educational vision, and during the 1830s the *Saturday Magazine* contained more than eighty extracts from the series, carefully selected to emphasize its agenda.

46. For an overview, see Dawson and Topham 2020.

47. Priestley 1908, 30; "Chambers's Edinburgh Journal," *Chambers's Edinburgh Journal*, 2 February 1833, 1–2; Bennett 1982.

48. Anon 1832, 18, 65, Topham 1992, 420–23; Clarke 1959, 181–83.

THE
# WEEKLY VISITOR.

NUMBER XLII.] TUESDAY, SEPTEMBER 10, 1833. { PRICE ONE HALFPENNY.

PHALENA PAPHIA AND JUJUBE TREE.

1. Eggs of the Phalena Paphia or Tusseh Silkworm.
2. The Newly Hatched Worm.
3. The same half-grown.
4. The same full grown, and ready to spin its cocoon.
5. The Male cocoon suspended.
6. The Female cocoon, which is always larger.
7. A Branch of the Jujube Tree in Flower, with some Fruit a little advancing.
8. The Ripe Fruit thereof.

THE SILKWORMS OF BENGAL.
*Tusseh.*

IN Bengal, and some provinces of India, is found a caterpillar, which, in its moth state, is called the phalena paphia, and which, in its caterpillar state, feeds on the leaves of the jujube or byer, and of the asseen-tree; and another, which is called phalena cynthia, which feeds on the leaves of the palma christi. From these insects, called, when in the caterpillar state, the tusseh and Arindy silk-worms, a coarse sort of silk is procured, called tusseh silk, though in reality it is not applicable to the purposes to which the more valuable and elegant product of the common silk-worm (*phalena bombyx mori*) is suited. A coarse dark-coloured cloth, however, is manufactured from the cocoons, spun by the insect, which is worn by some of the Hindoos, and is very durable; but as yet it has not become a valuable article of commerce, nor does it appear likely to be so. In fact, the whole class of caterpillars produce a silky thread, from which a cocoon is formed; but it is the silk-worm, properly so called, alone, which feeds only on mulberry leaves and twigs, that produces a thread of any important value.

The tusseh silk-worm is peculiarly interesting from the manner in which it suspends the cocoon or case in which it is to pass its chrysalis state of existence. The insect having attained the full maturity of its caterpillar state, is furnished with a glutinous filament, by means of

VOL. I.

THE
# Saturday Magazine.

Nº. 147. OCTOBER 18TH, 1834. { PRICE ONE PENNY.

UNDER THE DIRECTION OF THE COMMITTEE OF GENERAL LITERATURE AND EDUCATION APPOINTED BY THE SOCIETY FOR PROMOTING CHRISTIAN KNOWLEDGE.

THE WILD PALM OF THE DESERT.

VOL. V. 147

FIGURE 4.3 *Saturday Magazine*, 18 October 1834, and *Weekly Visitor*, 10 September 1833.

As an article in the magazine's first number spelled out, while knowledge might well be power, it was crucial to remember also another Baconian aphorism: "*That all knowledge is to be limited by Religion, and to be referred to use and action.*" For philosophers such as Newton and Locke, whose Christian belief kept knowledge in its proper channel, it had been "not only power, but virtue and happiness." It was above all important that the search for knowledge should not be allowed to supplant the Bible or its precepts. Knowledge of God—his attributes, purposes, and laws—was "beyond the scope of unassisted human inquiry" and to be found only in revelation. In acquiring knowledge of nature, it was necessary for the learner to be properly versed in such biblical knowledge and to avoid too exclusive a focus on secondary causes. Thus, readers of the *Saturday Magazine* were to expect their education to be liberally laced with references to divine action in the created order and to "*rise through Nature up to Nature's God*," but not by means of a rationally constituted natural theology.[49]

Many of the *Saturday Magazine*'s extracts from the Bridgewaters were among its notably large number of articles related to natural history, a subject often used to lead the reader's mind to the creator. In addition, other forms of religious allusion were also valued, including moralizing passages and scriptural references. In such circumstances, the treatise of High Churchman William Kirby was particularly useful, and many of the passages taken from his Bridgewater were those that emphasized the strongly hierarchical character of the created order and that moralized about the virtue of created beings fulfilling their place in the created order with appropriate "honest industry."[50] From the perspective of the SPCK, the Bridgewaters were valuable not just because they offered authoritative and contemporary science in a form that was free from dangerously materialist or transmutationist doctrine. Their text was also leavened with references to divine design, to sometimes very socially conservative Christian morals, and even occasionally to the Bible in such a way that made them highly attractive.

Evangelicals also responded with concern to the appearance of secular useful knowledge periodicals, although with much more limited success. The SPCK's evangelical counterpart, the Religious Tract Society (RTS), had already ventured into the production of cheap religious periodicals in the 1820s, but in January 1833 it responded to the new cheap weeklies by commencing a halfpenny *Weekly Visitor* (see fig. 4.3). Edited by Esther Copley, an Anglican evangelical whose husband was a Baptist pastor, this was a relaunch of the society's fourpenny quarterly *Domestic Visitor* that had been intended to promote the "spiritual instruction of families, particularly domestic servants." The new magazine was designed to combine Bible teaching with "matter instructive to the general reader," and its opening article on the crocodile thus not only began with a biblical quotation but also appeared in company with extracts on "The Deity of Christ" and "The Way of Salvation." Each issue sought to explain and illustrate both

49. "On the Right Use of Knowledge," *Saturday Magazine*, 7 July 1832, 2–3. For a partial listing of Bridgewater extracts in the cheap weeklies, see Topham 1993, 553–615; for further discussion, see also Topham 1992.

50. "On Wasps and Bees," *Saturday Magazine*, 17 December 1836, 238 (quoting from Kirby 1835, 2:338); Hinton 1979, 280; "On the Study of Natural History," *Saturday Magazine*, 7 July 1832, 3.

the "works and the word of God," ministering to "both the temporal and everlasting interests of the reader."[51]

Selective quotation from the Bridgewaters could nevertheless be made to serve these distinctive objects of the evangelical press. Unsurprisingly, the magazine's first lengthy extract came from Chalmers's Bridgewater and formed the basis of an article headed, "Is Man Accountable for His Belief." Another soon followed on "The Nature of Virtue." However, other extracts were also offered on a range of scientific topics, and these did not even necessarily make explicit reference to divine design, the editor clearly relying on the context that juxtaposed the "Utility of Water" from Kidd's Bridgewater with "A Dying Christian's Prayer" to draw out the religious meaning of science. Sometimes extracts were more vigorously worked into an evangelical perspective. Thus, while two extracted paragraphs on electricity from Whewell's Bridgewater carried a reference to God's goodness and wisdom, the text was run on indiscernibly from editorial matter ending in the biblical declaration, "Great and marvellous are Thy works, and Thy ways past finding out!" Another article drew on Whewell's account of Laplace's nebular hypothesis to offer an example of "the greatest philosophers" being, like Sampson, "weak as other men" (Judges 16:17) with the intention of destroying "that overweening deference to human authority, by which weaker minds are apt to be overpowered and misled."[52]

At the same time, however, the leading secular miscellanies to which the religious critics were responding also found that they could raid the Bridgewaters for material. It was *Chambers's Journal*, rather than the *Penny Magazine*, that made most use of the Bridgewaters. The magazine was a commercial venture by the Edinburgh publisher William Chambers, offering four newspaper-like pages of original and extracted articles, anecdotes, stories, snippets of advice, biographies, and poetry with the stated aim of providing the best possible food to satisfy the public appetite for instruction. Earlier publications, the opening editorial claimed, had failed to present knowledge "under its most cheering and captivating aspect," not least because of the political and religious interests of the "corporations" behind them—the useful and Christian knowledge societies. By contrast, the new magazine would be nonpartisan, and it would blend instructional

51. "Domestic," *Tract Magazine*, n.s., 4 (1833): 71–72, on 72; Jones 1850, 98, 134–36; Fyfe 2004, 63–64.

52. "False Views of Creation," *Weekly Visitor*, 19 November 1833, 423–24, on 423; "Electricity," *Weekly Visitor*, 4 February 1834, 42–44, on 44. See also *Weekly Visitor*, 19 November 1833, 419–20, 29 April 1834, 154–55, and 13 January 1835, 12–13.

articles and writings on morality with sketches and tales that would amuse readers and draw them in. In this it reflected not only Chambers's meliorist belief in the socially regenerative power of education but also a pragmatism about what the public might be prepared to purchase and read.[53]

For all that the new journal avoided sectarian religion, its editors—who soon also included William's brother Robert—were far from shy about introducing what they deemed to be nonsectarian religious matter. In particular, when they looked back over their editorial practice a few years later, they were happy that they had presented "exalted views of Creative Wisdom and Providential Care," and they certainly showed themselves by no means averse to drawing on the Bridgewaters. The process began in June 1834, when an entire chapter from Whewell's volume was squashed into just two columns of the journal, complete with its references to the design that had adapted the atmosphere and the human ear to each other.[54] This was followed by a number of further extracts over the next few years, although the magazine's emphasis on originality meant that such extracts were relatively infrequent. In addition, however, the series was also turned into source material for a wide range of original scientific articles on topics such as the history of life on earth, the nervous system, and infusoria. Despite the significantly transformed framing of the subjects in these original articles, the design perspective was often carried over in quotations, as was the Bridgewater imprimatur, in references.

The *Penny Magazine*, by contrast, drew relatively little material from the Bridgewater series, apparently reflecting the society's principles of not publishing on religious topics. While the SDUK's architect Henry Brougham had initially hoped that its publications would be laced with a discourse of divine design, more radical members had made clear their opposition to such an emphasis, causing Brougham to publish his new edition of Paley's *Natural Theology* independently. Moreover, in seeking to counteract radical street literature—the "blasphemous and seditious press"—it became ever-more necessary to avoid the appearance of political or religious partisanship. Thus, the *Penny Magazine* contained no religious articles and made relatively few references to divine design. It was these strikingly secular features that quickly attracted critical comment from religious leaders, and Charles Knight later felt it necessary to defend

53. "Chambers's Edinburgh Journal," *Chambers's Edinburgh Journal*, 2 February 1833, 1–2, on 1; Chambers 1832, 1; Fyfe 2012, 21–25. See also Chambers 1872, 220–48.

54. "Sound," *Chambers's Edinburgh Journal*, 7 June 1834, 150–51 (from Whewell 1833b, 117–25); "Fifteen Years Ago," *Chambers's Edinburgh Journal*, 2 January 1847, 15–16, on 16; Priestley 1908, 30.

his editorial practice, insisting that the magazine had not been as strictly secular as had been claimed. Some of the extracts from the Bridgewaters certainly did retain references to divine design. However, many others did not, offering an account of modern science that was "safe" insofar as it was not at odds with Christian doctrine but that equally made no connection to Christianity.[55]

The cheap miscellanies reproduced text from the Bridgewaters in strikingly varied ways to suit their strongly divergent ideas concerning what kind of scientific knowledge was "safe" for the populace to encounter. Certain features made the series attractive to them all. The treatises offered a commanding view of the sciences written by acknowledged authorities committed to maintaining Protestant orthodoxy, and they were also written in an accessible manner despite their high price. Moreover, while they were not appealing as works of natural theology, their discourse of design was not unwelcome to the SPCK and the RTS provided it was tempered by appropriately distinctive Christian references. Even the SDUK and, more particularly, the Chambers brothers were sometimes able to reconcile such references with their avowedly nonsectarian tone. As a result, albeit in different ways, the cheap miscellanies brought the Bridgewaters to the attention of millions of readers, conveying to a greater or lesser extent the impression that the sciences at the heart of the "march of mind" were in keeping with Christianity.[56]

❋

The cheap miscellanies were not the only periodicals that found the Bridgewaters useful vehicles of scientific enlightenment. From the mid-1820s, British readers were offered a rapidly growing range of more or less scientific magazines aimed at naturalists and gardeners, at mechanics and autodidacts, and at practical men in fields such as mining and veterinary surgery. At a time when the most studiously learned of the scientific magazines had been reduced to just two titles—the *Philosophical Magazine* and the *Edinburgh New Philosophical Journal*—these new journals addressed a much wider range of readers.[57] Moreover, as their editors cast about for

55. Knight 1864–65, 2:192. Compare, for instance, "The Dinotherium" and "Proof of Design in the Origin of Iron and Coal," *Penny Magazine*, 27 May 1837, 195–97 and 200. See also Topham 1992, 405–20.

56. "Preface," *Penny Magazine* 1 (1832): iii–iv, estimated that that journal alone had a readership of a million.

57. Dawson and Topham 2020.

copy, the Bridgewaters had some appeal as authoritative, accessible, and religiously safe guides to contemporary science.

Neither the *Philosophical Magazine* nor the *Edinburgh New Philosophical Journal* devoted much space in their admixture of short scientific articles and scientific news to reviewing, and neither reviewed any of the Bridgewaters, although readers soon began to respond to their scientific claims in original articles. By contrast, many of the proliferating natural history periodicals viewed their remit as being at least in part educational, and some of the Bridgewaters seemed especially useful in this context. In November 1834, for instance, Roget's account of "vegetable physiology" attracted a review in the cheap monthly *Magazine of Botany and Gardening*, probably written by the magazine's editor James Rennie, which approved of both its useful scientific exposition and its religious framing. Rennie was the professor of natural history at the Anglican King's College London, but he was not alone in giving his botanical magazine this religious cast or in using Roget's Bridgewater to do so.[58]

Kirby's Bridgewater was also of obvious relevance to the natural history magazines, but here there were problems. The first was its high price, as the inexpensive *Magazine of Natural History* tartly pointed out in a review in August 1835 that began, "We have not read this work." But there was also Kirby's idiosyncratic biblicist approach. The *Edinburgh Journal of Natural History and of the Physical Sciences*—a populist magazine begun in 1835 by William MacGillivray, former assistant to the Edinburgh professor of natural history—found the work "interspersed with much interesting detail, especially regarding the lower classes of living beings," but lamented the fact that it resembled "a lecture on divinity" more than a "philosophical treatise." The details nevertheless remained both interesting and useful, and MacGillivray later provided readers with an extract.[59]

This dual sense of the utility and idiosyncrasy of Kirby's work was evident in the much more learned quarterly *Entomological Magazine*, edited by Quaker manufacturer Edward Newman on behalf of London's recently founded Entomological Club as an outlet for the "scientific Entomologist." Part of the mix was to be "Reviews (with extracts) of all new Ento-

58. *Magazine of Botany and Gardening* 2 (1834): 165–66; *London and Edinburgh Philosophical Magazine*, 3rd ser., 5 (1834): 33–39; Prout 1834c; Wright 1838. See also "The Physiology of Plants," *Magazine of Botany and Gardening* 2 (1834): 172–74, "Preface," *Botanic Garden* 5 (1833–34): [iii–iv], and "Preface," *Botanist* 1 [1836–37]: i–iv (the last two journals were edited by the devout Bromsgrove bookseller Benjamin Maund).

59. *Edinburgh Journal of Natural History*, 24 October 1835, 4; *Magazine of Natural History* 8 (1835): 471.

mological works," and Kirby's Bridgewater was duly reviewed at length. The magazine had, the reviewer reported, been "sorely disappointed" with the Bridgewaters as "works of science." Yet while Kirby's contribution was "necessarily a work of compilation," containing little new "in the way of fact," it did contain a "highly valuable" body of material from reputable sources. Thus, the reviewer had "no hesitation in recommending it to the general reader," observing that "The aged cannot rise from its perusal without pleasure, nor the young without having received instruction." The only hiccup was Kirby's insistence on using the Bible to elucidate scientific questions.[60] Newman had adorned the title page of the magazine's first volume with quotations concerning the religious value of nature study, but Kirby's peculiar approach was altogether objectionable.

Kirby's idiosyncrasies aside, however, many of these natural history magazines clearly considered that several of the Bridgewaters were of considerable value as works of instruction and reference. Moreover, readers joined editors in that conviction. In 1837 one contributor to the *Analyst*—a quarterly journal of "science, literature, natural history and the fine arts" produced in the West Midlands—littered his article describing fossil Crustacea with references to the plates in Buckland's Bridgewater, observing that the work was "doubtless in the possession" of the journal's readers.[61]

Few scientific periodicals aimed at as wide an audience as the cheapest of the magazines of natural history and gardening, but those that did also saw the Bridgewaters as being of educational interest. The first such journal had been the threepenny weekly *Mechanics' Magazine*, founded in 1823, which discussed the sciences alongside technical developments, typically in contributions from readers. With its limited space, the magazine reviewed relatively few publications—mostly publications on manufactures and machinery, inexpensive books, or periodicals. Bell's work on the "mechanism of the hand" nevertheless seemed sufficiently "germane" to the interests of mechanics to warrant attention, with its "vast multitude of facts, of the most curious and important character." From the perspective of those wishing to educate themselves in science, however, its "arrangement" was notably "deficient." Moreover, the book prompted an artisanal critique of the scientific establishment as reflected in its management of patronage, leading one correspondent to offer an altogether more damn-

60. *Entomological Magazine* 3 (1835–36): 296–97; "Introductory Address," *Entomological Magazine* 1 (1833): 2–3.

61. W., "An Account of Two New Crustacea from the Transition and Carboniferous Strata," *Analyst* 6 (1837): 85–89, on 89.

ing indictment of the appointing of the authors and the quality of the works. The writer nevertheless exempted Bell's "mass of curious information" from the criticism together with Roget's "work beyond all praise."[62]

A similarly equivocal response came from the short-lived monthly *Magazine of Popular Science*, which was started in 1836 under the supervision of the Adelaide Gallery, a commercial gallery of practical science recently established in London by the American inventor Jacob Perkins. The magazine set itself to inspire and educate "the general reader" with a "popular view" of the "progress and condition of the Physical Sciences," particularly seeking to appeal to young people and practical men. Much of its content was introductory, and it promised short reviews of works that were of a "*popular cast*" or designed for "*elementary* instruction." Running to ten pages, the review of Buckland's Bridgewater that opened the issue for November 1836 was anything but short, however, reflecting the book's extraordinary public profile and the magazine's promise to examine the sciences in relation to their wider connections and above all "as connected with Divine Truth." This concern reflected the wider interests of the editor, Oxford's clerical geometry professor Baden Powell, and it seems clear that he was the reviewer of Buckland's Bridgewater.[63]

Already in October the magazine had used its "Popular Course of Geology" to expound the claim that geological discoveries were "not opposed to Sacred History." In the process it drew heavily on Buckland's Bridgewater to show that the Bible could be interpreted in ways that were consistent with modern geology and that even Buckland had now abandoned his claims concerning geological evidence of the flood. The following month, however, the magazine offered Buckland a sharp rebuke that reflected Powell's growing frustration with his colleague's more placatory approach to High Church critics. In general, the reviewer observed, the Bridgewaters had marked an "epoch" in the literature. The "agreeable and attractive manner" in which "knowledge of the Creator and his attributes" had been inculcated in the series had produced "as much pleasure combined with instruction as perhaps any equal number of volumes ever published." By contrast, Buckland had foolishly continued to mix his geology with revelation, indulging in "speculative hypotheses"—for instance, about the initial formation of the globe—that would leave the uninitiated with "erroneous notions as to the proper objects and limits of the science." More generally, his desire to make his work "popularly attractive"

62. *Mechanics' Magazine*, 30 August 1834, 376–79; 13 September 1834, 412.

63. Corsi 1988, 6, 153, 312; "Prospectus," *Magazine of Popular Science* 1 (1836): 1–4; Bathe and Bathe 1943, 155.

had damaged its usefulness, meaning that novices could not "acquire any connected outline" of the subject from it. The work provided a wonderful account of fossil remains but ultimately failed as a work of instruction.[64]

The reviewer in the *Mining Review* for February 1837 was rather more sanguine concerning the value of Buckland's Bridgewater in advancing geological understanding among practical men. Begun in 1830 as a quarterly by the stockbroker Henry English, this was one of Britain's earliest trade periodicals, initially filled largely with news of mining companies and Latin American mining districts. English hoped from the start, however, to reach the "scientific man" as well as the "practical miner" with original papers on practical mining and notices of new publications "connected with mining." By 1836 the review's more commercial and practical matter had been removed to a weekly *Mining Journal*, leaving space for reviews and original matter. The thirty pages devoted to the review of Buckland's Bridgewater was nevertheless quite unprecedented and offered lengthy quotations on the grounds that its price was "such as to prevent that general circulation which the work deserves." The reviewer thus clearly considered Buckland's work to be highly instructive, although much of his commentary was nevertheless devoted to repeating the author's arguments concerning the religious significance of the subject in a manner that echoed the review's increasingly learned emphasis.[65]

The belief that certain of the treatises would be useful for "practical men" was also evident in the *Veterinarian*, a monthly begun in 1828 with the intention of improving professional standards among veterinary surgeons. In this context, several of the Bridgewaters were deemed worthy of review by the coeditor and prominent veterinary surgeon William Floyd Karkeek on the grounds that they offered invaluable and accessible instruction concerning sciences that were ancillary to veterinary practice. First came Roget's, which was deemed to have met the long-felt need for a "popular treatise on Comparative Anatomy," and the review provided an account of some of the "interesting facts that most intimately concern us as veterinarians," while noting the work's religious purpose almost as an afterthought. The following April, Karkeek was excited by the prospects that Prout's volume offered of extending the lamentably poor state of education

64. *Magazine of Popular Science* 2 (1836): 337, 339, 340; "A Popular Course of Geology," *Magazine of Popular Science* 2 (1836): 267–78. At least one reader considered the reviewer to have gone too far. See [Smith] 1836.

65. *Mining Review*, n.s., 1 (1837): xx; "Reviews of Recent Publications Connected with Geology, Mining, Metallurgy, &c.," *Mining Review*, n.s., 1 (1837): 72; "Preface," *Quarterly Mining Review* 1 (1830): v–viii, on v–vi.

among veterinary surgeons in pharmaceutical chemistry and physiology, summarizing the part on the function of digestion and recommending the work to the "veterinary pupil." The next month the journal returned to Roget's treatise in similar vein, recommending that students and others should obtain it. When in July and August 1835 the magazine provided an account of Bell's Bridgewater, Karkeek felt "almost ashamed" of copying so much. These reviews repeatedly endorsed the design perspective, and Karkeek even reviewed the less practically relevant volumes by Kirby and Buckland, although with reservations about Kirby's scriptural interpretations. Yet the overriding concern was with how the Bridgewaters contributed to the advancement of "veterinary science" in a nascent profession.[66]

For these scientific and technical periodicals, the Bridgewaters were by no means an unmitigated boon. Their prices were high, they were not consistently successful as instructional works, and the religious views of Kirby and Buckland were potentially problematic. On the whole, however, their command and readability were more useful than otherwise, and their pious framing was welcomed even in occupational publications.

❋

Despite the fact that the medical profession had an increasingly well-developed system of education, the medical press also presented some of the Bridgewater series as being potentially useful for professional purposes, not least in offering practitioners wider scientific perspectives on medical subjects. In part the interest in the Bridgewaters reflected the fact that half of the authors were medical men and three were significant figures in metropolitan medicine, making their publications at the very least newsworthy. Since the founding of Thomas Wakley's reformist *Lancet* in 1823, news had become a prominent aspect of medical journalism, with two competing titles joining a weekly market that was strongly marked by the politics of medical reform. The reformist agenda entailed challenging the vested interests of the Anglican-dominated medical establishment, and the heated exchanges between the medical weeklies not infrequently touched on questions of religion. Yet the reaction to the Bridgewaters was largely appreciative.

Although most of the short book reviews that appeared in the medical weeklies were narrowly professional in focus, all three weeklies noticed some of the Bridgewaters in this way. As one of the leading figures in Metropolitan medicine, it was Bell whose Bridgewater first attracted attention

66. *Veterinarian* 8 (1835): 240, 283, 411; *Veterinarian* 7 (1834): 618.

at the start of September 1833. The moderately reformist *London Medical and Surgical Journal* welcomed his contribution in superlative terms, asserting that it was sufficient in itself to mark Bell out as "one of the greatest biologists of the past or present eras." The reviewer was probably the editor, the reformist private medical school teacher Michael Ryan, who claimed never to have read a work with greater pleasure. Indeed, in offering some specimen extracts he claimed that he would have been happy to "transcribe the whole of it" into the journal if only space had allowed. Nor did he shy away from Bell's religious position, welcoming his repudiation of phrenology and the materialism that "would make man a vile and perishable insect," as well as his emphasis on design.[67]

The *Lancet* followed with a largely positive review at the end of October. When attacking the Anglican establishment, the *Lancet* could be acid about the manner in which leading figures made medicine subservient to Anglican orthodoxy. This had been the context for its earlier satirical depiction of Charles Bell's tendency to refer students to the "winderful eevidence of *desin*." Yet in other moods the journal was highly respectful of Bell, adding him to its short-lived "Gallery of Medical Portraits" and publishing his latest course of lectures at the Royal College of Surgeons. The review of his Bridgewater acknowledged the authors appointed to be largely creditable and claimed to recognize the content of Bell's work from his recent lectures at the Royal College of Surgeons. Moreover, while the reviewer considered it "needless" to prove design from the human hand ("Who doubts design?"), he claimed that Bell was not seeking to prove a "*particular* design, directed to a *partial* end, but a design which forms part of a grand scheme, planned by a superhuman mind." The review proceeded to offer an appreciative overview of Bell's comparative anatomy, referring readers to further "highly interesting matter" in the original, including Bell's "vigorous attack on St.-Hilaire's theory of elemental parts." According to the *Lancet*, Bell was no narrowly Paleyan natural theologian but a comparative anatomist able to trace divine action in morphological laws.[68]

It was the conservative establishment weekly, the *London Medical Gazette*, that brought up the rear in November with a glowing review of Bell's "delightful work" that ran to several pages. For this reviewer, what was

67. *London Medical and Surgical Journal*, 7 September 1833, 180–81.

68. *Lancet*, 26 October 1833, 165–69; "Sir Charles Bell," *Lancet*, 7 September 1833, 756–61; Bell 1833–34; "Mr Green's Sky-Rocket Lecture," *Lancet*, 27 October 1832, 151–55, on 154. Cf. Adrian Desmond's reading of Bell's work as narrowly "Cuvierian" and as having "elicited a yawn from the Wakleyans." Desmond 1989b, 259, 116.

most impressive about Bell's Bridgewater was the author's "happy tone," which reflected his habit as a lecturer, disparaged by the *Lancet*, of finding design in everything. It was this habit, the reviewer claimed, that underpinned Bell's extraordinary discoveries in anatomy and physiology, and the *Gazette*'s review drew out the author's emphasis on functional adaptation and his associated opposition to Geoffroy Saint-Hilaire and Lamarck rather than the emphasis on a "grand scheme" of design found by the *Lancet*. The reviewer's parting shot was to regret that what he saw as anticompetitive reforms in medical education, such as the founding of London University, had resulted in the capital's students being deprived of Bell's "elevated views and philosophical spirit."[69]

These early responses were altogether encouraging, and there followed a steady flow of further reviews, not only in the weeklies but also in the quarterlies, which provided abstracts of or selections from new works for those too busy or too poor to be able to read the originals. On the whole, the reviews were appreciative of these "excellent treatises." However, some of the ways in which the Bridgewaters sought to render the sciences religiously safe proved controversial. Not surprisingly, for instance, the quarterly *Phrenological Journal* took issue in December 1833 with those of the authors who had attacked phrenology. Bell's views were critiqued on the basis of one of his older publications, while Kidd's Bridgewater was characterized as admitting the truth of phrenology's "general principles" but "*denying all its details*" and thus escaping from its applications. However, it was Chalmers's nonmedical Bridgewater that was made the subject of a lengthy review written anonymously by Britain's preeminent phrenologist, George Combe, who was outraged that it offered what he took to be a Christian gloss on his own phrenological *Constitution of Man*.[70]

Like Combe's *Constitution of Man*, the *Phrenological Journal* was intended to make phrenology respectable, and Combe's review began by warmly applauding the objectives of the Earl of Bridgewater. Moreover, Chalmers's work was treated with respect. However, the review made very vivid the difference between the vision of the moral position of the human species offered by Chalmers and that offered by Combe. By allowing his "preconceived scriptural opinions" concerning the "irremediable corrup-

69. *London Medical Gazette*, 16 November 1833, 253, 258.

70. Combe to R Chambers, [1833], in Gibbon 1878, 1:295–97; "On the Natural Provision for the Security of the Brain; with an Answer to Sir Charles Bell's Objections to Phrenology," *Phrenological Journal* 8 (1832–34): 332–38; *Phrenological Journal* 8 (1832–34): 354; *Medical Quarterly Review* 1 (1834): 358; Conolly 1835, 22.

tion" of humans to interfere with the inferences he had been appointed to draw by reason alone, the review claimed, Chalmers had betrayed his commission and gone awry. Nevertheless, Chalmers's analysis of the mutual adaptation of the laws of nature and human morality bore a certain resemblance to Combe's own. The reader must therefore decide between Combe's vision of human nature being progressively improved by being brought into alignment with natural and moral laws and Chalmers's vision of Adam's Fall and the need for God to accomplish spiritual regeneration through grace. Outlining Combe's principles, the review ended by asserting that only phrenology could properly manifest the divine attributes in the human constitution.[71]

The views of phrenological critics also made the pages of the *Lancet*. Reporting on the latest meeting of London's Phrenological Society, of which the journal's editor was a founding member, the journal related how the society's president, Wakley's friend and ally John Elliotson, had answered Kidd's critique. The Bridgewater of this "highly informed and excellent man," Elliotson had noted, made so many positive statements about phrenology that it was clear evidence of the progress of the science. A professor from Oxford, "the university of all that is established and orthodox," had agreed to all the basic principles of phrenology, showing "soundness of judgement and freedom from intellectual and moral prejudice." Only then had Kidd unaccountably demurred from phrenology's concrete scheme for the localization of brain function, and Elliotson answered his objections robustly, asserting the consistency of phrenology with true religion.[72]

Another aspect of the Bridgewaters that provoked controversy in the medical press was Kirby's riposte to Lamarck. The moderately reformist *Medico-Chirurgical Review*, a quarterly run by Ulster Protestant James Johnson, began a three-part review of Kirby's Bridgewater in October 1835 that critiqued and derided his Hutchinsonian reasonings. These were as ridiculous as the "absurd and atheistical hypotheses and speculations of La Place and Lamarck" that Kirby had attacked. The literalism of his account of creation was foolish and his miraculous providentialism "degrading alike to God and man." However, these criticisms were interspersed with appreciation both of Kirby's character and of his systematic natural history. His Bridgewater offered useful information concerning the sci-

71. *Phrenological Journal* 8 (1832–34): 340.

72. *Lancet*, 22 February 1834, 835–37; 1 March 1834, 861–63. On Elliotson and Wakley, see Winter 1998, 48–49.

ences "collateral" to the "severer studies of practical medicine." Complaining about the book's high price, the review promised to condense its contents to make them available to "all classes and ranks of our brethren."[73]

This review provoked an anonymous riposte by a correspondent of the *London Medical Gazette*, who admitted that Kirby's theology left something to be desired but feared that the reviewer's expressions might be taken to give countenance to the common accusations of irreligion against the medical profession. This did not deter the *Medico-Chirurgical* reviewer from further ridicule or from asserting the dangers of attempting to "unravel" final causes—"fruitless speculations" that the medical profession had abandoned—but he concluded by telling his critic that the strictures had been offered to "promote religion and piety." The sentiment was repeated when in the same issue the journal began a two-part review of Prout's treatise by insisting that it wished to second Lord Bridgewater's intentions by spreading knowledge of natural theology among members of the profession, too many of whom were accused of levity or irreligion. The series had done nothing to extend the arguments of natural theology and was "rather adapted to confirm faith than to create it," but then no one but an idiot or a "downright metaphysician" could be an atheist. Once again, however, a major incentive for reviewing the work was to provide an account of those parts of it that would be professionally useful to medical readers while also taking seriously the need to enforce "philosophical habits of thought" on the many readers of "imperfect education."[74]

That works like Kirby's offered medical men opportunities to flex their philosophical muscles was further emphasized by a respectful reviewer in the *Lancet*. The reviewer did not agree with Kirby on everything, he wrote in conclusion, but this was an original work of merit that readers would find offered valuable "materials for thought." Like Kirby, he began by discussing Lamarck's theory, claiming that it was well worthy of examination but agreeing that it was "not satisfactorily made out." Next came the "recapitulation" theory of German embryologist Johann Friedrich Meckel, discussion of which drew in some criticisms from Bell's Bridgewater, followed by a critical discussion of spontaneous generation. Of course it was a radical move to claim that Lamarck's theory was even "a bare possibility," but there were many questions to be answered, and the writer asserted that the origin of living animals from physical matter and the process of transmutation could not have taken place without "some cause" that was "not

73. *Medico-Chirurgical Review* 23 (1835): 400–401, 409, 413.

74. *Medico-Chirurgical Review* 24 (1835-36): 88, 364, 386, 391; 25 (1836): 115; *London Medical Gazette*, 28 November 1835, 318-19.

in operation now." Altogether, the discussion was strikingly unorthodox and significantly at odds with those in the Bridgewaters, but it was offered as a respectful response to questions raised by Kirby.[75]

The relatively large numbers of reviews in the medical press offered readers a sense that, even though some of them had significant failings, the Bridgewaters could be professionally useful to medical practitioners. In addition, they were works that medical editors considered to be of relevance to ongoing debates about scientific theory, including on such controversial topics as transmutation and phrenology. In the highly politicized context of medical journalism, they were also occasions to debate the place of Christianity in both medical teaching and scientific theory, as well as to comment on the management of patronage. The editors of reformist medical journals were certainly willing to question the Bridgewaters' religious positions, but even those who did so offered reassurance that medical science would always be found consistent with Christianity.

The number, diversity, and readership of periodicals offering scientific content in Britain during the 1830s was quite unprecedented, but the Bridgewater authors' readable and commanding accounts of the several sciences appealed widely across those periodicals, couched as they were in terms that appeared to offer no threat to Christianity. For High Church and evangelical editors keen to ensure that the sciences were encountered in a distinctively Christian framework, significant effort was required to cut and paste appropriately. For *Lancet* radicals, by contrast, the Bridgewaters could provide opportunities for speculative discussion about matters of high theory. Yet these works of safe science were generally welcomed as being of real value in offering trustworthy instruction to a range of readers, from impoverished workers to veterinarians.

## *Geological Controversy and the Press*

Over the space of four years, British readers encountered the Bridgewater Treatises through a huge range of periodicals in which they were typically portrayed as offering a religiously safe vision of the sciences. At the end of that period, however, Buckland's much-delayed treatise on geology followed a markedly different trajectory through the periodical press that turned its author into "quite as much a newspaper subject as would an horrid murder or a glorious victory."[76] During an interval lasting sev-

75. *Lancet*, 17 October 1835, 105–6, 109.
76. *Spectator*, 1 October 1836, 947.

eral months, the work became a cause célèbre of scientific progress and religious orthodoxy, with not only newspapers but also magazines and reviews commenting at length. Yet in spite of the fact that many commentators were fiercely critical of Buckland's reinterpretation of the Genesis creation narratives, numerous major conservative publications weighed in to defend it. The overall impression conveyed by the press was that the clerical professor had won the day with his new vision of harmony and that his critics were overzealous outsiders.

The peculiarly public controversy generated by Buckland's Bridgewater reflected several particular circumstances. First, geology was a science whose foundational principles had become a matter of growing dispute over recent years, with numerous religiously conservative writers offering alternative visions of the role that biblical evidence should maintain in interpreting the earth's past. The often accessible self-styled "scriptural" or "Mosaical" geologies that such authors offered achieved a significant degree of success, captivating readers with an alternative vision of both the history of the earth and the science of geology and increasingly alarming Buckland and his peers.[77] Several such works—notably Sharon Turner's very successful *Sacred History of the World* (1832) and George Fairholme's *General View of the Geology of Scripture* (1833)—were widely reviewed in the 1830s, especially in the religious press. Moreover, while scriptural geologies were not infrequently criticized even by conservative reviewers, the topic attracted lively debate, especially in evangelical periodicals.

A second factor affecting the response to Buckland's Bridgewater was the great delay in its publication and the growing sense of anticipation. While some of Buckland's friends feared that the delay might actually damage sales, the opposite proved to be the case, perhaps assisted in fashionable circles at least by the speculation concerning what he would do with respect to his earlier views concerning geological evidence for the biblical flood. In any event, Buckland was taking no chances, and a third factor behind the press coverage of his Bridgewater was the author's own active role in managing it. This was especially evident in the manner in which he set about securing supportive reviews, although here he experienced a touch of good fortune.

While touring in Germany in the summer of 1835, Buckland ran into John Murray, who told him that he was eager to get "the best possible Re-

77. O'Connor 2007a, 133–59, 202–3; O'Connor 2007b. See also the useful but polemical Mortenson 2004.

view" of his Bridgewater in time for the April 1836 issue of his *Quarterly Review*. Murray's enthusiasm partly reflected his friendship with the author, but it also made good business sense given how much the book had been talked about and given also that it was—as the review's editor declared—one of those rare scientific works that was "readable by all literary persons who can follow a logical chain of argument." Happy to capitalize on Murray's enthusiasm, Buckland set about ensuring that the review would indeed be commanding. The geologist advised that, since his work had a "3 fold Character," the review would need the involvement of two or three individuals. Lawyer and naturalist William Broderip, who had corrected the proofs of the paleontology sections of Buckland's Bridgewater, was drafted in on that topic, and Buckland hoped that botanist Robert Brown and stockbroker and geologist Charles Stokes, whom he had lined up to read the proofs on fossil plants, would prepare notes for the review of that portion. The editor agreed that Buckland's opposite number in Cambridge, Adam Sedgwick, should be asked to write the review of the "Pure Geology, and the Physico Theology" in the introduction and conclusion, and Buckland sent a sample to entice him to do so, although without the desired effect. In the event it was taken on by geologist and regular *Quarterly* contributor George Poulett Scrope.[78]

Appearing fully five months before the book itself, this carefully orchestrated review served "to convince the world that it had not perished in the gestation." It introduced Buckland's treatise to the public in grandiose terms, observing that this was a "work of a dignitary of the church, writing, *ex cathedrâ*, from the head-quarters of orthodoxy" to demonstrate that geology was consistent with Christianity and that it afforded new evidence for the arguments of natural theology. Scrope proceeded to outline Buckland's reconciliation scheme and general geology before handing over to Broderip to describe the most interesting part—"the consideration of fossil organic remains." The effect of the whole was to convey an overpowering sense of the support that geology offered to Christian faith. Buckland's treatise was not merely an invaluable "*reportorium palaeontologicum*"; it made the subject "an appeal to the better and nobler sentiments of our nature, in plain language, unincumbered as much as possible by the technical terms that deter too many from entering this most pleasant field of inquiry." So commendatory was the review that Buckland's friend

78. WB to Sedgwick, 28 October 1835, and WB to Featherstonehaugh, [25 April 1836], CUL, Add.7652, IB.44 and II.LL.32; WB to Murray, 6 January 1836, NLS, MS.40165, fols. 57–58.

Francis Chantrey thought it would provoke a "slashing article in the Edinburgh" that would "mummyfy" him.[79]

In fact, the *Quarterly*'s review provoked a rash of derivative reviews and extracts in newspapers from journalists who had not yet seen the book.[80] Such press coverage was doubtless to be expected given the anticipation regarding Buckland's Bridgewater. Much more unusual, however, was the press coverage that followed the author's public statements about his new book and its religious interpretations at the meeting of the British Association for the Advancement of Science meeting in Bristol in the last week of August 1836 (fig. 4.4). The association's proceedings were commonly recounted in local newspapers as well as in the metropolitan weekly literary journals, but when Buckland presented a copy of his Bridgewater to its president on the final Saturday, his words reverberated far more widely. In the middle two weeks of September, the following paragraph appeared in newspapers the length and breadth of the country:

> Age of the World.—At the meeting of the British Association on Friday, the only fact elicited through the evening, was the declaration of Dr. Buckland, that millions of years must henceforward be assigned to the age of the world, and that the best Hebrew scholars had lately given a new interpretation to the two first verses of Genesis. This announcement of the rev. doctor was received with an applause that lasted some minutes.

Apparently based on a local newspaper report, the paragraph quickly became the focus for widespread and often contentious commentary and discussion, including in the metropolitan dailies, giving Buckland's Bridgewater an exceptional amount of publicity before the book became available for purchase.[81]

New publications did sometimes warrant discussion in newspapers, but it was usually weekly newspapers that found space to offer comment. Whewell's Bridgewater, for instance, was noticed appreciatively in the

79. Chantrey to WB, 1 January 1836, BO, Ms.Eng.Lett.b.35, fols. 16–17; *Quarterly Review* 56 (1836): 31, 62; WB to A Irvine, 25 February 1837, CC, Ms.531.

80. The derivative review in the *Morning Post*, 7 May 1836, 6f, was widely reprinted (see app. B). One extract on "Formation of Coal and Iron," taken via the *Quarterly*'s review from Buckland 1836, 1:66–67, was especially widely reprinted between July and October 1836.

81. The earliest version of the paragraph I have located is in the *Belfast Commercial Chronicle*, 5 September 1836, 1b. For a list of some of the many papers that reprinted it, see appendix C.

BRITISH ASSOCIATION.

BRISTOL, FRIDAY, AUG. 26.

The sections met as usual this morning. Section C.—Mr. R. J. MURCHISON, V. P. for Geography, in the chair.

THE ANCIENT CITY OF MEMPHIS.

The Marquis Spinetto read a paper, entitled "A report of the attempts made to ascertain the latitude of the ancient city of Memphis." The details of this communication are of importance to geographers, as tending to elucidate a point on which Pocock, Shaw, Bruce, and other travellers have differed. The question may now be considered to be set at rest, it having been clearly ascertained that it was in the present bed of the Nile, in latitude 29, 46, north, and longitude 31, 30, east from Greenwich.

The CHAIRMAN congratulated the section on having heard these satisfactory details, and observed that the same process which had buried the ancient city of Memphis in the bed of the Nile—an accumulation of mud and drifted Lybian sands, in consequence of the demolition of the dikes, which once turned aside the water—had already sunk the beautiful beds of Purton beneath the Severn.

Dr. BUCKLAND then took the chair, and stated that he had received engravings, prepared under the direction of M. Agassiz, of some of the splendid fossils in the Bristol Institution; and he also placed upon the table a copy of his own work on geology, forming one of the Bridgewater Treatises.

The next paper was,

ON THE CHANGE IN THE CHYMICAL CHARACTER OF MINERALS INDUCED BY GALVANISM.

Mr. FOX mentioned the fact, long known to miners, of metalliferous veins intersecting different rocks containing ore in some of these rocks, and being nearly barren, or entirely so, in others. This circumstance suggested the idea of some definite cause; and his experiments on the electrical magnetic condition of metalliferous veins, and also on the electric conditions of various ores to each other, seem to have supplied an answer, inasmuch as it was thus proved that electro-magnetism was in a state of great activity under the earth's surface, and that it was independent of mere local action between the plates of copper and the ore with which they were in contact, by the occasional substitution of plates of zinc for those of copper, producing no change in the direction of the voltaic currents. He also referred to other experiments, in which two different varieties of copper ore, with water taken from the same mine, as the only exciting fluid, produced considerable voltaic action. The various kinds of saline matter which he had detected in water taken from different mines, and also taken from parts of the same mine, seemed to indicate another probable source of electricity; for can it now be doubted, that rocks impregnated with or holding in their minute fissures different kinds of mineral waters, must be in different electrical conditions or relations to each other? A general conclusion is, that in these fissures metalliferous deposits will be determined according to their relative electrical conditions; and that the direction of those deposits must have been influenced by the direction of the magnetic meridian. Thus we find the metallic deposits in most parts of the world having a general tendency to an E. and W., or N. E. and S. W. bearing. Mr. Fox added that it was a curious fact, that on submitting the muriate of tin in solution to voltaic

FIGURE 4.4 Daily newspaper report of the geological section at the British Association meeting, 26 August 1836, relating that William Buckland made a copy of his Bridgewater Treatise available for inspection by members. *The Times*, 30 August 1836, 4f.

radical weekly *Examiner*, while Chalmers's stature as a leading figure in national life ensured a lengthy review in the twice-weekly Whig-Radical *Scotsman*. It was much more unusual for the dailies to give column inches to books. However, Buckland's carefully staged statement about his potentially controversial new publication was perfectly timed for the moment when news was at its slowest, with Parliament in recess and the bon ton away from London, and it became a matter of discussion in newspapers up and down the land. The newspaper press was, admittedly, scandalously expensive at this period, but a landmark reduction in stamp duty on 15 September—the week before Buckland's Bridgewater hit the shelves—led many newspapers to drop their prices from 7d to 5d. Moreover, since many papers were available in reading rooms and public houses, they reached an enormous readership.[82]

The first comments on Buckland's reported claim came from two weekly newspapers noted for their often outrageous political interventions—the radical Whig *Satirist* and the ultra-Tory *John Bull*. The *Satirist* had made fun of the British Association meeting the previous week, suggesting that Buckland had received there a prize "for an essay proving that cabmen are naturally great *gee*-ologists, and that the Mosaic account of the creation of the world 6,000 years ago" tallied exactly with "what is observed of the strata of rocks which were formed fifty millions of years previous." This was now followed up by a satirical article purportedly warning the Archbishop of Canterbury that Buckland had committed a "Frightful Blasphemy" and suggesting that the geologist might next affirm the truth of eighteenth-century Scottish philosopher Lord Monboddo's suggestion that humans were descended from apes. Meanwhile, the somewhat more sober *John Bull*, which had also been ridiculing the pretensions of the British Association, archly claimed not to believe Buckland to be guilty of so serious a charge.[83]

These journals were designed to foment controversy, but more mainstream newspapers soon also joined in. At the forefront were two newspapers from the Tory stable of proprietor Charles Baldwin: the London evening *Standard* and the thrice-weekly *St James's Chronicle*, which was intended for the country market and widely read by the Anglican clergy. A letter signed "A Reader of the Bible" appeared in each, disputing Buck-

82. Altick 1957. Grant 1836 identifies the *Examiner, Atlas, Observer, Bell's Weekly Messenger*, and *Sunday Times* as having been especially attentive to new literature.

83. "Scientific Prizes," *Satirist*, 4 September 1836, 283; 11 September 1836, 293; *John Bull*, 12 September 1836, 293a–b.

land's long earth history as a matter both of biblical interpretation and of science. Geology and chemistry were such infant sciences, it claimed, that it remained perfectly possible that the creation might have taken place within the week described by Genesis. The author referred to the notion that electricity might explain how rocks came to be formed on so short a timescale—a view associated with Bampton Lecturer Frederick Nolan—noting in support Andrew Crosse's recent experiments on galvanic crystallization, which Buckland had publicized at Bristol.[84]

This letter was picked up in the provincial press, where editorial comment was added, but London's Whig papers quickly moved to forward the opposite view. The *Morning Chronicle* carried a letter welcoming and publicizing Buckland's announcement, while the most widely sold evening paper, the *Globe*, addressed a lengthy editorial to answering the "blundering fanatic of the *John Bull*." While admitting that neither scientific nor religious questions were usually suitable topics for newspaper discussion, the *Globe* wished to remind readers that all reputable philosophers agreed in the antiquity of the world, that religion and science were in no danger of being in "substantial opposition," and that it was would-be inquisitors who put the church in danger.[85]

The terms were thus set for a high-profile and politically charged public controversy. The *Globe*'s editorial prompted a reply from the *Standard* and the *St James's Chronicle* that was relatively temperate in tone but rather extreme in content. The objection to Buckland lay not in his chronology but in his denying the fact of the creation by insisting on its protracted nature. This, they declared, was tantamount to atheism. Buckland naturally bridled at such an accusation and, against his wife's counsel, he wrote an authoritative but conciliatory letter to the *Standard*, the *St James's Chronicle*, and *John Bull*, directing readers to his imminent Bridgewater. At the same time, the *Globe* was able to bolster its own restated editorial position with letters from correspondents, including one who signed himself as a Fellow of the Royal Society and member of the Royal Irish Academy. As the correspondence rumbled on in other newspapers, the *Globe* allowed itself to revisit the topic one last time, publishing a letter from Cambridge's Regius Professor of Hebrew, Samuel Lee, which argued that Buckland's notion of a hiatus in the Genesis narrative was unsustainable but that there was no reason why previous creations might not have taken place before

84. *Standard*, 14 September 1836, 2e; *St James's Chronicle*, 13–15 September 1836, 3d. On Crosse and his high profile in newspapers at this juncture, see Secord 1989b.

85. *Globe*, 17 September 1836, 2d–e; *Morning Chronicle*, 17 September 1836, 3f.

that narrative. Meanwhile, the *Scotsman* reported the London controversy through the *Globe*'s eyes, adding its own editorial comment.[86]

Such counterblasts only fanned the flames of controversy in the Tory press. Still without having seen Buckland's work, the *Standard* and *St James's Chronicle* devoted a two-column editorial to the question, associating Buckland's geology with Erasmus Darwin's rejected theory of transmutation and arguing that creation must always be miraculous, whether immediate or protracted. Over the following four weeks, further letters were published, all on the negative side of the question, since the editor viewed attempts to restate Buckland's case as tantamount to "rolling the road to make smooth the march of *Atheism*."[87] The Tory *Bell's Weekly Messenger* joined the fray, reprinting and expanding on articles from its contemporaries.

Such extended discussion added considerably to the already considerable notoriety of Buckland's work, emphasizing above all its controversial character. Such, indeed, was its public profile that in mid-November *The Times*—recently turned Peelite in its politics and still considered to be the nation's leading newspaper—devoted most of a column to an appreciative synoptic review from "a Correspondent well acquainted with the subject." Only such a book as Buckland's, the reviewer noted, could warrant "the least literary encroachment" on the pages of the self-styled "leading journal of Europe" at a time when the country required its "undivided and most strenuous support." It was, the reviewer assured readers, the "most valuable and eloquent discourse" on geology that had appeared since the seventeenth century "and the most complete summary of authentic facts relating thereto." Moreover, its statement of the "consistency of geological discoveries with sacred history"—quoted at length—effectually "cleared the ground of many popular errors and prejudices," leaving readers free to enjoy its authoritative and stirring account of modern geology.[88]

The peculiar profile that Buckland's Bridgewater had gained thus meant that it was placed at the forefront of public debate. And while it attracted vocal critics, its scheme of biblical reconciliation was seen to have the public backing not only of reform-minded commentators but also of *The Times*. Moreover, thanks to Buckland's clerical friend William Conybeare, the second substantial review to appear was also an overwhelmingly supportive one in a conservative quarterly. Impressed by what he read in the

86. For details, see appendix C. Mary Buckland's undated letter is reproduced in Gordon 1894, 196.

87. *St James's Chronicle*, 8–11 October, 2a.

88. *The Times*, 15 November 1836, 3d.

*Quarterly*, Conybeare asked Buckland for a copy of the book in order to write another early positive review. This was to appear in the High Church *British Critic*, which had previously been critical of scriptural geologists. Although the journal was now increasingly concerned by the degree of freedom claimed for science and turning toward the Tractarians, Conybeare was nevertheless able to ensure that his commendatory review appeared there within a fortnight of the book's publication.[89]

The *Critic*'s account applauded, without undue fanfare, Buckland's "very satisfactory system of conciliation between the results of science and the declarations of Scripture." However, its main emphasis was on the value of the work in giving a "popular view" of geology that fostered religious feelings. Commenting on the sublimity of Buckland's scientific findings, the review observed that the way in which the reader was "directed to look still higher" to see in them "proofs of the unity and attributes of the great designer of universal nature" promised "great benefit" from the "habit thus impressed of giving a religious association to our most interesting intellectual speculations." It was this, the reader was told, that would be found to be the "principal advantage" resulting from the Bridgewaters, although Conybeare did also welcome Buckland's extended but appropriately subordinate natural theology.[90]

This well-engineered review set a tone that the other High Church periodicals echoed. In November, the *British Magazine* carried a letter from Bedfordshire vicar William Balfour Winning that claimed the support of Buckland's new work for the interpretation of Genesis that he had previously outlined in letters to the magazine. The next issue carried a letter from a further correspondent supporting Buckland's interpretation on biblical grounds. Winning's views had previously generated some dissent from the magazine's correspondents, and there was a distinct degree of circumspection in the manner in which geology was handled by Rose, yet this was a welcome response. In December, moreover, the *Dublin University Magazine*, a publication associated with the city's Anglican Trinity College, greeted Buckland's work as offering an "abundantly satisfactory" account of the agreement between the Bible and geological findings.[91]

Although the *Christian Remembrancer* was a little slower to respond,

89. Corsi 1988, 52–53; WB to Featherstonehaugh, [25 April 1836], CUL, Add.7652, II.LL.32.

90. *British Critic* 20 (1836): 297, 300, 328.

91. *Dublin University Magazine* 8 (1836): 695; *British Magazine* 10 (1836): 554–56, 714. Note, however, that the critical [Carter] 1837 was based in part on an earlier letter to the *British Magazine* (Carter 1835).

the lengthy multipart review essay it devoted to Buckland's Bridgewater between January and May 1837 offered more commanding support. It began by asserting the ultimate congruence of geology and the scriptures and by answering the "unseemly and unchristian" attacks on Buckland in the *Standard* and the *St James's Chronicle*. Next came a lengthy extract containing the majority of Buckland's chapter on geology and scripture, followed by a learned discussion leading to the judgement that Buckland's reading of Genesis seemed likely to be correct, quite independently of geological considerations. The reviewer then proceeded to answer objections over succeeding numbers. Giving little attention to the detailed content of Buckland's treatise, he sought to vindicate its author from the charge of heresy and thus to remove "a great impediment in the way of the circulation of this interesting, able, eloquent, and learned addition to the evidences of Natural Theology."[92] Moreover, once the review was complete, an exchange of letters between the reviewer and two correspondents who responded to his invitation to outline serious objections kept both the topic and the magazine's party line before readers for a further four months.

Other religious periodicals also joined in, although the response of the evangelical magazines, several of which were well disposed to the self-styled "scriptural" geologists, was rather more muted. The editor of the *Christian Observer* had earlier engaged critically with literalist correspondents, including Fairholme and Henry Cole. Nevertheless, while Fairholme had pointed to Buckland's forthcoming Bridgewater as an intervention of crucial importance, the editor had apparently wearied of the subject by the time it appeared, allowing it to go unnoticed. In Scotland, however, the *Presbyterian Review* was entirely confident that geology would always agree with scripture "rightly understood," while the English *Wesleyan-Methodist Magazine* offered readers extracts from Buckland's work.[93] Moreover, the evangelical periodicals of "old" dissent (notably Congregationalists) were especially eager, as the champions of religious liberty, to stand up against Buckland's attackers.

Buckland's was the only one of the Bridgewaters that either the *Congregational Magazine* or the *Eclectic Review* brought to the attention of their readers, which they both did at the start of January 1837. The former argued that Buckland's was "decidedly the best" of the series but also claimed that it called for attention because of the severe attacks of "some

92. *Christian Remembrancer* 19 (1837): 2, 97.

93. *Presbyterian Review* 9 (1836–37): 242; [Wilks] 1834, 374n; [Fairholme and Wilks] 1834, 492; O'Connor 2007a, 210.

zealots" on its contents. The magazine lambasted the "spirit of sectarian hatred and ecclesiastical intolerance" that had assailed Buckland, casting it in a lineage of intrareligious persecution going back to Galileo. Science and revelation could never be at odds, the reviewer asserted, and he was proud to point to a series of essays on geology by Congregational minister John Pye Smith that had recently been published in the magazine's own pages that accorded with Buckland's reconciliation scheme. The *Eclectic Review* also approved of Buckland's geology, criticizing his handling of the scriptural issue only for being somewhat timid and incomplete. He should have "sternly grappled with the question, and pursued it, through all its ramifications, to a final adjustment."[94] Such views were also expressed in several of the Unitarian magazines.

The attacks on Buckland's Bridgewater in sections of the conservative press thus gave way to a remarkably broadly based endorsement of its message of biblical congruence. A year after the *Quarterly*'s early review, the *Edinburgh* summed up the positive reaction nicely. The essay began by reviewing the history of geology from the earliest days, when its laborers feared a "conflict between science and religion," to the present, when "in assemblies composed of Churchmen, and Dissenters, and Conservative statesmen, we have heard the walls ring with rapturous joy, when geology renounced her ecclesiastical tenure, and demanded a lease of MILLIONS OF MILLIONS of years for the range of their enquiries." Penned by Brewster, this was a nationalist story, with Scotland receiving the laurels for having embraced this viewpoint first, but it had two important lessons: that science must be unfettered by religion, and that natural and revealed truths would nevertheless always be found to be ultimately congruent. Buckland's work gave welcome, if rather belated English baptism to such views in geology.[95]

The widespread endorsement of Buckland's approach did not, of course, mean that his opponents ceased to be heard. The public controversy led *Blackwood's Edinburgh Magazine* to overcome its wonted indifference to science to include comments in a regular column devoted to "strange, touching, odd, melancholy, humourous, and terrible" passing events. This was by one of the magazine's most regular contributors, George Croly, the recently appointed rector of a notable church in the City of London and a fiercely conservative Irishman who had already published his own work of scriptural geology. In February 1837 Croly's column turned its eye on

94. *Eclectic Review*, 4th ser., 1 (1837): 25–26; *Congregational Magazine* 13 (1837): 42–45.

95. *Edinburgh Review* 65 (1837): 2, 13.

the "exceedingly trifling ambition of science" flourishing among "geologists, naturalists, and political economists," reserving particular contempt for the mischief done by the "presumption" of the geologist. Buckland's Bridgewater had, he sneered, had "the ill effect of exhibiting an English divine ranked on the side of the French geologists," with their "faith of chalk and lizards." Similarly, when *Fraser*'s final Bridgewater review appeared in December 1837, it was highly critical of Buckland's reinterpretation of the biblical text.[96]

Joining these two Tory magazines in attacking Buckland was a new *Church of England Quarterly Review*—a rabidly "*Church and State*" organ, bent on a "crusade against the triple alliance of infidelity, liberalism, and popery." In October 1837 the review subjected Buckland's work to a lengthy review essay accusing geology of being "infidelity in disguise." Where geologists—Buckland prominent among them—had not long before promised additional testimony from their subject in support of the Mosaic history, it was now clear that the doubts of many concerning this promise had been justified by Buckland's reinterpretation of the scriptures. The issue in the "contest of science with religion" had thus been clearly joined. Championing the scriptural geology of Fairholme and Granville Penn, the reviewer claimed that geologists were really covert materialists and infidels who would soon show their true colors. The kind of "concessions to the spirit of the times and to the march of intellect" demanded by Buckland in biblical interpretation would soon, like recent Catholic emancipation legislation, undermine the "whole bulwarks of religion and the state." Before long, England would go the way of France.[97]

The *Church of England Quarterly Review*'s response to Buckland's Bridgewater was fierce in the extreme, but the journal conceded that this was far from being the establishment view. The theological implications of the work's contents combined with the circumstances of its publication (including Buckland's stage management) meant that it became a matter for widespread and sometimes highly critical comment in the public press. Overall, however, the impression offered by many of the country's most sober and respectable periodicals—ranging from *The Times* to the *Quarterly Review* by way of the leading Anglican journals—was that right-

96. *Fraser's Magazine* 16 (1837): 719–24; [Croly] 1837a, 180–82; [Croly] 1836, 609; Croly 1834; O'Connor 2007a, 309–10. See also [Croly] 1837b, 690–92, and [Green] 1836, 589–91.

97. *Church of England Quarterly Review* 2 (1837): 452, 467; "Introduction," *Church of England Quarterly Review* 1 (1837): 1–34, on 1, 31.

minded people could be assured of the religious safety of modern geology. Oxford's professor of geology, speaking at the behest of the head of the church, had shown it to be so.

❋

While the high price of the Bridgewaters significantly limited the range of those who could read them, the phenomenal expansion of the periodical press in early nineteenth-century Britain meant that millions of readers had the opportunity to learn about the series or to read extracts from it. The writers, editors, and publishers of the various periodicals had different reasons for taking up the Bridgewaters and consequently presented them to readers in very distinctive ways. Yet as the tide of comment and quotation washed across the range of periodical publications, readers of many different hues encountered the Bridgewaters as scientifically instructive and as confirming the ongoing consistency of the sciences with Christian orthodoxy. Observations on natural theology in the periodicals ranged from strongly antagonistic to warmly appreciative. However, most commentators were agreed that this was not in any case where the importance of the Bridgewater series lay.

Few books of the age of reform went through such a process of description, discussion, and excerpting across the full range of periodical publications as the Bridgewaters. The public notoriety that resulted from their origins, the attractiveness of their "safe," authoritative, and accessible scientific contents, and their conspicuous religious framing made these works of wide appeal to those who produced such periodicals. But what of their readers? For all that those who produced them considered themselves to have their fingers on the pulse of their readers, periodicals—like books—were never experienced just as their producers hoped or expected. It is thus time now to turn away from the producers of printed matter and to begin to consider how readers actually encountered the Bridgewaters, whether in their original form or through periodicals and other media of literary replication.

# PART III

# *Reading*

✷ CHAPTER 5 ✷

# *Science and the Practice of Religion*

*It cannot be doubted, that great and extensive mischief may arise to Religion, and to the eternal welfare of mankind, should our general literature, and the various institutions of society, acquire a character and tendency decidedly contrary to the principles and practice of Christianity.*

Report of the Society for Promoting Christian Knowledge, *1832*[1]

In the new age of cheap print, the natural sciences intruded themselves relentlessly into the daily lives of Christians of all kinds and classes. As a growing supply of ever-more various scientific publications tumbled from the presses, their implications for the Christian religion became a topic for urgent examination. Many considered it to be a crisis, but if it was, it was at least as much a consequence of changes in print culture as of changes in the sciences. According to these critics, a terrible danger to religion lay in the growing availability of printed works discussing the sciences without reference to religious concerns—something that the publications of the Society for the Diffusion of Useful Knowledge made especially clear. Such works might contain erroneous claims at odds with Christian doctrine, but they were much more likely to do damage by undermining the religious sensibilities of their readers as a result of presenting the creation independently of its creator, his revelation, and his dispensation of saving grace. It was these concerns that bodies such as the Society for Promoting Christian Knowledge sought to address with their alternative enterprises in scientific publishing.

For many of those who read the Bridgewater Treatises, the critical concern was likewise with the relevance of the sciences to the everyday practice of religion. As the previous chapter showed, many commentators considered that the series was valuable precisely because it allowed religious readers to engage with the latest scientific discoveries in such a way as to

1. Anon 1832, 16.

protect and enhance their religious sensibilities. While some evangelicals and High Anglicans considered them to be insufficiently imbued with gospel truth or objected to their geology, Christians of various stripes were happy that the Bridgewaters could be turned to substantial good. The series provided an up-to-date survey of the several sciences, written by leading scientific men, and with the imprimatur of the Anglican primate as well as the president of the Royal Society. The treatises were, moreover, written in an accessible style and connected their scientific exposition throughout both to religious truths and to religious feelings. For many religious readers concerned that modern science should sustain rather than undermine their spiritual journey, the Bridgewaters bore real promise.

But how was that promise to be fulfilled? How were these new books to be incorporated into the daily lives of Christian believers in such a way as to bear spiritual fruit? To answer that question we have to enter into the quotidian practice of Christianity and reflect on how reading in general and scientific reading in particular fitted into the daily round of religious lives in the age of reform. Where historians of science and religion have commonly concerned themselves with public religious pronouncements—typically of a theological cast and often made by men—our concern here is with a much more private religious world of meditation and prayer, of family relationships, and of education and recreation in which feelings, habits, and dispositions were at least as important as beliefs. These concerns had a critical role in framing readers' encounters with books of all kinds, and if we are to understand how the Bridgewaters operated to reconnect Christianity and the sciences on the eve of the Victorian age, then we must follow them into that unfamiliar world.

Since reading was a matter of such consequence in religion, it was thankfully the subject of much published discussion. Many of the sources of religious reading advice have, however, been notable in modern libraries for their tendency to gather dust. This chapter disturbs some of that dust in order to make sense of the often passing traces that religious readers left of their encounters with the Bridgewaters. The spiritual manuals made clear that reading needed to evoke appropriately religious feelings if it was to sustain a life of Christian devotion, and we begin by exploring the most private and emotionally charged world of religious meditation before enlarging our focus to consider how family devotions could also shape the religious reader's experience of the sciences. Religious practice was often intimately connected with the domestic unit more generally, and we thus turn in the latter part of the chapter to consider the centrality of family relationships in managing the religious use of reading matter. In strikingly vivid ways, it turns out that it was in developing and strength-

FIGURE 5.1 Ellen Parry, aged seventeen, Summer Hill, March 1840. Engraved by Finden after M. Theweneti. From [Parry 1846?], frontispiece. Image: Bodleian Libraries, University of Oxford. CC-BY-NC 4.0.

ening marital intimacy and in navigating the pitfalls of child-rearing—as much as in meditation and prayer—that the Bridgewaters entered into the everyday practice of religion.

## *Private Devotion*

On 28 April 1841, the fashionable Bath physician and fellow of the Royal Society Charles Henry Parry faced the first anniversary of his seventeen-year-old daughter Ellen's death (fig. 5.1). Only now, after a year of mourning, could he finally bring himself to open her desk in his mansion Summerhill on the rural fringes of the town and look into her pocket books and other private papers. What he found there ravished his tender heart. Her fervent outpourings of devotion to God, her earnest meditations on her own shortcomings, and her desire for spiritual renovation reflected depths of piety that even her loving parents had not suspected. Everyone who had known Ellen Parry agreed that hers had been a life of true Christian faith. Her uncle, the Arctic explorer Sir Edward Parry, felt that every detail of her life confirmed that she was "another lamb added to the Saviour's fold." It was an exemplary story, worthy of being better known, and her father hoped that his "excellent friend" Wordsworth—who had been "much struck" with Ellen on a visit the previous year—might turn it into pious verse. But while the poet thought such a life of youthful devotion was worthy of bringing before "adults of all ages," he expressed himself unequal to "anything so holy." Now, with the evidence of all her papers in hand, the grieving father sought to weave them into an account for circulation among family and friends that ran to more than two hundred and fifty pages.[2]

As for countless of his contemporaries, Charles Parry's experience of death was pervaded by an overriding concern for the eternal welfare of his loved ones. In the case of Ellen, all was well. "Oh! what a privilege is this," Parry's brother comforted him, "to be the parent of a child in heaven!" But such an outcome could not be taken for granted. As Ellen's personal papers showed, true religion was the product of an active process of daily devotion to which she had devoted her best efforts. Her parents and especially her mother, Emma, could take comfort that they had, by training, precept, and example led her "into the ways of eternal life." When Ellen took stock of her daily routine in June 1839, it was permeated with pious activities. On a weekday, family prayers were supplemented by three hours of reading and by four hours spent on "Sundries" and "Extras" that included transcribing sermons from memory, writing in her pocket book, "self-examination," and "meditation." On a Sunday, the fare was "Attending Divine service," "Studying the Bible," "Reading serious and devout books," "Meditation," "Self-examination," "Secret prayer and praise," "Holy *conversation*," "Learning and repeating chapters of the Bible, and Hymns," and "Writing out Sermons."[3]

In Ellen's regulated life of Christian devotion, both reading and exploring the natural world had a prominent place. Emma Parry's annual reports on her daughter's progress gave details of her reading, starting at the age of three with her first primer and catechism. From the age of twelve, Ellen's

2. [Parry 1846?], 9, 13, 240, 243.
3. [Parry 1846?], [iii], 177–78, 243–44.

journals listed the dozens of books she read each year, including histories, biographies, travels, natural history, poetry, and occasional novels, but also a heavy dose of religious fare, including theological works, histories, sermons, and devotional guides. In among these works, Ellen found time to read several of the Bridgewaters: Roget's (twice), Prout's, Kidd's, and Chalmers's. As she faced her death, she was hoping to read Buckland's *Geology*. She had manifested a love of the natural world from an early stage, and from the age of seven her reading included works of natural history. By the time she was nine she could broadly identify most of the plants she came across and was assembling a collection of dried specimens. She became an avid gardener and spent hours painting flowers. In addition, she read a range of scientific works, attended exhibitions at the Bristol Literary and Scientific Institution, and wrote out transcripts of courses of lectures she attended on botany and chemistry. Yet as her journals made clear, it was essential for Ellen that such activities be incorporated into the daily practice of Christian devotion, which was the purpose and measure of her existence.

Ellen Parry's devotional life was notable in its intensity, but it was far from being idiosyncratic. Her great grandfather had been a leading dissenting minister, and while the family had subsequently turned to Anglicanism, their evangelicalism reflected a rich religious heritage. Ellen had learned to regulate her religious life by careful scrutiny of some of the most successful of the myriad works intended to assist readers in the daily practice of Christian living. Works of this sort had a long pedigree, becoming especially popular in the evangelical tradition, and they help us to locate Ellen's practice within a wider religious culture.[4]

Among Ellen's reading were two standard works that had gone through dozens of editions: the Puritan Richard Baxter's *Saints Everlasting Rest* (1650) and the dissenting minister Philip Doddridge's *Rise and Progress of Religion in the Soul* (1745). She first read Baxter's work at the age of fourteen, and it soon became one of two favorite books. However, she was worried that she was not using it correctly and that her Christian profession was not yielding the necessary fruit in overcoming her besetting sins of pride, false shame, self-love, and vanity. Aged sixteen, she abridged

4. Rivers 1982; Rivers 2018, chaps. 3 and 9. There is no nineteenth-century equivalent to Rivers's excellent recent study, although see Ledger-Lomas 2009 and, for the parallel US history, Brown 2004. On the Parry family, see Glaser 1995. Charles Parry's manuscript autobiography and volumes of early letters detail his religious development, including his adoption of Anglicanism; BO, MS.Eng.Misc.d.613, p. 148.

Doddridge's chapter on "maintaining a devout character" and Baxter's "directions for self-examination" with a view to implementing them more successfully. Doddridge enjoined regular habits of meditation, prayer, and self-examination that would yield "continual communion with God" and leave the reader "in his fear all the day long." The upshot would be that every aspect of life was made holy, including "worldly business" and "seasons of diversion." Ellen Parry attempted to live this way. In her times of self-examination she upbraided herself for being "too much engrossed by worldly cares and pleasures." Her prayers and praises had been "cold and heartless," and she had misused her time and talents.[5]

How did this relate to Ellen's love of reading? The devotional guides were agreed that inappropriate reading might contribute to a spiritual malaise but that, done correctly, reading might equally be made part of the cure. In her *Practical Piety; or, The Influence of the Religion of the Heart on the Conduct of the Life* (1811), Anglican evangelical Hannah More had identified two indispensable aspects to the "cultivation of a Devotional Spirit." A regular practice of "retirement and recollection" was to be accompanied by "a general course of reading" that would "never be hostile" to, even if it did not actively promote, "the spirit we are endeavouring to maintain." The pious reader should not only avoid "corrupt writings which deprave the heart, debauch the imagination, and poison the principles" but also those "insipid," "idle," and "frivolous" books that waste time and "gradually destroy all taste for better things." Why be satisfied with fiction when the faculties "might have been expanding in works of science?" Above all, what were called for were books that would be morally purifying and "raise a devotional spirit" without disordering the affections. This would also add to the believer's "materials for prayer."[6]

Ellen was among the many readers of More's book, which reached its seventeenth edition by 1838, and her father recalled that she "read to be made wiser and better, and not for amusement alone." In March 1837, for instance, she penned reflections on reading Paley's *Natural Theology* in which a breathless summary of the argument was followed by an attempt to turn it to practical effect in her own life:

> The existence of a God, as proved by the works of nature, which is, in fact, only another name for God, I firmly now believe, and that I live in his constant presence, and that I am at His merciful disposal. I shall henceforward, take much greater pleasure in examining His works, and

5. [Parry 1846?], 165, 172, 174; Doddridge 1745, 175, 180.
6. More 1811, 1:126–30, 121–22.

in seeking for striking proofs of design, as they will strengthen my *practical* belief in His existence.

According to Charles Parry, this kind of devotional use of the creation was common among the young, prompting in them a "glow of grateful acknowledgement" to God.[7] For his daughter, certainly, reading about God's involvement in the creation and studying the natural world in the right spirit formed part of her practice of piety. To her, the Bridgewaters furnished valuable materials for both rational recreation and pious reflection, fostering her love of nature and of reading within the higher demands of her Christian devotion.

Ellen had thus made good use of the Bridgewaters, but the great joy of her father was that she had gone further. Writing to his daughters as they prepared for their confirmation in December 1838, he acknowledged the importance of their "employments and accomplishments," which allowed them to fill their time innocently, offering practical benefits and contributing to the training of both mind and character. They were, however, "secondary to the high objects of religion." While most young people were caught up with their experience of the world around them, Ellen had pushed on to consider her eternal destiny. She could engage in "rational accomplishments" while at the same time being prepared—through "meditation on her own state," "habitual self-examination," and "practical resolutions of improvement"—to "abandon all these pleasures, and meekly submit to the decrees, however afflicting, of her God and Saviour."[8] It was this unusual capacity for separating herself from the transient pleasures of the creation that had made hers such an exemplary life.

But for all that it was exemplary, how typical was Ellen Parry's experience as a reader? Certainly it was by no means exceptional. In our profoundly secular age, it is easy for us to overlook the extent to which Christian devotion was an overriding concern in the lives of many readers in the 1830s. Advice concerning how reading could be turned to spiritual good in this way was ubiquitous, coming not only from Calvinist dissenters and Anglican evangelicals but also from High Anglicans, as in the case of William Law's hugely successful *Serious Call to a Devout and Holy Life* (1729), and from Methodists, for whom John Wesley's many publications continued to be important. Moreover, the myriad religious periodicals offered counsel concerning appropriate reading matter. As the previous chapter showed, those that discussed the Bridgewaters tended to emphasize that

7. [Parry 1846?], 5, 84–86.
8. [Parry 1846?], 8, 151.

their value was primarily devotional or affective, but with such scientific fare readers often had to work out for themselves how they should use them based on more general principles.

What we have of Ellen's devotional outpourings is obviously somewhat unusual in its extraordinary detail. This was perhaps in part a reflection of her age. Young people with so many developing thoughts and feelings to examine and evaluate for the first time were doubtless particularly inclined to put them down on paper, and pious middle-class families certainly encouraged the keeping of a diary or commonplace book for such purposes. In the case of Ellen, however, her early death prompted her family to take the unaccustomed step of publishing extensive extracts from her notebooks, in confirmation of her eternal destiny, where most such records perished. A similar situation pertained with Frederic Post, the only son of an Islington merchant and evangelical Quaker, who kept extensive diaries and notes up to his death, aged sixteen, in 1835. Like Ellen, Frederic had kept the contents of his diaries private, and his parents were able to take comfort in what they found to be an unexpected depth of religious experience recorded there. The diary showed that he had been "ardent in the pursuit after knowledge," but also that he had breathed "throughout a spirit of piety, and a desire for the attainment of heavenly wisdom, with a fixedness of purpose." The dead boy's father soon published extensive extracts, again comforting himself by making his son's life an exemplary one.[9]

As with Ellen, the Bridgewaters had featured in Frederic's early reading. Educated at home from the age of eleven as a result of lung disease, he had always been passionate about books, and they became his "companions." His favorite reading, other than on religion, concerned the arts and sciences, and he read many of the volumes of Lardner's Cabinet Cyclopædia as they appeared through the early 1830s. In addition, following the founding of the Islington Literary and Scientific Society in 1833, he became a regular attendee at monthly meetings, lectures, and classes, his father paying for him to become a subscriber at the age of fourteen. His diaries were full of detailed notes and thoughts concerning what he learned there, and he engaged in electrical experiments and started an entomological collection. Throughout, however, these entries sat alongside extensive notes and essays on theology together with his intense spiritual reflections. Like Ellen, Frederic "had a great regard for the duty of frequent retirement for meditation and prayer," and his diary entries indicated a similar concern with "*self-examination*." He reflected that "intellectual acquirements"

9. [Post] 1838, ix. The Religious Tract Society produced a much abbreviated edition in 1850.

seemed to be of no use, even to a pious man, in the world to come. Above all it was piety, not "metaphysical speculations," that anchored young souls.[10]

In the last year of his life, Frederic more than once read the *Essay on the Habitual Exercise of Love to God, Considered as a Preparation for Heaven* that Norwich banker and evangelical Quaker Joseph John Gurney had just published. To allow the Holy Spirit to transform the soul ready for heaven, Gurney's book made clear, the believer should seek to develop appropriate habits, including the regular "contemplation of God, in nature and providence." God could be little known from nature alone, but contemplating God "*in his works,* under the beaming light of the religion of the Bible," was "one of the most profitable exercises of the human mind" that could not fail to generate "filial love and gratitude." Gurney was a great admirer of both Chalmers and Whewell, and he proceeded to show that the discoveries of modern science could combine in furthering this important aspect of Christian piety. For a scientifically minded young Quaker like Frederic, this gave religious purpose to his studies, and it is thus no surprise that Bell's Bridgewater was also among the books he read in his final months. Where local evangelical critics of the Islington Literary and Scientific Society had claimed that its secular principles undermined religion, Frederic's experience suggested that the society's scientific fare was entirely compatible with evangelical piety provided that one's private devotions were conducted appropriately.[11]

Adults were perhaps generally less inclined than youths to give such detailed accounts of their private religious meditations in diaries and commonplace books. At any rate, their biographers were rather less inclined to publish such extensive extracts. However, one mode of religious reflection that has tended to leave more permanent traces is devotional poetry. Verse, whether in the form of poetry or hymns, was widely recognized across denominational boundaries as an important means by which religious feelings might be evoked and expressed. High Church cleric and professor of poetry in Oxford, John Keble, made the point very clearly in 1827 in his *Christian Year,* a volume of devotional poetry that ran to sixteen editions in a decade and was probably the most widely circulated poetical publication of the century. The object of the work was, he claimed, to

10. [Post] 1838, 10, 46, 70, 78, 86, 179.

11. Cromwell 1835, 282–88; Lewis 1842, 42–46; [Post] 1838, 31, 435; Gurney 1834, 13–14. On Chalmers, see Gurney 1853, Braithwaite 1854, 2:406–22, 462–63, 497–502, and Gurney 1832, i–iv. On Whewell, see Gurney to WW, 31 October 1833, TC, Add. Ms.a.205[63].

assist readers in attaining a "sober standard of feeling," something that in practical religion was second only to a "sound rule of faith."[12]

Devotional verse appeared in a range of contexts, including not only hymn books and published volumes of poetry but also the religious magazines, where much of it was penned by readers themselves. For a proportion of people, perhaps especially the well educated, poetic composition was a significant means of religious meditation even if it was intended solely for private use. Such personal devotional poetry has received even less attention than devotional verse in general. However, poetry could have great value for those seeking to explore the religious dimensions of scientific developments, enabling them to give vent to newly generated feelings of awe, gratitude, or praise and to engage imaginatively and emotionally with potentially disruptive implications. As the young Scottish geological lecturer Daniel Mackintosh put it in a series of short religious reflections reprinted from his lectures as a *Supplement to the Bridgewater Treatises* (1843), "the noblest field for the exercise of poetic emotion" lay "in the revolutions of the works of the Creator."[13] Certainly, a number of readers engaged with the Bridgewaters in just such a way.

One was the evangelical curate in charge of Purleigh in Essex, Robert Walker, whose response to Kirby's Bridgewater took the form of a panegyric of its author that expressed his own devotional response. What Walker appreciated in the treatise was its "*pious* worth," yielding "a warmer and more genial glow" than could a mere work of scientific erudition. Kirby's heart spoke to his own, and what Walker most appreciated was the way in which the author had made the Bible the key to seeing God in nature. Graced with the Bible, "fair science"

> Springs into life immortal, lives indeed;
> Borrows from Heav'n all help for time of need;
> Lures to a fount where mortals thirst no more,
> Points to a realm, for souls in spirit poor,
> Where smiles a home, to faith's far-seeing eyes,
> Not made with hands, eternal in the skies!

12. [Keble] 1827, 1:[v]; *ODNB*, s.v. "Keble, John"; Tennyson 1981. Anonymous commonplace books from the 1830s with extracts from the Bridgewaters are to be found in the Beinecke Rare Book and Manuscript Library, Yale University, Osborn d223, and in the Brotherton Library, University of Leeds, BC MS Lt/132.

13. Mackintosh 1843, iv. See also Inkster 1975, 463–66, and Astore 2001, 130–32. On devotional poetry, see Rivers 2018, chap. 11, and Brown 2016. On the private composition of poetry see O'Connor 2007a, 81, and Brown 2012.

As he later explained to Kirby directly, he found his Bridgewater a great help in seeing "those ways of Providence" that offer an endless subject "for adoring and grateful praise." The poem ended with an exhortation to both Kirby and himself to remain rooted in this faith.[14]

There is no indication that Walker wrote the poem for presentation to Kirby. It came to the author's attention incidentally, after the younger man visited the rectory of retired headmaster and canon of St. Paul's Cathedral James Tate in July 1838 and found him deep in meditation on Kirby's book. But the poem was part of Walker's established practice of piety. He had schooled himself in Doddridge's *Rise and Progress* while at Oxford and had long been in the habit of maintaining a diary of religious mediation and self-examination. Surviving extracts indicate that he was not unaccustomed to meditating on the natural world to religious ends. "Let me retire into myself," he wrote one summer's day, "and meditate upon the goodness and loving-kindness of my God and Saviour. 'I am fearfully and wonderfully made: marvellous are thy works.'" Moreover, while their spiritual intent was strikingly evangelical, these meditations sometimes drew on scientific expositions. The habit of expressing his feelings in poetry had likewise long been a feature of Walker's devotional life. He reminded Tate that, according to the Swiss pastor and poet Johann Kaspar Lavater, poetry was the "language of the heart." Walker could engage with the sciences theologically, as well as devotionally, when he chose. An accomplished linguist, his manuscript observations on the early part of Genesis referenced Buckland's Bridgewater as he strove to show that geology was not at odds with the Bible. However, it was in poetry that his pious heart found expression.[15]

The private devotional poetry of published poets had, of course, a higher chance of surviving to posterity. One such poet was John Edmund Reade, who read more than one of the Bridgewaters while traveling in Italy in the early 1830s and found himself moved by them to versification. In his thirties, and the grandson of a baronet, Reade saw poetry as his vocation and had already published two volumes. The second of these, *Cain the Wanderer* (1829), made very clear his inclination to emulate other poets, especially Byron, and it had received severe censure as well as significant praise. Now, as he read Whewell's account of the "resisting medium" through which heavenly bodies passed and its consequences in the finite duration of the universe, Reade was moved to render it into blank verse in

14. Pyne [1855], 113–15; Freeman 1852, 473–76.

15. Pyne [1855], 3–5, 41, 113, 129–33, 181, 194n. A panegyric on Chalmers's Bridgewater by the Quaker poet Amelia Opie is reproduced in Brightwell 1854, 305–6.

his journal. Reaching the end of the passage, where Whewell declared that to "dwell on the moral and religious reflections suggested by this train of thought" was not his "present purpose," Reade went on to develop those reflections for himself, writing,

> And who art thou, weak man! who dar'st complain,
> Least part in this great chain of life? . . .
> . . . Strive to rule thyself,
> Thy passions, and thy luxury, and pride:
> Be humble, meek, and ignorant; and know
> That resignation to the will of God
> Is the true magnanimity . . .

In seeking to approach their maker, Reade reflected, wretched humans were called on to engage in self-purification and love.[16]

That Reade's reflections survived was due to his reproducing them in a note to his narrative poem *Italy* (1838), which bore striking similarities to Byron's *Childe Harold's Pilgrimage*. There, as the Spenserian stanzas moved from the destruction of Pompeii to the transient fortunes of humankind, Reade's notes referred readers first to his private verse response to Whewell's treatise and then to Buckland's Bridgewater, which offered "a most eloquent and beautiful illustration of the text." Now writing for public consumption, he reflected that the fossil records of past ages were as if intended "to show vain men what dusty fame is worth!" Reade's fifth canto—its Neapolitan moment—reached its climax in the following stanza, reflecting the humble appeal of humans to God. This marked a contrast with Byron that was clear to reviewers who applauded his wholesome religious sentiments and the way in which he rendered the sublime "perfectly *safe*."[17]

An altogether more notable poetical reader was Alfred Tennyson, concerning whose engagement with the sciences much scholarly ink has been spilt. Tennyson did not respond to the Bridgewaters in the direct manner of Walker and Reade. He did, however, undertake a program of scientific reading in the 1830s and, while there are few explicit traces of his having read the Bridgewaters, the circumstantial evidence is strong. He certainly

16. Reade 1838, 500. Cf. Whewell 1833b, 202–3.

17. "Reade's Poems," *Dublin University Magazine*, 13 (1839): 734; "Modern Poetry of Remote Ages," *Eclectic Review* 5 (1839): 516; Reade 1838, 283–86, 501. Ironically, a quotation that Reade gives in his note is not from Buckland, as claimed, but from James Hutton's *Theory of the Earth* (1795).

read Chalmers's treatise, and in addition he acquired Roget's for his library (together with Babbage's unofficial contribution to the series) and sought to borrow the treatise by his former Trinity tutor Whewell through a friend. His brother owned Kirby's volumes, and his sister acquired Buckland's treatise while they were living together in the family's Lincolnshire home at Somersby in 1837. This was just at the time that Tennyson was also reading Lyell's *Principles of Geology*, and there is every reason to suppose that he read and was moved by Buckland's work.[18]

Not that the poet was convinced by the religious message of the Bridgewaters. As a member of a select student society at Cambridge, Tennyson had voted against the proposition that an "intelligible first cause" was "deducible from the phenomena of the universe," and much later in life he recalled finding Chalmers's Bridgewater disappointingly question-begging.[19] For Tennyson, there was no easy assurance of divine goodness in creation, and the findings of modern science, especially of geology, were hardly reassuring in this regard. Yet while his response as a reader of the Bridgewaters did not involve the kind of devotional spirit found in Reade or Walker, it is clear that the works furnished some of the materials for his meditations concerning fundamental existential questions that famously coalesced in poetic form.

Wrestling with the death in 1833 of his beloved friend Arthur Hallam, Tennyson gave vent to his thoughts and feelings in a series of "elegies" that were later shaped into his master work *In Memoriam*. In two elegies apparently written in the early 1840s, his emotional responses to the findings of the sciences as detailed by the Bridgewaters and similar works, and to their failure to offer unequivocal reassurance of providential care, were melded with his experience of personal grief:

> Are God and Nature then at strife,
> That Nature lends such evil dreams?
> So careful of the type she seems,
> So careless of the single life;
>
> . . . . . . . . . . . . . . . . . . . . . .
>
> "So careful of the type?" but no.
> From scarpèd cliff and quarried stone

18. Tennyson 1897, 1:124, 145; Campbell 1971; Lang and Shannon 1982, 1:128, 145; 3:467. I possess a copy of volume 2 of Buckland's second edition that is inscribed "Emily Tennyson ¦ Somersby. ¦ 1837."

19. Tennyson 1897, 1:44n; Lang and Shannon 1982, 3:467.

She cries, "A thousand types are gone:
I care for nothing, all shall go."[20]

The manner in which the published poem ultimately resolved these feelings in an optimistic vision of faith in divine love endeared it to the British public, but Tennyson's exploration of these spiritual concerns is indicative of the way in which poetry enabled the wrestling soul to engage with such questions. Whether through Whewell's vision of a dying universe or through Buckland's vision of a succession of extinction events, the Bridgewaters engaged spiritually questing readers with the emotional charge of modern science.

Alfred Tennyson and Ellen Parry were, of course, exceptionally wealthy and well-educated readers, but concern with the role of print in fostering spiritual development was by no means restricted to such people. The evangelical revival of the eighteenth century had been notable for its particular appeal to working people and for its inculcation among them of devotional habits rooted in printed matter, and these developments only escalated in the new century. By the 1830s, tract organizations such as the Religious Tract Society (RTS) were moving on from their formerly somewhat narrow fare to offer workers guides to devotion such as the *Fire-Side Piety* by Massachusetts pastor Jacob Abbott—one of Ellen Parry's favorite books. Likewise, cheap evangelical periodicals frequently offered advice about suitable devotional reading matter. For workers, however, the high price of the Bridgewaters did not make them particularly appealing for devotional purposes. Moreover, as we have seen, the cheap evangelical magazines found other books potentially far more helpful in private devotions not least because they made reference to specifically Christian sentiments. Works such as Thomas Dick's seven shilling *Christian Philosopher* (1823) were widely approved by evangelical reviewers and were often acquired by libraries attached to churches and chapels, making them available for devotional use by working people.[21] Nevertheless, the reprinting of passages from the Bridgewaters in cheap miscellanies such as the *Saturday Magazine* did make them available

20. *In Memoriam*, LV and LVI. See Shatto and Shaw 1982, 216–22; Ricks 1987, 2:370–74. On Tennyson and geology see also, for example, Gliserman 1975, Rupke 1983, Secord 2000, and Anderson and Taylor 2015–2016.

21. Astore 2001. On the Bridgewaters and religious libraries, see Topham 1993, 297–312.

to poorer readers for devotional purposes, and as the following section shows, the series also lent itself to more systematic repackaging by Christian ministers wishing to furnish families with anthologized devotional materials.

## *Family Prayers*

Family devotions played a significant role in the lives of many Protestant Christians of the early nineteenth century, typically taking the form of prayers in the morning, the evening, or ideally both. According to Abbott's *Fire-Side Piety*, family prayer had "perhaps more influence than any other one thing in bringing a household under the control of Christian principles." As with personal devotions, family devotions were enjoined in and supported by a plethora of publications emanating from a range of Protestant denominations and parties and produced for a variety of purses. Once again, the most successful of these passed through dozens of editions. Anglican rector Benjamin Jenks's *Prayers and Offices of Devotion for Families* (1697), for instance, had reached its twenty-ninth edition by 1816, and Bishop of London Edmund Gibson's *Family Devotion* (1705) had reached its thirty-sixth by 1828.[22]

Such practical guides usually envisaged family prayers being led by the paterfamilias, who was typically expected to read from the family Bible, possibly with some comments of his own, to lead the family in prayer (extempore or read), and perhaps also to lead them in the singing of a hymn. In practice there was no doubt considerable variety, and it is clear that in some families, just as for some individuals, the Bible was not the only book introduced into devotions. In the small rural village of Ruthwell in Dumfriesshire (fig. 5.2), for instance, Church of Scotland minister Henry Duncan regularly enlivened family worship with a range of extracts "illustrative of the Bible, or of the character and works of God," and for a spell many of these came from the Bridgewaters.[23] Duncan was somewhat unusual both in the extent of his scientific interests and in his desire to introduce the works of God into religious worship. Nevertheless, he turned his own readings from the Bridgewaters and other works into a highly popular devotional guide that was published to considerable acclaim.

22. Rivers 2018, chap. 5; [Abbott] 1835, 10.

23. Duncan 1848, 92; [Abbott] 1835, 10; McDannell 1994, 81–85. Little historical attention has been given to the topic of petitionary prayer and science, although see Ostrander 2000.

FIGURE 5.2 Henry Duncan's manse at Ruthwell. Engraved by W. Richards after W. W. Duncan. From Duncan (1848), frontispiece.

As a young minister at the start of the nineteenth century, Duncan's practice had smacked of the Moderatism then prevalent in the Church of Scotland, and his overwhelming emphasis in preaching had been on "Natural Theology, Virtue, and Morality." At that time, he had also unsuccessfully sought to interest his parishioners in Sunday lectures on the works of God, encouraged by David Brewster. Subsequently, however, he had been drawn by degrees into the evangelicalism that came to dominate the Scottish church under the leadership of Chalmers, and his preaching came to emphasize more the particular importance of the gospel message. Duncan nevertheless continued to believe that scientific knowledge supported and enlivened Christianity, and he again tried to interest his parishioners in scientific lectures. Later, he lent his support to the nascent mechanics' institute movement and (through his university friend Brougham) the useful knowledge society. Employing a tutor to teach his own children and paying pupils, he enjoyed enlivening their education by constructing scientific apparatuses and discussing scientific matters. His scientific reading was extensive, and in 1827 he followed up local reports of the existence of footmarks in the New Red Sandstone at a local quarry, establishing the veracity of the claims to the satisfaction of William Buckland, who visited Ruthwell to see the specimens and who discussed them in his Bridgewater.[24]

Duncan's household included his own three children and a regular stream of pupils, all of whom were expected to attend the regular routine of twice-daily family worship. This was "so arranged as to engage the attention and interest of the young," with Bible reading and prayer being supplemented by additional readings and the minister's own "questions and remarks." For Duncan, the works of God were a fit topic for contemplation on such occasions, and he found a useful source for appropriate readings in the *Reflections on the Works of God in Nature*, translated from the German of evangelical pastor Christoph Christian Sturm. This work contained short mediations for every day of the year, designed to educate the reader about the natural world but also to show how "wisdom and virtue" might derive from the contemplation of such knowledge. According to one of the work's translators—the prominent Methodist minister and theologian Adam Clarke—it was a work "in which sound philosophy and pure practical piety" went hand in hand. Indeed, Clarke recommended the work to families, suggesting that the reflections be read daily with a portion of the Bible.[25]

It is not clear how typical Duncan was in using Sturm's work in family prayers, but this was certainly one of the many uses that made the *Reflections* one of the great best sellers of the British book trade in the first third of the nineteenth century. First translated in 1788, a torrent of editions, reprints, and retranslations issued from presses up and down the land over successive years, numbering well in excess of fifty by the 1830s. There were also popular abridgements, including some for children. Moreover, while Sturm's *Reflections* was relatively untypical among the popular devotional works in its focus on the natural world, it was not alone. Most notable was the even more successful *Meditations and Contemplations* (1746–47) by Anglican evangelical James Hervey, which saw more than sixty printings in the first three decades of the nineteenth century, although it was both less scientific and less exclusive in its focus on nature than Sturm's

24. Duncan 1848, 35–39, 66–68, 87–90, 151–57, 178–83, 210–12, 244–46; Hall 1910, 40–41, 87–88, 96–100, 111–14. See Duncan's letters in OUMNH–BP, Footprints/1, and Buckland 1836, 1:258–63.

25. Clarke 1810, 1:v, xi; Duncan 1848, 92, 204–5.

work.[26] In addition, some families—like the Duncans—were evidently in the habit of turning even more scientific books about the natural world to good use in their devotions.

Having used Sturm's *Reflections* and other books to introduce the natural world into family worship, Duncan found himself in the middle of 1836 contemplating the publication of his own devotional work on the same model. The first six-shilling volume of his *Sacred Philosophy of the Seasons* was published in Edinburgh in October, containing readings for the winter months, and Duncan exhausted himself producing three further quarterly volumes in rapid succession. Drawing on his own knowledge and on a variety of scientific readings, among which the Bridgewaters were very prominent, he thus offered readers a full set of daily reflections illustrating God's attributes from the creation.

In his preface, Duncan related his work to those existing works of natural theology that had "enlightened and delighted the pious mind," but he also drew attention to its distinctiveness in focusing on the seasons and, more fundamentally, in connecting natural theology with the distinctive claims of the Christian revelation. To Duncan's mind, it was crucial to explore the analogies between God's two books, since the book of nature was dark and obscure without that of revelation. When he declared that the "attention of scientific men" had recently been very successfully but too exclusively focused on natural theology, the reference to the Bridgewaters was obvious. Without assistance drawn from revelation, the readers of such publications might overlook God's moral attributes and the whole divine dispensation toward humans. Duncan set out to counteract this and to show that the God of nature could only be properly known "when regarded as the God of Grace." When creation was studied in the light of revealed truth, then it not only served to educate readers but would also "improve the heart." Like Sturm, his purpose in giving daily readings was to lead the reader to "sanctify each day," although he wished to provide the readings according to a more "systematic method," justifying the title "sacred philosophy."[27]

Duncan promised to draw together readings on natural theology for those without access to a large library with a view to making them suitable for "family reading," and he often gave extensive quotations from the Bridgewaters. However, the different context in which these passages now appeared transformed their significance. In its enthusiastic review of the middle two volumes of Duncan's work in July 1837, the *Edinburgh Chris-*

26. Rivers 2018; Duncan 1848, 205.
27. Duncan 1836–37, 1:i–iv.

*tian Instructor* juxtaposed them with the Bridgewaters. The latter had been "eminently beneficial" in marshalling prominent scientific men to make "all the different departments of science tributary [ . . . ] to natural theology." They constituted "a vast magazine of facts of the most wondrous and authoritative nature." Ultimately, however, natural theology was of limited value. Duncan's aim had been less "ambitious," but it was altogether more important. He aimed "to lead his readers forth amidst the scenes of nature which lie spread around them, to seize on striking facts and phenomena as they occur; to bring the light of science to bear upon them, and to consecrate them by applying them to the purposes of an enlightened and scriptural piety."[28] This positive assessment of Duncan's work was shared by others, and the four-volume work passed through five editions in fifteen years.

Another indication that the Bridgewaters were considered potentially useful in the practice of family devotions comes from Edward Bickersteth's *Christian Truth: A Family Guide to the Chief Truths of the Gospel; with Forms of Prayer for Each Day in the Week, and Private Devotions on Various Occasions*, compiled between 1834 and 1838. A leading Anglican evangelical, the author had worked for many years for the Church Missionary Society and was widely known for his many practical publications on such subjects as prayer, Bible reading, and communion, which passed through dozens of editions. On becoming rector of a small village thirty miles north of London in 1830, he found many of his new parishioners ignorant "of the first principles of divine truth," and he soon set about preparing a series of tracts designed to bring it before them "in a simple, easy and devotional form." Each quarterly contribution to what he thought of as his "Cottager's Guide to Christian Truth" consisted of short theological readings of a page or two followed by a meditation or prayer. Having produced eighteen such tracts, Bickersteth reissued them as a volume, and while he suggested that it might be used weekly by cottage or district visitors or by parents in educating their children, its primary intended purpose was in family devotions. Moreover, earnest Christians of various stripes agreed that it made a "very useful family-book."[29]

Bickersteth's topics were purposefully aligned with the first six sections of his own spectacularly successful hymn book, *Christian Psalmody* (1833),

28. "Duncan's Sacred Philosophy of the Seasons," *Edinburgh Christian Instructor*, 3rd ser., 2 (1837): 463–74, on 465.

29. "Select List of Books," *Wesleyan-Methodist Magazine* 18 (1839): 231–37, on 233; "Notices and Reviews," *British Magazine* 14 (1838): 694–707, on 700–701; Birks 1851, 2:86; Bickersteth 1838, vii–viii.

which he had compiled for "public, social, family, and private worship." He began with a section on the Bible before moving on to God and the creation. Knowledge of God comes primarily from the Bible, he explained. While the creation "declares God in his works," humans are "alienated from him" in their sinful nature. However, with the aid of the Bible, God's works could become a "looking-glass" in which the believer might "contemplate continually" the creator's perfections and "rise to a constant communion with him." In evoking such contemplations, Bickersteth quoted at length from Kirby's Bridgewater, explaining some of the more difficult terms, but leading the reader on to the author's pious ejaculation, "I am lost in the depths of the unfathomable deity." As with many conservative Christians, he also found Kirby's politically charged vision of every rank of creature promoting "the good of the whole system" particularly attractive in engaging with the poor. The meditation that followed gave the humble parishioner a secure place in the paternal care of a transcendent deity: "O Lord, our Lord, how excellent is thy name in all the earth, who has set thy glory above the heavens! And yet thou condescendedst to be my God and my portion for ever."[30]

Bickersteth's treatment of creation soon moved on to its fallen state as outlined in the Bible, but even here Kirby's Bridgewater was useful. According to Bickersteth, while the Bible made "plain the spiritual meaning of earthly things," the "book of nature" was also "a picture to discover to us the book of grace." This was a view that Kirby shared, and Bickersteth was able to draw on his account of how the system of predation and parasitism preached the "great doctrine of *vicarious suffering*." Bickersteth's meditation emphasized this distinctively symbolic reading of the natural world: "O my soul! Use the helps which thy God has given thee in creation for learning those things which belong to thy everlasting salvation!" Kirby's Bridgewater was uniquely useful for such purposes, but Bickersteth looked more generally for knowledge of the creation to stir up a devotional spirit within the believer, and the Bridgewaters in general—books he recommended for the minister's library—were included in his short list of useful books at the end of the section.[31]

For many Christians, the main concerns of the religious life were the urgent priorities of spiritual renovation that could only be achieved through an active daily practice of individual and family devotion. The history of

30. Bickersteth 1838, 25, 50, 52–54, 58. Bickersteth's *Christian Psalmody* sold over 220,000 copies by the 1850s. Ledger-Lomas 2009, 350.

31. Bickersteth 1838, 64–66; Bickersteth 1844, 383.

such devotional practice has barely begun to be addressed, and the vast numbers of associated publications have moldered unnoticed. However, for devout readers like Henry Duncan and Edward Bickersteth, the Bridgewaters were valuable inasmuch as they could be made subservient to the purposes of the spiritual life. Both, indeed, made the series more available for such purposes by repackaging them within popular devotional works. For all the reservations of evangelicals, such uses—properly moderated—could make the Bridgewaters highly attractive and useful.

## *Love and Marriage*

The practice of Christianity was a product, in the main, of family life—of precept and example, of discussion and engagement, and especially of education. Underpinning such Christian family life was the marriage relationship, and few decisions were more consequential for the daily practice of Christianity than the selection of a marriage partner. On that decision rested the harmony of domestic relations, the conduct of personal and family devotions, the upbringing and education of children, the engagement with public worship, and the involvement of the family in work, social intercourse, and community action. Not surprisingly, the topic of identifying an appropriate marriage partner was frequently discussed in devotional guides as well as in novels and advice manuals. Hannah More, for instance, saw the "indiscreet forming" of marriage connections as a leading cause of the decline of Christian piety, noting that the irreligious more frequently drew the religious "to their side" than the other way around. Similar advice came at a lower price from Esther Copley, an evangelical Anglican turned Baptist whose *Female Excellence* (1839) was published by the RTS and sold over twenty-one thousand copies in a dozen years. The "moral and religious character" of a potential husband, urged Copley, was "the foundation of happiness in the married life."[32]

Given the high stakes, the question of how to identify and secure an appropriate life partner was a challenging one. For all that family was expected to be able to offer advice and direction, particular responsibility was placed on the couple themselves by the increasingly individualist conception of marriage and all the more so as marriage became ever-more associated with the notion of romantic love. For those earnest about their Christian religion, romantic love was not expected to develop independently of

32. Copley [1839?], 205; Jones 1850, app. V; More 1811, 2:84. See also Gisborne 1794, 600, and Gisborne 1797, 233–42. Copley's clerical husband seems to have been an alcoholic (*ODNB*, s.v. "Copley, Esther").

religious fitness. Hannah More made the point in her *Cœlebs in Search of a Wife*, a religious novel of 1808 that in thirty years sold twenty-one thousand copies. Her eponymous narrator explained in the preface that in his own exemplary case, love had not been an "ungovernable impulse" but rather a "sentiment arising out of qualities calculated to inspire attachment in persons under the dominion of reason and religion." More generally, with "personal morality" increasingly taking the place of "external constraints" in managing relationships during this period, courtship became fraught with danger in new ways, especially for women, who were ever more seen as the bearers and defenders of virtue and sexual innocence.[33]

Conversation regarding reading could prove an invaluable aid in this courtship minefield. The prevalence of advice about the religious merits of appropriate reading meant that reading was a useful signifier of the disposition of a potential spouse. The point was made explicit in the advice manuals that became increasingly common during this period. Such manuals were notably popular in the United States, and one very successful American import was the *Young Lady's Friend*, written anonymously by Eliza Ware Farrar (whose husband was a Harvard mathematics professor) and republished in Britain in 1837 by High Church publisher John William Parker. In Massachusetts, Farrar appeared "essentially of English breeding," having been brought up in London, and while she had since turned from Quakerism to Unitarianism, her advice to young ladies was strongly rooted in what she called "vital piety." Thus, marriage was represented as very much secondary to that "great end of existence," the "preparation for eternity." Farrar's advice was that in conversation with gentlemen, young ladies should talk "as one rational being should with another," never reminding the men of their status as "candidates for matrimony." They should demonstrate elevated sentiments and a cultivated intellect, conversing of "books, pictures, and the beauties and wonders of nature" rather than about people. Familiarity with the "works of great minds" or of "nature and scientific researches" would displace any tendency to gossip.[34]

Farrar's advice was echoed by Quaker turned Congregationalist Sarah Ellis in her successful *Daughters of England*—one of a suite of books she produced for middle-class readers from the late 1830s that did much to codify Victorian gender norms. Like Farrar, Ellis addressed herself to women who were sincere Christians, dealing with questions of love and courtship while also reflecting on how scientific reading might contrib-

33. Coontz 2006, 158; More 2007, 38; Thompson 1838, 244. For general discussions, see also Davidoff and Hall 1987 and Macfarlane 1986.

34. [Farrar] 1837, 183, 185, 188, 198; Schlesinger 1965, 152.

ute to that process. A "general knowledge of science," Ellis considered, made women "more companionable to men." Not that women needed to "*talk much*." She pointed out that men usually just wanted an "attentive listener" in conversation, but their female interlocutors needed "considerable understanding" in order to offer this. Moreover, women might thus learn from men, as also from the study of scientific books and nature itself, about how "the wisdom and goodness of God pervade all creation," a topic of real spiritual importance. For men and women to talk on such sublimely religious topics accorded perfectly with Ellis's vision of Christian courtship.[35]

In the 1830s the Bridgewaters were an obvious recourse for men and women wishing to develop their courtship in this way. However, their utility was constrained by an emerging Victorian conception of the lady's sphere of activity as exclusively domestic. The encroachment of such views had made scientific discussion increasingly uncommon in the many ladies magazines, so that only one (the *Ladies' Cabinet*) offered reviews of the Bridgewaters.[36] There was, commentators made clear, a real danger in becoming the kind of "literary young lady" who only read the scientific books on her frequent visits to the circulating library, pontificating indiscriminately about their contents. Such a figure was an easy target for satire, as vicar's son, Oxford student, and Tractarian Edward Caswall made clear in his pseudonymous *Sketches of Young Ladies*. This was a "philosophical enquiry" into the natural history of young ladies that was published by Dickens's publisher and illustrated by his new illustrator "Phiz." It proved to be one of the most popular books of the 1837–38 season, although its initial droll conception as a supplement to the Bridgewater Treatises was tactfully removed before publication. Caswall retained his "Linnæan" classification, identifying all young ladies as domestically cloistered "Troglodites" in spoof correction of Buckland's alleged claim that they were "Ichthyosauri." In addition, he characterized his female literary bore as having lately become interested in political economy and geology, telling all comers of her emphatic approval of Cuvier. Such behavior, Caswall made clear, was hardly conducive to courtship or good company.[37]

Courting couples certainly encountered pitfalls in discussing the

35. Ellis [1842], 77–78. On Ellis, see Davidoff and Hall 1987, 180–85.

36. Shteir 2004a, 2004b; Beetham 1996. Other titles examined include *Blackwood's Lady's Magazine*, the *Christian Lady's Magazine*, *Court Magazine*, *Lady's Magazine*, and *Royal Lady's Magazine*.

37. [Caswall] 1837, v, 21–26; de Flon 2005, 22–23. Caswall also sketched out a comic Bridgewater Treatise on lying in his private notebook. See also Schilke 2012.

Bridgewaters. Take, for example, the love letters of John Torr of Exeter and his Islington-based cousin Maria Jackson. Interspersed between occasional family visits, the early affectionate correspondence between these two middle-class youths was often about their reading. In the spring of 1835, however, their intimacy hit the rocks. Maria was initially unhappy that her cousin, now a sixteen-year-old articled solicitor's clerk, had written warmly to her about politics and had accused her of being a "Novel reader." After defending novel reading, she sought to demonstrate that her reading soared much higher. She had lately been reading "some of the Bridgewater Treatises" she told him and had been "highly entertained" by them. John should take a look at them himself—especially at Whewell's. Apparently feeling patronized by his older cousin and piqued by her report that her cousins at Oxford were good company, John was not entirely placatory in his reply. His parting shot had its own sharply patronizing edge. "I am much obliged to you for recommending one of the Bridgewater Treatises," he wrote. "I shall take the first opportunity of perusing it." He had heard the series highly spoken of and was happy to find that she had been employing her time "so usefully" in reading them. Maria was in no mood to brook such a high-handed tone, and their correspondence was abruptly terminated.[38]

For this young couple, the Bridgewaters served as a signifier of moral and religious earnestness, and when John Torr reopened their correspondence three and a half years later with a bold declaration that he had always loved his cousin, her initial discouraging reply placed great weight on the difference in their "religious sentiments"—a subject "of the first moment in marriage," without which there could be no "perfect union." This was a reference to the fact that Torr had been brought up Unitarian and sometimes treated religious matters with levity, while Jackson was a devout Anglican. The two continued to discuss the matter over their prolonged courtship before finally entering into a marriage that was soon ended by her tragically early death.[39]

Another courtship between cousins was that of Scottish baker's son James Young Simpson and his second cousin Jessie Grindlay. In 1835 Simpson was beginning to make his way as assistant to the professor of pathology in Edinburgh, and he finished a summer medical tour of London and the Continent by calling on his father's cousin, a Liverpool merchant. The ensuing correspondence with Jessie included discussions of reading mat-

38. Carritt 1933, 13–17.
39. Carritt 1933, 23. See also Torr 1918–23.

ter alongside details of his improving professional position and flirtatious talk of possible love rivals. By the summer of 1837, James was lending his cousin Lyell's *Principles of Geology* and Whewell's Bridgewater—his favorite of the series. "You will find," he told her, "one or two chapters near the beginning, and most of the second part of the book, on the 'Stability and Vastness of the Universe,' etc., very beautiful and interesting." Later in life Simpson underwent an evangelical conversion, but his upbringing had been earnestly religious, and he had studied natural theology as an undergraduate, so it was easy enough for him at this moment to offer the religious sublimity of scientific study to his cousin. For a man of science like himself, such exchanges were important in establishing common interests.[40]

As Simpson's comments indicate, the Bridgewaters were more than merely sober and improving reading matter: readers often found them sublime, both in the striking views they offered of the wonders of the universe and in their inspiring visions of divine power, grandeur, and care. Such material was especially suitable in mixed company and not least in the context of courtship, where a shared experience of higher feelings established a common bond. The vision of the immensity of time and space offered by many geological and astronomical works was especially awe inspiring, engrossing the feelings and subtly changing the practice of etiquette. When, for instance, Buckland gave a domestic lecture on his forthcoming Bridgewater to a private party at the Cornish house of Quaker geologist Robert Were Fox in the autumn of 1836, his seventeen-year-old daughter Caroline recorded that they had sat "with great and gaping interest" as he described his frontispiece, with its astonishing vision of earth history.

These effects could, of course, be used with dubious intent, as George Eliot's novel *Mill on the Floss* (1860) made clear. Eliot had read Buckland's Bridgewater with "much pleasure" on its first publication, finding its subject matter sublime. Moreover, as scholars have recognized, *Mill on the Floss* was a novel that drew heavily in its plot and narrative on the geology of both Buckland and Lyell.[41] However, the one explicit appearance of Buckland's Bridgewater in the novel occurred in relation to a pivotal

40. Duns 1873, 79. Simpson's biographer, Free Church of Scotland minister and professor of natural science John Duns, emphasized the lack of feeling in his early religion (125–27).

41. Eliot to M Jackson, 4 March 1841, in Haight 1954–78, 8:7–8; Shuttleworth 1984; Smith 1991; Buckland 2013.

scene of polite but ultimately disastrous lovemaking between the novel's heroine, the mill owner's daughter Maggie Tulliver, and a wealthy businessman's son, Stephen Guest.

The two met for the first time in the drawing room of Maggie's uncle, Mr. Deane, a junior partner in Guest and Company. However, while Deane's daughter Lucy had an unspoken understanding with Stephen, he was captivated by Maggie's dark eyes and her rebuttal of his rather flirtatious compliments, and he wished to make her look at him again. His opportunity came as he talked lightheartedly to Lucy of "impersonal matters." Addressing the question of what she might recommend to the next meeting of the ladies' book club in the neighboring town, he recommended Southey's *Life of Cowper* or one of the recently published Bridgewater Treatises. Responding to Lucy's request to know more, Stephen became "quite brilliant in an account of Buckland's Treatise."

> He was rewarded by seeing Maggie let her work fall, and gradually get so absorbed in his wonderful geological story that she sat looking at him, leaning forward with crossed arms, and with an entire absence of self-consciousness, as if he had been the snuffiest of old professors, and she a downy-lipped alumnus. He was so fascinated by this clear, large, gaze, that at last he forgot to look away from it occasionally towards Lucy.

Only when his "stream of recollections" ran shallow and Stephen offered to bring Maggie the book, did she become conscious of, and embarrassed by, the intimacy into which they had entered.[42]

Lucy vetoed the suggested loan, protesting that she would never get Maggie away from books if she were once plunged into them. In fact, Maggie had, since her father's bankruptcy, entered into a religiously colored renunciation of her passionate love for demanding reading matter similar to that engaged in by the then-evangelical Eliot during the 1830s. Of late, Maggie had sought to satisfy herself with the Bible, Keble's *Christian Year*, and *The Imitation of Christ* of Thomas à Kempis in keeping with more conservative gender norms. What Stephen offered her was an account of a learned book—mediated to the cloistered domestic sphere of the ladies by a man of the world—that was suitable in its purity and sublimity even for the chastened Maggie. Strikingly, the only other moment in the conversation when Stephen was similarly able to engage her unselfconscious attention was when he recounted the good deeds of a lo-

42. Eliot 1981, 378, 380–81.

cal clergyman, prompting Maggie to let her needlework fall and exclaim, "That is beautiful."[43]

The irony of Eliot's narrative is that this conversation was the dawn of a passionate bond between Stephen and Maggie that rapidly led to their elopement and Maggie's disgrace—the pivotal point of the novel's plot. In a culture in which women sequestered in domestic purity were viewed as being likely to have their feelings stirred by sublime subjects, they were naturally also viewed as being in danger of having that propensity turned to other ends. As a droll essayist in *Fraser's Magazine* observed, minds moved by the scientific sublime were "more open to the impressions of *la belle passion*."[44]

The consequences of this did not need to be as disreputable as in Maggie's case. Indeed, sometimes scientific reading was seen as working in quite the opposite way. After Byron's eighteen-year-old daughter Ada attempted to elope with her tutor early in 1834, the counsel of her mother's medical and religious adviser, the physician and cooperativist William King, was that she should learn to control her imagination and thoughts (fig. 5.3). In response, Ada engaged in a program of "study and intellectual improvement." She found that only "very *close & intense* application to subjects of a scientific nature" kept her imagination "from running wild" or filled the "void" left in her mind by "want of excitement." King approved the plan: mathematics and natural philosophy had "no connexion with the *feelings* of life" and could not therefore "lead to objectionable thoughts." And when Ada was soon lapping up Whewell's Bridgewater ("How interesting it is!"), King was glad to hear it, merely advising her to make it a subject of study and avoid "loose reading."[45]

The role of reading not only in binding intimacy but also in channeling sexual desire is strikingly illustrated by the role of Buckland's treatise in the courtship between the young vicar's son Charles Kingsley and the wealthy MP's daughter Fanny Grenfell. They met in 1839 while Kingsley was staying with his family in Oxfordshire during his first summer vacation from Cambridge. The two entered into a correspondence that over the course of the following year developed into an amorous one. Kingsley had developed doubts about aspects of Christianity while embracing an almost pantheistic love of nature, and during his second year in Cambridge his life there became somewhat dissipated. But by the start of his third

43. Eliot 1981, 293, 379. On Tulliver as a reader, see Golden 2003.

44. [Mitchell] 1838, 294.

45. Stein 1985, 42–45. On King, see also Moore 1977, 39–41.

FIGURE 5.3 Ada Lovelace (née Byron). Stipple engraving by William Henry Mote after Alfred Edward Chalon, published 1839. Image: © National Portrait Gallery, London.

year he was finding in Christianity an escape from his growing despair, and the pious and theologically well-read Fanny had a significant role in helping him to a Christian faith. This journey to faith went hand in hand with their growing intimacy. Fanny, a keen follower of the nascent Tractarian movement, was attracted by the notion of joining a celibate order, and Charles also wrestled with the question of celibacy as he struggled to control his sexual desires. However, the pair came to a vividly religious understanding of sexuality that became an increasingly important part both of their correspondence and of their religious practice as their courtship developed. When they met in January 1841, Charles took her in his arms and kissed her, and when they met again in March, Fanny confessed her love for him. Three days later he wrote that his faith was secure, and by the time of his twenty-second birthday in June, he had committed himself to becoming a clergyman.[46]

The course of their courtship was not, however, smooth. Fanny's wealthy siblings objected to Charles's lack of financial resources and sent her abroad for the summer. The following summer, after graduating, Charles took up a curacy in Hampshire and, with the prospects of their marriage seeming bleak, the two lovers wrote intimately confessional letters to each other as "husband" and "wife." Then Charles suggested that they demonstrate the depth of their love by undertaking a year's abstinence from correspondence. The prospect horrified Fanny, and Charles offered lengthy advice to help her to sustain the trial. They must not look on it in a negative light and as a punishment but, through prayer, in a positive light and as a mercy. He referred Fanny to Jeremy Taylor's seventeenth-century devotional classic *Holy Living*—now past its thirtieth edition—for practical advice about loving God and working to his glory and then proceeded to offer his own very detailed guidance. She must avoid Taylor's Popish notions about religious duties and celibacy and avoid mystification, aiming instead at depth and simplicity of thought and the cultivation of religious habits of mind. Above all, she must engage in "objective" studies and practical action that would take her out of herself and keep her from reveries.[47]

The proper study of mankind was not so much man as God. "He is the only study fit for a woman devoted to Him," Charles advised, and he should be studied in three ways. First, "From His dealings in History" (here, Thomas Arnold's recently published lectures would help). Second, "From His image as developed in Christ the ideal, and in all good men" (here, works by Frederick Maurice and Thomas Carlyle that Fanny had previously lent him were the key). And third, "From His works." Here, Charles became rhapsodic:

> Study nature—not scientifically—that would take eternity, to do it so as to reap much moral good from it. Superficial physical science is the devil's spade, with which he loosens the roots of the trees prepared for the burning! Do not study matter for its own sake, but as the counte-

46. Chitty 1974, 51–62; [Kingsley] 1877, 1:43–62.

47. Kingsley to Grenfell, August 1842, in [Kingsley] 1877, 1:86–90; Chitty 1974, 71.

nance of God! Try to extract every line of beauty, every association, every moral reflection, every inexpressible feeling from it.

Yet for all that this was to be a nonscientific engagement with nature, Charles concluded, "Read geology—Buckland's 'Bridgewater Treatise' and you will rise up awe-struck and cling to God." Charles had studied geology with Sedgwick in Cambridge, riding out with him on field trips. But, for all that Buckland's work was scientific, it offered an encounter with the natural world that could be profoundly devotional; here was an account of geological phenomena that would rouse the emotions. Even a little child could study nature in such a way. Fanny was to use her senses much and her mind little: "Feed on Nature, and do not try to understand it," he advised. "It will digest itself." She should "Think little and read less," keep herself busy, "pray and praise," and fill her commonplace book with observations and sketches of nature.[48]

There is something strikingly visceral about Charles's advice regarding the observation of nature that reflects the intensity of the physical passion between them. Indeed, alongside this advice went the introduction of a weekly ritual that the pair were to share during their enforced separation. On Thursday nights, they were to engage in a festival in which they imagined lying together in bed, while on Fridays they were each to fast in penance, Charles scourging his naked body. The pair suffered intensely from the separation, Fanny becoming increasingly ill, so that at last her family agreed to the resumption of their relationship. Indeed, the wedding was brought forward when she began to suffer "spasms which the family doctor declared only married love could cure."[49] Yet in all this sexual frustration and in the couple's determination to find a spiritual depth to their sexual desires, a sensual engagement with nature, facilitated by Buckland's Bridgewater, seemed to Charles to offer an important release, reconnecting their feelings with the divine.

Even on a more mundane level, works like the Bridgewaters offered important means of cultivating the kind of intimate marriage relationship commended by Christian writers. In an age in which middle-class gender roles and spheres of operation were increasingly being differentiated, such means were more necessary than ever. One of the most forthright supporters of the notion of separate spheres was Sarah Ellis, but her *Women of England* (1838) elaborated on the important role that women had in making the domestic sphere one that was appealing to men. Through the

48. Kingsley to Grenfell, August 1842, in [Kingsley] 1877, 1:86–90.

49. Chitty 1974, 74, 83.

careful management of conversation, women could wean their husbands from the temptations to which the world exposed them and bring to bear a weighty moral and religious influence. Science had an important part to play in this if handled aright. It was easy to talk of astronomy in technical terms, but what was needed was a "general survey of the laws of the universe, and to bow before the conviction that all must have been created by a hand divine." This was all the more necessary in a world where many of the middle-class women Ellis was addressing lived in "tumultuous" industrial cities. As "commerce, and arts, and manufactures" endlessly increased, it was the role of women "to invest material things with attributes of mind," especially so that they might "assist in redeeming the character of English *men* from the mere animal, or rather, the mere mechanical state, into which, from the nature and urgency of their occupations, they are in danger of falling."[50]

Ellis would have been well pleased with the household of the Manchester cotton manufacturer's wife Mary Greg. Just three weeks after the publication of Buckland's treatise, the couple were "reading him, in an evening, aloud." Greg's husband, Robert, had public opportunities to develop an interest in the sciences, being involved in the founding of both the Mechanics' Institute and the Royal Institution in Manchester. However, this domestic reading of Buckland offered the couple an opportunity to draw together around the higher meaning of the sciences. And, as devout Unitarians, they were able to discuss what they considered the negative implications of Buckland's scheme of biblical interpretation.[51]

Even among the gentry, however, such scientific reading played a valuable role. When, in May 1836, admiral's daughter Juliana Bond traveled from Exeter to visit her married sister on the edge of Dartmoor, she was pleased to observe the state of Sophia's "*domestic* life." It was four years since Sophia Bond had married Captain Edward Baring-Gould, nephew of Arctic explorer Sir Edward Sabine, following his injury-enforced retirement from the East India Company's army at the age of twenty-eight. But Juliana was delighted by how "kind and attentive" her sister's husband was. He did not, like many gentlemen, find such a retired life "very wearisome," craving "Society" or hunting and shooting but instead amused himself with "reading and gardening." At that moment, his reading was rather "deep" and "dry": he was to be found "in the agonies of mastering a passage of De la Bèche, or Whewell's *Treatise on Nature*." Yet his wife joined him in the discipline: "Every evening after tea we sit down at the table,"

50. Ellis [1838], 114–17, 160–62, 324, 329–30.

51. RH Greg to J Phillips, 14 October 1836, OUMNH, Phillips Papers.

Juliana reported, "Sophy with her work, trying to understand and acquire a taste of Whewell's *Treatise*." Sophia herself reported that the pleasures of being "a-courting" were as nothing compared with "sitting down at a table in the act of writing to a dear sister, with a dear husband sitting with his book before him, like a good boy close to your side, one, I should specify, who makes you love him more and more every day, by his kindness and affection."[52] For this devout if theologically ignorant Anglican couple, sharing the demanding reading of Whewell's Bridgewater helped to make the domestic sphere suitable to the ongoing challenge of Edward's enforced retirement even if, within a year, they had given up their sequestered existence for travel on the Continent.

Given the religious importance accorded to reading, it comes as little surprise that it also played a key role in the making and conduct of many marriages. In the perilous context of courtship, devout reading was a valuable sign of character, but when that reading was also scientific, it could prove particularly invaluable in cultivating the marriage bond at a time of growing gender differentiation. Here was a point at which the baser concerns that were increasingly associated with the public world of male action could be refined by association with the religious and moral sublimity of the feminized domestic sphere, carrying the couple to a condition of devout domestic harmony. Few books were better suited to such purposes than the Bridgewater Treatises.

### *Christian Parenting*

It was not only between husband and wife, however, that the Bridgewaters came to play a significant role at the heart of the Christian family. They also featured heavily in relationships between parents and children. Edward Baring-Gould was a strong believer in corporal punishment and had his own particular take on the subject. His son, later the clerical author of the hymn "Onward, Christian Soldiers," recalled him saying that "a very convincing Bridgewater Treatise" might be written on the evidence of divine design to be found in the provision of a "portion" of the "person" of boys suitable for benign chastisement. Other parents, however, found more conventional ways of using the series. The role of science in middle-class domestic education had been growing steadily since the latter part

52. Sophia Baring-Gould to a sister, 28 August 1832, Juliana Bond to unknown, 23 May 1836, in Baring-Gould 1923, 9–10.

of the eighteenth century, accompanied by a growth in the number of science books for children. Rational dissenters (notably Unitarians) had been in the vanguard of this development, advocating the advantages of a scientific education, while more orthodox Christians, such as Kirby's cousin Sarah Trimmer, were concerned that scientific education should be rooted in the primacy of scriptural religion.[53] As the range of science books for children grew apace in the early nineteenth century, such concerns persisted, and evangelical and High Church parents alike sought to manage their children's scientific education so as to foster rather than to hinder their spiritual progress.

Ellen Parry's parents offer a good example. Her father came from a family at the heart of late eighteenth-century rational dissent, where the education of girls was especially promoted. The educational scheme that he and Emma devised for their six daughters was relatively liberal, affording many of the opportunities to study the sciences deemed suitable for girls of their social standing. Alongside learning to read, Ellen's earliest training involved learning prayers, hymns, and the catechism. Thereafter, she learned arithmetic, grammar, geography, French, English and Scripture history, and natural history—at first with her governess, Miss Huntley, and from the age of eight with her mother. Natural history was common fare for girls in such circles, but Ellen's exposure to the subject was notable. By her fifth birthday she knew "the names of the grasses, and a few shells," and by her sixth she was reading Sarah Trimmer's sequel to her *Easy Introduction to the Knowledge of Nature,* having been much pleased with "an exhibition of wild beasts" that included an elephant. The following year, she was reading "a book of Natural History," and by her eighth birthday she was beginning to make a collection that included dried flowers and shells.[54]

While Ellen's parents were pleased with her intellectual accomplishments, they were also always keen to encourage her religious training, as Emma's annual memoranda and letters make clear. In 1838, when she was working her way through several Bridgewaters, Emma's birthday letter dwelt on the "pains" Ellen was taking to cultivate her mind "by reading on instructing subjects, and especially by the study of the scriptures, from whence all our happiness must be derived." Since her father was a keen supporter of phrenology, Ellen had the additional benefit of phrenological readings by Britain's foremost practitioners to assist her self-

53. Fyfe 2000; Baring-Gould 1923, 108.
54. [Parry 1846?], 26–28.

examination. When her father's friend Johann Gaspar Spurzheim gave a reading of Ellen's character in February 1831, he noted her "great talent for natural history, in all its branches." But as Ellen grew older, phrenology became useful for developing her religious habits. Following a later visit from George Combe, a fifteen-year-old Ellen was, like her sisters, enjoined to address her deficiencies of character and presented with a diary for daily self-examination to be reviewed weekly by her parents.[55]

Within this rigorously managed religious upbringing, Ellen's parents evidently considered the Bridgewaters consistent with their ambition of educating their daughters in a spiritually healthy manner. And while the Parry family were hardly representative, it is clear that the series was introduced into other devout middle-class families in a similar way. One commentator looking back from the turn of the century reflected that the "good parent" of "Evangelical Clapham of sixty years ago" interested his children in the wonders of the universe by having them read the Bridgewaters and introducing them to the microscope and telescope.[56] Nevertheless, there was a persistent undercurrent of fear across Anglican and nonconformist evangelicalism and among High Anglicans about the possible dangers of such reading for one's children.

The point was made forcibly in 1836 by John Angell James, a prominent Birmingham minister who had recently been one of the architects of the Congregational Union, in a work dedicated to his twenty-two-year-old daughter, Sarah Ann. She had been impressed by the account of the piety of Clémentine Cuvier, the French naturalist's daughter, that was sent to the *Evangelical Magazine* in 1828 by the British minister of a leading Protestant congregation in Paris. James now offered her an enlarged reprint, together with some reflections on Clémentine's early death, under the title *The Flower Faded* (fig. 5.4). The work was aimed at girls and young women—drawing out fearful lessons about the vanity of the world, the nature of true religion, and the mortality of youth—and offered warnings about treacherous false religions. Young people might be snared into "educational reverence" for God or "mere religious sensibility," but especially specious was "*that admiration of the power, wisdom, and beneficence of the Creator*, in which science indulges as it surveys the proofs of benevolent intelligence with which the universe is replete."[57]

James declared himself fully aware of the advantages of science, includ-

55. [Parry 1846?], 29–30, 129–41, 148; see also 65–67, 87–88.

56. "Short Notices," *Church Quarterly Review* 65 (1908): 228–30, on 229.

57. James 1844, 77–79. See W[ilks] 1828; another account of Cuvier (from the *Archives du Christianisme*) appeared in the *Christian Observer* 28 (1828): 531–34.

FIGURE 5.4 Clémentine Cuvier. Engraving by Dick. From John Angell James, *The Flower Faded: A Short Memoir of Clementine Cuvier* (New York: D. Appleton, 1838), frontispiece.

ing in prompting religious adoration. Books such as Paley's *Natural Theology* and the Bridgewater Treatises could be "read with instruction and advantage by all." But it should "never be forgotten," he continued, "that it is by *revealed* and not by natural religion that the sinner is to be saved. It is Christianity, and not deism, that will take us to heaven." Without "penitence and faith in Christ, and love to a holy God," the sublimity of nature was "all mere poetry, not piety." And the point was well made by the case of Clémentine Cuvier. For while she had been exceptionally well educated in "profound science," she knew that religion was the only education relevant for heaven. Her father, by contrast, seemed likely to have risen no higher than the "altar of natural religion," and his dying moments shared none of the religious hopes of his daughter's.[58]

James was already a popular author, and by midcentury his *Flower Faded* had passed through seven editions. Moreover, its viewpoint was evident in his other publications. His *Christian Father's Present to His Children* (1824), which reached twenty editions in four decades, urged children to study diligently and assured them that natural history and experimental philosophy offered inspiring displays of God's existence and attributes. At the same time, however, it made clear in the strongest terms that such knowledge was not the "great end of life." More popular still was his *The Anxious Inquirer after Salvation Directed and Encouraged* (1834), which, once taken on by the RTS, sold two hundred thousand copies in five years and more than half a million by 1861. Once again, the sciences were shown to be a potential snare. If his readers could, "by the most splendid discoveries in science, or the most useful inventions in art," fill the earth with their fame, still, he thundered, their lives would nevertheless have been in vain if they had "lost the salvation" of their souls.[59]

Such priorities affected the practice of many parents, such as Irish landowner Charles Cobbe and his wife Frances of Newstead House, ten miles to the north of Dublin. In contrast to the Calvinist James, the couple were remembered by their daughter, the activist and writer Frances Power Cobbe, as moderate evangelicals exhibiting the "mild, devout, philanthropic Arminianism of the Clapham School" that then "prevailed among pious people in England and Ireland." Nevertheless, they were both actively involved in teaching her religious lessons from the earliest age, maintaining a strict routine of family prayer and Bible study. Frances became a passionately religious child, writing devotional reflections and poring over the classic devotional guide *The Whole Duty of Man* (1658) before undergoing a conversion experience in 1839 at the age of sixteen.[60]

Largely educated at home by governesses, Frances Cobbe increasingly used the large family library to inform herself about a range of subjects, developing a passion for astronomy and becoming "very familiar" with all of the Bridgewaters. Soon after her conversion, however, she started to question Christianity, doubting the testimony of the gospels and the truth of

58. James 1844, 79–81, 93, 97, 115–16. On Cuvier's religious views, see Outram 1984, chap. 7.

59. James 1838, 8; Dale 1861, 288, 308; James 1825, 192–93, 354–60.

60. Cobbe 1894, 1:51–52, 81–82, 86, 88. On Cobbe, see Mitchell 2004; on her religious views, see Peacock 2002.

miracles. After four years of struggle she privately abandoned Christianity, moving quickly from what she later thought of as an "Agnostic" moment to a deism rooted in the arguments of Paley and the Bridgewater Treatises and in the Enlightenment tradition. As it proved, this was only the beginning of Cobbe's pilgrimage to a much richer theism, but it was precisely the awful destiny that James had conjured up. Indeed, when Frances finally told her father about her rejection of Christianity, it came as a "terrible blow" to him, and he exiled her to her brother's house for many months.[61]

There was a growing body of printed matter designed to help wealthy and middle-class parents to avoid such painful outcomes—notably books and magazines for children that showed by precept and example how scientific matter should be incorporated into the Christian life. One author very active in writing for this market was Bourne Hall Draper, a Baptist pastor in Southampton, who in the 1820s and 1830s produced, as one obituarist had it, "a great number of little books, adapted to interest children and promote their spiritual welfare." Having served his time as an apprentice printer at Oxford University Press, Draper studied for the pastorate at a time when the new head of the Baptist Academy in Bristol, John Ryland, was expanding its curriculum to cover a broad range of subjects, including the sciences. Draper acquired a considerable interest in scientific subjects, and many of the books for children that he wrote between 1825 and his death in 1843 concerned the natural world.[62]

Most of Draper's earliest children's books were Bible related, and his *Bible Story Book* (1827) remained in print for half a century, reaching a seventeenth edition. However, even his *Stories from Scripture* (1827) included scientific information in its retelling of the biblical narrative of creation and ended with a meditation that ran from the "striking display" that creation gave of God's power, goodness, wisdom, and glory to the human need for divine grace in Christ. In March 1826 Draper issued a work that made more explicit his view of the "great importance" of leading young people "early to see the hand of God, as it is displayed around them in the creation." His two-shilling *Conversations on Some Leading Points in Natural Philosophy* was offered in place of elementary works that were either too expensive or too secular. And gospel truths were never far away in Draper's conversations. The opening discussion on matter, graduating to one

61. Cobbe 1894, 1:47–48, 52–53, 69–75, 88–95, 100.

62. Moon 1979, 27–30; "Rev. B. H. Draper," *Baptist Magazine* 35 (1843): 585. For Draper's publications, see Moon 1976, 40–41, Darton 2004, 76–77, 384–86, and Peddie and Waddington 1914, 170–71. For his unspoken engagement with phrenology, see Van Wyhe 2004, 172–73.

on the human soul, ended with the biblical observation, "It shall profit a man nothing, if he gain the world and lose his own soul."[63]

Several other works for both youths and children followed in a similar vein, published by leading children's publishers John Harris and, subsequently, William Darton. These became more elaborate, and his final two such works—*Juvenile Naturalist* (1839) and *Stories of the Animal World* (1841)—offered very detailed information about natural history in fictional conversations taking place between Edward and his father Mr. Percy while on country walks, although always in such a way as to link it to Christianity. As a title-page quotation from Paley made clear, Draper's object was to create in his readers the most morally desirable "train of thinking," namely, "that which regards the phenomena of nature, with a constant reference to a supreme intelligent Author."[64] Within the safe confines of evangelical family conversation, there was no question but that the Bridgewaters could be made a useful resource in establishing such a pattern of thought, and Draper drew on the series in both books, with the fictional Mr. Percy offering seasonable quotations for Edward's edification and recommending him to further reading.

Of all Draper's publications, his *Stories of the Animal World* had the highest scientific pretentions, being "arranged so as to form a systematic introduction to zoology," but the religious references continued unabated, and the second conversation began with Edward asking for an account of the Bridgewater Treatises that his father had lately been reading. Mr. Percy suggested that his son should read the books for himself when "a little older" before drawing from them some brief but quite technical instances of design in nature. Edward's reply modeled the appropriate response: "How wonderful are the works of God! I am never tired of hearing you discourse of them." Since the Bridgewaters were "illustrative of the works of God," further references to them were entirely in harmony with the purpose of the walks. Moreover, among the passages that Draper's paterfamilias was able to draw on were some that touched on the distinctive doctrines of Christianity—such as Kirby's reflections on vicarious suffering and Chalmers's on the human conscience.[65]

Such practical training of young people to find the deeper Christian meaning in nature, and to incorporate works such as the Bridgewaters in doing so, met with warm approval by reviewers ranging from High Anglicans to Wesleyan Methodists. Moreover, it was not merely in expensive

63. Draper 1828a, [iii], 16; Draper [1828b], 15.

64. Draper 1839, 1:[iii], v.

65. Draper [1841], iii, 44–48, 137–38, 299–304. See also Draper 1839, 1:123–24.

books like Draper's two *6s 6d* volumes that such advice was available. Among the most widely circulated periodicals of the period were monthly religious magazines for "youths" costing just a few pennies, and these also offered parents and their offspring guidance about how such books as the Bridgewaters could be turned to good.[66]

Britain's first lastingly successful periodical of this sort was the Sunday School Union's *Youth's Magazine*, which was aimed primarily at middle-class readers and by the 1830s had a circulation of around ten thousand. Its editors saw its main purpose as promoting "Real Religion" and "pointing the youthful mind to the WORD OF GOD as the *only pure source* of Religious Knowledge," but they also sought to contribute to readers' intellectual development, cultivating "a taste for reading" and self-improvement and providing regular and varied scientific fare.[67] The way in which the magazine presented this material, however, was calculated to instill in young readers an essential habit of associating knowledge about nature with scriptural truth. It offered implicit advice about reading through the introduction of extracts and references in factual articles, but it was especially the moral tales and homilies that were meant to promote the practices considered to be necessary to create appropriate religious associations in relation to secular reading, including conversation and family interaction as well as suitable patterns of study and note taking.

Thus, while the July 1834 issue of the *Youth's Magazine* carried a lengthy extract from Bell's Bridgewater—offered by a contributor with the comment that it was impossible for the mind to survey the human body "without acknowledging God as its author"—it came after a brief moral tale offering exemplary guidance about how the series should be read. The tale recounted the visit of Miss Henley to Mrs. Harcourt and her daughter Maria while they were in the garden one June morning. Miss Henley related that she had been reading some of the Bridgewater Treatises, and, as her mother had enjoined her not to "neglect any domestic duty," even for mental improvement, she had consequently had no time for visiting. Now, however, she was glad of the opportunity to admire the natural world itself, and Mrs. Harcourt joined in, remarking, "All nature seems to rejoice in her

66. "Notices of Books," *Christian Remembrancer*, n.s., 1 (1841): 123–37, on 137; "Select List of Books," *Wesleyan-Methodist Magazine* (1841): 397–401, on 400. A number of small penny magazines for younger children, including the *Children's Friend, Child's Companion, Child's Magazine*, and *Teacher's Offering*, were even more popular, but the Bridgewaters were deemed too advanced for such a readership.

67. These comments come from a prospectus to the third series in my copy of the *Youth's Magazine* for 1828. On the *Youth's Magazine*, see Topham 2004a; for an index of its scientific content, see Cantor et al. 2020.

Maker's works; may we be excited to elevate our hearts, from 'Nature, up to Nature's God!'" The ensuing conversation between the two focused on the divine care taken in the production of even the smallest of the garden flowers, with both quoting from a variety of books, including the Bible, concerning the lessons thus to be learned concerning God's love for humans. Moreover, when an envious Maria complained, after Miss Henley's departure, that there had been "too much of display" about their visitor, Mrs. Harcourt rather applauded her quotations for having enriched the conversation with pious reflections.[68]

Boys' conversations with their fathers were sometimes portrayed as more determinedly studious, but they were also intended to inculcate the habit of pious association in using books such as the Bridgewaters. This point was exemplified at length in a series of fictional "Conversations at Carringford Lodge" contributed by Cornish Congregationalist minister Richard Cope over the second half of 1837, at a time when to the *Youth's Magazine* featured numerous extracts from Buckland's Bridgewater. In the conversations, the fictional representative of a landed family, Mr. Ravenstone, set about educating his eldest son Edwin in knowledge that would make him a "useful member of society" and furnish his mind with "a rich supply of topics for profitable contemplation."[69] The overarching framework for their informal and "natural" conversations was provided by the questions, "What am I?" "Where am I?" and "Why am I here?" As Mr. Ravenstone sought to teach his son about the opportunities that the universe offered for observation, he did so in relation to these larger questions.

Offering written comments on an essay concerning observation that he had assigned his son, Mr. Ravenstone included a lengthy quotation from Prout's Bridgewater about the minuteness and complexity of animalcules. The information amounted, he explained to Edwin, to a "delightful illustration" of "the special providence of God," and it evoked from Mr. Ravenstone a medley of biblical texts that exemplified appropriate habits of religious association. Moreover, his example rubbed off on his son, whom he was delighted to find increasingly attentive to the "paramount question of his eternal happiness." It was, to Mr. Ravenstone, confirmation that "sanctified knowledge of history and the sciences, instead of diminishing the love which the Christian feels to God, increases it, and enables him to perceive God in all things, and to find all things in God." Devout learning required appropriate habits of association and reflection, and Mr. Ravenstone emphasized that "all subjects should be guided by a pious re-

68. M. W., "Envy," *Youth's Magazine*, 3rd ser., 7 (1834): 229–31.

69. C[ope] 1837, 160–61.

flection into a holy and practical tendency." In such a way, study of the creation would lead on "to the new creation, the necessity of that new birth effected by the same author." Thus, when the series of tales climaxed in Edwin writing a pleasing essay on preparing for a future state, Mr. Ravenstone reflected, "Nature and providence improved and sanctified, lead us to grace; and grace points to eternal glory." He closed with the benediction "Go then, my son, go, and continue your researches into the kingdoms of nature, providence, and grace."[70]

A third story in the *Youth's Magazine*, this time featuring Buckland's Bridgewater, again made much of the importance of giving appropriately religious associations to the knowledge of nature. Taking a mixed family party on a rural walk, Mr. Halesworthy sought to impress on his children the importance of such associations. He told his daughter Charlotte that if she had read the introduction to Buckland's *Geology*, she would have been able to understand how the wild flowers they encountered reflected the "very structure of the world itself." But there were deeper associations to develop. The hidden parts of the earth, he pointed out, had all been "weighed and measured, and adjusted to their proper ends by God." As the "happy family group" reflected on these and the other providentialist associations that Mr. Halesworthy suggested, they fell into a sublime and distinctively Christian reverie from which the paterfamilias roused them with the observation that "the Christian will never rest satisfied with any associations that stop at second causes; he must carry every thing up to God himself."[71]

According to the *Youth's Magazine*, the fear of some devout parents that scientific reading might damage their children's religious training was thus to be addressed by ensuring that children were taught how to associate what they read with the distinctive outlook of Christianity. This was a view that was widely shared, albeit with some variations in emphasis between Christians of different hues. While the Bridgewaters could certainly be misused, they nevertheless readily lent themselves to such devout usage, combining authoritative and accessible exposition with repeated reference to religious perspectives.

❋

70. C[ope] 1837, 298–99, 344, 350, 415–16. Cf. Prout 1834a, 24–25.

71. S., "The Afternoon Walk," *Youth's Magazine*, 4th ser., 3 (1840): 190–98, on 194, 197.

As the previous chapter showed, most religious reviewers of the Bridgewaters exhibited disdain for a rationally constituted natural theology, especially in the context of attempts to convince those who doubted the truth of Christianity. Yet many reviewers had also been sanguine that the series could do good by visibly reconnecting the sciences with the religious concerns of the Christian. This was certainly at the forefront of the minds of devout readers as they engaged with the new works. Properly used, many considered, the Bridgewaters could evoke suitable feelings toward God while developing an enlarged but theologically orthodox understanding of the creation. Such a purpose was important in a new way in the age of cheap print, and its management was not undemanding. But in the typically domestic context of personal and family religion, the emphasis was much more on the management of religious habits, associations, and feelings than it was on theological reasoning or argument. Whether in daily habits of religious devotion, in the relating of courting or married couples, or in the upbringing of children, earnest Christians were anxious to ensure that their encounter with the sciences, managed through the Bridgewaters, would sustain a life of religious faith. Moreover, as the next chapter shows, these concerns were also at the forefront in public religion, where preachers who drew on the series did so largely in the context of the same kind of interest in evoking religious feelings that would prompt congregations to action. Only when Christian protagonists sought to use the Bridgewaters to respond to the "infidelity" that was becoming a significant presence in Britain's towns and cities did theological arguments come to the fore, and there, their effectiveness was equivocal.

✹ CHAPTER 6 ✹

# *Preachers and Protagonists*

*During the first decades of my scientific life, science was rarely, within my experience, heard of from the pulpits of these islands: during the succeeding, when the influence of the "Reliquiæ Diluvianæ" and the Bridgewater Treatises was still felt, I often heard it named, and always welcomed.*

JOSEPH HOOKER, *August 1868*[1]

Looking back in the 1860s, the president of the British Association was nostalgic for a time when the Bridgewater Treatises had led Christian preachers to offer their parishioners a positive vision of the sciences. This, he claimed, stood in stark contrast to more recent years, when the majority of clergy named science only to revile it. Such a change mattered, he observed, because of the immense power of the pulpit. According to Hooker, the country clergy were especially negative, but it was from them alone that an "overwhelming proportion of the population" ever heard the name of science. And while clerical commentators disagreed as to whether preachers were as negative as Hooker suggested, they certainly agreed that the clergy were "looked up to in thousands of parishes as the natural leaders of opinion" whose authority gave an "almost oracular dignity" to their utterances.[2]

The power of the Victorian pulpit and its potential for affecting attitudes to the sciences is undeniable. Nearly eleven million attendances at public worship were recorded in England and Wales on census Sunday in 1851. This was equivalent to more than 60 percent of the total population (including children), although repeat visits by some meant that that the proportion attending worship was probably closer to 40 percent.

1. Hooker 1869, lxxiii.
2. Farrar 1868, 600; Hooker 1869, lxxiii–lxxiv. See also Hannah 1867, 1868, 1869.

Moreover, while the various forms of public worship typically combined prayers, Bible readings, preaching, singing, and the celebration of Holy Communion, it was chiefly within sermons that the sciences featured, if at all. A conservative estimate suggests that over twenty thousand sermons were delivered in Britain on a weekly basis, giving preaching the largest audience of any form of communication in the period.[3]

Some enthusiasts for the sciences in the age of reform certainly viewed the pulpit as a "powerful engine"—comparable in influence to the press—that was "capable of producing on the mass of mankind, a tone of thinking, and an enlargement of conception," in relation to the sciences that no other means could "easily effect."[4] Yet in producing such effects, preaching did not necessarily entail theological arguments or oracular statements concerning science and religion. On the contrary, while there were some contexts—notably educational—in which theological analysis of the religious tendency of the sciences might be appropriate from the pulpit, the great bulk of preaching was addressed to matters of practical religion. Most sermons were designed primarily to evoke in the Christian a practical response at the level of religious devotion, personal morality, or social duty. This chapter explores how preachers incorporated the Bridgewaters into this weekly sermonizing, examining further the connections that many Christians considered necessary to be developed between secular learning and the daily practice of religion.

While the Bridgewaters could thus be useful adjuncts to a religious life for those already committed to Christianity, this was nevertheless an age increasingly concerned about the repudiation of Christian faith, or "infidelity," and the latter part of the chapter moves on to investigate such concerns. Most writers in the religious magazines were, as we have seen, signally unimpressed by the potential usefulness of works of natural theology in addressing religious unbelief. Irrespective of whether they were prepared to admit that anything could be known of God independently of his self-revelation, evangelicals and High Anglicans alike were sure that the problem with infidels lay in their hearts. The Bridgewaters had little to offer such skeptics. Yet as the tide of infidelity rose in the 1830s and as the churches began to pay it more concerted attention, there were some activists—especially Protestant dissenters—who were prepared to enlist the Bridgewaters to answer the arguments of atheist disputants. As they did so, however, some of their opponents began to see the series as

3. Gibson 2012, 7; *Census 1851—Religious Worship*, 155–57. Among the few studies of science in sermons see Sell 2016, Francis 2012, Cantor 2011, and Brooke 1991a.

4. Dick 1824, 461–62.

a powerful resource on their own side, or at least as a useful foil for their opposing views.

## *Sermons in Stones*

How is the historian to weigh Joseph Hooker's rhetoric concerning churchgoing in the 1830s? So much of what was said from the pulpit, let alone of how listeners responded to it, has left no trace. Certainly, many sermons were published—indeed, the published sermon, or volume of sermons, was a staple of the nineteenth-century book trade. Numerous others survive in manuscript. Yet these are the tip of the iceberg, and they are inevitably unrepresentative. Of those that survive, a disproportionate number are "occasional" sermons prompted by the opening of an institution, an anniversary, a death or departure, or a special invitation, which often lent themselves to reflections on themes or topics of the day. By contrast, the routine cycle of congregational sermons was much more tied to the seasons of the Christian calendar, to the quotidian concerns of Christian morality with an emphasis on exhortation, and to the inculcation of Christian doctrine.[5]

One way of charting the vast and largely unknowable ocean of sermons is through the guides that conscientious clergy drew on for advice. In an age in which formal training of the clergy for their pastoral duties was still very limited, such published guides were an increasingly common recourse, with several titles running through numerous editions. One especially important guide was the *Essay on the Composition of a Sermon* by the seventeenth-century Huguenot Jean Claude. First translated by a Baptist minister in 1778, it was reissued in 1796 by leading Anglican evangelical Charles Simeon. A notable advocate for teaching the clergy how to preach, Simeon offered formal training to his students at King's College Cambridge and published over two and a half thousand sermon outlines in twenty-one volumes. His edition of Claude's work ran to six editions by 1833, but the work was also incorporated into Thomas Hannam's *Pulpit Assistant* (1799) and dissenter Edward Williams's *Christian Preacher* (1800), and it formed the framework for Congregationalist Saunderson Turner Sturtevant's *Preacher's Manual* (1828), all of which went through many editions.[6]

5. Francis and Gibson 2012, and esp. Gibson 2012. See also Ledger-Lomas 2009 and Rivers 2018.

6. On the training of the clergy, see, for example, Jacob 2007 and Brown 1988. On Sturtevant, see Anon 1842, 25, and the Surman Index, https://surman.english.qmul.ac.uk/.

According to Claude, the purposes of preaching were diverse: "to instruct, solve difficulties, unfold mysteries, penetrate into the ways of divine wisdom, establish truth, refute error, comfort, correct, and censure, fill the hearers with an admiration of the wonderful works and ways of God, inflame their souls with zeal, [and] powerfully incline them to piety and holiness." Insofar as it should inform the understanding, it should do so in a manner "which *affects the heart*; either to comfort the hearers, or to excite them to acts of piety, repentance, or holiness." Early nineteenth-century commentators agreed. Preaching was intended to result in a practical outcome in the form of changed behavior brought about by addressing the affections. Claude's account thus shared with many others a thoroughgoing commitment to applying the principles of classical oratory to Christian preaching. And the point was particularly evident in his prescription for the sermon's conclusion, which was meant not only to summarize the message but also (in accordance with the classical *peroratio*) to "move Christian affections—as the love of God—hope—zeal—repentance—self-condemnation—a desire of self-correction—consolation—admiration of eternal benefits—hope of felicity—courage and constancy in afflictions—steadiness in temptations—gratitude to God—recourse to him by prayer—and other such dispositions."[7]

Such a view was not confined to evangelicals. In 1835 High Anglican William Gresley's treatise on preaching claimed that its object was to "*make men real Christians*" and "to keep them so." This required the preacher to make a "lasting impression on the heart, and effect a corresponding change in conduct"—a task that would require the preacher's "whole power of persuasion" and the conscious adoption of the "principles of rhetoric." Gresley devoted two entire chapters to the question of "How to Move the Passions," observing, "Persuasion is the end of all preaching." Convincing one's hearers of what they ought to do was not enough; they must be persuaded to actually do it, and this was to be done "principally by moving the passions, or feelings." Having attended to this priority throughout the sermon, the preacher should make a "last vigorous effort" in the conclusion to "stir up, or raise to the utmost" an appropriate feeling, "whether it be of love, gratitude, zeal, courage, faith, hope, and charity; or of sorrow, shame, self-condemnation, resolution to amend, [or] repentance."[8]

The authors of preaching handbooks frankly acknowledged that in

7. Simeon 1838, 2, 5, 118. On Claude, see also Edwards 2004, 452–55, and Ellison 1998, 18.

8. Gresley 1835, 15, 23, 25, 33, 273, 464.

practice, preaching was often far removed from these ideals. Nevertheless, the advice they offered reveals a shared understanding that sermons were not merely intellectual exercises. Rather, the great bulk of sermons were straightforwardly practical in purpose, seeking to enjoin or encourage patterns of moral or religious behavior, often through religious instruction, or to arouse and develop religious fervor. The kinds of topics covered are well illustrated by such catalogs of sermons as the *Churchman's Guide*, published by John W. Parker in 1840, in which, for example, the entry for "E" encompassed "Earnestness in Religion—Elect—Enemies, love of—Enquiry in Religion—Envy—Evidences—Evil Example—Excuses—Expediency."[9]

Within the primarily pastoral context of Christian preaching, the sciences were nevertheless potentially relevant in a range of ways, and increasingly so as they became a more prominent aspect of public culture from the 1820s. As Claude's manual makes clear, inspiring congregations with a feeling for God's "wonderful works" had long been considered to be part of the preacher's remit. Exploring the creation was also sometimes important as the preacher sought to illuminate the "ways of divine wisdom." In particular, as the sciences came to the attention of an ever-larger number of people during the age of reform, preachers had new incentives to discuss the relevance of scientific claims to the vision of divine providence that they outlined. In some cases, of course, this became a matter of controversy, as preachers sought—in Claude's terms—to "establish truth" and "refute error." More generally, knowledge of the creation could be invaluable in contending for or confirming belief in God, even for the many preachers skeptical of natural theology, strictly defined. Finally, scientific knowledge could be introduced by way of illustrative color, although Claude was not alone in warning preachers to avoid "philosophical" observations and "passages from Profane Authors" that were "a vain ostentation of learning."[10]

This was the context in which published guides to preaching began to list the Bridgewaters as books suitable for clerical libraries. Some of the Bridgewater authors were themselves preachers of renown, and William Gresley, who elsewhere applauded the Bridgewater writers for elucidating "the agreement between Science and Revelation," drew attention to the famed Chalmers's Bridgewater as offering the preacher a cautionary example of "excessive prolixity." Of all the series, however, Chalmers's was probably the Bridgewater that clergy were most likely to encounter in

9. Forster 1840, [xiii].
10. Simeon 1838, 8.

their education. That was certainly the case in Edinburgh, where many trainee ministers read it repackaged as part of Chalmers's textbook of evidences. But earnest evangelical Presbyterians, such as aspiring ministers James Halley and James Hamilton, also read others of the series, recommending them to friends as part of their theology course. Hamilton, who was a friend of "Joe Hooker" from days of botanizing together as Glasgow students and soon after became minister at the National Scotch Church in London, even contributed a lively review of Buckland's treatise to the *Presbyterian Review*.[11]

The Bridgewaters informed the approach to the sciences of a generation of such student ministers. Halley and Hamilton's Edinburgh contemporary Joseph Taylor Goodsir was still reaching for the series when he dusted off one of his early sermons for republication in response to John Tyndall's address on "Scientific Materialism" at the 1868 British Association meeting. Nor was it only in Edinburgh that young men intended for the church became familiar with the Bridgewaters. As the next chapter shows, many students in Oxford and Cambridge encountered them even though they were not part of a formal curriculum. For instance, Cambridge's twentieth wrangler in 1830, Thomas Henry Steel, read Bell's and Chalmers's works shortly after graduating master of arts, while he was a fellow and assistant tutor at Trinity College. What he learned was put to good use in later life, when his sermons as assistant master at Harrow School drew on "copious stores of scientific knowledge in illustration of the wonderful works of God."[12] The Bridgewaters were clearly a valuable part of the preacher's tool kit. The critical question, however, was how to use them to good effect.

We might expect that the series was most useful to preachers in relation to filling their hearers, in Claude's phrase, "with an admiration of the wonderful works and ways of God," but this appears not to have been the case. There were certainly preachers who, in an increasingly scientific age, were keen to advocate the use of the sciences in such a way, but they did not generally find a very receptive audience. Indeed, when Scottish Secessionist schoolmaster Thomas Dick's popular *Christian Philosopher* urged the need for ministers to take up the sciences in their preaching, critics argued that scientific matter was a distraction from the main purposes of religion.

11. Arnot 1842, 78–79, 85, 260, 272; Arnot 1870, 83, 89–98; Gresley 1835, 148; Austen 1879, 24; Bickersteth 1844, 383; Simeon 1853, 428. See also Arnot 1877.

12. Steel 1882, xxiv, xli; Goodsir 1868, 22. A numbered ranking in the first class ("wranglers") in Cambridge's main bachelors of arts examination (the "mathematical tripos") was a mark of considerable academic distinction.

One of Dick's key objectives was to show that "the Teachers of Religion" ought to include a "wider range of illustration, in reference to Divine subjects, than that to which they are usually confined." Mirroring the error of those who reduced religion to natural theology, he claimed, was the error of "the greater part of religionists" who tended to "treat scientific knowledge, in its relation to religion, with a degree of indifference, bordering on contempt." Dick urged repeatedly that Christian ministers should be in the vanguard, in their pulpit ministrations, in helping Christians to see how valuable scientific knowledge was in learning about and worshipping God. Yet while his book was generally well regarded by evangelicals, reviewers resisted his suggestions for preachers, echoing the "zealous outcry" Dick had already heard "against every discussion from the pulpit, that has not a *direct* relation to what are termed the doctrines of grace."[13]

One forthright commentator was the prominent Wesleyan Methodist minister Richard Watson, whose *Theological Institutes* (1823–28) systematized Wesleyan theology for the first time. Reviewing Dick's book anonymously for the *Wesleyan-Methodist Magazine*, Watson claimed that his criticism of ministers was ill founded. Every Sunday the works of God were used in a thousand pulpits to provide "interesting allusions, impressive figures, and means of illustration" that enlarged religious knowledge and deepened religious feelings by their "*proper connexion* with other truths." But this was the key point: the impressions produced by the natural world were not "*in themselves* moral or religious impressions." Preachers should, as St. Paul directed, "know nothing" except "Jesus Christ, and him crucified." Displaying God's attributes in the creation was good only insofar as it served that purpose. Moreover, scientific discussions were actually less effective at producing such impressions than the "general *observation* of nature."[14]

Watson elaborated on these views when he addressed the 1827 Wesleyan Methodist conference. Speaking as president on the occasion of the ordination of new ministers, he reviewed the characteristics that their office required—courage, love, and a sound mind—stressing that the last of these involved prioritizing the truths that they were "appointed exclusively to teach." Referring back to Dick's book, he emphasized that while ministers should not be ignorant of scientific subjects (which might be useful in illustrating truth or adorning eloquence), extensive scientific knowledge was simply not helpful to the preacher. Watson himself had interests in botany and astronomy, and he was not averse to using his knowledge of

13. Dick 1824, [v], [17], 33; Astore 2001, 63–68.
14. [Watson] 1824, 35. Cf. 1 Cor. 2:2.

nature in the pulpit. Moreover, he was an able preacher who was considered to have contributed to improving the standard of Wesleyan preaching. However, his message was clear: the use of scientific knowledge in preaching required great circumspection.[15]

Preaching excessively about the works of God, or at least without due reference to distinctively Christian concerns, could easily cast doubt on a preacher's religious standing. In his *Metropolitan Pulpit* (1839)—an ostensibly impartial review of "the most popular Preachers in London"—Scottish Calvinist and journalist James Grant offered a damning indictment of Joshua Frederick Denham for his propensity to preach on the "perfections of the Divine Being" as evidenced in the "volumes of creation and providence." Denham—the Cambridge-educated curate of St. Mary le Strand and evening lecturer at St. Bride's Fleet Street—had formerly been classed as an evangelical preacher, but now his sermons usually had a "closer connection with natural religion than with the theology of the Bible." He could offer extended quotations from Paley's *Natural Theology* and display scientific erudition, but, with little reference to the Bible or Christian doctrine, his discourses might as well be moral lectures at a mechanics' institution. His preaching was hardly likely to effect "the conversion of sinners or the edification of saints." There was nothing more "frigid and soulless," one Oxford-educated vicar later reflected, than the "discourses of those who adopt the moral essay and Bridgewater Treatise style of preaching." The preacher needed to "preach the Gospel of Jesus Christ" even if he also needed to respond to scientific developments.[16]

Despite these dangers, preachers found the Bridgewaters useful in a distinctively Christian fashion. Indeed, in the writing of their treatises, certain of the most active and able preachers among the authors—notably Whewell and Chalmers—had made it especially easy for other preachers to draw on their works. The point is well illustrated by a parish sermon "on Natural Theology" delivered in February 1834 by the vicar and archdeacon of Brecon Richard Davies. Davies had been a close friend of John Kidd as a student in Oxford, but it was his reading of Whewell's Bridgewater that provided the materials for his sermon. Indeed, when the sermon was later published, fewer than two of the fifteen pages were original to Davies, with the remainder consisting of a slightly abridged excerpt of the first eighteen pages of the third book of Whewell's Bridgewater, his "Religious Views," together with some additional excerpts from the preface and the chapter

15. *ODNB*, s.v. "Watson, Richard (1781–1833)"; Jackson 1834, 72, 110, 153, 167, 174, 639; Watson 1827, 27–29.

16. Zincke 1866, 101; Grant 1839, 257–66.

on "inductive habits." These were presented as Davies's own "reflections," and when his parishioners petitioned him to publish them, he did so unabashedly, dedicating the sermon to Kidd and Whewell. But while the decision to publish seems outrageous, Davies's wholesale and not fully acknowledged quotation from a published source in an oral sermon was not unusual, as both anecdotes and preachers' guides make clear.[17]

Davies found Whewell's treatise so congenial to his purpose not because it offered "philosophical arguments for religion," which were hardly suitable topics for extended treatment in "mixed" company, or because it provided "abstract explanation of deep scientific investigation," which would inevitably be obscure to many. The Bridgewaters themselves offered those with both leisure and the "preparatory requisites" the opportunity to pursue such aspects further. Instead, Davies's objective was to draw from the series a pastorally appropriate message of divine superintendence. His text was thus a verse from Psalm 8, "O Lord our Governor, how excellent is thy name in all the world!" and his sermon followed Whewell's emphasis on the need to reconnect natural knowledge with knowledge of God as a moral governor. The bulk of the sermon—drawn from Whewell's chapter "The Creator of the Physical World is the Governor of the Moral World"—not only did this effectively but it did so in preacherly style. Like Whewell, Davies considered there to be a profound pastoral significance in interpreting the findings of the sciences in relation to "religion, as it affects the actions of man, *by reforming men's lives—by purifying and elevating their characters—by preparing them for a more exalted state of being.*"[18]

It is striking that a sermon ostensibly on "Natural Theology" and taken almost entirely from Whewell's treatise in this way should be so clearly in keeping with the practical, pastoral concerns of Christian homiletics. Yet it reflects the close connections we have already explored between Whewell's preaching in the 1820s and the writing of his treatise. Not surprisingly, others, including some preachers of a more strikingly evangelical cast, also found that they could draw on Whewell's text, albeit in a less wholesale fashion. Among these were two honors graduates from Cambridge well versed in mathematical natural philosophy who had gone on to be successful evangelical preachers in London: Henry Melvill and George Smith Drew.

Melvill had graduated as second wrangler in 1821, becoming a fellow

17. Some commercially available sermons were even printed to look like they were handwritten. See Francis and Gibson 2012, 23–24, and Jacob 2007, 258–59.

18. Davies 1834, 6, 17–18. The added italics indicate a direct quotation from Whewell 1833b, vi.

and tutor at Peterhouse, but in the 1830s he was in the process of becoming London's most popular preacher. Viewed by contemporaries as a model for evangelical homiletics, his style was reportedly modeled closely on that of Thomas Chalmers—so much so that some claimed that he plagiarized Chalmers's sermons. According to James Grant, while Melvill's orations were "often abstruse," they appealed, like Chalmers's, not only to "a large intellectual audience" but also to working people. Thus, his two-thousand-seat Camden Chapel in Camberwell was usually crowded out by a further four or five hundred people in the aisles, and many others read his sermons in the weekly *Penny Pulpit* or in his own often-reprinted volumes.[19]

Melvill turned Whewell's Bridgewater to homiletic ends soon after it was published. Working from the text "But Jesus answered them, My Father worketh hitherto, and I work" (John 5:17), his theme was that the Christian should have as much confidence in the ongoing redemptive power of Christ as in the ongoing providence of God the father. The first half of the sermon was given over to elaborating on the second of these points, beginning with an account of God's continuing role in upholding the natural order. Here the perspective on natural laws as manifestations of God's ongoing action in the universe drew explicitly on Whewell's Bridgewater, giving authority to Melvill's claim that any other view was as "unscientific" as it was unscriptural. Moreover, this was not Melvill's only sermon reference to Whewell's work. His sermon on "Heaven," for instance, made passing reference to Whewell's account of the human constitution in order to offer a better exposition of a biblical text concerning the excellence of the afterlife (Rev. 22:5).[20]

That Melvill's intellectually robust yet staunchly evangelical preaching could blend well with a reading of the Bridgewaters is further illustrated by the experience of a young midshipman, John Irving. Having been brought up in the Church of Scotland, Irving soon found naval friends who shared his evangelical earnestness—"saints"—including an admiral's son, William Malcolm. After Malcolm left the navy to study for the church, however, Irving found it difficult to maintain his old ways. In February 1837 he visited Malcolm at Trinity College Cambridge, and his old friend took the opportunity to draw him back to his evangelical roots, taking him to hear one of Melvill's university sermons as Select Preacher. Irving was deeply moved by Melvill's account of "the unnaturalness of disobedience to the

19. A range of cheap sermon periodicals existed, the most successful of which were widely circulated. Titles included the threepenny *Pulpit*, *British Pulpit*, and *Preacher*, and the twopenny *Scottish Pulpit*.

20. Melvill 1838, 2:43, 324.

gospel," which offered reassurance, in passing, that acceptance of the gospel was rational and well grounded. His response to his visit to Cambridge was to return to regular Bible reading and immediately to start reading the Bridgewaters, which he was excited to remind Malcolm included one by Trinity fellow, Whewell.[21]

A similar emphasis was to be found in the sermons of George Smith Drew, who graduated as twenty-seventh wrangler in 1843 and immediately became curate at the impressive neoclassical parish church of St. Pancras that had only recently been built to serve the affluent area of Bloomsbury. Drew was active in the British Association for the Advancement of Science, and one of his earliest sermons was on the text "Stand still, and consider the wondrous works of God" (Job 37:14). Yet, while the first part of his sermon emphasized the spiritual value of taking time to reflect on the "number, variety, and magnitude" of God's works, he was very careful to emphasize the distinctive character of such meditation for the Christian. Above all, Christian reflection on the "wondrous works" of creation should highlight that they were "nothing less than manifestations of the attributes and the energy of an Omnipresent God." Only meditations of this character came "well and wisely" from a Christian minister or were "listened to with propriety and with profit" by a congregation, and Drew went on to emphasize how such meditations fed the Christian's attachment to a divine father.[22]

In laying out this view of the matter, Drew contrasted the Christian's sense of God's constant activity with the "enormous error" of some "self-styled philosophers" of allowing the laws of nature to substitute for the divine will. Here, or course, he was drawing on Whewell's Bridgewater, and in the published version of the sermon, a note offered extended quotations from Whewell's discussion of the religious tendency of inductive and deductive habits of mind. Drew had not finished, however. Having outlined the Christian's proper mode of meditation on the works of God, he sought to address the case of those who did not approach them with a sense of God's immanent activity. Such "practical" atheists, he assured his congregation, would never be recalled to the truth by argument but only by a conviction of the sin that blinded them and a true repentance. Only then would they apprehend the "true lesson and meaning" of nature.[23]

Cambridge clergy well educated in mathematical natural philosophy might have particular cause to draw on Whewell for evangelical purposes,

21. Bell 1881, 47–50; Melvill 1837, 10.

22. Drew 1845, 57, 62.

23. Drew 1845, 64, 69–72.

but they were by no means the only ones who did. In the 1820s John Hartley had gone straight from Oxford to work for the Church Missionary Society in developing a mission to the Eastern Mediterranean. Now in a new role as British chaplain in Nice, his sermon to promote a London collection in support of "Protestant efforts" in France used Whewell's Bridgewater for its illustrations. Preaching on the verse "Heaven and earth shall pass away; but my words shall not pass away" (Matt. 24: 35), Hartley asserted the importance and enduring truth of the Christian revelation before urging his auditors in a lengthy conclusion to support the promotion of Protestantism on the Continent. But, while the "internal evidence" of the Bible was strong enough not to require support, Hartley used Whewell's Bridgewater to show that his text received "confirmation from modern science." "Astronomers inform us," the preacher declared, "that they have discovered causes in operation, which will eventually destroy the world, and all worlds by fire." Hartley felt it was hardly his place to judge such claims, but a footnote referred readers of the published sermon to his source.[24]

While preachers thus found the Bridgewaters helpful in illustrating the "ways of divine wisdom" and even in "establishing truth," their uses little resembled either a mere descant on the creation or a reasoned natural theology. Rather, the series was brought in to evoke distinctly Christian sentiments, and this was a purpose to which Chalmers's Bridgewater lent itself even more naturally. Not only did its subject matter include the "moral constitution of man," allowing Chalmers to develop his popular argument for belief grounded in the supremacy of the human conscience, but in addition his approach was decidedly evangelical. Equally significantly, Chalmers was one of the most noted preachers of the age (fig. 6.1). He took London by storm in 1817 to such an extent that William Wilberforce found himself having to climb into a church window in order to hear him. With volumes of his sermons running to numerous editions and individual sermons being widely republished, preachers were used to finding in him an exemplar, and his Bridgewater offered another sample of his distinctive style. Indeed, Chalmers himself led the way in appropriating it in the pulpit. When asked to preach at the National Scotch Church in London's Regent Square shortly after the publication of his treatise, he drew on his account of the supremacy of conscience to urge the practicability of evangelism in the context of "Indian Missions."[25]

24. Hartley 1840, 194–95. On Hartley, see Stock 1899, 1:227, 231, 264, 350.

25. "A Sermon, Delivered by the Rev. Dr. Chalmers, at the National Scotch Church, Regent's Square, Monday, Jul 15, 1833, on Behalf of the Indian Missions," *Preacher*, 25 July 1833, 97–111, esp. 98–100; Wilberforce and Wilberforce 1838, 4:324.

FIGURE 6.1 "Rev. Thomas Chalmers, 1780–1847. Preacher and social reformer." Cut paper on watercolor background, by Augustin Edouart, ca. 1830. Image: National Galleries of Scotland.

Admirers of Chalmers again included some of the leading preachers of the age. Among these was James Anderson, the Oxford-educated incumbent of St. George's Chapel Brighton. Situated close to the Royal Pavilion, this privately owned chapel had become fashionable as Queen Adelaide's place of worship in the town. Anderson became her chaplain in ordinary and, following the king's death in 1837, that of the new Queen Victoria, who thought him a good preacher, if a little "theatrical" and long winded. His sermons were indeed both long and erudite, but they were firmly directed toward the practical concerns of the Christian minister, and few topics were more practically religious than that of conscience. Chalmers rapidly became one of Anderson's standard sources on the topic, worthy of quotation, although references were only given in the footnotes to the published sermons. Anderson's cultured auditors were expected to be familiar with the works of "easy access" that explored the evidence of the moral attributes of the creator offered by the supremacy of conscience, and the flow of the sermon was not to be diverted from the "immediate application of the subject to our own hearts." Thus, in a sermon explicitly "On Conscience," Anderson's concern was not with Chalmers's evidential question but with urging his auditors not to deify their own consciences and instead to allow God's grace to make their consciences responsive to his edicts.[26]

Anderson's cultured preaching style meant that Chalmers's was not the only one of the Bridgewaters he referred to in his sermons. Especially notable was a series of sermons delivered in the late 1830s on the great "hymn of faith" in the book of Hebrews. Preaching on the text, "Through faith we understand that the worlds were framed by the word of God, so that things which are seen, were not made of things which do appear" (Heb. 11:3), Anderson began by elaborating on the inadequacy of natural theology to make God the creator known independently of revelation, referring to many of the recent works on the topic. In the second part of the sermon, however, he urged the "duty" of "tracing out" the many corroborations of God's activity as creator to be found in the natural world. Here he reviewed the science of geology from the perspective of Buckland's Bridgewater, urging that once mature, geology would offer renewed confirmation of the truth of revelation. But even in so "recondite and philosophical" a sermon, refuting error and confirming truth, Anderson ended

26. Anderson 1837, 8–9; Anderson 1835, 323n; Dale 1989, 42–45; Queen Victoria's Journals, 8, 22 October, 25 December 1837, 30 December 1838, 6 January 1839, 6 March 1842, RA VIC/MAIN/QVJ (W), by permission of Her Majesty Queen Elizabeth II (http://www.queenvictoriasjournals.org).

in his "characteristically practical" mode, reminding auditors that the ultimate purpose of faith in revelation was to secure salvation through belief in God's redeeming grace. These were thus "edifying" sermons that reviewers expected would do good among the nation's elite.[27]

It was not only noted fashionable preachers who drew on Chalmers's Bridgewater. Within months of publication, John King—the evangelical incumbent of a recently built chapel of ease designed to accommodate the growing population of the important Yorkshire port of Hull—was using Chalmers's work to illuminate a sermon on "the peace of God in the heart of man." Urging that God's peace guarded the heart "by inclining the affections to their proper objects," King quoted at some length Chalmers's assertion that God had constituted human nature such that the "exercise of the good affections" brought pleasure (the "oil of gladness"). The published version of the sermon referenced Chalmers's Bridgewater in a footnote, but the manner in which the quotation flowed within the preacher's oratory demonstrates how easily Chalmers's writing could be appropriated for pastoral purposes. Silently, Chalmers's account contributed to the sermon's grand practical purpose of urging its auditors to seek the "spiritual blessings" that derive from a "devotional spirit."[28]

At the opposite end of the country, in the Devon market town of Tavistock, the young curate at the parish church also found Chalmers's Bridgewater useful. Preaching on "Human Excuses," Whittington H. Landon—brother to the fashionable poet Laetitia—emphasized that it was human depravity that blinded reason to the many "witnesses" to himself that God had implanted in human nature and in "all creation," so that people were "without excuse" for their disbelief. Here there was no direct quotation from Chalmers, but while he felt that the "countless tokens of design" in the natural world required no further elaboration, Landon referred the readers of the printed sermon to Chalmers's Bridgewater for the "full and powerful development" of the argument from the moral constitution of man. The work was a useful adjunct in the context of stirring up a congregation to a more fervent religious devotion. Moreover, when Landon's vicar, the poet and writer Edward Atkyns Bray, soon after referenced the Bridgewaters in a sermon on "Divine Protection," the emphasis was similar. The series confirmed, he assured his congregation, that knowledge

27. "Anderson's Cloud of Witnesses," *Christian Remembrancer* 22 (1840): 332–38, on 337; "Notices and Reviews," *British Magazine* 16 (1839): 561–62; Anderson 1839–43, 1:50–94; see also Anderson 1839–43, 1:161n (on Kidd) and Anderson 1842, 15n–16n (on Bell).

28. King 1833, 293, 296; Allison, ed. 1969, 213, 239.

of nature contributed to spiritual wisdom and the understanding of the creator's ways. But while genuine science was thus "perfectly compatible" with religion, the purpose of Bray's sermon was far more practical: to urge his listeners to hide themselves under God's wings.[29]

An altogether more sharp-edged use of the series appeared in an "occasional" sermon delivered in February 1834 by William Daniel Conybeare, Rector of Sully, near Cardiff. Yet even here, with a preacher who was notably positive regarding natural theology, the tenor of the sermon was decidedly practical. The Oxford-educated Conybeare came from a prominent ecclesiastical family, but he was also a notable geologist and a friend of Kidd and Buckland. He also showed Buckland's enthusiasm for natural theology and, as we have seen, greeted his Bridgewater with approbation in the *British Critic*. As "Visitor" to the new Bristol College—opened in 1831 as a nondenominational alternative to England's universities—he had been called on to offer a course of theological lectures to the Anglican students and had commenced with "the evidence and doctrine of natural religion." He published the results in 1834, and when a second edition was called for in 1836, he took the opportunity to annotate the lectures extensively with references to the Bridgewaters, which offered, he told readers, accounts that were "very similar" to but "far more elaborately developed" than his own.[30]

Conybeare's 1834 sermon, however, originated in a very distinct context and offered an intensely practical religious lesson in reactionary social morality. He had been asked, in his role as chaplain to the High Sheriff of Glamorganshire, to preach at Cardiff parish church on the occasion of the spring assizes (a system of legal administration only recently introduced in Wales). In that context, his theme—"The Origin and Obligations of Civil and Legal Society"—was highly topical. South Wales had witnessed considerable unrest during the crisis preceding the passage of the 1832 Reform Act, including the Merthyr Rising of 1831 that had constituted the business of a recent Cardiff assizes. His sermon—based on the text "Wherefore ye must needs be subject, not only for wrath, but also for conscience sake" (Romans 13:5)—began by arguing that "the whole order of justly-constituted political communities" was divinely instituted. Here, Kidd's Bridgewater was useful. His friend had outlined Galen's argument that human supremacy depended on the hand and intellect rather than any natural strength, and Conybeare now argued that the development of these attributes required the prolonged nurturing only achievable within

29. Bray 1860, 1:118; Landon 1835, 83, 85.
30. Conybeare 1831, vii; Conybeare 1836, 17n.

a settled society. From this, he moved rapidly to assert that, to the "philosophical eye," the "*apparent inequalities*" of modern society in fact reflected "the regulated workings of a wonderfully and wisely constituted machine." The poverty of the poor was "not caused by the wealth of the wealthy, but in reality alleviated by it." However, the whole system needed "a spirit of religion" to operate as intended.[31]

The sermon then proceeded into Chalmers's territory as Conybeare sought to demonstrate that the institution of civil law was rooted in the "moral constitution" of humanity and especially in the conscience. He preferred, however, to refer readers of the published sermon to his own Bristol lectures and to the recent *Discourse on the Studies of the University* of his friend, Cambridge geologist Adam Sedgwick, rather than Chalmers's Bridgewater. Defending Britain's laws on the grounds that they originated in "general consent," he devoted the conclusion of his sermon to arguing for the value of "a truly religious education" in ameliorating crime, enjoining his audience to support such endeavors. This was a religious justification for Britain's constitution and laws that was bound to attract the approbation of the local conservative press. Moreover, the judge and jury who were in attendance requested the sermon's publication, and Conybeare now doubled its length with notes that included several pages of extracts from Kidd's Bridgewater.[32]

The providentialist emphasis of the Bridgewaters made them very well suited to political-theological applications of the sort made by Conybeare, but his was a theme much better adapted to occasional sermons than to regular parochial preaching. Indeed, he soon afterward incorporated material from his published sermon in an appendix to the second edition of his Bristol lectures in illustration of "the application of natural theology to political philosophy."[33] Much more commonly, the Bridgewaters were used from the pulpit in urging and inspiring habits of Christian living in an altogether less politicized and more devotional fashion. In the weekly delivery of congregational sermons, preachers typically found the series useful in stirring religious feelings and actions, whether in engendering faith in and love for God as creator and redeemer or in developing an awareness of the human condition. Moreover, such preachers were strikingly little attracted by the claims of a rationally constituted natural theology or by

31. Conybeare 1834, 1, 4–7.

32. "Cardiff Spring Assizes," *Glamorgan, Monmouth, and Brecon Gazette*, 1 March 1834, 2f.

33. Conybeare 1836, 138.

the possibility of discoursing at length on the wonders of the created order as revealed by modern science. In addition, while the overwhelming impression conveyed by those mentioning the Bridgewaters from the pulpit was that modern science confirmed the Christian faith and offered useful materials for its practice, the religious tendency of scientific knowledge did not frequently become a major focus of attention.

### *The Religious Tendency of the Sciences*

Given the practical, affective purposes of congregational sermons, it is no surprise that preachers seem generally to have considered public worship an inappropriate context for extended theological consideration of developments in the sciences. To the extent that clergy felt drawn on to comment publicly on such matters, they generally did so in special religious lectures, discourses, or prelections or in published pamphlets or works of theological controversy. When, for instance, Thomas Chalmers wished to discuss "the Christian revelation, viewed in connection with the modern astronomy" in 1817, it was in a series of hugely popular discourses delivered on weekday lunchtimes from the pulpit of his church in the heart of Glasgow rather than in Sunday sermons. Similarly, the extended theological critique of science offered in the spring of 1833 by Frederick Nolan was delivered in a series of lectures, rather than sermons, albeit delivered from the pulpit of Oxford's university church.

One minister who did directly address the "religious tendencies of modern scientific discoveries" in a congregational sermon was Henry Acton of Exeter, and his sermon perhaps naturally drew on the recently published Bridgewaters. It is clearly significant, however, that, unlike the preachers considered so far, Acton was a convert from Anglicanism to Unitarianism. Working class in background, he had risen to the task of offering a learned ministry at the George's Meeting in Exeter in succession to the prominent Unitarian theologian Lant Carpenter. His Sunday evening lectures—including one in the winter of 1832–33 on the devout Unitarianism of Milton, Locke, and Newton—had been well attended by a diverse audience. However, his ordinary Sunday-morning preaching reflected his attachment to the growing emphasis in Unitarianism on religious sentiment and his practical interest in "the duties of the Christian life." Thus, when he preached on scientific discoveries on a Sunday morning, Acton promised only a "general and desultory" treatment.[34]

Taking as his text a prayer of the first Christian believers acclaiming

34. Acton 1846, lxxvi, 166; Acton 1833.

God as the creator (Acts 4:24), he divided his sermon into four heads. Scientific discoveries, he told his congregation, confirmed belief in the existence of God and in his "*beneficent purposes*" in creation. It was on the latter point that Acton offered a quotation from Whewell's Bridgewater, announcing "*This* is the lesson of modern science, on the authority of its most illustrious cultivators." He next proceeded to affirm that scientific developments confirmed "our belief in the near, immediate, constant superintendence and agency of God," drawing out Whewell's emphasis on laws as the "settled modes of operation" of divine providence. The final part of his sermon promised an even "stronger feeling" of God's presence. Acton asserted that modern science confirmed the recent creation of humans and other "comparatively recent acts of creating power." Here, he followed the Bridgewater line that the origin of humans fell beyond the domain of natural laws, pointing out that theories of "gradual transformation" were considered ridiculous by philosophers. His grand conclusion was that faith in the "great principles of religion,—the existence, the providence, and the moral government of God, his especial care for the human race, and his designs for their final elevation"—was confirmed by modern scientific discoveries. These principles, he claimed, underpinned a true religion of the understanding, the heart, and life. Thus, while there were lessons to be learned about maintaining a liberal attitude to modern science and to biblical interpretation, the sermon's burden was still to sustain practical faith.[35]

Similarly, while England's Anglican universities were among the few places where sermons devoted to the religious tendency of the sciences were felt to be appropriate, the emphasis was typically practical. University preachers were properly concerned with how students should respond religiously to what they learned about the sciences. That had been the case with Rose and Whewell in the 1820s, and the themes were revisited in relation to the Bridgewaters by a number of preachers in Cambridge and Oxford in the 1830s, as the following chapter shows. Yet while clergy might use such opportunities to inculcate in their congregations appropriate habits of thought, feeling, and practice, they usually reserved substantive theological controversy for other occasions. The preacher's role as outlined in the manuals included establishing truth and refuting error, but the preacher was to do this above all in relation to the daily practice of religion.

The point is nicely illustrated by the activities of William Cockburn, who was by far the most notable of the several clergy who took exception

35. Acton 1846, 170, 172–73, 176, 179, 181, 186–87.

to the scriptural implications of Buckland's Bridgewater.[36] Following its publication in 1836 he spent a number of years issuing vituperative pamphlets on the subject of geology, some of which were notably popular, but his theological dispute seems not to have seeped into his preaching. Cockburn was a man of standing. He was the son of a Scottish baronet, dean of York Minster, and brother-in-law to the recently ousted prime minister, Robert Peel. As a college fellow in Cambridge in his younger days, he had also been the first to hold the university's office of "Christian Advocate," the duties of which were to deliver addresses in answer to anti-Christian views. It was in a similar vein that he mounted a spirited attack on what he saw as Buckland's errors. He had easy access to the Bridgewater series, since the York Minster library had spent a large part of its restricted budget in the 1830s buying both them and Nolan's Bampton Lectures. Volumes from the series were borrowed not only by Cockburn but also by his wife and her friend, two of the canons, and the clerical librarian. However, the dean did not respond to Buckland immediately, having more pressing matters with which to contend.[37]

In a scenario of which the chronicler of fictional Barchester would have been proud, Cockburn had come into dispute with one of York's canons, the archbishop's son and vicar of Bishopthorpe William Vernon Harcourt. A former student at Oxford of both Kidd and Buckland, Harcourt had been the first president of the Yorkshire Philosophical Society in the 1820s and the principal architect of the British Association for the Advancement of Science in the 1830s. Over succeeding years, his dealings with Cockburn became increasingly troubled, resulting in an unprecedented archiepiscopal visitation of the minster by Harcourt's father in 1841 to inquire into its affairs. Cockburn was found guilty of selling church livings (simony) and removed from office before being reinstated on appeal. Not surprisingly, Harcourt and his association stood rather low in the dean's estimation.[38]

Cockburn's initial salvo against Buckland's Bridgewater consisted of

36. Most of the several pamphleteers who engaged with Buckland's biblical interpretations were negative. See Best 1837, Brown 1838, [Carter] 1837, and De Johnsone 1838. On self-styled "scriptural" geologists, see O'Connor 2007a, 133–59, 202–3, and O'Connor 2007b. Mortenson 2004 is also useful but polemical.

37. Loan Registers, DCY. For details of the Minster library, see Barr 1977; for more about loans of the Bridgewaters, see Topham 1993, 299–301. On Cockburn, see Chadwick 1977, 272–94.

38. See Chadwick 1977, 282–87. Cockburn's assault on the British Association is discussed in Morrell and Thackray 1981, 240, 242–44, Rupke 1983, 216–18, and Orange 1981, 21–25.

two pamphlets published in 1838. The first addressed Buckland directly, arguing that the facts of geology described in his work could not be accounted for by his own account of progressive change but that they could reasonably be explained by following literally the narrative in the first chapter of Genesis. Cockburn's second pamphlet was a "remonstrance [ . . . ] upon the dangers of peripatetic philosophy" addressed to the Duke of Northumberland, who was shortly to be president of the 1838 meeting of the British Association in Newcastle, warning that "these annual assemblies of Thespian orators" were "injurious to religion." Cockburn claimed that a "favourite subject" at the upcoming meeting would be "the theory of the creation of the world, many ages before the birth of Adam," and he cited Buckland's work in evidence. Fierce as these attacks were, however, it is notable that they took place outside of Cockburn's pulpit, and while his comments were reportedly echoed in a sermon by the minister of Belgrave Chapel Pimlico, it was "only a sentence or two" that geologist Roderick Murchison considered "unworthy" of notice.[39]

Cockburn's general attack was answered at the Newcastle meeting by the bishop of Durham, Edward Maltby, who was vice president for the year. However, this did not restrain the dean from issuing further controversial pamphlets over succeeding years, causing the association much concern. Returning to York for its 1844 meeting, the men of science had to face Cockburn's attack in person (fig. 6.2). One unsympathetic commentator observed that the only reason Cockburn gained a hearing for his paper titled "Critical Remarks on Some Passages in Dr Buckland's *Bridgewater Treatise*" was because of his "local importance" and for fear of being accused of refusing to meet "such an opponent." The dean merely reiterated the points he had made before at greater length: that Buckland's theory did not account for geological facts and that those facts supported instead a literal interpretation of Genesis. In the absence of Buckland—one of whose sons had just died—Cockburn was answered authoritatively and at length by Cambridge geologist Adam Sedgwick. Nevertheless, his comments received enormous publicity, both through newspaper coverage and in the printed version, published the following day under the title *The Bible Defended against the British Association.*[40]

Cockburn's pamphlet passed through five editions within a year, and

39. Murchison to Harcourt, 11 May 1839, in Morrell and Thackray 1984, 313; Cockburn 1838b; Cockburn 1838a, [5], 6.

40. Cockburn 1844a; "The Scientific Meeting at York," *Chambers's Edinburgh Journal*, 23 November 1844, [321]–24, on 322.

FIGURE 6.2 Members of the British Association for the Advancement of Science visiting Bishopsthorpe, the seat of the archbishop of York. *Illustrated London News*, 5 October 1844, 221.

it is clear that his chosen mode of attack allowed him to make his mark. By contrast, when he rose to preach in the pulpit of York Minster on the Sunday in the middle of the British Association meeting, he had nothing to say directly about Buckland, geology, or the association. His sermon was on "the evils of education without a religious basis," and his text was certainly pointed: "If any man among you seemeth to be wise in this world let him become a fool that he may be wise. For the wisdom of this world is foolishness with God" (1 Cor. 3:18–19). However, the short sermon was distinct in tone from Cockburn's earlier controversial interventions. In the face of the educational ambitions so characteristic of the age of reform, Cockburn insisted that unless knowledge of "earthly things" was pushed on into the spiritual, it was "likely to produce more evil than good" by filling the heart with pride, and he urged his congregation to contemplate the creator, the afterlife, and the need for redemption. Where "worldly studies" were "subservient to the higher interests of eternity," they could contribute to developing the understanding and enable it "more thoroughly to appreciate the works of the great Creator." Too often, however, such studies led to the arrogant rejection of Christianity, as exemplified by Laplace. Cockburn concluded by cautioning those new to science to beware lest the study of it undermine and destroy their Christian hope while warning those directing the studies of the young to give Christian education primacy.[41] In short, this pugnacious controversialist had turned practical pastor in the pulpit.

Matters sat similarly with those clergy who publicly argued for a positive view of modern geology. Pamphlet protests against Buckland's Bridgewater had come mostly from evangelical Anglicans, but a very different response was offered by the prominent Congregationalist minister John Pye Smith. Smith had a long-standing interest in geology. At the turn of the century he had lectured students on the subject and on other sciences as a tutor at the Congregationalists' Homerton Academy in North London. He developed a renewed interest in the 1830s, and in November 1836, following the publication of Buckland's Bridgewater, he became a fellow of the Geological Society of London. Troubled by religious opposition to geology, he announced his belief in the harmony of geology and the Bible in the winter of 1837, first in a lecture to young men at Weigh House Chapel on Fish-Street Hill in the City (part of a series organized by the Congregationalist Christian Instruction Society) and then in a letter to the *Congregational Magazine*. By the following winter, he was planning to write a pamphlet or small book, possibly to be called *A Vindication of Geology and Geologists*, and in January 1839 he delivered "a few familiar Lectures" on the subject in the vestry of his Old Gravel Pit Chapel in Hackney.[42]

It was at this juncture that Smith was invited to deliver the "Congregational Lecture," an annual series of lectures instituted at the establishment of the Congregational Union in 1832. Taking place at the Congregational Library, just off Finsbury Circus, these were designed to be "Academic prelections," rather like the Bampton, Boyle, and Warburton lectures of the Church of England. Smith had been invited to lecture in 1837, when his chosen subject had been "On the Divine Attributes," a topic on which he had also given congregational lectures. However, illness had forced him to withdraw, and when the invitation was renewed in February 1839 he took the opportunity to propose a series of lectures on "Revelation and Geology, or the Relations between the Holy Scriptures and Some Parts of Geological Science." The eight lectures offered a thoroughgoing and learned account of how modern geology could be understood to be in accordance with the Christian scriptures, drawing to a significant extent on Buckland's Bridgewater and seeking to answer many of Buckland's critics, including

41. Cockburn 1844b, 5, 7.

42. Smith 1837; Smith 1839b; Medway 1853, 412–20. See also Smith 1834 and [Smith] 1836. On Smith and geology, see Medway 1853, 406–36, and Helmstadter 2004.

Cockburn. Yet while Smith began each lecture with a biblical text, they were still unmistakably lectures and had nothing like the rhetorical structure of a sermon or the intention to rouse hearers to feel or to act. Detailed discussions of the Christian bearings of modern science needed to take an entirely different form from regular preaching.[43]

Preachers were generally agreed that extended discussion of the substantive theological issues raised by the modern sciences was incompatible with the practical purposes of public worship. Important as it might be for some, such discussion was best handled in controversial pamphlets, theological treatises, or congregational lectures. By contrast, when preachers drew on the Bridgewaters in sermons, it was largely to confirm, clarify, and enliven perceptions of God's agency in human lives with practical ends in view. Moreover, while evangelical preachers frequently sought to convince the heedless, wavering, or skeptical hearer of the need for an active faith, they mostly shared the widespread conviction among religious reviewers that a rationally constituted natural theology had nothing to offer in relation to such purposes. There were, however, some preachers who did turn to the Bridgewaters in their attempts to defend and promote Christianity. In particular, a number of those who found themselves in the thick of growing opposition to Christian belief in Britain's industrial towns found the series useful in answering such antagonists in public discourses and addresses, and it is to these that we now turn.

## *Science in the Spiritual Battleground*

In the age of reform, the specter of unbelief had taken on a new and, to many, horrifying aspect. There had long been religious skeptics among the wealthy, but the new "infidelity" was associated above all with the ever-growing working class of the country's burgeoning industrial towns. Since the first appearance in 1794 of Thomas Paine's *Age of Reason*, with its attack on Christianity, the radical reformism so distinctive of the new urban working class had been associated to some extent with religious radicalism. For many, this went no further than criticism of Anglican privileges or of the power of the clergy, but a significant minority adopted and advocated more skeptical views, including deism and, less commonly, atheism. This religious radicalism became increasingly visible in the decade

43. Smith 1839a, [vii]; Medway 1853, 410, 420. The lectures were published in much-expanded form and achieved a significant reputation among more educated dissenters, passing through five editions.

FIGURE 6.3 In the aftermath of the Peterloo Massacre, George Cruickshank depicted the Prince Regent at the head of a "Holy and Compact Alliance" (involving lords, bishops, military officers, foreign powers, and the devil) all dancing around a pyre consuming a printing press and the figure of liberty. From *The Man in the Moon, &c. &c. &c.* (London: William Hone, 1820), [13].

that followed the Peterloo Massacre of 1819, when draconian measures had suppressed political radicalism to a significant extent (fig. 6.3).[44] Self-styled "infidels" began to engage in their own "missionary" endeavors, and by the time the Bridgewaters started to appear, a growing number of religious activists—many of them dissenters—found themselves called on to engage in a new kind of spiritual warfare in which the souls of workers were at stake. As Christian protagonists sought to answer the claims of atheists in particular, some reached for contemporary works of science in

44. For an overview see Royle 1971 and Royle 1974, chap. 1; see also Wiener 1983. On the history of atheism in this period, see also Berman 1988 and Priestman 2000.

order to counter the often crude Enlightenment materialism of their adversaries or to offer more positive arguments from the natural world for the existence of God. In such a context, the Bridgewaters naturally lent themselves to apologetic use.

The figurehead of infidelity in these years was Richard Carlile. An itinerant tin worker turned radical journalist, Carlile had been present at Peterloo and made his reputation through publishing cheap editions of Paine's works, playing a major role in making them accessible to working-class readers. In October 1819 his publication of Paine's *Age of Reason* and the deist *Principles of Nature* (1802) by Elihu Palmer resulted in a conviction for blasphemy, and he spent the next six years in Dorchester Gaol. This imprisonment marked Carlile out as "a standard-bearer of free expression," with a wide range of reformers rallying to his cause. Ironically, prison also contributed to his intellectual development, giving him the leisure to read the infamously atheist and recently republished *System of Nature* (1780) of Baron D'Holbach alongside the deist *Ruins of Empires* (1793) of François de Volney, David Hume's *Dialogues Concerning Natural Religion* (1779), and a host of other skeptical works. A growing number of such titles appeared among the pamphlets that Carlile published, and he also gave wide publicity to his developing views in his serialized anthology, the *Deist* (1819–20), and his weekly journal, the *Republican* (1819–26).[45]

By 1822 Carlile had abandoned deism in favor of atheism. His views drew heavily on the crude materialism of D'Holbach, of which he became a leading promoter. He also sought, however, to give his views the authority of modern science. Thus, as we have seen, his 1821 *Address to Men of Science* used William Lawrence's notorious lectures at the Royal College of Surgeons to support the claim that the sciences were, at heart, materialist. In practice, however, Carlile's atheist strategy was eclectic, pragmatic, and somewhat out of touch with recent authorized science. "We like all the sciences," he wrote in 1828, "or every system that bears the name of science, and that is available wherewith to attack the current superstition." Thus, alongside D'Holbachian materialism, he drew on the ultramaterialist anti-Newtonian theory of erstwhile radical and publisher Sir Richard Phillips and on the highly contested but potentially materialist science of phrenology.[46]

Despite these limitations, Carlile made converts, establishing Britain's

45. Wiener 1983, 33, 63–64, 67–68.

46. Carlile 1828, 481; Carlile 1821; Carlile 1823, 397. On the roots of Carlile's atheism, see Wiener 1983, 109–12, Cooter 1984, 201–23, Berman 1988, 201–6, and Topham 1993, 427–39.

first "cohesive 'republican and infidel' movement," numbering several thousand individuals. Many of these were artisans in the rapidly growing towns of the nation's industrial heartland in the North of England who not only found common cause in the pages of Carlile's periodicals but also formed local societies that were often termed "zetetic" societies in allusion to the ancient skeptical school of philosophy. Such disciples caused considerable disquiet in particular neighborhoods. Carlile himself continued to provoke alarm despite his religious views evolving in the later 1820s toward a form of allegorical Christianity. His new collaborator was the deposed Anglican priest Robert Taylor—the self-styled "Devil's Chaplain"—whose deism and allegorical interpretation of Christianity chimed with Carlile's developing views. The pair undertook repeated lecture tours, culminating in a high-profile "infidel mission"—the first of its kind in Britain—starting with a challenge to the University of Cambridge in May 1829.[47]

Not surprisingly, the manner in which the Carlileans increasingly sought to make their presence felt in a number of towns began to prompt action from the churches. Of course, there were well-established means of engaging with the impious poor. The industrial age had seen the old established Anglican Society for Promoting Christian Knowledge joined by new ventures, including the Religious Tract Society, that were designed to reach those dismissive or careless of the churches' claims by means of distributing tracts that often featured emotive scenes of deathbed terrors. Yet while the Christian knowledge society formed an "Anti-Infidel Committee" in 1819, issuing numerous new tracts, there is no evidence that these societies mounted a direct riposte to the claims of the new infidel movement.[48] In particular locations, however, dissenters undertook to engage more directly.

One such dissenter was Thomas Allin, a former president of the egalitarian Methodist New Connexion, who in 1826 was appointed minister of the burgeoning Yorkshire mill town of Huddersfield. This, he was shocked to discover, had become a place in which "the grossest atheistic materialism prevailed" in infidel clubs. Many boasted that "the great standard both of their disbelief and faith," D'Holbach's *System of Nature*, was "unanswered, and unanswerable." Allin rose to the challenge, delivering three congregational discourses on "the character and folly of modern atheism"

47. Wiener 1983, 101, 112–15, 130, 141–63; Berman 1988, 201. On the "infidel mission," see also Carlile's periodical the *Lion* (1828–29).

48. Allen and McClure 1898; Clarke 1959; Knickerbocker 1981. See also Topham 1992 and Fyfe 2004.

and one on the "necessity of a divine revelation." With his High Street chapel filled to the aisles, he concentrated the bulk of his fire on what he called the "atheist's bible," attacking D'Holbach's materialist system of science as inadequate, especially in its promise to "explain the origin and appearances of things, without first having a recourse to an intelligent cause." The proffered explanation was, Allin claimed, utterly insufficient and, from the perspective of modern science, laughable. Notably poor, he claimed, was D'Holbach's attempt to account for the origins of living creatures by spontaneous generation, and Allin also briefly dismissed the idea that organisms were eternal (which was at odds with modern geology) or had progressed naturally over time.[49]

A "large edition" of Allin's published discourses sold quickly. He was asked to deliver them again in the barns and mill rooms of neighboring villages and in many other northern towns that were plagued by infidelity, and he became a magnet for visits and letters from skeptics. Allin had been stationed in Liverpool by the time Carlile and Taylor took their infidel mission there in August 1829, and he preached against the missionaries, reluctantly being dissuaded from debating with them directly. The following year he was sent back across the Pennines to Sheffield, England's cutlery capital, which had witnessed disputations between dissenters and Carlileans over a number of years. As early as 1824 Abel Bywater—a Congregationalist awl-blade maker—had responded with pamphlets to the atheism of local Carlilean grinder Thomas Turton and the leader of the Edinburgh Zetetic Society James Affleck. Bywater elaborated at greater length, in an address at the Scotland Street Chapel of the New Connexion Methodists in March 1830, on the second anniversary of their tract society. Like Allin he attacked the naturalistic account of the created order found in works by D'Holbach and Volney, focusing above all on the origin of the human species. Carlile subsequently became a target for humor in Bywater's very successful comic publication, *The Sheffield Dialect*, first published in parts in 1830. However, when the archinfidel visited the city in October 1833, it was Allin whom he sought out in public debate, and while the latter declined the invitation, he republished his discourses with additional notes referring to Carlile.[50]

By this date, the spiritual battle had reached a crisis in another north-

49. Allin 1828, iii, 9–10, 34; Hulme 1881, 72–83.

50. Carlile 1833a, 1833b; Allin 1833; Hulme 1881, 85, 108–10; Bywater 1830. For more on Bywater, see the memoir in Bywater 1877, vii–xi; on Affleck, see Wiener 1983, 119n47.

ern mill town, and here the first of the newly minted Bridgewaters began to be deployed. Bradford was one of the most rapidly growing of Yorkshire's wool towns, in which "all was bustle and activity." Carlile's followers there had begun to make their presence felt in the mid-1820s. Prominent among them was Squire Farrar who, as a bookseller and vocal materialist in nearby Otley, had organized support for the imprisoned Carlile. By 1825 Farrar had become an attorney's clerk in Bradford, where he had been involved in an attempt to found a mechanics' institute under working-class control. However, after the nascent institute collapsed later in the year during the notoriously bitter strike of the Bradford wool combers and handloom weavers, contemporaries attributed its failure to the way in which the skeptical reputations of Farrar and his collaborators had alienated potential middle-class patrons.[51]

Shortly before Farrar had arrived in the town, Benjamin Godwin had come to take up an appointment as tutor at Horton Academy, a Baptist ministerial college. Also engaged as pastor of the new centrally located Sion Chapel, Godwin became aware that there were considerable numbers of infidels, including atheists, who held regular discussion meetings in the town and sought to spread their opinions in the mills. There were also "many works of infidelity" circulating in the city, including D'Holbach's *Système de la Nature* and Carlile's *Deist*. Yet while Godwin suggested to friends of "scientific attainments" that they might halt the progress of such views by "a reference to nature, and an appeal to reason," none took the hint.[52]

Events such as a planned anniversary dinner for Paine at the start of 1833 provoked pulpit remonstrances, but matters were brought to a head at the end of that year. John Matfin, a newly appointed Primitive Methodist minister, had caused controversy by declaiming against infidelity from several village pulpits and by threatening anyone who sought to dispute his assertions in chapel with the ruinous forty-pound penalty for disturbing religious assemblies. Here, indeed, was a preacher whose pulpit endeavors to refute error and establish truth had sharp teeth, but as the engagement

51. Godwin 1839–55, 365. On Farrar, see Farrar 1823, 1826, 1832a, and 1832b, "The Celebration of Paine's Birth-Day in Leeds," *Republican*, 27 February 1824, 275–85, on 277–78, Baines 1822, 1:232, and Prothero 1979, 209. On Bradford's first Mechanics Institute see James 1841, 248–49, Godwin 1839–55, 537, Farrar 1889, 44–46, Fraser 1905, 219–31, Harrison 1961, 61–62, and Koditschek 1990, 308–11. On the wider context, including the atheist controversy, see Morrell 1985, esp. 9–13.

52. Godwin 1834, vi. On Godwin, see Steadman 1838, Godwin 1839–55, and Sell 2004, 170.

intensified it became a matter for learned discourses rather than sermons. When leading infidels posted placards around the town challenging the advocates of Christianity to establish its truth on the basis of reason and argument, Godwin considered himself bound to answer. Accordingly, he delivered a series of six lectures from his pulpit—each one lasting around two and a quarter hours—with the chapel nevertheless "crowded to excess" and a courteous phalanx of infidels in attendance.[53]

In preparing his lectures, Godwin obtained a reading list from the town's leading skeptics, who helpfully lent him copies of their core texts—the *System of Nature*, Hume's *Dialogues*, Palmer's *Principles*, and Carlile's anthology *The Deist*. Alongside these he studied works from his own library and others that he bought or borrowed. Godwin had long lectured to his students at the academy on mathematics, the globes, and experimental natural philosophy. In addition, just the previous January, he had joined together with other leading dissenters in the town to form the Bradford Amicable Book Society. This met quarterly in the members' houses to select books and discuss topics of the day, forming "a bond of union and the means of a profitable exchange of thoughts and feelings" that soon resulted in the establishment of a liberal dissenting newspaper, the *Bradford Observer*. Members subscribed a guinea per year for the purchase of books, and starting with their second meeting they began purchasing the Bridgewaters, as well as other scientific books.[54] At the same time, Godwin had become a figurehead and guarantor of the religious orthodoxy of the newly founded Bradford Mechanics' Institute, an organization whose library grew rapidly, purchasing the Bridgewaters as they appeared.

Armed with these sources, Godwin's approach drew heavily on science. His first lecture was largely devoted to clearing the ground. The pastor explained that he would not presume the truth of God's existence but rather seek to offer evidence for a creator in opposition to the atheist attempt to explain the phenomena of the universe independently of such a being, as exhibited above all in the *System of Nature*. In the next lecture, he examined atheist explanations of the universe, becoming inexpertly metaphysical as he dissected arguments for the eternity of the system of nature. But he was more successful in borrowing from Lyell and Sedgwick to argue for the recent origin of humans and from Whewell to argue that the uni-

53. Godwin 1834, iii–x; Godwin 1839–55, 577–78; "The Late Dr Godwin," *Bradford Observer*, 22 February 1871, 4a–c; Farrar 1832a, 28.

54. Godwin 1839–55, 369–72, 573–74, 578; Minute Book and Borrowing Register, Bradford Amicable Book Society, WYAS-B, DB17/C4/2.

verse must have an end.[55] The following three lectures were altogether more positive, offering "proofs of the existence of God from the works of nature" and detailing the divine attributes as thus inferred. A concluding lecture brought the arguments home by comparing atheism and Christianity intellectually and practically.

The central portion of Godwin's lecture series offered a fairly well-constructed natural theology drawing on a range of reasonably current and authoritative works. These included a number of medical works, including John and Charles Bell's *Anatomy of the Human Body*; Antholme Richeraud's *Elements of Physiology*; two of the Society for the Diffusion of Useful Knowledge treatises (including Bell's on *Animal Mechanics*); and such high-profile scientific works as Lyell's *Principles of Geology*, Herschel's *Treatise on Astronomy*, and Thomas Thomson's *History of Chemistry*. There were also a number of references to the Bridgewaters—specifically those of Whewell, Kidd, and Roget. Yet while Godwin found them useful alongside other sources and developed many of the same arguments, his lectures were rather different in tone. In particular, they were much more clearly directed at answering D'Holbach's *System of Nature* in general terms, although there was also some brief discussion of Lamarckism based on Lyell's *Principles*.[56]

As Godwin prepared his lectures for the press, he amiably furnished his opponents with the proof sheets in order that they might reply. The task was undertaken in part by Farrar but more particularly by a bookseller, Christopher Wilkinson. However, their "examination" of Godwin's arguments was striking in its lack of purchase on the scientific details. Notwithstanding the dramatic failure of Farrar and Wilkinson's 1825 mechanics' institution, Wilkinson had been appointed to the committee of the new one in 1832. Nevertheless, with the weight of gentlemanly science arrayed against the prospect of offering an atheist account of the universe, Bradford's infidels struggled to offer a sophisticated rebuttal of Godwin's core arguments. Moreover, the local publishers soon took fright, withdrawing the infidels' reply from sale with only about 150 copies in circulation.[57] For Godwin at least, the battle had been decisive and the lesson clear: the weight of modern science as exhibited in works like the Bridgewaters

55. Godwin 1834, 44–46.

56. Godwin 1834, 175–79.

57. [Wilkinson and Farrar] 1835; Wilkinson and Farrar 1853, viii–ix; Fraser 1905, 229–30; Godwin 1839–55, 605. On Wilkinson, see Wheeler 1889, 334, and Harrison 1961, 61.

could crush the materialist claims of atheist infidels, affording ample materials for a rational justification of the creator's existence.

### *Bringing Christianity into Disrepute*

Godwin's use of the Bridgewaters in countering militant infidelity was relatively unusual in the early 1830s. Atheism continued to be the position of only a small minority among working-class radicals. Many viewed it as an unhelpful distraction from the campaign for political reform, and many actually rooted their radical political analysis in Christianity or at least in deism. Moreover, as Allin had shown, the lack of sophistication of Carlilean materialism did not require an extended scientific rebuttal. Over the course of the 1830s, however, both of these aspects of the spiritual battleground changed. As Carlile's reputation and following declined in the early 1830s, that of the former mill owner turned prophet of "socialism" Robert Owen rose rapidly, spawning a well-developed movement. In addition, some of Owen's followers sought actively to appropriate the sciences as they encountered them in the new popular publications of the age of reform with a view to making the case for atheism.[58] And while Christian protagonists reached for the Bridgewaters in reply, Owenite atheists found the series could also be repurposed for their ends.

Owen was the son of a saddler whose rise to become owner of New Lanark mills at the start of the century enabled him to put into action his developing vision of radical social reform rooted in an Enlightenment-inspired belief in environmental determinism. He became increasingly known to working-class radicals after 1817, when he publicized his proposals for cooperating communities based on labor-value exchange, and, while his own attempt to found such a community in the United States proved unsuccessful, he found on his return to Britain in 1829 that his cooperative philosophy had been adopted by others. By August 1834 Owen's attempts to coordinate and develop these cooperativist initiatives and the associated trades unions had all failed, but he now shifted the focus of his activities more toward a secular millenarianism in which society was to be transformed by the organized promotion of his principles. He started a weekly journal, the *New Moral World*, and an organization that in 1835 became the Association of All Classes of All Nations, and by 1837 the self-styled "socialists" massing under his banner had begun to organize on

58. On Owen and Owenism, see Harrison 1969, Royle 1974, 43–53, and Cooter 1984, 224–55.

a provincial basis, with local branches building "Halls of Science" and employing social missionaries to propagate Owen's social gospel.

Owen's critical views of the immoral and divisive effects of all existing religions had long been on record, but it was not until his return from America that his religious teachings attracted extensive comment or concern. Even then, it was his social teaching that proved of particular concern. Especially abhorrent were his environmentally determinist doctrine of moral nonresponsibility and, from 1835, his denial of the sanctity of marriage and advocacy of artificial contraception, but his whole social philosophy was subjected to religious attack. Indeed, prominent Owenites were sure that the critical mention of "Cosmopolitanism" in Chalmers's Bridgewater was directed at them. Chalmers had criticized those who sought to "substitute a sort of universal citizenship, in place of the family affections," considering the latter a carefully designed part of the divine social mechanism that Owenite measures would disrupt. Significantly, the response of Owen's supporters was religious in character. The editor of Owen's journal, *The Crisis*, was a millenarian Universalist—the self-styled "Rev." James Elimalet Smith—who charged the Bridgewater author with heresy for preferring the family ties of "corrupt" nature to the renovated morality of the Owenite system. More conventionally, Chalmers's Bridgewater also prompted criticism from wealthy Owenite philanthropist and publicist John Minter Morgan in a pamphlet reply to the critics of his philosophical novel *Hampden in the Nineteenth Century* (1834). Morgan, an Anglican, criticized the Scot for censuring "the struggle of humanity to advance in conformity with 'the matchless wisdom of nature's God'" and for failing to appreciate the divine imperative for reform.[59]

When Owen did elaborate on his religious beliefs, he made clear that they were broadly deistic rather than atheistic. In November 1835, having been asked to give his opinion concerning "that, at present, to us, mysterious Power 'which directs the atom, and controls the aggregate of nature,'" he offered readers of the *New Moral World* a series of deistic or pantheistic "conjectures" to the effect that the universe was pervaded by an "eternal, uncaused, omnipresent Existence" that was beyond human comprehension but was best understood through the laws of nature. He asserted,

59. [Morgan] 1834, 46–56; V.C.L., "The True Philosophy of Mysterious Dogmas," *New Moral World*, 10 January 1835, 85–86, on 85n; *Crisis*, 11 January 1834, 154–55; Chalmers 1833, 1:225. On Morgan, see Harrison 1969, 33–35; on Smith and his attempt to "spiritualize" socialism, see Harrison 1969, 109–22, and Smith 1892, v, 101–13.

moreover, that knowledge of this power was not essential to human flourishing while disagreements on the subject were a major obstacle to it. This statement of the "Religion of the Millennium" was adopted the following May by Owen's Association of All Classes of All Nations and was distributed in bulk as a tract.[60] Yet, while Owen found phrenology very useful to support his environmentally determinist views of human character, and while his magazine carried articles and extracts about science that supported materialist or antiscriptural views, neither showed any interest in offering a naturalistic account of the history of the universe that would support atheism. For instance, the extensive account given in the magazine of the anti-Newtonian "electrical theory of the universe" formulated by Owenite lecturer Thomas Simmons Mackintosh eschewed any atheist intent, supporting Owen's general position on religion. Indeed, its author soon afterward wrote in support of the existence of God.[61]

The one Owenite who did offer a detailed atheist account of the universe at this time met with little encouragement from the movement's leadership. This was Manchester radical Robert Whalley, who issued the first of three planned parts of his *Revolution of Philosophy: Containing a Concise Analysis and Synthesis of the Universe* in 1835. Having been brought up an Anglican, Whalley had grown skeptical while a Sunday school teacher, becoming a deist in the early 1820s. Paying sixpence a fortnight as a member of the Carlilean Miles Platting Reading Society, he now gained access to a range of radical works, such as D'Holbach's *System of Nature*. And, having previously believed the existence of God secure on the grounds that no atheist could otherwise explain the phenomena of the universe, Whalley abandoned that belief as he developed his own revolutionary theory of the universe.[62]

Whalley's account was as elaborate as it was unconventional. Running to ninety-five pages, it offered a systematic materialist interpretation of the universe drawing heavily on Daltonian chemistry and adding a crystallization-based explanation of species origins to a nebular theory of the origins of the universe. It was published for a shilling by his radi-

60. [Owen] 1835; [Owen] 1836; *New Moral World*, 5 March, 21 May, and 4 June 1836, 152, 236, 256. See also Owen 2005, 84–89.

61. Mackintosh 1837–38, 3:380, 4:154. See also Mackintosh 1842 and Mackintosh [1845?], 24. The book version of the theory made anti-Newtonian use of Whewell's Bridgewater. Mackintosh [1845?], 263–64; Mackintosh [1846], 236–37. On Mackintosh, see Morus 1998, 135–39, and Secord 2000, 325.

62. Whalley 1835, v. On his upbringing, see Whalley 1825b. On the Miles Platting Reading Society, see Harper 1823 and Wiener 1983, 114. See also Whalley 1825a and Topham 1993, 448–63.

cal neighbor, the bookseller Abel Heywood, and advertised through the radical network. However, it made little impression and was later repudiated by Whalley. The only immediate public attention came from Owenite Universalist James Smith, who described it in his penny weekly *Shepherd* as "only the old French materialism, which is far past its prime, and is fast decaying." Owen's *New Moral World* paid the work no attention until 1840, when, after a change of heart, Whalley published a *Philosophical Refutation of the Theories of Robert Owen, and His Followers*. The Owenite journal now naturally criticized Whalley's attack on socialism at length, but at the same time it congratulated the author on his escape from atheism. It disparaged his earlier scientific claims as outdated and asserted that "the appearances of design in the adaptation of means to an end through the universe, prove the existence of intelligence in arranging that disposition of things." Whalley's atheistic system of science was thus far from representing an Owenite orthodoxy.[63]

By the time of this review, however, the religious orientation of the Owenite movement had in any case moved on significantly. From 1837 the Association of All Classes of All Nations began to organize in a much more developed manner, rooting itself in the activities of local branches and "Halls of Science" and appointing "Social Missionaries." As this suggests, of course, the Owenites had learned about effective organization from Christian missionary and tract societies and from the various Methodist denominations. In May 1841 the annual congress heard that in the preceding year there had been nearly 1,500 lectures delivered by eighteen social missionaries in Britain's largest towns and cities, one-sixth of which were on religion. Between 1839 and 1841, two and a half million tracts were distributed. Such a scale of activity brought significant concern, not least from clergy, who were to be found speaking against the immorality and irreligion of Owenites across the industrial heartland.[64]

One prominent organization that set itself to oppose the new wave of Owenite infidelity was the London City Mission. Founded in 1835, this collaboration between evangelical Anglicans and dissenters was prompted by a shared disquiet that—largely unbeknownst to the wealthy—more than a quarter of London's two million inhabitants were living in "the

63. "Reviews," *New Moral World*, 16 January 1841, 38–39; Whalley 1840, 3; [Smith] 1835, 244. For advertisements, see, for example, *Poor Man's Guardian*, 20 December 1834, 367.

64. Royle 1974, 47–51; "Proceedings of the Sixth Annual Congress of the Universal Community Society of Rational Religionists," *New Moral World*, 29 May 1841, 331–39, on 331; Harrison 1969, 231. On the example of Bradford, see Morrell 1985, 13–19.

most wretched, ignorant, and depraved state" (fig. 6.4). The mission's founder, the Scot David Nasmith, had over the preceding decade established similar missions in Glasgow, Dublin, and the United States. He had learned the basic principle of putting lay people in charge of visiting homes in small urban districts from the innovative visitation scheme previously forged in Glasgow by Thomas Chalmers. Nasmith, however, paid his visitors as full-time lay missionaries, employing working men who stood a chance of being tolerated in crime-ridden and dangerous neighborhoods. They were to visit the poor in their homes, conversing and praying with them, reading the Bible, and distributing tracts with a view to converting them to Christianity. The plan rapidly drew support, and within five years the number of missionaries had reached fifty-eight. Between them they claimed to have made 871,891 visits and distributed 1.2 million tracts.[65] In addition, the committee had organized the distribution of over thirty-five thousand Bibles and a quarter of a million tracts against intemperance.

This avowedly "aggressive" approach to working-class irreligion soon came up against the opposing effort of Owen's followers. Standing up to give the annual report in Exeter Hall in May 1839, Congregational minister Robert Ainslie described the socialists' program with horror, asking "What is to be done?" He continued, "Never before did men calmly and openly unite together, organize institutions, frame laws, and employ missionaries to overturn the constitution of society, destroy the social relations, abolish marriage, and blot out from the mind the belief and love of the one living and true God." This was too much for busy ministers to grapple with, Ainslie observed, and it was down to institutions like the London City Mission to circulate tracts and deliver lectures. Initially, the mission planned to hire one of London's Owenite Halls of Science in order to lecture directly to their target audience, but the socialist leadership insisted on having a right of reply. It was consequently in the London Mechanics' Institution that the mission made its stand over nine nights in January and February 1840, addressing "overflowing audiences" with "thousands obliged to leave for want of room."[66]

65. "The Necessity of Missionary Effort in London," *City Mission Magazine* 1 (1836): 3; *London City Mission Magazine* 5 (1840): 81–82; Ainslie 1836, 3, 41. On Nasmith, see Campbell 1844; on Chalmers, see Brown 1982. On the founding of LCM, see also Noel 1835; the best general history is Weylland [1884].

66. "Socialism," *London City Mission Magazine* 4 (1839): [165]–67; Fleming 1840, 1110; *Annual Report of the London City Mission*, 1839, 12–14; "Proceedings at the Fourth Annual Meeting of the London City Mission," *London City Mission Magazine* 4 (1839): 101–16.

FIGURE 6.4 The realities of urban poverty. "Down the Court." John Matthias Weyland, *Round the Tower; or, The Story of the London City Mission* (London: S. W. Partridge, 1875), opposite 108.

The lectures were intended to cover all the main points of Owen's system so that much attention was given over to moral responsibility and marriage. Yet while there was admitted to be a "great variety" of religious sentiment among the socialists, some were avowedly atheists, and the first lecture, by Ainslie himself, was given over to the question, "Is there a God"? Ainslie's response to Owen and his followers' stated views was to offer a wide range of standard lines of reasoning, including the argument from universal consent, the standard arguments of natural theology, and, more briefly, arguments from the Christian revelation that supplemented natural theology and answered certain socialist objections. In offering this argumentation, Ainslie not only quoted from Bell's Bridgewater but directed his hearers to the elaborate proof of the existence of design provided by the series. Yet while there was an embarrassment of riches in regard to such evidence, he conceded that it was of little use with the socialists, since they accepted the existence of design and merely quibbled about its interpretation. After contending with their objections, Ainslie's climax was to point out that the corollary of disbelief in God was the absurd belief that the entire created order—from the cosmos through the whole variety of living beings to the human mind—was the result of matter in motion.[67]

Ainslie's use of the arguments of natural theology found a more ponderous echo in the longest of the lectures, by Congregationalist minister and professor of philosophy and logic at University College London, John Hoppus. Hoppus apparently spoke for over three hours on "The Province of Reason in Reference to Religion," urging the importance of religious reasoning in both natural theology and the evidences of Christianity, while also claiming that, just as in science, so in religion some matters were mysterious. Despite such prolixity, the hall was reportedly crowded and attentive throughout the lectures, which had been widely publicized, with free tickets sent to Halls of Science, advertised in the *New Moral World*, and distributed by the missionaries. Owen himself sat on the platform for one lecture. The mission was delighted with the result, publishing the lectures at a shilling each and sixpence more in the case of Hoppus. Moreover, the society's missionaries now lent copies out. Later in the year, the missionary in the wretchedly poor Old Pie-Street neighborhood of Westminster reported having spoken to a visiting woman from the East End "who had been a great sinner," urging her to seek salvation. She was concerned because her husband was a socialist, but the missionary lent her Ainslie's

67. *London City Mission Magazine* 4 (1839): 167; Ainslie 1840b.

pamphlet and the one on the Bible, reporting that the husband had found them convincing and had given up his socialism in consequence.[68]

Robert Owen was not slow to respond to the London City Mission's assault, delivering three lectures at the London Mechanics' Institute—again "crowded to suffocation"—in late March and early April in direct response to the mission and a range of other critics. However, in the face of a tidal wave of religious opposition at a time when funds were needed for his community schemes, Owen's religious statements were becoming more conciliatory. The previous year he had instructed his missionaries—in what was renamed the Universal Community Society of Rational Religionists—to cease "contending with religious error" and thereby arousing "insane anger" among the irrational. Having faced prosecutions for blasphemy, the society was now claiming legal protection as a "religious body." Thus, when in the spring of 1840 Owen responded to the London City Mission, he offered a detailed restatement of the socialist scheme but had little to say about his religious views except for a brief and diffident concluding assertion in the first lecture to the effect "that there is a Power which directs the atom and controls the aggregate of nature; a Power, in fact, which governs the universe as it is governed: but *what* this Power is [ . . . ] no man knows." He also referred in passing to the "most evident harmony and unison of design" shown in the formation of humans while asserting that the power behind that design was a "mystery."[69]

Even by 1840, Owen's unwillingness to offer a more forceful riposte to the kind of religious argumentation offered against his movement was causing outrage among some adherents, and this soon spilled over into action. In the vanguard was former bookseller and social missionary Charles Southwell, who in November 1841 began the penny weekly *Oracle of Reason* as a vehicle for his fiercely atheist views. From the outset, the new journal promised to undermine theism with a serious, scientifically credentialed alternative to the Christian account of creation. Most notably,

68. Wilson 1840, 182; "Proceedings at the Fifth Annual Meeting of the London City Mission," *London City Mission Magazine* 5 (1840): 82–96; Weylland [1884], 47; G—— H——, "Meeting of the London City Mission," *New Moral World*, 30 May 1840, 1261–62; *New Moral World*, 18 January 1840, 1040; "The Lectures against Socialism," *London City Mission Magazine* 5 (1840): 49–54, on 49. Hoppus drew on Whewell's account of "deductive" habits in the published version of his lecture; Hoppus 1840, 92.

69. Owen 1840, 14, 31; 1839, 593; Royle 1974, 66–68; Ainslie 1840a, ii; *New Moral World*, 28 March 1840, 1208; R. P., "Progress of Social Reform," *New Moral World*, 18 April 1840, 1250–54, on 1254.

Southwell's partner in the enterprise, the Bristol printer William Chilton, began in the first issue a long-running series outlining a theory of species transmutation. As Chilton later observed, this was a subject of "vital importance" for materialists eager to show that they could explain the origin of living forms from "purely *natural* causes" independently of divine creation. Much more than their predecessors, these new-style atheists plundered authoritative and recently published scientific works to give their naturalistic theory credibility. Thus, when Southwell was imprisoned for blasphemy in January 1842, he requested help from supporters in obtaining copies of Lyell's *Principles of Geology* and Buckland's Bridgewater. These were works he could get past the censorship of the prison governor and local magistrates but still put to use in the cause of atheism. Chilton likewise used Buckland's Bridgewater in arguing the case for "progressive development," reinterpreting its history of life on earth naturalistically.[70]

This radical riposte to those who sought to deploy the Bridgewaters in the spiritual battleground of working-class atheism entailed a new and direct engagement with contemporary science underpinned by the explosion of cheap scientific literature over the preceding decade. Not only did the *Oracle* reengineer works such as Buckland's Bridgewater for atheist purposes but it also offered an explanation of why they had to be reengineered. Cowardly and dishonest scientific men had, Chilton claimed, hidden their own infidelity because of their vested interests in Christianity. This was perfectly characterized by the Bridgewater bequest: Buckland's "apology" for Christianity was merely "a sop for the dragon." Moreover, another writer argued, the strained and ridiculous reasonings about design in nature that such works contained had brought the Christian God into contempt. The blame for the spread of atheism lay "almost solely upon the advocates for god's cause," and the likes of Buckland ought consequently to be indicted for blasphemy. The atheist had only been "the servant of the Bridgewater Treatise authors." Who, after all, could be held responsible for the manner in which they were affected by reading a religious book? Notwithstanding the Bridgewater Treatises, the *Oracle*'s third editor Thomas Paterson claimed, "god-belief" was "on the wane."[71]

The rapidly expanding freethought press of the early 1840s offered a

70. C[hilton] 1842, 194; Southwell 1842, 79; C[hilton] 1845, 9. On Southwell, Chilton, and the *Oracle*, see Royle 1974, Desmond 1987, and Southwell [1845]. See also Topham 1998.

71. W., "The New Argument 'A Posteriori' for the Existence of God. I," *Oracle of Reason*, 10 September 1842, 317–18, on 317; P[aterson] 1843, 217, 236; [Chilton] 1842, 194.

newly sophisticated critique of the classic arguments of natural theology not least as instantiated in the Bridgewaters. Especially important were the publications of onetime Owenite lecturer George Jacob Holyoake, who was ultimately to steer the remnant of the movement into what he termed Secularism. Holyoake had offered to edit the *Oracle* following Southwell's imprisonment only to be jailed for blasphemy in turn. He used his six months in Gloucester Gaol well. Having been presented with a copy of Paley's *Natural Theology* by a magistrate, he prepared *Paley Refuted in His Own Words* (1843), which reduced the design argument to an infinite regress, as one of Hume's protagonists had done. The work naturally included the Bridgewaters in its sights, since their writers had promenaded up and down the "walk of design, which Paley made, and gravelled, and rolled," adding new illustrations, but they were a sideshow, since they had added "*no new principles.*"[72]

Southwell greeted Holyoake's book with enthusiasm in the latter's new 1½*d* weekly *Movement and Anti-Persecution Gazette*. The article had been intended for Southwell's now defunct *Investigator!*, and it reviewed *Paley Refuted* alongside a penny reprint of a pamphlet by the aging Irish landowner, radical political writer, and freethinker George Ensor. This was his *Natural Theology: The Arguments of Paley, Brougham and the Bridgewater Treatises on This Subject Examined*, which had been published privately in 1836 "for the author's friends." It adopted a skeptical rather than a strictly atheist position, with the Irishman claiming in distinctly Humean terms that design arguments "implied transcendental egotism and impertinence" and using examples from the Bridgewaters to argue that natural theology was liable to encourage disbelief by its false presumption. Ensor also expressed his skepticism about the creation of the universe and the existence of the soul. However, while he had invited others to reprint his pamphlet, it was not until August 1843 that it was taken up by the radical press, reflecting the new mood. Southwell in particular was thrilled by it, considering its account of the Bridgewaters unparalleled.[73]

Ironically, Southwell was able to draw such appraisals together with the criticisms of natural theology emerging from those he dubbed "Sen-

72. Holyoake [1843?], 9–10; Holyoake 1892, esp. 1:166. On Holyoake, see Royle 1974.

73. *Oracle of Reason*, 19 August 1843, 288; Southwell 1844, 82; Ensor 1836, 15. On Ensor, see *ODNB*, s.v. "Ensor, George," Topham 1993, 479–83, Robertson 1929, 1:84–85, Budge et al. 2002, 1:363, and *Dublin Evening Post*, 12 July 1834, 4b. Ensor's circle included Florentine author and former British soldier André Vieusseux, who also ridiculed the Bridgewaters. See [Vieusseux] 1836–37, 96–97, Wheeler 1980, 336, Boase 1892–1921, and Allibone 1870–71.

timental Theists," for whom faith, and not reason, was the key to religion. Describing the *Argument from Design Equal to Nothing* (1842) by "Fidian Analysis" (actually the Scottish chemist Samuel Brown), Southwell listed its author among those "most able opponents" of atheism "who broadly tell us that no proof can be furnished of God's existence." Of course, Southwell had no truck with the author's alternative emphasis on faith, but the pamphlet helped him to establish his claim that natural theology "was fast falling into contempt." Thus, he was able to quote the work to the effect that the "illustrious argument of Paley and the Bridgewater Treatises" was "inconclusive" and that it was "now fraught with danger to the success of that gospel of faith which its promulgators profess their anxious readiness to defend." From Southwell's perspective, this new Christian questioning of rational arguments and the emphasis on fideism among "leading theologians" was a clear result of the fear generated by the "amazing spread of Atheism" in 1841 and 1842.[74]

According to the leaders of the burgeoning atheist movement of the early 1840s, the use of the Bridgewaters in the spiritual battleground of Britain's industrial towns had only helped Christianity's opponents. To some of their protagonists—especially Congregationalist and Baptist clergy—the Enlightenment materialism of the infidels had seemed an obvious target in the 1830s that was easily overcome by the securely Christian vision of the creation offered by the Bridgewater series. Yet if, at first, they had been confident of receiving no substantial answer, that confidence was short-lived. Alongside other newly accessible, authoritative, and up-to-date scientific works, the Bridgewaters soon provided increasingly outspoken atheists with the weapons that they needed to challenge Christian orthodoxy scientifically, even to the extent of constructing a naturalistic history of the creation. Moreover, atheist radicals made the series a focus for attacks on the arguments of natural theology, contending that they demonstrated the facileness of claims about divine design.

❋

There were, of course, plenty who disagreed with this characterization of the Bridgewaters as failing to answer atheism, and the series continued to be used as a resource in seeking to counter infidelity. Even in the 1850s, Charles Kingsley sought to convince the skeptical Chartist leader Thomas Cooper of the truth of Christianity by having him read the Bridgewaters

74. Southwell 1844, 35, 44, 82. The quotation is from [Brown] 1842, 22–23.

(although it was Cooper's supposedly providential deliverance from death in a train crash that actually returned him to the fold). Others sought to use them in Christian missions overseas, and by 1840 Alice Lieder of the Society for Promoting Female Education in the East reported that she had persuaded the Pasha of Egypt, Muhammad Ali, to translate one of them for use in Arab schools. However, the utility of natural theology in making the case for Christianity continued to be questioned by many ministers active in the spiritual battleground as well as by atheists. When, for instance, the evangelical minister of St. George's parish in Glasgow gave the first of a course of lectures on infidelity in 1842, he was happy to affirm that the "proofs of wise and benevolent design" had satisfied the "most rigorous enquirers" of the existence and agency of God, quoting from Whewell's Bridgewater in support. Nevertheless, he claimed that it was neither necessary nor appropriate to dwell on such matters, since no amount of evidence would overcome the resistance of a "corrupted heart."[75]

This concern with the emotions and will rather than with the intellect was quite characteristic of public religion in Britain during the 1830s, with preachers seeking to use the Bridgewaters to evoke an emotional response from their congregations that would contribute to the development of Christian dispositions, habits, and behavior. Moreover, it was not only clergy who were concerned with the capacity of the sciences to evoke such responses. Faced with criticism of the religious bearing of the sciences, men of science also found the Bridgewaters to be useful in pointing up the ways in which their work sustained the characteristics that lay at the heart of true religion. Historians have often pointed to the ways in which scientific practitioners at this time sought to demonstrate that their work contributed to arguments for Christian belief. Perhaps easier to miss are the ways in which they fashioned an identity for themselves as Christian gentlemen whose labors contributed both to their own religious character and to the religious character of British culture. In this context, as the next chapter shows, the Bridgewaters offered a great resource, enabling scientific readers to enact and reproduce a Christian persona that justified the growing place of the sciences in a Christian nation.

75. Smyth et al. 1842, 1–6; Timpson 1841, xix–xx; Cooper 1872, 262–63, 371, 375–78; [Kingsley] 1877, 1:380–83. The Arabic translation seems not to have taken place (Perron 1843; Heyworth-Dunne 1940); on missionary pedagogy in the Middle East, see Sedra 2011 and Elshakry 2013. On Cooper, see Larsen 2006, 72–108.

✷ CHAPTER 7 ✷

# *Being a Christian "Man of Science"*

*On the Continent, where there is far less religion than in England, a man who cultivates Natural History, who studies only the works of his Maker, is highly considered and raised by common consent to posts of honour [ . . . ] while, on the contrary, in England, a man who pursues science to a religious end (even who writes a Bridgewater Treatise) is looked upon with suspicion, and, by the greatest number of those who study only the works of man, with contempt.*

MARY BUCKLAND, *November 1845*[1]

During the age of reform, the number and public profile of the all-too-gendered "men of science" in Britain rose to unprecedented levels. There had been expanding opportunities for men actively engaged in the natural sciences since the late eighteenth century, with the founding of literary and philosophical societies in the new industrial towns of Northern England and the Midlands, the formation of specialist scientific societies in London and Edinburgh, and the establishment of new medical schools, colleges, and even universities. The activities of these men were publicized through an ever-increasing array of periodicals and the new, widely circulated reflective works that discussed the distinctive contribution of the sciences to knowledge. Moreover, as Babbage's *Reflections on the Decline of Science in England* made particularly clear, the social role and financial remuneration of scientific savants were becoming important topics of debate within a changing political environment.

The status and standing of scientific men nevertheless remained somewhat precarious. In particular, as the scientifically adept Mary Buckland knew only too well, the "men of science" were sometimes objects of religious "suspicion." Even writing a Bridgewater Treatise exploring the re-

1. Gordon 1894, 220.

ligious tendency of the sciences had not secured her clerical husband against a degree of "contempt." Nevertheless, at this pivotal moment in British science, the Bridgewaters offered men of science an invaluable opportunity to cultivate an image of themselves as respectable figures who were well fitted to take a leading role in a Christian nation and to educate its youth. In itself, the production of these works from the heart of British science but with the full approval of the Church of England hierarchy signified the close connection between the sciences and the Christian establishment. In addition, several of the treatises set out a vision of how scientific practice supported Christian piety, engendering qualities that were to be looked for in the leaders of a Christian land. For scientific savants seeking to establish the legitimacy of their work and the propriety of their social role, the series had much to offer.

This chapter explores how the men of science involved in Britain's learned societies and universities read and deployed the Bridgewaters in relation to these considerations. It begins with the way in which certain practitioners, including the authors themselves, deployed the series to publicly validate a particular identity for themselves as Christian men of science. This was especially evident at meetings of the British Association for the Advancement of Science, where the Bridgewater authors and others drew attention to the new works as signifiers of the moral and religious characteristics that would make members of the association dependable servants of a Christian nation. Moreover, such public performances were paralleled in the more private encounters between the savants and leading figures in the ruling class, where the identity conveyed by the Bridgewaters could open the door to patronage.

The remaining sections of the chapter focus on the role of the series in arguing for and carrying out scientific education at Britain's universities. This was one of the most heated subjects of the day, and it was closely related to the question of the role of men of science in Britain's future progress. Within the nation at large, England's historic universities of Oxford and Cambridge were coming under increasing criticism for their comparative neglect of the sciences in a curriculum primarily suited to the needs of the Anglican clergy. At these institutions, attempts to use the Bridgewaters to defend scientific education met with a rather mixed reception, especially from High Anglicans unconvinced that the men of science gave sufficient credit to the authority and doctrines of Christianity. If anything, it was at institutions that did not impose religious tests, such as the University of Edinburgh and the newly founded London University, that the broad religious framing of the series offered a more acceptable justification for scientific study. Here, however, teachers and students alike found the series useful more as foils for their developing scientific views than as statements of authority. Nevertheless, the Bridgewaters colored the education of a generation of university students whose developing notions about what it was to be a "man of science" grew out of an engagement with them.

## *The Virtues of the "Man of Science"*

The question of what it was to be a man of science was nowhere more publicly aired in the age of reform than at the annual meetings of the British Association for the Advancement of Science. The association represented a coalition of clerical and nonclerical university professors, medical men, and others cultivators of science in various professions and trades as well as the privately wealthy and aristocratic patrons. Its peripatetic meetings, held the length and breadth of the country, offered a platform on which such cultivators of science could advertise the value and meaning of their work to audiences far beyond the narrow confines of London's specialist societies (some of which included women), and the extensive press coverage accorded to them carried the message to an even greater audience (fig. 7.1).

The face that the association thus presented to the world suggested the emergence of a newly important figure in civil society who was devoted to the pursuance of natural science and was committed to making his distinctive knowledge serve the national interest. As one critic pointed out, the British Association had sought to give the term *science* a radical redefinition such that it referred exclusively to natural knowledge and excluded the hitherto highly regarded sciences of "morals, dialectics, and psycology, along with politics and divinity." As men of science in this new, more restricted sense, the association's members felt the need for a general term to characterize their role. At the third meeting, held in Cambridge in 1833, Whewell famously suggested that they use the term *scientist* by analogy with *artist*, but the idea met with little favor. Yet there was not much doubt that a new kind of public figure had risen to prominence, and the association did much to give form to the identity of the man of science.[2]

2. [Bowden?] 1834a, 402 (for the authorship see Morrell and Thackray 1981, 233). On "scientist," see [Whewell] 1834b, 58–61, and Ross 1962. On the British Association and the rise of the "man of science," see Morrell and Thackray 1981 and Ellis 2017.

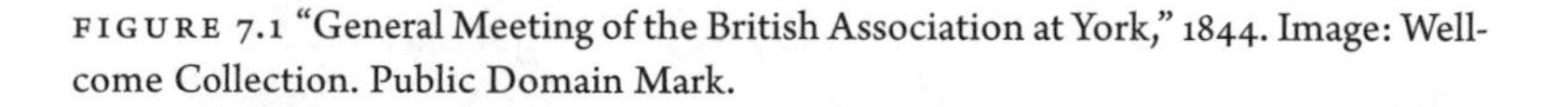


FIGURE 7.1 "General Meeting of the British Association at York," 1844. Image: Wellcome Collection. Public Domain Mark.

There were, of course, many facets to the question of the role that scientific savants should have in society. One particular concern of the early British Association was to recast its members as men of action—bringing the outcomes of their researches to bear on the needs of the nation—rather than merely cloistered scholars. At the same time, however, they were represented as retaining religious and moral qualities associated with the long-established ideal of the devout and humble scholar who served the cause of divine truth. In a Christian nation where cultural leadership was dominated by the power of the established church and its clergy, the men of science felt both the importance and the difficulty of demonstrating that they were properly devout and would not lead the nation astray. Moreover, the view that correct understanding was rooted in virtuous character was deeply entrenched in British learned culture, so that the British Association found it necessary not only to demonstrate the moral and religious tendency of its work but also its moral and religious grounding.[3]

In this context, the Bridgewaters—four of them written by clergy of the established churches of England and Scotland—were a godsend. These works by leading men of science offered confirmation that while scientific truths must be sought unfettered by religious constraints, they would ultimately be found to be both consistent with and supportive of religious truths. More than this, however, the treatises symbolized the Christian character and virtues of the man of science and offered testimony to the moral consequences of studying science. The Bridgewater authors were all to some degree involved in the early meetings of the association, which coincided with the years in which their several treatises first appeared in print. Indeed, several of their number—most notably Whewell and Buckland—were heavily involved in its organization and public presentations. It was natural, therefore, that contemporaries should read the Bridgewaters in relation to the British Association and vice versa. Moreover, the connection went further than this. The authors and their peers actively used the Bridgewaters at association meetings, both implicitly and explicitly, to convey a positive message about the moral and religious consequences of scientific knowledge and above all to represent symbolically the moral and religious virtues of the man of science.

Concern about such issues was especially evident in the two meetings of the British Association that followed its inauguration in York in 1831. These were held in the Anglican universities of Oxford and Cambridge, where three of the Bridgewater authors were based, and those three naturally took prominent roles. Buckland was the obvious choice as president of the second meeting in Oxford in 1832, with Whewell acting as his vice president and both Kidd and Prout serving on sectional committees. As Oxford's leading scientific professor over the preceding decade, he had an established reputation as an advocate for the value of scientific inquiry in supporting Christian belief, and he now reiterated that connection. But he also emphasized its role in fostering moral character. As the meeting reached its climax on Saturday 23 June, Buckland delivered an illustrated evening lecture on the recently imported fossil remains of the giant sloth *Megatherium*. In a less formal setting, with women present, Buckland had an opportunity to range widely concerning the figure of the man of science. The lecture rapidly turned into a panegyric on Cuvier's exemplary comparative anatomy, which proved that all living and fossil species were

3. These complex issues are explored in much greater depth in Dewitt 2013, Bellon 2015, and Ellis 2017.

products of God's design. More especially, in Buckland's hands the Frenchman became a "great and good man" to whom "the higher science of morals" owed a "debt of deep and everlasting obligation" for his new proofs of God's unity and goodness.[4] In his own novel account of the *Megatherium*, later to take such a prominent place in his Bridgewater, Buckland emulated the character attributed to the masterful Cuvier in drawing out fresh evidence of design.

Others were rather more forthright on the subject. Right at the start of the meeting, the outgoing president, Whig MP Viscount Milton, had connected the moral and religious significance of the association with the character of its members. The more a person knew, the more aware he would be of his own limitations, leading to "a deeper feeling of religious awe" and a "stronger sense of the reverence and duty which he owes to the power, the wisdom, and the beneficence of the Creator." This was echoed the day after Buckland's lecture by professor of moral philosophy William Mills, who delivered a sermon in the university church "on the duty of Christian humility as opposed to the pride of science." With many of the British Association's members in his congregation, Mills claimed that, far from the pride of knowledge being characteristic of eminent philosophers of the present day, their successes were dependent on "patient observation and habits of tranquil and deep reflection" that checked such pride in a way that contrasted with those in more active life. The greatest philosophers stood out for their "humble and religious spirit." Each new discovery only gave them a stronger sense of their own imperfection. It was this spirit that was responsible for the more rapid progress of science in modern times. For Mills the challenge was to ensure that each new generation of philosophers was schooled in such Christian humility.[5] Strikingly, however, the preacher's most particular warning was directed at theological inquiries; it was in such studies rather than in physical science that the age's predilection for speculation and novelty was especially threatening.

Buckland encouraged Mills to publish his sermon, but he was naturally less happy with the series of discourses that the British Association meeting elicited the following spring from the Irish-born but Oxford-educated vicar and High Church theologian Frederick Nolan. Invited to deliver the eight Bampton Lectures on divinity in the university church, he took the opportunity to address what he considered the dangerously and insidi-

4. "Proceedings of the General Meeting, 1832," *Report of the* [. . .] *British Association for the Advancement of Science* (1833): 95–110, on 105.

5. "Proceedings of the General Meeting, 1832," *Report of the* [. . .] *British Association for the Advancement of Science* (1833): 96; Mills 1832, 10–16.

ously anti-Christian tendency of modern science, with its ambitions for a naturalistic synthesis. Much of the content of the lectures was occupied with opposing modern geology on scriptural grounds, but Nolan did not see himself as opposing science so much as reforming it. Indeed, he took the trouble to be elected fellow of the Royal Society before delivering his lectures, nominated by his patron the Archbishop of Canterbury William Howley. In his sermons he set himself the task of reconciling science and religion by demonstrating, as the title of the published lectures had it, *The Analogy of Revelation and Science*. Both, he claimed, depended on faith as well as reason. However little scientific men might wish to admit supernatural causes, he claimed, they were obliged to do so at least in acts of creation, and their science was baseless without them. This reliance confounded the besetting pride and vain pretensions of science. The proper lesson of modern science was a "mortifying sense of our weakness" and a renewed submissiveness to God.[6]

The terms of the dispute were clear. Was the modern man of science arrogant in his sense of his own knowledge and importance? Or was he humble in his awareness of his ignorance and dependence on God? Nolan had suggested that the British Association was motivated by an arrogant and dangerous desire to wrest authority from the church with overblown promises of power reminiscent of the *philosophes*. Other High Anglicans were less outspoken, but they were still critical. The *Oxford University Magazine* claimed that while true philosophy could "never be at variance with true religion" and while the British Association had certainly exhibited a "tone of piety," its meetings were nevertheless more productive of "vanity" than knowledge. Buckland's presidency had not especially helped the cause, given his jocular comments, his antediluvian reenactments, and his playing up to the female portion of the audience.[7] As the first four of the Bridgewaters appeared in the spring of 1833, however, they more visibly placed the men of science on the side of the angels. Whewell's treatise, in particular, responded to the concerns of High Church critics (including his friend Rose) by robustly characterizing the admirable moral and religious qualities of scientific men. Moreover, the endeavor received a significant impetus as the scientific savants gathered in Cambridge in the last week of July for the third British Association meeting.

Among the five Bridgewater authors present, it was Whewell's turn to act as host as one of the secretaries. As the meeting opened in the Senate

6. Nolan 1833a, 36; Election Certificate, RSL, EC/1833/03.

7. [Bowden?] 1834a, 407–8. See also "Scientific Meeting at Oxford," *Fraser's Magazine* 5 (1832): 750–53.

House, the new president, Cambridge geologist Adam Sedgwick, immediately paid tribute to his close friend, readers of whose Bridgewater would have recognized his reference to the manner in which the "very constitution" of the mind led humans to consider the idea of natural laws as the "annunciation of the will of a Supreme Intelligence." Sedgwick emphasized that the association's meeting was not motivated by "personal vanity," and Whewell's opening review of its first two meetings more generally picked up where the Oxford meeting had left off. Some had charged that modern science made those involved in it "confident and contemptuous, vain and proud," but a brief glance at the history of science made clear how little had been achieved, how much was owed to past discoverers, and how humble such individuals had been. The triumphant successes of the mechanical arts might lead to a "natural exultation," but the small scientific advances on which they were based were not such as to lead to "any extravagant estimate of what man has done or can do." Seeking to explain how some scientific men had come to lose sight of God, Whewell's Bridgewater argued that deductive reasoners failed to recognize their debt to others, acquiring "an exaggerated feeling of the amount and value of their labours," but Whewell now made clear that such men should not be taken to be the norm.[8]

Whewell's statements concerning the "morals of science" were widely reported in the press, and they were also speedily put on sale in pamphlet form. Moreover, they set the tone for the whole meeting. As proceedings were brought to a close back in the Senate House on Friday, the theme was repeatedly revisited. Knowledge of nature's laws gained its "moral worth" as the mind turned to the creator, Sedgwick reported, and he concluded the meeting by reminding his auditors of their mortality, entreating "God grant that all our attainments in science may tend to our moral improvement." The same topic continued to dominate in an interminable number of speeches and toasts during the course of the afternoon when almost six hundred members of the association had assembled at Trinity College for a cold collation. Among those to return to the theme was Chalmers—a late but much-feted arrival at the meeting. Asserting that "Christianity had every thing to hope and nothing to fear" from the advancement of science, Chalmers pointed out that "humility of mind" was the "offspring of true philosophy, which manifested itself in well constituted minds." It was only arrogant second-rate philosophers who sought to attack true religion.[9]

The ritualized enactment of the piety and humility of the inductive

8. Whewell 1833b, 335; Whewell 1833a, 19–21; Anon. 1833, 65, 69.

9. Anon. 1833, 88, 94–95, 97–98, 108.

man of science in which the Bridgewater authors took a prominent part at these early British Association meetings carried a positive message both to those with scientific pretensions and to the wider public. The Cambridge meeting was attended with enthusiasm by the veteran leader of the university's evangelicals, Charles Simeon, who reported it had "gone off well." He was clear that "one atom of heavenly science" was preferable to all that the philosophers had achieved, but their transactions were in no way at odds with evangelical piety. Indeed, when Simeon was ill the following year, Whewell's Bridgewater was read to him more than once and was, next to the Bible, his favorite reading matter. By then the association had moved on to Bell's native city of Edinburgh, where the same message was repeated. With his Bridgewater fresh in readers' minds, Bell lectured to a crowded and enthralled room on the "proper method of studying the Nervous System." Emphasizing the importance of functional analysis, Bell concluded by suggesting that a man could not be an anatomist without assuming that animal morphology had a divinely ordained final cause. His observations were echoed by the section president, Edinburgh's leading physician, the evangelical John Abercrombie, who closed the session by denying that the pursuit of science was "hurtful to the higher interests of man considered as a moral being." It was presumptive ignorance that led to irreligion.[10]

In addition to this positive message, however, a negative one was never far from the surface. As the Oxford meeting had shown, there was no shortage of religious conservatives who considered that pious humility meant placing ultimate trust in a literal interpretation of the Bible. This, of course, touched Buckland's interests most particularly, and Nolan's attack on the British Association, together with a growing range of works on scriptural geology, led him to urge the association's clerical general secretary, his former pupil William Vernon Harcourt, to take action. The time had arrived when the school "must be put down," he claimed, and if Harcourt and Sedgwick contributed articles on the subject to the leading quarterlies, that would nicely prepare the way for Buckland's Bridgewater. Buckland considered that he "must of course say a little" on the matter in his introduction, but that he "must be very brief," focusing instead on "the facts." Sedgwick addressed the topic in his *Discourse on the Studies at the University* (1833), but it also became the subject of the presidential address in Dublin in August 1835, when natural philosopher and university provost Bartholomew Lloyd directed the majority of his address to those fearful of

10. "Fourth Meeting of the British Association [ . . . ]," *Athenæum*, 20 September 1834, 694–99, on 698, 699; Bell 1870, 425; Bell 1835; Carus 1847, 722–23.

geology. What such individuals required, Lloyd emphasized, was greater humility in the interpretation both of scripture and of geological discoveries. In the study of God's word, as of his works, "patient self-distrust and humble reverence amounting to religious awe" were required.[11]

Lloyd had made due reference to Buckland as an "ornament" of the association and, as we have seen, the never-understated Buckland used the Bristol meeting in August 1836 to publicize his newly published treatise to the nation's largest scientific audience. In fact Buckland's notorious buffoonery meant that there was "immense laughter" when he referred to the "peculiar modesty which is characteristic of geologists," and Sedgwick and Roderick Murchison were both scandalized by his behavior. Nevertheless, both Buckland and others made reference to his "new work" in the geology section during the course of the week, and on Friday morning he placed a copy of it on the table for members to consult (see fig. 4.4). At the meeting's "theatrical finale" on Saturday he made a spectacular splash. Following the president's closing speech, Buckland rose to present him with a copy of his treatise in the name of the association with which it was "connected in no inconsiderable degree." Reports differ about just what words accompanied the presentation, but the gist of it was that the world had existed many millions of years before the Mosaic account of the creation and that this was consistent with the biblical interpretations of Luther and other leading theologians. Buckland had already rehearsed the theme of the "probable age of the world" at the Edinburgh meeting, observing that the opening verse of the Bible implied "an indefinite period of time, in which geological phenomena of the most extensive description might have taken place." Now, however, with the long-awaited Bridgewater in print, his statements spawned the huge newspaper storm previously discussed.[12]

Buckland's performance in Bristol had multiple audiences in view. While the men of science certainly wished to demonstrate to the nation at large, including its clerical leadership, that they were patient, humble, and devout, it was above all important to convince those who exerted political power and patronage that they combined these qualities with an authoritative command of the sciences that would serve the nation's practi-

11. Anon. 1835, 25, 27; WB to Harcourt, 20 November 1833, in Morrell and Thackray 1984, 181–83, on 182.

12. "Fourth Meeting of the British Association [ . . . ]," *Athenæum*, 20 September 1834, 694–99, on 699; "British Association," *Literary Gazette*, 1 October 1836, 634–35, on 634; "Age of the World," *Belfast Commercial Chronicle*, 5 September 1836, 1b; "British Association [ . . . ]," *Bristol Mercury*, 27 August 1836, 2a–6e, on 4a; Cannon 1961, 312n. See also "British Association," *Bristol Mercury*, 3 September 1836, 4a–d, on 4b, and C Lyell to G Mantell, 19 September 1836, in Lyell 1881, 1:471–72.

cal and economic interests. As with earlier British Association meetings, the Bristol meeting was attended by several peers and MPs, including a quarter of the cabinet—chancellor of the exchequer Thomas Spring-Rice, home secretary Lord John Russell, and president of the Board of Control for India Sir John Cam Hobhouse. Until illness intervened it had been intended that the association's president would be the scientific Whig and lord president of the council the Marquis of Lansdowne, but in the event it was Buckland's geological friend the Marquis of Northampton.[13] And Buckland took the opportunity to attach the additional cache of aristocratic support to his controversial claims.

Northampton had begun the week with some fairly forthright statements about the religious consequences of such assemblages, observing that the search after truth led inexorably to reflections on God. "Every true philosopher," he reported, was a "religious man." Taken together with Buckland's presentation, these comments opened him up to newspaper attack by one of his political opponents. In response he felt that he had to take the unusual step of writing to the Northamptonshire press to defend several of his statements before the association, including, most particularly, that the study of the creation was "entirely consistent with the character of a Christian." As to the details of Buckland's interpretation of Genesis, he left his critic to take up the matter with the author, who was "himself a Clergyman of the Church of England, and Canon of Christchurch, as well as Professor of Geology; and with Dr. Pusey, who is Professor of Hebrew in the same University." Buckland's ecclesiastical position provided him with a good deal of authority, but here was public confirmation by a leading aristocrat that the religious identity of the man of science was sound. Indeed, Northampton's successful performance at the Bristol meeting marked him out a couple of years later as the aristocratic champion best suited to take over from the queen's uncle, the Duke of Sussex, as president of the Royal Society, in which role he brokered considerable government patronage.[14]

Among those who crowded into the theater to hear Northampton and Buckland was the Irish poet Thomas Moore, who was much feted in Whig political circles. He had agreed to accompany his patron Lord Lansdowne

13. Morrell and Thackray 1981, 115–16; "British Association—Bristol Meeting," *Athenæum*, 20 August 1836, 587.

14. Hall 1984, 73–89; [Compton] 1836; "Sixth Meeting of the British Association [ . . . ]," *Athenæum*, 27 August 1836, 593–616, on 603; *Report of the* [. . .] *British Association for the Advancement of Science* (1833) 109. Northampton's critic responded to the report given in "Bristol Association," *The Times*, 24 August 1836, 3b–c; see also "The Marquis of Northampton [ . . . ]," *The Times*, 11 October 1836, 3e.

to the meeting and enjoyed himself making up satirical verse in which he imagined the association's meeting of a thousand years thence to involve discussion of fossil specimens of the extinct "Aristocratodon" and "Episcopus Vorax" in a poem that was soon after published in the Whig *Morning Chronicle*. On the last evening, however, Moore was surprised by Northampton with a request to offer a vote of thanks to the Bristol Society of Artists. Standing to rapturous applause, he went on to express the wish that "knowledge" would be found to be "virtue" as much as "power." The Bridgewater authors were clearly in view as he observed, "Some of the eminent men now before us have shewn, in most able and luminous treatises, that Science, so far from being the enemy, is the hand-maiden, or, if I may so say, the torch-bearer of Religion." It was consequently not too much to hope, he continued, that a "like good effect" would "flow from the study of science, in the paths of moral conduct." Unravelling the "mysteries of the material world," Moore continued, was likely to make men more capable of understanding and regulating themselves.[15]

As the nation watched, its leading poets and most senior aristocrats thus confirmed the testimony offered by the Bridgewaters and their authors of the moral and religious tendency of the sciences. Moreover, it was not only in its formal proceedings that the British Association gave the scientific savants opportunities to demonstrate their wholesomeness to the wealthy and powerful. The peripatetic nature of the meetings and their occurrence during the summer, when the wealthy were at their country estates, meant that the men of science were able to join contiguous house parties and confirm their standing as cultural leaders.

Whewell and Roget, for instance, arrived at Bristol from Laycock Abbey—the house of the well-connected Henry Fox Talbot—from where Lord Lansdowne's seat was in easy reach. After the meeting, Buckland left to stay with the family of wealthy geologist Robert Were Fox at his house Penjerrick, near Falmouth, where a wet day had him offering an impromptu lecture on his Bridgewater. After returning to Oxford for the birth of his latest child, he was soon in North Wales at Kinmel Park—the seat of Whig politician and mine owner Lord Dinorben—for a visit of the Duke of Sussex. Buckland was the lion of the season, and John Cam Hobhouse was doubtless not the only member of the large party delighted by his "agreeable mixture of sense and simplicity." Indeed, when the Duke of Sussex announced in his anniversary address to the Royal Society just a few weeks later that Buckland's long and "anxiously expected" treatise had

15. "British Association," *Bristol Mercury*, 3 September 1836, 4b–d, on 4c; Pym 1882, 4; [Moore] 1836; Russell 1853–56, 7:165; O'Connor 2007a, 258–59.

finally completed the Bridgewater series, he commented that a list beginning with the name of Whewell and ending with that of Buckland could "hardly be considered as an unworthy representation of the science and literature of this country." Shortly afterward, Buckland was back in Bristol conversing about his book "in a most agreeable and sensible manner" with the family of Unitarian physiologist William Benjamin Carpenter.[16]

Such conversations continued back in London as the season got under way. Like most of the authors, Buckland was absent from London the majority of the time. There were, however, many other men of science who were well placed to discuss the significance of the Bridgewaters with aristocrats, politicians, writers, and churchmen at fashionable dinners and soirees. The majority of scientific savants had little social standing, and their involvement in such circles depended on their knowing their limits. While they might be called on to express views about matters of scientific fact, they were not at liberty to range widely across philosophical and religious terrain, although scientific clerics such as Whewell and Buckland naturally had greater license.[17]

Charles Lyell had a conversation of this sort early in 1837 while dining with the poet Samuel Rogers and his sister. Among those present was cabinet member Lord Holland, whose grand Kensington house had long been a leading center of Whig society. After dinner, once the ladies had withdrawn, Holland asked Lyell "about Buckland's book, and whether he knew much of geology." It appeared to Lyell that the peer had not formed a high opinion of the work, so he drew on his specialist knowledge to speak up "in favour of the body of the work, on fossils." The conversation moved on to "a talk on new species, and that mystery of mysteries, the creation of man." However, in the account of the conversation Lyell gave to his sister, the only voice was that of Holland, who concluded that modern science had progressed no further on the subject than the Epicurean philosopher Lucretius. Here was the kind of radical comment that a religiously skeptical aristocrat might make in private. Lucretius's atomist work *De rerum natura* might, Holland observed, have supplied mottoes for the three volumes of Lyell's own *Principles of Geology*. But of course Lyell could never have risked such impropriety.[18]

It was not only in London, however, that men of science continued

16. Carpenter 1881, 48; *Proceedings of the Royal Society* 3 (1837): 433; [Carleton] 1909–11, 5:63; Boylan 1984, 213; Pym 1882, 9; Gordon 1869, 161–62; Russell 1853–56, 7:165. On the private hospitality surrounding the meeting see [Prestwich] 1899, 49.

17. Secord 2000, chap. 12.

18. C to S Lyell, 19 March 1837, in Lyell 1881, 2:7–9, on 8.

to exploit the Bridgewaters in securing their reputation with society's wealthy elite. In fashionable Brighton, where the royal party annually wintered, Buckland's work offered surgeon and geologist Gideon Mantell further opportunities to seek the aristocratic patronage that he desperately craved. The son of a Lewes shoemaker, Mantell's work on local geology—especially the giant saurians of Tilgate Forest—had brought him to the attention of the hugely wealthy Earl of Egremont. The earl's patronage soon took Mantell and his growing collection to nearby Brighton, and by the spring of 1836 a second thousand pounds had enabled the surgeon to establish the Sussex Scientific and Literary Institution in association with his "Mantellian Museum." From the outset, however, Mantell sought to shore up the new venture against potential criticisms of the religious implications of his geology, drawing heavily on Buckland's work to do so.

In a lecture on geology at the Old Ship Hotel in December 1835, Mantell had found himself having to respond to Kirby's Bridgewater, with its biblical literalism and its explicit repudiation of Mantell's "age of reptiles." Buckland's Bridgewater, Mantell promised, would completely refute Kirby's views. The following March he began a lecture on fossil crocodile remains by addressing religious prejudices against geology, drawing on the credit of Sedgwick and Buckland as clerical geologists. When the *Quarterly*'s review of Buckland's treatise appeared, a report of it in the *Brighton Gazette*, quite possibly by Mantell himself, pointed out that it confirmed Mantell's earlier claims. These efforts prompted attacks on both Mantell and Buckland in a letter to the newspaper and a pamphlet by local cleric Robert Fennell. Mantell was not, however, discouraged. In October, at the institution's first lecture of the winter season, he addressed some four hundred people, including the Earl of Munster (the king's illegitimate son) and the Earl of Rosse, who had recently penned a work of Christian evidences. Taking the opportunity to return to the question of religious prejudices against geology, he referred "with much liberality of sentiment" to Buckland's newly published work.[19]

Despite his elaborate efforts, Mantell's plan to secure his future as a geologist through aristocratic and even royal patronage ultimately proved ineffectual. In 1838 he sold his collections to the British Museum and purchased a medical practice in London before publishing a series of popular geological works. The first of these, *Wonders of Geology*, was based on his final course of lectures in Brighton, and its frequent recourse to Buck-

19. "Sussex Scientific and Literary Institution [ . . . ]," *Sussex Advertiser*, 24 October 1836, 3e; Dean 1999, 150, 152, 153; "To the Editor," *Brighton Gazette*, 26 May 1836, 3d–e; "Dr Buckland's Bridgewater Essay," *Brighton Gazette*, 28 April 1836, 2c–d.

land's treatise shows how much his presentation of the subject had come to owe to geology's theological guarantor.[20] More generally, the Bridgewaters allowed men of science to perform the role of humble and pious interpreters of nature, posing no threat to the religious safety of the nation. Given their often socially vulnerable position, such a role, when performed before the wealthy and powerful, had the potential to deliver important benefits, corporately or individually.

Two of the Bridgewater authors had particular occasion to witness this potential. When in 1841 Whewell was appointed Master of Trinity College Cambridge—a Crown appointment—his Bridgewater was to the fore in making the case. This was Robert Peel's first university appointment, but he received advice from the Bishop of London, to whom Whewell's work had been dedicated. According to Blomfield, Whewell not only had "extraordinary powers of mind" but also had sound religious views, as manifested by his Bridgewater. Concerned with combating "the extravagances of the Oxford Tractarians," the bishop expressed the hope that, if appointed, Whewell would devote himself "somewhat less to science and more to theology." In his nomination to the queen, however, Peel observed that Whewell's name was "probably familiar" to her as the author of one of the Bridgewaters. He was chiefly eminent as a writer on "scientific and philosophical subjects," Peel reported, but, while he was "not *peculiarly* eminent as a divine," his publications displayed "a deep sense of the importance of religion, and sound religious views."[21]

The pattern was repeated in 1845 when Peel recommended Buckland to the queen for the deanery of Westminster as a "divine of irreproachable life and sufficient theological attainments" who was "at the same time eminent as a man of science." He was successful at the second attempt and told Buckland that he felt that he was "adding strength to the Church" by fostering the appointment of one who united "a pure and blameless character and a kind and generous heart" with "distinguished intellectual attainments." This was, however, a bold move on Peel's part. Mary Buckland considered the appointment of a man of science bound to "raise a clamour, and among good people too."[22]

As the appointment of the Bridgewater authors had shown in the first place, this kind of connection between patronage and support for Chris-

20. Mantell 1838.

21. Blomfield to Peel, 21 September 1841, Peel to Victoria, 16 October 1841, in Parker 1891–99, 3:422–23.

22. M Buckland to P Egerton, November 1845, in Gordon 1894, 219–20, on 220; Peel to WB, 9 November 1845, in Parker 1891–99, 3:417–18.

tian orthodoxy seemed highly suspect to many reformers determined to achieve support for men of science independent of church and aristocracy. In the spring of 1845 Lyell was to be found retelling an anecdote he had heard from John Murray, one of whose friends had taken his pencil to his copy of Buckland's treatise, enclosing each of the concluding theological paragraphs in brackets with the marginal comment "As per contract." Lyell, however, had a rather higher view of Buckland's sincerity and had earlier been sorry to hear him described as "trading in humbug" for personal gain by a "great gun of the old Wernerian school," Jean-François Daubuisson. Likewise, Harriet Martineau considered Bell to have been perfectly sincere in his religious outpourings, although she relayed a variant of the Buckland anecdote, observing, "There was much covert laughter about this among the philosophers, while they presented a duly grave face to the theological world."[23]

Questions concerning sincerity and "humbug" are strikingly intractable for the historian. What is clear, nevertheless, is that the vision of the man of science as pious, patient, and humble became the predominating vision. This was accomplished in part through the manner in which the men of science, including the authors themselves, represented the Bridgewaters in key social settings, including the British Association meetings as well as the more select gatherings of high society. Such performances underpinned the growing access to patronage and social influence secured by the scientific savants within a reformist society in which powerful aristocrats saw them as having the potential to contribute to national wealth and progress. Those with dissident views had to manage their public profile with great care in relation to this social reality or risk the consequences in terms of professional failure and financial ruin.[24] Moreover, the connection between pious humility and scientific learning was something that those who learned about the sciences in Britain's universities could hardly avoid encountering. There, too, the Bridgewater Treatises played a significant role, and it is to these contexts that we now turn.

## *The Anglican Heartland*

At the heart of the changing status and identity of the scientific savants were the related questions of how men of science were to be educated and

23. Martineau 1877, 1:361–62; C to M Lyell, 9 July 1830, in Lyell 1881, 1:275–76; [Bunbury] 1890–91, 2:46.

24. These issues are explored further in Winter 1997, Dawson 2007, and White 2014.

how they might themselves contribute to educating the nation. Of course, education was one of the great topics of the age of reform. Alongside the Whig campaign for the "education of the people," concerted efforts were made from the mid-1820s to reform the nation's universities and to overhaul medical education and certification in London, fomenting controversy as long-established Anglican institutions came under challenge. Most of the Bridgewater authors were to a greater or lesser extent involved in these events, and their treatises were in part written with a view to addressing educational concerns. Not surprisingly, the series was read and used within these educational contexts, but in an age of heated controversy and rapid change, views about how they should be used reflected divergent visions of the proper character of scientific education.

The question of what a university education was intended to achieve was central to the different ways in which rival factions envisioned the future of Britain. In England in particular, reformers at the new London University sought to supplant the exclusively Anglican universities at Oxford and Cambridge, many of whose graduates became clergy, with a more inclusive university that would produce graduates better suited to take the lead in an industrial nation. Yet despite profound disagreements about both the most appropriate curriculum and the unique position of the Anglican Church in English universities, protagonists shared many common convictions. Before it became professional, most agreed, a university education should be liberal, offering some breadth of knowledge but especially training the mind. Moreover, whether or not religious education was to form a compulsory part of the curriculum, the general presumption was that religious teaching would in any case take place at home and in church and be reinforced by university study, producing Christian gentlemen.

There was widespread agreement, at least in public, that if the sciences were to find a place in the university curriculum, it should be within such a framework. When it came to the detail, however, there was much disagreement. England's historic universities, the new competing institutions in London, and the commanding University of Edinburgh offered strikingly different visions of the role of scientific education and its relation to Christianity. The Bridgewaters were nevertheless used by teachers and students in all of these universities. Historians have tended particularly to associate the series with the Anglican universities of Oxford and Cambridge, where three of the authors were based. Yet while it is certainly the case that scientific professors at both these universities sought to use the Bridgewaters to justify the place of scientific education in the heavily religious undergraduate curriculum, they also encountered significant opposition. In Oxford, in particular, where the series was used in attempts to justify an extension of the meager scientific curriculum, the effort was met with considerable hostility and ultimately proved unsuccessful. If anything, it was arguably at the universities of London and Edinburgh that use of the Bridgewaters was less controversial and more straightforward.

At both Oxford and Cambridge, the sciences continued to lie very largely outside of the curriculum. Both universities were wedded to a particular vision of liberal education that drew heavily on philosophy, the classics, and Anglican theology. In Cambridge this involved a significant emphasis—at least for honors candidates—on geometry and mixed mathematics, but as critics observed, the university's emphasis on mental training meant that students were left ill informed concerning the mathematical analysis that had proved so productive in the Laplacian school of physics. Even providing this limited amount of scientific training generated opposition within the university, as it also did in Oxford. Moreover, as previously discussed, natural theology was by no means an obvious means of responding to such critics. Nevertheless, there were those in both universities, especially among the science professors, who found the Bridgewaters useful with both critics and students alike in symbolizing the positive religious tendency of a scientific education.

In Cambridge, an important restatement of the value of science in the education of the Anglican gentleman was offered on the eve of publication of the Bridgewaters by professor of geology Adam Sedgwick. Invited to deliver the sermon at Trinity College's annual service for the commemoration of benefactors at the end of the Michaelmas term in 1832, Sedgwick explored themes discussed by Whewell in recent sermons. Specifically, he set out to show how the three branches of "human" (as opposed to divine) learning in the Cambridge curriculum—inductive philosophy, ancient literature, and moral and political philosophy—developed habits of Christian virtue. In words that were echoed at the British Association meeting the following summer, he emphasized that inductive philosophy inculcated habits of self-control and intellectual humility. In addition, he claimed, it taught individuals to perceive divine action in the universe, preparing the mind for revealed truth. In terms familiar from the manuscript of Whewell's Bridgewater, Sedgwick especially emphasized the close connection between knowledge of natural laws and the perception of a divine creator.[25]

When Sedgwick published an expanded version of the sermon as

25. Sedgwick 1833, 10–12; Clark and Hughes 1890, 1:401–2. For an excellent analysis of Sedgwick's address, see Bellon 2012.

*A Discourse on the Studies of the University* in December 1833 the scientific part was largely unchanged except for the addition of footnotes, one of which made due reference to the "very important and elaborate treatises" on natural theology that had appeared in the intervening months and the others that were still in progress. Sedgwick's view had been that students at the university should learn the habit of perceiving God's agency in natural laws, and he urged those starting out on their studies to make a habitual study of Paley's *Natural Theology* to this end. Moreover, while it was impossible that the Bridgewaters should individually or collectively "supersede" Paley's work, he suggested that they might expand on his argument using "new and pregnant" illustrations and address some "deficiencies." For instance, while Paley had not grappled with "the adaptation of the mechanical laws of the universe to each other, and to the wants of man," Whewell had done just that. Sedgwick was sure that he would be in good company in the university in rejoicing "at the appearance of a work" that fell in "admirably with the course of reading of our best class of students." Whewell's Bridgewater, born out of sermons delivered to defend the religious safety and value of scientific studies within the university, was ideally suited to inculcate Christian character in the small number of undergraduates pushing beyond the regular curriculum to study the natural sciences in more depth.[26]

Sedgwick was altogether less impressed by Buckland's geological treatise when it finally appeared three years later. Scientifically, he valued certain parts of it very highly, especially the paleontology. By contrast, the "moral and theological part" was a "great failure." This was in part a matter of writerly skill. While the descriptive style was "good," the "moral and didactic parts" were "sometimes mouthy and turgid." The argument was too much broken up, and the concluding chapter offered an incoherent string of quotations in place of a "grand outpouring" of Buckland's "own soul, pregnant with high thoughts." More substantively, Sedgwick considered natural theology a matter of laws, but Buckland had only offered "minute details." Furthermore, although God's laws established "wisdom and benevolent design" in general terms, their outworkings involved "much particular and individual evil," which needed a separate theodicy. Worst of all, Buckland on occasion exhibited serious "bad taste." In an age of revolutionary fears, his contrast between the "fleeting perishable" lives of even the most powerful human rulers and the permanent record of the fossilized footsteps of tortoises in the New Red Sandstone left Sedgwick almost throwing the book down "in a passion." While he was eager not to

26. Sedgwick 1833, v, 88; see also 85n, 107.

offend a "most kindhearted, and a most valuable friend," the hot-blooded Yorkshireman found Buckland's treatise much less useful than Whewell's in inculcating religious and moral habits in undergraduates.[27]

Sedgwick's view of the religious value of the scientific studies of the university was shared by other scientific professors, such as botanist John Stevens Henslow and anatomist William Clark. Moreover, his published discourse was seen as an important public defense of the university's curriculum in the face of continued scrutiny, passing through four editions in two years. The view did not, however, go unchallenged. On the one side, the first quarterly number of the philosophical radicals' new *London Review* sneered at Sedgwick's "trite commonplaces" and clerical wonderment concerning proofs of divine greatness. On the other, an approving editorial in *The Times* prompted a declamatory riposte from Anglican priest and staunch evangelical Henry Cole.[28] Cole was a former student at Clare Hall who ran a private school in Islington, where he was also the minister of a proprietary Anglican chapel. In a work pitched as an answer to Sedgwick's discourse and titled *Popular Geology Subversive of Divine Revelation!*, he highlighted the dangers of geology for true biblical religion. And, while the bulk of Cole's fire focused on Sedgwick's geological assertions, he also attacked his "erroneous and anti-scriptural substitution of *natural religion*" for "real, true, and saving" religion. Denying that knowledge of God was possible independent of revelation, he claimed that the "academics" and public of the "art-and-science-illustrious" nation of Britain had been led to worship the creation more than the Creator. Their "moralizing and naturizing system" was one of the devilish "signs of the times," visible in the Bridgewater Treatises and similar "moral philosophications" and tending to undermine true, scriptural religion.[29]

For all that Cole was later to preach in the university church against both natural theology and the religious tendency of modern science as taught in the university, his was hardly the predominating theological viewpoint. Indeed, Sedgwick's critics in Cambridge were much more concerned about his leading role the following spring in preparing a highly divisive petition to Parliament to allow non-Anglicans to be admitted to degrees, a position that Whewell opposed.[30] Nevertheless, there were cer-

27. Clark and Hughes 1890, 1:469–71.

28. Cole 1834a; *The Times*, 10 January 1834, 4c–d; [Mill] 1835, 100–101; [Lord] 1834; Jenyns 1862. On Clark, see Secord 2000, 248–52, and Weatherall 2000.

29. Cole 1834b, 96, 110–11, 129–30; Clarke 1988, 41–44; Lewis 1842, 265.

30. "The Admission of Dissenters to Degrees in English Universities," *British Critic* 16 (1834): 186–99; "The Universities," *British and Foreign Review* 2 (1836): 483–95; Searby 1997, 492–503; Todhunter 1876, 1:91–92; Cole 1853.

tainly voices of concern within the university about the religious tendency both of the sciences and of attempts to shore them up through natural theology. The most powerful of these was Whewell's friend Hugh Rose, whose concerns about science and rationalism had not abated in the years after 1826 and whose wider concerns about how to defend the church had involved him in a foundational meeting of the Tractarians in 1833. Rose's principles were outlined in his lectures as the inaugural Regius Professor of Divinity at the new Anglican University of Durham, where he argued for the superiority of theology over modern learning. When he appeared back before his alma mater on Whitsunday 1834, preaching on "the duty of maintaining the truth," he reaffirmed his earlier skepticism about the religious value of science. The sermon, published "by desire of the Vice Chancellor and the heads of houses," was a doctrinaire restatement of High Church theological conservatism that proved controversial in the midst of the debate about the admission of dissenters.[31]

While Rose's views had considerable currency in Cambridge, they represented one extreme. A more circumspect response was offered from the university pulpit in the following spring by Richard Parkinson, a graduate of St. John's College and fellow (canon) of Manchester's collegiate church (cathedral). A rising cleric of the High Church party, Parkinson had been appointed to deliver eight discourses on the "evidence for revealed religion" and the "truth and excellence of Christianity" as Hulsean lecturer for the year. His distinctive approach was to focus on what he identified as the prevailing error of the age—the doubting of the necessity of scriptural truth. The error was prevalent among the "educated classes" he encountered in Manchester and was fueled by the engrossing character of trade and the "new lights of natural and moral philosophy." No longer confined to "real philosophers," such knowledge was now disseminated "in small treatises among the common people," and it spawned practical unbelievers. Often, such individuals were externally moral and made no profession of disbelief in scriptural truth, and it was thus of no value to direct them to Paley's *Evidences of Christianity* or Butler's *Analogy of Religion*. But the modern "free-thinker" made "the constitution and course of nature his paramount creed—the very God of his idolatry," taking from it a belief in human perfectibility.[32]

31. Rose 1834a, esp. 12; Rose 1834c, esp. 32; Rose 1834b. On Rose, see Valone 2001, Thompson 2008, 56–61, and Bennett 2018. See also Rose to WW, 27 March 1833, TC, Add.Ms.c.211[143].

32. Parkinson 1838, xliii–xliv, 10–17. For an appreciative High Church review, see "Parkinson's Hulsean Lectures," *Christian Remembrancer* 20 (1838): 573–79.

The Bridgewater Treatises, Parkinson continued, had ably added the "force and eloquence" of modern science to the proof of God's existence and attributes, but such an exercise did not necessitate the acceptance of scriptural truth. Parkinson's objective was to show that the "moral and intellectual constitution of man, *coupled* with the power, wisdom, and goodness of God" were the strongest proofs of the necessity of revealed truth. The Bridgewaters might seem to offer succor to the rationalist's obliviousness to revelation, but Parkinson promised to turn such endeavors to good. Indeed, he quickly drew directly on Chalmers's Bridgewater as he developed his first argument concerning the testimony of the human conscience to the truth of revealed religion. At the end of the lecture series, Parkinson concluded that he had shown convincingly that "all human knowledge and all legitimate philosophy" led to the gospel. The duty of the clergy, he now urged, was likewise to appeal to both the reason and the affections, and he exhorted those in the congregation preparing for a clerical career to use their time in Cambridge to acquire human knowledge as well as divine, sanctifying the "discoveries of science" by connecting them with the priorities of the religious life.[33]

Thus, while some followed Rose in considering the effect of mathematics and the sciences in Cambridge's curriculum to be baneful, others were prepared to accept the justification offered by Sedgwick and to embrace the Bridgewaters as useful to students. In the modern world, Parkinson made clear, an appropriately pious education in the sciences was an essential prerequisite even for those intent on a clerical career. Moreover, while the university's theological educators generally gave little space to natural theology, not all were as negative on that subject as Rose. Indeed, one of the theological professors was drawn into discussing the subject at length by the appearance in May 1835 of Henry Brougham's *Discourse of Natural Theology*. This was Thomas Turton, who as professor of divinity was responsible for examining students for divinity degrees, which were taken after completion of the bachelor of arts degree. Finding what he considered inaccuracies in Brougham's account, he wrote up his notes for publication, correcting Brougham on a number of points but ultimately remaining positive about the contribution that natural theology made to Christianity.

Turton had previously been the Lucasian Professor of Mathematics, and he claimed to have reflected on natural theology from his "earliest days." He did, however, consider that Brougham had blundered in trying to give the subject the semblance of an inductive science in order to

33. Parkinson 1838, 18–20, 175–87, 202–14.

make it appeal more to scientific men who doubted its relevance. Turton was shocked by Brougham's claim that many "scientific men" of his acquaintance relied little on natural theology because they viewed it "as a speculation built rather on fancy than on argument—or, at any rate, as a kind of knowledge quite different from either physical or moral science." Brougham's ill-conceived endeavor to make natural theology a science was, in Turton's view, unnecessary given that modern science had already given a "new stability" to the argument from design. This was where the Bridgewaters came in, since they had brought forward a "variety of important information" in the "most recondite departments of science." According to Turton, by making clear that the argument from design had a settled position in relation to the sciences, the series rendered Brougham's overreaching ambition of making natural theology itself into a science entirely unnecessary.[34]

Natural philosophy had a relatively settled, if limited, place within the Cambridge honors curriculum, but the university's scientific professors were glad of the opportunity that the Bridgewaters provided of restating the religious tendency of scientific education in the face of High Anglican and evangelical disapproval. Yet while they convinced some theological educators, they by no means convinced all. Nor were the arguments of natural theology especially likely to ease such tensions. Although some within the university were happy to endorse such arguments, the "dim and deceitful light" of natural theology appeared to others to be a misleading distraction from revelation.[35] To the extent that the Bridgewaters seemed to develop this program, such readers were liable to find them harmful.

❋

In Oxford, there was a rather more pronounced conflict regarding the Bridgewaters and what they represented. While over the preceding two decades Kidd and Buckland had justified a place for the natural sciences in the university's studies partly through an appeal to their religious tendency, the age of reform brought a chill wind for such endeavors.[36] With the growing cultural authority and pervasiveness of scientific knowledge

34. Turton 1836, 2, 7, 41; Brougham 1835, [1]. On Turton see Yule 1976, sec. 6c, Schaffer 2003, 246–47, and Thompson 2008, 61–62.

35. Sedgwick 1833, 87.

36. For a good overview, see Corsi 1988, chap. 9. See also Rupke 1997, Fox and Gooday 2005, and Topham 2013.

and the rising challenges to the privileges and doctrines of both the church and the university, many of the university's clerical leaders became less open to the sciences. Even those who were reform minded generally found themselves too occupied with larger battles to make the sciences a priority. Thus, while the scientific professors saw in the new works another opportunity to demonstrate the beneficial tendency of the sciences, with Buckland's teaching in particular being remolded by the writing of his Bridgewater, the prevailing mood in the university was distinctly against the attempt.

The situation is well illustrated by the unsuccessful endeavors of the professor of geometry Baden Powell to improve the position of scientific studies in the university. Those studying for the bachelor of arts degree were obliged to learn parts of Euclid alongside their concentrated diet of theology and classical literature, and they had, since 1807, been able to compete for honors in mathematical and physical sciences as well as classical literature. However, only a small proportion did so, and the standard was not high. In February 1827, shortly before Powell was elected to his chair, the university's governing body had agreed to the formation of a separate board of mathematical examiners, which promised higher standards. However, the new professor's attempts to increase the independence and competitiveness of mathematical examination further over the next few years ultimately failed (fig. 7.2). Powell's proposals crucially lacked the support of his reformist former teachers at Oriel College—dubbed the "Noetics" (i.e., "intellectuals"). In addition, as the clamor for reform in the country at large grew, they also began to make him a target for conservatives who saw in the sciences and in competitive examination a marker of reformist danger.[37]

Resigning in disgust as an examiner at the start of 1832, Powell's subsequent attempts to garner external support for educational reform only served to open him to a charge of disloyalty to his university. The "public lecture" with which he began his geometry course in the Easter term of 1832 cleverly argued that it was a betrayal of the university's "long established principles" not to include the sciences as core to its scheme of liberal education. By neglecting the sciences in a world where even workers were scientifically educated, he concluded, the university was at risk of endangering its leadership role in the nation and even its own survival. The lecture was published in time for the visit to Oxford of the British Asso-

37. Corsi 1988, chap. 9; Brock and Curthoys 1997, 370; *Oxford University Calendar*, 1810, 55–58.

FIGURE 7.2 Cartoon by Baden Powell concerning the debate over the Oxford examination system, ca. 1831. An emaciated don representing "Disciplinæ Mathematicæ et Physicæ" and a corpulent star-strewn one representing "Litteræ Humaniores" look on as a series of competing systems for ranking candidates follow behind a dragon bearing the name of the planned statute *De examinandis graduum candidatis*, who gives an apparently fatal blow to the figure of logic (signified by the mnemonic "Barbara Celarent Darii"). Baden Powell 31(16). Image: Bodleian Libraries, University of Oxford. CC-BY-NC 4.0.

ciation in June, which senior members of the university had hoped would defuse some of the recent external criticism of its scientific failings, and it clearly aligned Powell with such critics. Moreover, Powell subsequently publicized his failed attempt to make science compulsory in anonymous articles for the Society for the Diffusion of Useful Knowledge's reformist *Quarterly Journal of Education*, coming by 1834 to argue that Parliament should intervene.[38]

38. Powell 1832, 25, 27; A Master of Arts 1832; Morrell and Thackray 1981, 386–93; Corsi 1988, 115–19.

These plans for reform were clearly politically charged, but they also had a growing religious significance. Powell was the scion of a well-connected High Church family, and his first book, published in March 1826 while he was a young vicar, had offered a critical examination of the "rational religion" of Unitarians, taking a highly circumspect view of the value and validity of natural theology. As he applied the lessons of his Oriel teachers to changing times, however, Powell's perspective developed. In April 1826 he preached a sermon before the Archdeacon of Rochester designed to show that Christianity had nothing to fear from the "advance of knowledge," and when Hugh Rose delivered his alternative perspective in Cambridge shortly afterward, Powell sent a polite critique to a leading High Church magazine. He also took the opportunity, when invited to preach the university's Easter sermon for 1829, to defend the religious safety of science before his Oxford peers. Selecting as his theme "Revelation and Science," he argued that conflicts between the biblical text and the findings of science did not undermine either the Bible or Christianity, since the Bible was not designed to offer scientific instruction but, rather, spiritual instruction alone. As a result, he claimed, there was no cause for hostility to science, but neither was it necessary to seek a reconciliation, as Buckland had done.[39]

This strikingly liberal position did not win over Powell's Noetic patrons, and it was not until 1833, when Nolan's High Church attack on the sciences issued from the same pulpit, that Powell decided to publish "the substance" of his sermon in reply, adding additional illustrations. He considered his sermon a satisfactory answer to Nolan's basic principles, but he felt it necessary to elaborate in an added note on an error that he felt was more general, namely, the complaint that the exclusive focus on secondary causes in the sciences led to a neglect of the first cause or final causes. In terms that echoed Whewell's recently published Bridgewater, he argued that, on the contrary, the independent search for ever-more general laws was the source of science's capacity to "support the doctrine of a first cause"—something that his sermon had already claimed was fundamental to establishing the truth of revelation. Any suggestion that scientific pursuits had an irreligious tendency, Powell urged, was unjustifiable in a period that had "given birth to the Bridgewater Treatises."[40]

Powell's supposed "heterodoxy" over revelation brought down on him the severe censure of the High Church *British Critic*, where John Henry Newman's close friend John Bowden noted that it clearly set him apart

39. Corsi 1988, esp. chap. 10; Powell 1826a, 1826b, 1833.

40. Powell 1833, 32–34.

from Buckland's earlier reconciliatory ambitions. Moreover, the alignment between the sciences and natural theology on the one hand and religious, political, and university reform on the other seemed to many to be as clear as it was objectionable. In the midst of the great public debate about the admission of dissenters to the Anglican universities, popular college tutor and High Church stalwart William Sewell made the point explicit in a pamphlet concerning the reformers' "attack upon the University of Oxford." Responding to Sedgwick's *Discourse* and his role in the Cambridge reformist petition Sewell observed, "We have no wish to be encumbered with the help of a natural theology, unconnected with the great facts of Christianity. We do not require to be taught, as science has been lately teaching us, that there is a God, and that we are his creatures." In fact, he told prime minister Earl Grey, it might be necessary for true Christians in the age of reform "to protest solemnly and anxiously" against a natural theology that, rather than assisting Christianity, was "more than irrelevant to its truth." There was nothing of value in the Bridgewaters, Sewell observed in the newly Tractarian *British Critic* in October 1838, since they had "studiously dropt from their pages all mention of the Gospel, and confined themselves to illustrate the existence of [a] vague undefined power above us."[41]

Despite such fierce opposition, other Oxford scientific professors in the 1830s also continued Buckland's earlier approach of using natural theology to justify science's place in a liberal education. Powell's friend, the college tutor Robert Walker, made just such a point in an anonymous 1832 pamphlet supporting Powell's campaign for the expansion of scientific education. Rather more vocal was Kidd and Buckland's protégé, Charles Daubeny. Elected Kidd's successor as professor of chemistry in 1822, Daubeny had argued in his inaugural lecture for the value of his subject within a liberal education not least because of its value in mental training. Moreover, his *Introduction to the Atomic Theory* of 1831 showed how modern chemistry naturally intersected with ancient philosophy and with questions concerning God's agency in the universe.[42] By the early 1830s, however, Daubeny was increasingly concerned about the opposition of religious conservatives. When Nolan launched his attack in 1833, it was Daubeny—who had been the one to invite the British Association to Oxford—who now anonymously took up the cudgels in the *Literary Gazette*.

41. [Sewell] 1838, 306; Sewell 1834, 21–22; [Bowden] 1834b, 431.

42. Daubeny 1823, 1831; [Walker] 1832, 20. On Daubeny, see Oldroyd and Hutchings 1979.

While Daubeny represented Nolan's as an extreme voice in Oxford, he also acknowledged that Nolan did not "stand altogether alone" in his anxieties about the tendencies of the sciences. Elected additionally to the chair of botany in 1834—an opportunity for him to eke out his meager living—Daubeny's inaugural lecture this time made much more of the religious safety and value of scientific knowledge. The natural sciences were not, he insisted, especially productive of intellectual pride. Nor did they alienate the mind from religion. And in such a scientific age, it was especially important, if the clergy were to maintain their control over education, that scientific "supplementary studies" should be encouraged within the university. To remain a "national establishment," Oxford had to turn out graduates capable of scientific study who were not "prejudiced against those pursuits."[43]

In such a context, the Bridgewaters seemed again to offer welcome confirmation of the religious tendency of the sciences, and Daubeny saw the series as directly relevant in baptizing the curriculum. Much of his inaugural lecture was taken up with an account of what he took to be the foundational principles of natural classification as developed in Swiss botanist Augustin De Candolle's *Théorie élémentaire de la botanique* (1813). De Candolle had argued that apparent irregularities in floral anatomy were to be accounted for by abortive, degenerative, or connective deviations from the type. Daubeny considered that this view needed to be defended from the imputation that it excluded the idea of design. One must, he argued, understand the larger plan of divine design without expecting every detail of plant anatomy to be adaptive. Yet the subject of "abortive or rudimentary organs" in plants or animals had not yet been "satisfactorily discussed." Daubeny looked to "some one of the Bridgewater Treatises, to whose province such a subject more properly belongs," to explain the phenomenon, and of course Roget's work on animal and vegetable physiology, published a few weeks later, did not fail to rise to the challenge.[44]

The presence among the Oxford science professors of two of the Bridgewater authors left few in doubt about the relevance of the series to scientific studies there. Buckland particularly made it felt. While his Bridgewater grew out of his existing lectures, his work on the book led to his giving a shorter course in 1832 and 1834 that was focused more specifically on "organic remains." Then, with the book in print, he cashed in

43. Daubeny 1834b, 37–39; *Literary Gazette*, 7 December 1833, 769.

44. Daubeny 1834b, 22–26. See also Daubeny's use of the Bridgewaters in Daubeny 1840.

during the Michaelmas terms in 1836 and 1837 with a course offering a "Demonstration of Organic Remains Figured in [the] Bridgewater Treatise." This was intended quite literally. The professor had numerous of the illustrations from his book reproduced (and in many cases enlarged) on large sheets of paper for use in his lectures. At a time when attendance at Oxford's scientific lectures was rapidly falling, the Bridgewater course bucked the trend, with forty-nine in attendance in 1836.[45]

The high profile of Buckland's publication certainly generated a degree of excitement. First on the lists for Buckland's lecture courses in 1837 was a young John Ruskin. Brought up in a devoutly evangelical and increasingly wealthy middle-class household in London, Ruskin had been educated privately by his parents, then at a private evangelical school, and briefly at King's College London. His parents had nevertheless encouraged his youthful interest in geology. They were familiar with Buckland's *Reliquiæ diluvianæ* and eagerly purchased Buckland's Bridgewater in December 1836. Thus, it is no surprise that one of the first things Ruskin did on his arrival in Oxford at the start of 1837 was to sign up for Buckland's mineralogy lectures, thereafter eagerly attending each one of Buckland's courses. Within weeks, Ruskin was being invited to Buckland's lodgings to meet visiting geologists (talking "all the evening" to Charles Darwin), and he was soon on friendly terms with the family. The Bucklands also appreciated his artistic talent, and he was drafted in to provide lecture illustrations as well as to offer advice to Mary.[46]

For Ruskin, Buckland's lectures were not only useful in learning about the science for which he had a passion. Together with a sermon Buckland delivered at Christ Church in January 1837, in which he sought to align the new geology with the doctrine of the Fall, they provided a means for Ruskin to square his scientific interests with his evangelical faith. Nor was Ruskin the only such student. Banker's son John Charles Ryle attended the lectures in 1836, shortly before taking first class honors in humane literature and undergoing an evangelical conversion. Yet as a rising leader in Anglican evangelicalism fifteen years later, he recalled his Oxford education. The Bible was not "written to explain geology or astronomy," he told his readers, and those who wished to know about such subjects should "study

45. Attendance at Buckland's lectures is detailed in the surviving registers (OUMNH-BP, Misc.Mss./13); see also Rupke 1997, 547. For the reuse of his illustrations, compare, for example, OUMNH-BP, D-82, D-98, D-101, and Buckland 1836, vol. 2, plates 15, 35–36, 44.

46. Burd 2008; Burd 1973, 2:463; Cook and Wedderburn 1908, 198, 385.

Herschel and Buckland," while those who wanted to learn how their souls could be saved "must study the written Word of God."[47]

Within the university more generally, Buckland's geology continued to have its adherents. Even the imperious dean of Christ Church and professor of Greek Thomas Gaisford told Ruskin—on his way to Buckland's lecture in the spring of 1838—that he liked students to "attend to Sciences," although he warned that Ruskin should not let geology occupy him too much. More positive support came from some of the Oriel Noetics. One example was Powell's contemporary, Renn Dickson Hampden, who returned to Oxford from various curacies in 1829, using his role as examiner to encourage the study of Butler's *Analogy of Religion*. Invited to give the Bampton Lectures in 1832, Hampden used them to develop his distinctively empirical science of the scriptures, and when he was himself examined for the bachelor of divinity degree in May 1833, he chose as one of his two theses the claim "The Scriptures have not been designed to convey philosophical instruction." Buckland filleted away the obligatory printed notice of the examination in his files with suitable underlining—proud of his former student.[48]

Hampden, however, was in no position to support the cause of the sciences in Oxford. In his introductory lectures as professor of moral philosophy in 1835 he rejoiced in the growing public attention to "Physical Science," seeking the same for "Moral Science" and offering an account of the evidences of natural theology in the human moral constitution. But when he was appointed to the Regius Chair of Divinity the following spring, his reformist views and Whig patronage made him a target for the Tractarians, who organized a concerted if unsuccessful campaign to remove him from office. Another of Powell's contemporaries, the reforming headmaster of Rugby School, Thomas Arnold, was among those who came to Hampden's defense—writing critically in the *Edinburgh Review* of the "Oxford malignants." Like Hampden, Arnold had attended Buckland's geology lectures as an undergraduate, and his early career as a schoolmaster had been in partnership with Buckland's brother John, who was Arnold's brother-in-law. Enthused by Buckland's geology, he had in turn encouraged its study by individual pupils at Rugby within his general scheme of a modernizing classical education. Moreover, he brought this enthusiasm back to Oxford in 1841 as Regius Professor of Modern History, making the case that his-

47. Ryle 1852, 18–19.

48. Hampden 1837, liii; OUMNH-BP, Lecture Notes 1/7; M to JJ Ruskin, in Burd 1973, 1:590. By the 1850s Gaisford had wearied of Buckland's geology. See Gordon 1894, viii.

tory was "closely connected" to geology and drawing on the "very highest authority" of Buckland's claims in his Bridgewater about coal measures to make the point.[49]

Within the year, however, Arnold was dead, and such interventions had hardly counteracted the prevalent mood at Oxford respecting the sciences. Many of the increasingly dominant Tractarians—notably the vicar of the university church, John Henry Newman, the professor of Hebrew, Edward Bouverie Pusey, and the nonresident professor of poetry and rural clergyman, John Keble—had enjoyed Buckland's lectures in the 1820s but were now resistant to his natural theology and discomforted by the professor's backtracking on his earlier flood geology. Indeed, one young High Church acolyte, the newly graduated curate William Josiah Irons, mounted a lengthy and learned attack on natural theology and its support for "modern deism," pointing out profound limitations in its arguments that even "the whole Library of 'Bridgewater Treatises'" could not answer. Of the Tractarians, the best disposed toward science in general and the Bridgewaters in particular was consumptive Oriel fellow Hurrell Froude, who was pleased enough by Whewell's opposition to the admission of dissenters to read and enjoy his Bridgewater. But even Froude considered that "every new step in science" would likely weaken Whewell's argument. By contrast, Keble reportedly once argued geology with Buckland "on a coach-top all the way from Oxford to Winchester."[50]

The Tractarians saw no religious value in the sciences, and to a large extent Oxford agreed. While several of its scientific professors sought to use the Bridgewaters to make the case for the continued and increased inclusion of the sciences within the university's core curriculum, the appeal to natural theology in particular proved more controversial than effective. Dominated by High Anglicans and increasingly by the Tractarians, the university was ever-more antagonistic to the claims of Powell and Buckland about the theological value of the sciences in grounding the truth of Christianity. To such critics the Bridgewaters appeared to be part of a wider reformist and scientific assault on the rightful territory of the Anglican Church in educating Christian gentlemen.

49. Stanley 1844, 1:19, 123; Arnold 1842, 161–62; [Arnold] 1836; Hampden 1835, vii–viii. On Arnold's vision of history, see Forbes 1952.

50. Froude to Newman, 1 February 1835, in Gornall 1981, 20; Mozley 1882b, 1:178–79, 2:429–34; [Mozley] 1882a, 20–24; Irons 1836, 133n. On the Tractarians and science see, for example, Newman 1843 (and the commentary in Newman 1872), Newman to Pusey, 11 and 21 April 1858, in Liddon 1893–97, 4:78, Morrell and Thackray 1981, 161–63, 230–33, and Rupke 1983, 136–37, 267–74.

## *The Great Metropolis*

The agitation concerning the admission of dissenters to Oxford and Cambridge was only the latest challenge to Anglican control over English university education. It followed in quick succession from the establishment by leading Whigs and dissenters of a competing self-styled London University, founded in 1826 as an unchartered joint-stock company. Drawing inspiration from Scottish and German universities, the new institution offered the opportunity to study a much broader curriculum than England's medieval universities, giving the sciences a far more significant role. Unlike at the Anglican universities, students were not subjected to religious tests at matriculation or graduation, and there was no requirement to undertake doctrinal education or attend public worship. This was above all designed to navigate around the religious differences of students, but it led to the charge that the university was undermining religion. The Anglican establishment was not slow to respond, founding King's College London just two years later with a similar curriculum but with the conventional Anglican elements retained both in what was taught and who taught it. The battle was thus joined in the great metropolis.

These developments represented only one element in a wider process of diversification. Numerous medical schools were established in English provincial towns in the years after 1815, when new legislation was introduced that regulated the education required by the bulk of practitioners. In addition, a wide variety of new colleges sprang up across the nation in the early nineteenth century. One such was the Royal Belfast Academical Institution, founded in 1810 by businessmen and professionals to offer a comprehensive nonsectarian education, including in the higher branches of study. Another was the nonsectarian Bristol College, opened in 1830 by a group that included Anglican clergy but also the prominent physician James Cowles Prichard. Within a few years, many of these institutions had become involved in the preparation of students for degrees. When in 1836 a new University of London was chartered as an umbrella organization examining and awarding degrees for the two London colleges, it also rapidly came to ratify the activities of other new institutions. Within five years, sixteen such establishments and a long list of medical schools were recognized by the university.[51]

51. Hearnshaw 1929, 137; Harte 1986, 95–98; *London University Calendar*, 1844, 57–60; Jamieson 1959; Latimer 1887, 140–42; Butler 1981, 17; Loudon 1986, 49; Bonner 1995, 168.

As these new institutions thrashed out their differing visions of education, religion, and the sciences, the Bridgewaters once again came into play. Yet for all that historians have been inclined to expect the nonsectarian University College (as it soon came to be known) to be more resistant to the message of the series, it was arguably the Anglican King's College that was the less receptive. The nonsectarianism of University College was grounded in what many considered a profoundly troubling separation of public education in secular subjects from private education in religion. However, Tory claims that it was an "infidel college" or "godless school" were a form of abuse that its many Whig and dissenting advocates roundly rebuffed. A few of the college's supporters—most notably James Mill and his acolyte, the banker and MP George Grote—were Benthamite "philosophical radicals" who repudiated Christianity. A few years earlier, Grote had even published from Bentham's notes a pseudonymous pamphlet attacking natural religion, and he took active measures to try to keep the Congregationalist minister John Hoppus out of the chair of philosophy. However, as his ultimate failure to do so perhaps suggests, Grote's atheism was not generally shared.[52]

For most of the college's promoters, the question was not whether young men needed religious education but whether it should be provided within the university curriculum or at home and through public worship. Moreover, while formal theological education was carefully excluded from the curriculum, a certain amount of religious common ground was presumed to exist. This was symbolized—when the king's brother laid the first stone of its building in Gower Street—by the prayer of Anglican cleric and prominent Whig Edward Maltby, which referred to the way that the study of God's works simultaneously advanced "human science" and led on to divine truth (fig. 7.3). This kind of rather general commitment to divine creation also found expression in the lecture courses on natural philosophy, comparative anatomy, physiology, and other sciences taken by general and medical students.[53] Even in this most secular of establishments, students could expect to encounter the sciences in relation to divine design.

52. On Grote see *ODNB*, s.v. "Grote, George," Bellot 1929, 108–11, 288–89, Berman 1988, 191–201, and [Grote and Bentham] 1822; see also Hoppus 1830 and 1840. For an early example of the abuse, see the *Standard*, 19 June 1828, 2e; for religious defenses of the university's secular policy, see [Macaulay] 1826 and especially [Brougham] 1828.

53. University of London 1827, 52. On the curriculum, see also University of London 1828.

FIGURE 7.3 The main building of the London University in Gower Street. Engraving by W. Wallis after T. H. Shepherd, 1828. Image: Wellcome Collection. Public Domain Mark.

It was a mark of Charles Bell's eminence that he was the professor whose address opened both the new university and its medical school on 1 October 1828. He began by unfurling a reformist vision of how the new university would allow for the development of a more practical and scientific approach to medical education than had been possible within the medical schools of the metropolis. Rather more space was devoted, however, to Bell's concern for the moral and religious welfare of medical students, as he welcomed the opportunities offered by the new university for professors to notice such matters in contrast to existing medical schools. Medical students had a lack of respect for authority that was born out of a sense of the progressive character of scientific knowledge and consequently had "peculiar" needs that were most properly met by a scientifically educated clergy, by parental tutelage, and by the "voluntary labours" of professors. Accordingly, Bell followed his introductory oration with a lecture on the circulation of the blood illustrating "the constantly recurring proofs of design in the animal frame," before ending with a promise that the next lecture would expand on the theme. According to the *Lancet*, these comments were greeted with "Loud applause." Moreover, this continued to form the opening theme of Bell's course in succeeding years.[54]

Bell's comments about the religious needs of medical students were echoed on the university's second day of teaching by John Conolly, professor of the nature and treatment of diseases. Conolly's subsequent observations on the importance of "character" were seen to have a political edge that angered those professors committed to more radical medical reform, especially of the Royal Colleges and teaching hospitals. Yet these political differences, which probably contributed to Bell and Conolly's early departure from the university, did not entail repudiating the religious context and significance of scientific study. When the reformist professor of zoology and comparative anatomy Robert Edmond Grant delivered his inaugural lecture, it likewise outlined the value of comparative anatomy in enlarging the understanding of divine design, thus laying "the most rational and lasting foundations of piety and virtue" and strengthening "the best principles of morality and religion." For all that Grant rejected Bell's Cuvierian emphasis on functional adaptation in favor of transcendental anatomy and privately endorsed species transmutation in a law-bound universe, he nevertheless echoed Bell in casting his course in terms of the confirmation of divine design.[55]

Other medical professors offered similar observations. When James Bennett was appointed as a second anatomy professor in 1830, he focused on the vast importance of recent developments in comparative anatomy on the Continent but couched the work of Cuvier, Geoffroy Saint-Hilaire, and Meckel in relation to the "laws which the creator has ordained" and the additional evidence offered of "the creative powers of the Deity" before ending with religious reflections on death. By the following year, Bennett was himself dead, and he had been succeeded in what was now a com-

54. "London University," *Lancet*, 4 October 1828, 8–10; University of London 1828, 13–14; Bell 1831, 311. See also Bell's endorsement of secular education in Bell 1829.

55. Grant 1828, 7; Conolly 1828, 31–34; "The London University," *Lancet*, 11 October 1828, 50–52; "Dr Conolly's Lectures," *London Medical Gazette*, 4 October 1828, 568. On Grant, see Desmond 1989b but also the useful corrective in Secord 2000, 62–65. On the medical reform agenda at London University, see Desmond 1989a, 1989b and Berkowitz 2015; for criticisms of it as a "politico-medical establishment," see "Memorial of the Medical Teachers," *London Medical Gazette*, 17 May 1834, 241–44 on 241.

bined chair of anatomy and physiology by the Irishman Jones Quain. But, while Quain had also learned his comparative anatomy in Paris, he, too, wished to emphasize to students the "arrangement and order" pervading the "Animal Series" and the "wondrous adaptation of means to end—of structure to function,—furnishing such incontestable evidence of design and contrivance, that the mind cannot but ascend at once to the contemplation of a designer, and soar from nature up to nature's God."[56]

Even Conolly's successor as the professor of the principles and practice of medicine, John Elliotson, who was later described by the professor of mathematics, Augustus De Morgan, as one of the "strongest" materialists he had encountered, joined in. In opening the medical session of 1832–33, he explained that the university's nonsectarian and consequently secular principles did not "interfere with the duty of teaching all that can be known of the Creator from the works of creation, nor all that can be known of our moral obligations from the principles and laws of human nature." He observed that "a philosophical reason *may*, indeed, be given for every moral duty, though enforced by Christianity." While doctrinal religion should be taught by parents and ministers, he insisted, "Natural religion and morality, in which all sects agree," were "no doubt, as fully taught in this as in any university."[57]

In the university's general department, the Anglican professor of natural philosophy Dionysius Lardner gave a thorough justification of the value of his subject for the purposes of a liberal, as well as a professional, education in terms of the enlarged religious perspectives it offered. Lardner was an ordained clergyman who had been educated at Trinity College Dublin. Moreover, he collaborated with his fellow cleric, the evangelical professor of English Thomas Dale, to offer religious services and lectures on divinity for Anglican students, commencing with lectures on natural theology. This was done with permission of the university council alongside a similar venture by dissenting clergy, although neither lasted long. The professor of chemistry, Scottish evangelical Edward Turner, took a different approach. Actively engaged with Prout's atomic hypotheses, he doubtless referred students to the chemical Bridgewater as part of his prescribed "course of reading." Yet Turner was more especially concerned with "imparting moral and religious instruction to such of the students as might desire it." He was, Thomas Dale emphasized to students following the chemist's untimely death in 1837, a perfect exemplar of "The Phi-

56. Bennett 1830, 8–11, 21–22; Quain 1831, 12.

57. Elliotson 1832, 5; De Morgan to J Smyth, 1 September 1864, in De Morgan 1882, 324–25, on 325.

losopher Entering, as a Child, into the Kingdom of Heaven." What this meant was that Turner knew from the inadequacies of natural theology that human reason was clouded and that true religion depended on faith. Moreover, his was not a religious "example" for the few. In this most secular of universities, about three hundred students turned out on a Saturday in February to hear Dale urge them to follow in Turner's footsteps.[58]

As one critic observed, it was "almost impossible" for London University's professors to avoid their religious views bleeding through, since scientific discussions always bore in some way on natural or revealed religion. Yet by no means all the scientific professors presented their subjects in a religious context. It is notable, for instance, that professor of botany John Lindley, who was determined to remodel his subject as manly and professional, made no reference to religious subjects in his inaugural lecture or even in the notes he supplied for Brougham and Bell's edition of Paley. Mathematician Augustus De Morgan avoided religious references for different reasons. Rejecting his evangelical Anglican background, he developed his own "unattached" Christianity, shaped by Unitarianism, but was passionate about making religion a private matter, fully separated from the curriculum. However, this was no easy matter, and as scholars have shown, the mathematics he developed continued to bear marks of his religious quest.[59]

The overall impression given by the scientific and medical professors at University College was that their subjects contributed to a sense of divine design. Indeed, even the college's more religiously unorthodox supporters argued as much. Speaking at its annual meeting in February 1834, the wealthy philanthropist and advanced liberal Robert Fellowes condemned the religious prejudices that had thwarted the college's purposes, claiming that rather than being irreligious, it was "a sanctuary of science" in which science elucidated "the Divine agency," making clear God's laws and attributes. He continued: "For do not all our scientific classes partake, more or less, of a theological character? Do they not all refer, more or less, to the acts, the volitions, the laws and ordinances of Deity in the moral, the intellectual, and the material universe? Nor can any student well attend any one lecture in any one of those classes without knowing more of the

58. Dale 1837, esp. 10, 19, 23; Terrey 1937, 146; Brock 1985, 161–70; Dale 1829, 33–38; Bellot 1929, 56–58; Lardner 1829, 17–22, 36.

59. Lindley 1829; Lindley 1830b; Brougham and Bell 1836, 2; Kennedy 2017; Stearn 1998; J. G. "Strictures on the London University," *Imperial Magazine* 9 (1827): 1098–1102, on 1101. There are, however, references to design in Lindley 1830a, 322, and Lindley 1835, 253. On De Morgan's teaching see De Morgan 1830, Anderson 2006, and Rice 1997, 1999; on his religious views, see Richards 1997, 2002, 2007.

Divine agency than he did before." Fellowes went on to claim that the college's approach would "give more purity and simplicity" to religion and (despite having been ordained as an Anglican priest in his youth) he elaborated on this point shortly afterward in a deistic work titled the *Religion of the Universe*.[60] This, of course, was precisely the kind of outcome that many Anglicans expected from a scientific education that made reference to divine design but offered no grounding in Christian theology. It was nevertheless to a significant extent consistent with the account of the sciences offered in many of the Bridgewaters.

While the experiences of students in such a plural establishment must have been quite diverse, there were sufficient references to God's design for them readily to perceive it to be a standard feature of scientific discourse. Moreover, whether professors directly recommended the series or not, the Bridgewaters consolidated such views in ways that were pedagogically useful, as the scientific and medical press had made clear. The collection at the library of the new university was notoriously thin, and with only £49 spent on it between 1832 and 1835, we may doubt that the Bridgewaters were purchased. However, there was a range of ways for ambitious students at the university to lay hands on them, and the series made it into several of London's medical school libraries, ranging from the conservative and patronage-riddled St. Thomas's Hospital to the radically reformist Webb Street School. Other students doubtless purchased individual Bridgewaters of interest.[61]

One such student was William Benjamin Carpenter, the son of a leading Unitarian minister, who had been apprenticed to a local surgeon in Bristol and studied at the town's medical school before arriving in London in 1834, aged twenty. Carpenter attended as many as thirty-five lectures per week and found himself especially excited by Grant's lectures on comparative anatomy—so much so that he began to contemplate writing a "little work" on the "philosophical study of Natural History." Before leaving London at the end of the session to continue his studies in Edinburgh, Carpenter put pen to paper, producing an article on the "structure and func-

60. [Fellowes] 1834; Fellowes 1836.

61. On the library, see Bellot 1929, 417–26. The library's first printed catalog (1879) listed early editions of all except Chalmers's Bridgewater, but the online catalog suggests that these may have been later donations. There are early editions of the Bridgewaters in the library of King's College London identified in the online catalog as being from St. Thomas's Hospital Medical School Library (Bell, Whewell, Kidd, Prout, Roget, and Kirby) and the Webb Street School of Anatomy and Medicine (Bell and Kirby).

tions of the organs of respiration in the animal and vegetable kingdoms" for a minor journal associated with the Bristol College. This drew on his London lecturers and key Continental authors, but Carpenter was especially taken with Roget's Bridgewater. Indeed, he cast this early account of the laws of vital action, in which he demonstrated the analogies in form and function between respiration in plants and animals, in relation to Roget's vision of the "unity of design," bookending it with quotations from Roget's Bridgewater.[62]

There was no shortage of places in which Carpenter might have learned about Roget's newly published Bridgewater (and that of Prout, which he also quoted). Perhaps it was from a grumbling Grant, who was hardly reticent about the use Roget had made of his lectures. In any case, Carpenter's experience provides a revealing insight into the world of the devout and scholarly dissenter sent to study at the misnamed "infidel college." Reflecting on his time there, he agreed with his parents about its "valuable influence" on him, adding that the high intellectual standards he had set for himself had been of substantial "assistance" to his religious principles. Most helpful from a "moral point of view" had been reading Herschel's *Preliminary Discourse.*[63] But, as we shall see, his engagement with the Bridgewaters continued to inform his thinking in significant ways as he left for Edinburgh and worked on the book that became his *Principles of Physiology*.

In a college in which Unitarians such as Carpenter rubbed shoulders with Anglicans, Congregationalists, and even deists, the Bridgewaters had the potential to support a range of views concerning the religious significance of scientific education. Their emphasis on the appearances of design in nature and on the consistency of scientific inquiry with Christian orthodoxy was well suited to support the college's widely publicized claims to strengthen, rather than to undermine, the Christian views of many of its students. At the same time, however, the absence of any distinctively Christian theology from many of the treatises was consistent with the college's nondoctrinal stance. The generality of the notion of divine design, which disquieted more conservative Anglicans, was precisely the quality that suited it to the college's desire to draw together its disparate com-

62. [Carpenter] 1835; Carpenter 1888, 7–11, 17–18. On Carpenter, see Secord 2000, 63–66, Desmond 1989b, 210–22, Rehbock 1983, 61–68, and Winter 1997, 35–43.

63. Carpenter 1888, 13.

munity around a common commitment to demonstrating the religious safety of the sciences.

❋

Leading Anglicans meanwhile wasted no time in procuring support from the Tory government to found a competing London college that combined a curriculum in the sciences and modern languages with religious education. Enjoying royal patronage, the new King's College opened its doors on the afternoon of 8 October 1831, only hours after the rejection by the House of Lords of the second attempt to pass the Reform Bill. Notoriously, the bishops' votes had carried the day, causing such anger that later in the month the bishop's palace in Bristol was burnt down during riots, but the ceremonies at the new college were presided over by Archbishop Howley, with Bishop of London Charles Blomfield preaching the sermon. A year earlier, of course, the pair had joined the president of the Royal Society in appointing the Bridgewater authors, but while Blomfield acknowledged that scientific study could confirm and enlarge the understanding of divine design, he asserted that God's nature and purposes could only be known through his self-revelation. In the new college, religious knowledge must and would have primacy. This set the tone for education at King's College and was reflected in how the Bridgewaters were appropriated within it.[64]

Although students were not expected to submit to religious tests in order to study at the college, those taking the general course (rather than just individual lecture courses) were required to attend daily prayers and Sunday worship as well as the principal's Monday lunchtime lectures on natural and revealed religion in the Anglican tradition. The science professors also knew what was expected. Even the professor of geology Charles Lyell, in private a deist, rose to the challenge. Lyell's unwillingness to relate his geology to the Bible had left college council member Bishop Edward Copleston seeking reassurances that he would not cast doubt on Noah's flood. Lyell was fully conscious of the context in which he was operating and "worked hard upon the subject of the connection of geology and natural theology" as he prepared his second course for a select audience in May 1832. But there was no compromising his principles. In his opening lecture he quoted "a truly noble and eloquent passage" from Blomfield's inaugural discourse to the effect that "truth must always add to our admiration of the works of the Creator" and that "one need never

64. Blomfield 1831; Hearnshaw 1929, 93–96.

fear the result of free inquiry." This reinforced his claim that his geological outlook, with its vast tracts of time and no "traces of a beginning," offered the "most sublime" vision of the divine creator. Thus, while friends assured him that he had not compromised "the utmost freedom of the philosopher," Lyell's opening lecture pleased the college's hierarchy, who recommended its publication.[65]

Most of the scientific professors were more orthodoxly Christian and readily connected their studies to the college's Anglican program. One after another they outlined how scientific studies enabled students to enlarge their conceptions of the creator's role in the universe. In such a context, the Bridgewaters were invaluable. One of the most forthright in his religious statements was the Glasgow-educated professor of zoology James Rennie. Rennie had written several works on natural history for the useful knowledge society's Library of Entertaining Knowledge that had emphasized evidence of design in nature, and he even prepared an elementary *Alphabet of Natural Theology* in 1834 for those who might be overwhelmed by such elaborate works as the Bridgewaters. With his professorial lectures suffering low attendance, and his income from fees consequently low, he continued to expand his literary output, commencing a *Magazine of Botany and Gardening* in 1833. There, as no doubt in his lectures, he recommended Roget's treatise warmly as a consummate "help" in the study of God's living creation. As with every branch of study, scientific researches should lead to "moral results," and "religion should be taught by science." Often, Rennie claimed in another review, distinguished scientific men had been opposed to gospel truth, but there had recently been a sea change. With his eye firmly on his own role at King's College, he noted that there were now many "teaching us the true philosophy" in such a way as to make the creator more manifest. Moreover, the Bridgewater authors had "acted effectually" in supporting such endeavors.[66]

Rennie was not the only lecturer to recommend the Bridgewaters to students. Indeed, Lyell's successor at King's College, John Phillips, was still recommending *Geology and Mineralogy* to his geology class (now in Oxford) some twenty years later. Furthermore, the college bought the Bridgewater series for its library as the volumes appeared, including three copies of Chalmers's treatise, and loan records show that they received significant use from both professors and students. At the time of his res-

65. Lyell 1881, 1:381–83; Hearnshaw 1929, 50–51, 99.

66. *Magazine of Botany and Gardening* 2 (1834): 165–66, 172, Rennie 1834, vii, 69; Page 2008. See also, for example, Burnett 1832a, 36–37, Burnett 1832b, 26–27, Daniell 1831, 31–32, and Daniell 1839, 4–11.

ignation in 1836, the college principal and Anglican cleric, William Otter, had Kidd's and Bell's Bridgewaters sitting in his office, quite possibly for use in preparing his theology lectures.[67]

Otter's successor as principal, Hugh Rose, promptly returned the Bridgewaters to the library. Otter had been theologically and politically liberal, but Rose was much more in harmony with the ethos of the ruling council. Indeed, his 1826 Cambridge Commencement sermon that had caused Whewell such disquiet had been a rallying point in the High Church riposte to the new London University. As Professor of Theology at the new Anglican University of Durham in 1834, Rose had managed ingeniously to weave his friend Whewell's new Bridgewater into a justification of High Church views of ecclesiastical history, and at King's he continued to give Christianity primacy. This he did in difficult circumstances. Almost immediately following his appointment, the Whig government granted a charter for a new University of London to act as a degree-awarding body for both King's College and the rebranded University College as well as for the rapidly growing list of other colleges across the nation. Against all Rose's protestations, theology was excluded from the curriculum for the bachelor of arts examination and while exacting questions about natural theology appeared in the philosophy examinations for the master of arts degree, such religious education fell far short of what Rose expected. In consequence, King's directed students to Oxford or Cambridge to take degrees.[68]

Rose's concerns are made particularly evident by developments in the medical school. The school had made a poor start, and Rose worked with professor of physiology and morbid anatomy Robert Bentley Todd to put it on a secure footing. However, there were not merely practical failings to deal with but also concerns regarding medical students' moral and religious education. The medical curriculum was heavily determined by the requirements of the licensing bodies, and medical students were not required to undergo the same religious observances as other students. This required active mitigation. Thus, an early benefactor had endowed a prize to encourage medical students to attend the chapel services and the principal's theological lectures. In addition, medical professors cast

67. KCL, Author Catalogue, 1841–43 (KAL/CA17), Senior Library: Books Missing or Out (KAL/CA20), and Loans Register, 1845–54 (KAL/BR1); Morrell 2005, 253 (see also 133–38). Henry Moseley's astronomy lectures (1839) included many references to design but not specifically to the Bridgewaters; see also Moseley 1847.

68. Hearnshaw 1929, 33–34, 127–37; *London University Parliamentary Returns* 1840, 304–6.

their studies in relation to religion, and for some, such as the professor of surgery Joseph Henry Green, this entailed elaborating a vision of the role of medicine within an Anglican polity. Green saw the profession as a key element of the "National Church, or Clerisy" of his friend Coleridge, offering leadership that was rooted in the education of England's Anglican universities.[69]

This emphasis was redoubled under Rose and Todd, but they were assisted in 1838 by a donation of £1,000 from High Church philanthropist and cleric Samuel Wilson Warneford to found a prize to encourage the study of theology among the college's medical students. Warneford's wealth had come largely from his family's involvement in the druggist trade, and over time he dispersed vast sums to medical and other charities, including the Oxford Lunatic Asylum and the Queen's Hospital in Birmingham, with its newly founded medical school. In the Birmingham medical school, Warneford used the considerable leverage his money gave him to ensure that medical students were given a distinctively Christian education, emphasizing the importance of the Bible and the need for salvation. Alongside chaplaincies and books for the library, in 1838 he endowed an essay prize there on a medical subject treated in a "practical and professional manner" but with a view to "exemplify or set forth by instance and example, the Wisdom, Power, and Goodness of God, as revealed and declared in Holy Writ." In its emphasis on scripture, Warneford's essay prize offered a very distinctive view of how evidence of design should feature in medical education. Some of the Birmingham school's supporters had welcomed the Bridgewaters as a boon to medical education. Warneford, by contrast, had a "*Bridgewater-phobia*." His concerns were well set out by his loyal lieutenant Vaughan Thomas, who argued that natural theology as found in works like the Bridgewaters was too rationalist to be useful or even safe. What students needed was an incentive to develop their religious qualities through biblical piety before bringing them to bear on their professional medical studies.[70]

Warneford wanted the Birmingham prize to form the model for that at King's, but he was ultimately dissuaded by his High Church advisers, including the Archbishop of Canterbury, his Hackney friend Joshua Watson, and Rose himself. They were fully in sympathy with Warneford's strong emphasis on revealed religion. Watson, for instance, thought it fair to say

69. Green 1832, [iii]; Mayo 1834; Hearnshaw 1929, 113–17, 138–42; Todd 1837. On Green, see Desmond 1989b, 260–75.

70. Thomas 1843, 35; 1855, 10–15, 39–54, 123–39; Cox 1873, 69–81, 88–91, 221, 228; Law 1873, 69. The first seven winning essays were all published.

that "Treatises built on Lord Bridgewater's foundation" could never "*make* a Christian." But it was a question of practicalities. Rose was skeptical of the ability of students to apply biblical views in the way Warneford sought to prescribe, and his alternative suggestion was that they write on the evidences either of natural or revealed religion or on their harmony. Warneford was unpersuaded, and in the end, their compromise was that candidates for the prize should be examined in the Bible and Joseph Butler's *Analogy of Religion*, followed by two branches of medical science.[71]

Although Rose did not adopt Warneford's rather exacting position on the place of natural theology in medical education, his own position stood in clear contrast to the very general design theology found at University College. Like their peers in that secular institution, the scientific and medical professors at King's College made reference to the evidence that the sciences offered of divine design, and the Bridgewaters were perhaps more prominently brought to students' attention at King's. Yet the college's Anglican ethos, especially under Rose, meant that it was the students at King's, rather than those at University College, who were most frequently reminded of the limitations of natural theology and its dangers in the absence of a vigorous engagement with the Christian scriptures. Significantly, it was there, rather than at the "infidel college," that there was talk of "*Bridgewater-phobia*."

## *The Athens of the North*

The new London colleges somewhat diminished the striking contrast between university education in England and that in Scotland, which had five historic universities located in Edinburgh, Glasgow, St. Andrews, and Aberdeen (where there were two). Writing in 1827, Charles Lyell had highlighted the oddity of England's Anglican universities in the wider European perspective. They offered, he observed, a preliminary education that was both extraordinarily protracted and excessively narrow in its focus on the classics and theology while offering no opportunity to proceed to a "regular professional course of study." By contrast, Scottish universities began with a "liberal education" that was much broader than at Oxford and Cambridge, including the "elements of mathematics and natural philosophy," while allowing students to proceed to professional studies in medicine, law, or divinity in an expeditious manner.[72] Partly in consequence,

71. Thomas 1855, 52; King's College London 1845, 15.
72. [Lyell] 1827, 218, 221.

the University of Edinburgh had long been Britain's favored center for medical education, with a first-rate course of study and with no religious restrictions. Moreover, the latter circumstance made Edinburgh a magnet for English dissenters more generally.

In the early nineteenth century, the University of Edinburgh's medical dominance was somewhat undermined by the development of English medical schools. The actions of the ruling town council to address the ensuing concerns resulted in 1826 in the initiation of a royal commission to inquire into the state of the Scottish universities. Nevertheless, Edinburgh remained by far the largest and most highly reputed of Scotland's universities, located in a city whose topography, neoclassical architecture, and above all intellectual renown had recently led to its being dubbed the "Athens of the North." Like others in Scotland, the university was ultimately rooted in the Presbyterian tradition, but for many years it had not submitted its students to religious tests—or even its professors, despite official rules. Moreover, it no longer operated strict moral and religious oversight like that found in Oxford and Cambridge. Indeed, the 1826 commissioners reported that some students were unhappy about the lack of religious provision, with seventy-three medical students petitioning them for more adequate accommodation for Sunday worship.[73]

A similar laissez-faire ethos pertained to the course of studies. A standard curriculum existed for the master of arts degree, but it did not include Christian theology except to the extent that, as at Scottish universities more generally, the course on moral philosophy typically included natural theology. Moreover, few formally completed the course and graduated. Those wishing to study divinity required certificates of attendance from key courses, but most others merely chose the courses that interested them, paying the necessary fees to the relevant professors in a highly competitive marketplace. Student experiences were consequently very varied. Those proceeding to professional studies in divinity, law, and medicine experienced more constraints, but even the medical students, whose studies were to some extent prescribed by the requirements of licensing, could choose between professorial and extramural teachers.[74]

Although it was not compulsory, the fact that systematic teaching on natural theology was part of the arts curriculum meant that Edinburgh

73. *Report into the Universities of Scotland*, 194–95, 273–74, 284–85; Bonner 1995, 173–75.

74. *Report into the Universities of Scotland*, 275; Morrell 1972. For useful accounts of the state of scientific and medical education in early nineteenth-century Edinburgh, see Gray 1893, 1:21–22, 97–106, and Jenkins 2019.

arguably gave more attention to the subject than any English university. In the early 1830s, however, the tone was notably circumspect. The professor of moral philosophy was Oxford-educated John Wilson, famous as "Christopher North" of *Blackwood's Magazine*, who owed his appointment to Tory interests. As was expected, his lectures covered the classic arguments for God's existence, refutations of skepticism, the problem of evil, and the question of free will—all considered as "*metaphysical questions*" and consequently constituting part of moral philosophy according to the Scottish conception of the subject—but he gave them a conservative twist, telling students forthrightly that much of what was taken to be known by reason was only known by the aid of revelation. These lectures were mainly heard by divinity students, who required a certificate of attendance to enter the Divinity Hall. However, while such students accounted for two-thirds of the hundred and fifty or so taking Wilson's class in 1826, there was also a variety of others attending his course. These even included some of the medical students who still dominated the university. The medical professors—very conscious of the need not to alienate potential students within an increasingly competitive educational market—disagreed with the commissioners' view that medical students should complete the arts curriculum before proceeding to their professional studies. Nevertheless, many of them, including extramural teacher John Thomson, were keen to see medical students receive a more liberal education. Thomson's son Allen took notes on Wilson's natural theology lectures as a student in the late 1820s, and some of what he learned was woven into his own lectures on medicine, delivered in the early 1830s as a professor in Glasgow.[75]

Edinburgh's laissez-faire curriculum offered students other formal settings in which to study natural theology. As we have seen already, Chalmers based his Bridgewater on his lectures as Edinburgh's professor of divinity and subsequently made his treatise into a "text-book" for the use of students. Moreover, while divinity lectures in Edinburgh were primarily intended for students preparing for a clerical career, there was nothing to stop other students from attending. In the autumn of 1830, for instance, Chalmers's class included James David Forbes, the son of a well-connected Tory banker who hoped he would become an advocate. Forbes was a devout Episcopalian and had considered entering the Anglican clergy, although his real interests lay in natural philosophy. His attendance at Chalmers's lectures came after four academic sessions in which

75. Jacyna 1994, 170–71; [Thomson] 1824, 1826a, 1826b; *Report into the Universities of Scotland*, 240–41, 297–300; Swann 1934, 154, 184; *Evidence on the Universities of Scotland* 1837, 35:156, 158, 164.

he had followed the arts curriculum at length—including classes on Latin, Greek, chemistry, moral philosophy, and natural philosophy as well as natural history, civil law, and Scottish law—and he was "strongly of the opinion that to hear such masterly lectures by Chalmers upon Natural Theology and the Evidences" was a "most fitting conclusion to a course of liberal education, and singularly well calculated to prevent injury from the skeptical insinuations of Laplace and other modern philosophers" whose works were "oracles" for those engaged in the exact sciences.[76]

Like Allen Thomson, Forbes went on to incorporate what he learned in his own professorial lectures. In 1833, when he was just twenty-three, family and political interests helped him to secure the Edinburgh chair of natural philosophy in succession to his own teacher Sir John Leslie. Leslie was widely considered to have been an atheist. His first election as a professor in 1805, in spite of his denial of certain claims of natural theology, had been a striking sign of the growing power at Edinburgh of the Evangelical party in the Church of Scotland, many of whom considered natural theology to be unnecessary as a foundation for Christianity. One clerical witness in 1826, however, reported of Leslie's class that natural phenomena were presented as "little more than proofs of the capabilities of nature," and Leslie's critic argued that a "regular exercise" on Paley's *Natural Theology* was needed to deepen "the impression of the existence and superintendence of a First Cause." Like earlier natural philosophy professors, Forbes adopted this more religious approach to his subject. Moreover, he was delighted with Whewell's Bridgewater and used it in the process. The Cambridge man had "charmed" him on their first meeting in 1831, and after supporting him for the Edinburgh chair, he had been Forbes's main adviser in preparing his course. Forbes followed his mentor in seeking to teach natural philosophy in a Christian context, rooted in moral and religious discipline, and he considered Whewell's chapter on inductive and deductive habits especially helpful in enforcing the mental discipline that would preserve students against the religious dangers of superficial learning.[77]

Other scientific professors were perhaps not so liable as Forbes to connect their studies with religious concerns. Long-standing professor of natural history Robert Jameson had earlier cast Cuvier's geology in biblical terms, but he had preceded Buckland in abandoning flood geology, and his well-attended lecture course was heavily technical. Yet he was eager

76. Shairp, Tait, and Adams-Reilly 1873, 66.

77. Forbes 1849, 12, 50–51, 63–66; Shairp, Tait, and Adams-Reilly 1873, 72, 95–98, 140, 147, 167; [Forbes] 1832, 2–4; *Evidence on the Universities of Scotland* 1837, 35:469; Morrell 1971, 1975; Topham 1999.

to encourage divinity students to attend his class, claiming that it would provide them with pulpit illustrations. Moreover, his *Edinburgh New Philosophical Journal*, which committed students consulted, announced the publication of the Bridgewaters in highly approving terms. Extramural lectures on natural history were also offered by Jameson's longtime assistant William MacGillivray, who was now the conservator at the Royal College of Surgeons. His publications from this period cast the study of natural history in devout terms, drawing on several of the Bridgewaters, although he was critical of Kirby's peculiar biblicism.[78]

The Bridgewaters were also pertinent in the medical school. Not surprisingly, Bell was to be found pointing out proofs of "design and benevolence" to students after becoming Edinburgh's professor of surgery in 1836. Another pious Episcopalian, professor of the institutes of medicine William Pulteney Alison, also repeatedly introduced divine design in the *Outlines of Physiology* based on his lectures—notably in preliminary remarks concerning the independence of the evidence of design in vital phenomena from the question of their underlying causation—and he soon introduced additional evidence from several of the Bridgewaters into subsequent editions.[79] In addition, however, just as at University College in London, some of those developing more innovative approaches to anatomy and physiology, notably in the extramural medical schools, also sought to root their teaching in an enlarged discourse of design.

This was notably the case with Alison's leading extramural competitor John Fletcher, whose highly regarded lecture course at the Argyle Square medical school won him a reputation as one of the leading exponents of the latest work in philosophical anatomy. Fletcher's lectures were nevertheless cast firmly within a religious framework. Whereas physiology had sometimes been charged with leading to irreligion, he told students, the "excellent Paley" and "numerous other authors" had shown that the "strongest proofs" of God's existence and character were afforded by the subject. Irreligion was not the sign of a "daring hero," he warned in no uncertain terms, but of "a half-witted and half-educated, dastardly, and vulgar reptile." Expanding on his lectures in print, Fletcher again observed that the

78. *Edinburgh Journal of Natural History*, 24 October 1835, 4; *Edinburgh New Philosophical Journal*, 3rd ser., 15 (1833): 403–4; *Evidence on the Universities of Scotland* 1837, 35:187, 795–98. On Jameson's teaching, see Morrell 1972, Secord 1991, Jenkins 2019, and [Jameson between 1830 and 1839?]; on MacGillivray's devout presentation of natural history, see MacGillivray 1834, 1836, and MacGillivray and Thomson 1910.

79. Alison 1833, 2–3, 216, 244–45; Alison 1839, 4, 13–15, 288, 364; Bell 1838, 2:303.

religious fears generated by advances in physiology, as in the sciences more generally, were ill founded. His hope was that the "recent admirable *Bridgewater Treatises*" would help finally to remove prejudices against physiology while also freeing religion from criticism of the "ill-judged zeal of its false friends." He elaborated on the point yet further when in 1836 he began an illustrated course of "popular" lectures on physiology for an audience of over three hundred "ladies and gentlemen." The Bridgewaters were a reassuring emblem of the religious safety of the sciences, Fletcher declared, but while natural theology might strengthen faith, it was revelation alone that made Christians.[80]

The Bridgewaters were not merely useful at this emblematic level, however. Fletcher liberally laced the published version of his technical physiology lectures with references to the relevant treatises. He drew especially on Roget's Bridgewater in his account of the new anatomy, partly in support of his opposition to the associated "wild" transmutationist "speculations." At the same time he used his former teacher Bell's criticisms of Geoffroy Saint-Hilaire's views as a foil in introducing them. To Fletcher, like Roget, the "doctrine of the unity of organic structure" provided evidence of design that was far superior to the narrow functionalism that was so commonplace. Fletcher similarly saw Roget as in ally in his controversial claim that life was properly defined as "living action" resulting from organic structure rather than as a "vital principle" that caused such structure—a view that he claimed was fully consistent with Christianity. Moreover, while the physiologist lamented Bell's, Prout's, and Kirby's recent reassertion of the vitalist position, he found much of use in their discussions.[81]

Not all the medical lecturers were so deferential to the Bridgewaters. Another Geoffroyan extramural lecturer, Robert Knox (fig. 7.4), offered a rather different view at the anatomical school in Surgeon's Square that he had taken over from the devout John Barclay. In 1828 Knox's extraordinary popularity as a lecturer had led him into notorious and damaging dealings with the body snatchers turned murderers Burke and Hare, and lecture attendances had tailed off sharply in the aftermath. He was, like Fletcher, a knowledgeable advocate of philosophical anatomy, but as a deist he com-

80. Fletcher 1836–37, 11:190–91; 1835–37, pt. 1, 64n–65n; pt. 3, xix–xx. On Fletcher, see the life in Fletcher 1835–37, pt. 3, xi–xxiii, Desmond 1989b, 71–79, Secord 2000, 62–65, and Jenkins 2019.

81. Fletcher 1835–37, pt. 1, 11n, 17n, 38n, 51n, 62n, 64–72; pt. 2a, 19n, 22n, 28n, 38n; Fletcher 1834–35, 6:712n.

FIGURE 7.4 Robert Knox. Line engraving. Image: Wellcome Collection. Public Domain Mark.

bined these views with a hearty contempt for Christianity that he did not hide. Knox was also much more outspoken than Fletcher about the failings of what he called the "coarse utilitarianism of Paley," which reached its ne plus ultra with Bell. In January 1830, for instance, he attacked the failings of Cuvier's principles of functional adaptation before the Royal Society of Edinburgh in a memoir on the Peruvian llama. The existence of design in animal bodies was "too obvious to require a thought," he claimed, but the problem was in "particularizing" it. In his later years, he looked back on the period of his teaching in the 1830s as an era in which "much baleful influence was cast over English science" by what he was by then fond of calling the "Bilgewater Treatises." Yet one former pupil and colleague later recalled that even Knox believed in a "perfection in creation," albeit that, in private moments, he "expected the advent" of an improved understanding of the "Creative scheme."[82]

Students were thus faced with a range of responses from lecturers, but Edinburgh's many societies also presented them with opportunities to try out their own views. By the mid-1830s the once vigorous Plinian Natural History Society—where students in Darwin's time had sometimes discussed materialism—was now in decline, but the long-standing Royal Medical Society and Royal Physical Society were both still active. One keen member, and soon president of both, was William Benjamin Carpenter, who arrived in Edinburgh in the autumn of 1835 armed with letters of introduction to his father's old college friend John Wilson and other professors. He found the university very congenial, especially since the Moderates in the Church of Scotland were so liberal they seemed like Unitarians at heart. The Royal Medical Society proved particularly helpful in completing his education, and he was soon planning to present a paper in which he would "bring forward a number of most beautiful analogies" that Roget had missed in his Bridgewater.[83]

By April 1837 Carpenter was in the president's chair at the Royal Medical Society, and he gave a talk "On the Unity of Function in Organized Beings" that was intended to be the counterpart of a recent paper on the unity of structure by his predecessor, the Continentally trained and independently wealthy Martin Barry, which had also framed the new anatomy in relation to the notion of design offered by Roget. Shortly afterward Carpenter won a student essay prize on the theme set by his physiology lecturer William Alison, "On the Difference of the Laws regulating Vital and Physical Phenomena." His argument (like Fletcher's) was that there *was* no fundamental difference between the two and that, in time, both sets of phenomena would probably be found to result from a higher law. Emulating Fletcher, whose lectures he greatly admired, he adroitly used the vitalist Prout's Bridgewater findings to support his opposing view.[84]

By this point Carpenter was rapidly transitioning from student to teacher, and in January 1839 his nonvitalist physiology was combined with an account of the "Unity of Design" in animal and plant morphology in his *Principles of Comparative Physiology*. Four of the Bridgewaters were listed among Carpenter's key sources in this important treatise, and the

82. Blake 1870, 334; Lonsdale 1870, 243–48, 402–10; Knox 1850, 169; 1831, 486; Dawson 2016, 271–79.

83. Carpenter 1888, 13–20; Desmond and Moore 1992, 31–34, 38; Jenkins 2019, 61.

84. Carpenter 1838b, 339; Carpenter 1837, 92; Carpenter 1888, 20–22; [Carpenter 1838a], 100; Barry 1837, 116, 141.

whole was framed in clearly theistic terms, climaxing in a chapter on the "evidence of design," which used Whewell's Bridgewater (rather than his more recent contrasting writings) to argue that understanding God's action as being expressed in natural laws offered a "far higher and nobler conception of the Divine Mind." When Carpenter approvingly reviewed a new edition of Prout's Bridgewater a few years later, he observed that, in general, the Bridgewaters had laid too much emphasis on "individual instances of design" in comparison to the "general plan." Bell, in particular, had offered a "sneer" to philosophical anatomy.[85] Yet between Roget's, Prout's and Whewell's contributions, there were significant materials for him to work with.

For a Unitarian such as Carpenter, whose religion soon afterward resulted in his being rejected as Alison's successor as professor, being able to enlist the support of the Bridgewaters for his scientific views was invaluable in managing his reputation. Indeed, a year later he was having to answer charges in David Craigie's *Edinburgh Medical and Surgical Journal* that his *Principles of Physiology* was unsafe because it denied God's continual agency in the universe and tended to materialism. Carpenter argued in a brief pamphlet that his views were fully consistent with those of reputable authors, including Alison, Fletcher, and Roget. Moreover, he was able to include testimonials concerning the orthodoxy of his position from numerous savants who included Anglican clerics as well as Alison and Roget. Developing his radically new perspective on physiology and morphology in dialogue with the Bridgewaters and under the tutelage of Alison and Fletcher, Carpenter had matured a religious understanding that he was confident could meet the expectations of orthodox Christians.[86]

By comparison with the discussions Carpenter engaged in at the Royal Medical Society, those that took place at Edinburgh's newest student society were significantly more lively. Founded by medical student Edward Forbes and three friends in 1834, the Maga Club was at first preoccupied with producing a spirited weekly magazine that poked fun at the professors, but by the following spring they were conceiving of themselves in grander terms as a quasi-Masonic brotherhood devoted to "wine, love, learning." In 1838 they took the title "Universal Brotherhood of Friends of Truth," calling themselves a "Union of the Searchers after Truth, for the glory of God." Membership was select in its high intellectual and moral standards, and for young men with scientific aspirations—including

85. [Carpenter] 1846, 116; Carpenter 1839, 464 (from the riposte to Whewell 1837a in [Carpenter] 1838c, 342).

86. Carpenter [1840]; Carpenter 1888, 31. See Winter 1997, 35–43.

chemists Samuel Brown, George Wilson, and Lyon Playfair; anatomist John Goodsir; and geologists Andrew Ramsay and David Page—the brotherhood offered the prospect of mutual support ennobled with Forbes's Platonic mysticism.[87]

In between the drinking of "whiskey toddies" and the performance of initiation rites, the "fraters" discussed all manner of topics, often with a religious cast. Samuel Brown later recalled in his private journal that, while Forbes was not much of a churchgoer, he was a "kind of half-intellectual, half-æsthetical believer" whose religion was "sincere rather than earnest." He was not much given to theology, but he never "talked infidelities," even in their "rash youth." Indeed, the 1838 yearly meeting found him quoting Saint Paul's famous verse that the "invisible things" of God were "clearly seen" in the "things that are made." There was not, however, a party line on religion. Rather, the fraternity offered a secure and tolerant but strictly self-regulating environment that allowed ambitious young men the opportunity to explore new ideas beyond the control of their seniors. This they did with spirit.[88]

Samuel Brown offers a striking example. It is not clear whether his *Lay Sermons on the Theory of Christianity*, published pseudonymously in 1841 and 1842, were ever delivered to the Universal Brotherhood, but they were presented as emanating from "a company of brethren" and bore the brotherhood's mystic letters "οεμ." In the two "sermons" Brown expounded his "fidianism," arguing that Christianity was rooted in faith rather than reason and repudiating the natural theology of Paley and the Bridgewaters as unsound and dangerous. Brown had been brought up in the evangelical United Secession Church, but whereas his devout father and clerical grandfather had found a limited place for natural theology as a prelude to the Christian evidences, he drew on his "great master" Coleridge (from whom he borrowed the term "fidianism") to argue for a radical alternative emphasis on faith. Only to one armed with such faith could the world become the "sublimest commentary" on God's attributes. For this reason, while the Bridgewaters tended "to rob the religion of faith of its essential character," Brown considered his own developing scheme of exploring the unity of God in the unity of matter entirely legitimate.[89]

87. Gay and Gay 1997, 428–31; Wilson and Geike 1861, 188–202; Wilson 1860, 225–31; Turner 1868, 1:58–62; [Brown] 1857, 380–83.

88. Gay and Gay 1997, 442–45; Wilson 1860, 228; [Brown] 1857, 381; Reid 1899, 39. The quotation is from Romans 1:20.

89. On the Bridgewaters, see [Brown] 1841, 18, and [Brown] 1842, 22, 31, 38. On Coleridge, see [Brown] 1857, 384. On the views of Brown's family, see Brown 1782 and [Brown] 1856, 34 (cf. the "allegorical biography" in [Brown] 1842, 5–8).

Brown's "transcendental chemistry" and his career soon broke on the rocks of his unreplicable conversion of carbon into silicon. Forbes went on to an altogether more successful career before his early death, but he also viewed nature in terms radically differently from the dominating religious views of many of his elders. A Platonist, looking at nature as "a visible manifestation of the ideas of God" rather than as mere mechanism, he became indignant at "what he called 'Bridgewater writing,'—well-meant, but foolish expositions of the argument from design." Like his great friend and fellow transcendentalist John Goodsir, Forbes had been greatly impressed by Knox's lectures at the start of the 1830s, and he shared the lecturer's sense of the smallness of functional design. Devout Baptist George Wilson, though more open to the arguments of natural theology, also found the Bridgewaters shockingly narrow in their conception of divine action.[90] Thus, while many of their teachers saw in the Bridgewaters a means of baptizing the sciences they taught, this coterie of religiously minded students found themselves questioning the approach taken and striking out in a new direction.

For all that Edinburgh University had none of the religious tests, observances, or compulsory teaching of Cambridge or Oxford, the education it offered in science and medicine was significantly colored by Christian concerns. Moreover, for those who wished it, this was a university at which a particularly extensive course of study concerning natural theology was available. Its lectures on secular subjects and its vibrant student societies also offered opportunities to explore the issues further, not least in relation to the sciences. In the Athens of the North, so important in training the next generation of scientific practitioners, the Bridgewaters thus came to have a significant role in how that generation understood the religious bearings of their studies.

❋

Reviewing a new edition of Buckland's Bridgewater twenty years after its first publication, an anonymous writer in the *Saturday Review* reflected that the work had filled "a very important place in geological education" and had, for all its failings, "prepared another generation to expand its faith so far as to recognise that all true knowledge, whithersoever it may seem to lead, is a righteous pursuit, and cannot in the long run be found

90. Wilson 1862; Rehbock 1983, 68–73, 87–96; Wilson and Geikie 1861, 546–47; Knight 1978.

inconsistent with that divine truth of which it needs must form a part." Of all the treatises, Buckland's was perhaps the one that most lent itself to formal educational purposes. Indeed, while religious views had changed by then, Boyd Dawkins was still using the revised edition of the 1850s as a "class-book" at Owens College in Manchester in the 1890s.[91] However, others of the series were also widely used and referred to in teaching the sciences at Britain's universities and in justifying that teaching to skeptics. These were works that introduced a generation of students to a vision of the sciences as embedded within Christian orthodoxy and as inculcating Christian habits of mind.

Historians have tended to assume that the Bridgewaters were most useful within the context of Anglican education. However, while they were certainly used to defend the religious tendency of a scientific education at Cambridge, Oxford, and King's College London, there were many in those institutions who viewed them with suspicion. By contrast, in the less theologically exacting context of University College London, their broadly based references to divine design contributed to a more general strategy by which scientific education was represented as supportive of students' preexisting religious views. In Edinburgh, too, the series could be appropriated to the diverse religious views of professors and students alike, offering opportunities for both to explore how the sciences should be related to religious concerns. Of course, not all such appropriations were equally legitimate. In the late 1850s the professor of chemistry at Owens College, Henry Roscoe, was shocked to find that one of his students had not only forged his scientific results but that he had also passed off, both on him and on the monthly *Chemist*, "a prize essay cribbed from the Bridgewater Treatises and Dr Chalmers's sermons."[92] Yet the Bridgewaters had proved pivotal in elaborating a vision of the intellectual virtues and social role of the Christian man of science and in securing patronage for some. Moreover, in a momentous decade, they helped to shape how existing practitioners as well as neophytes understood the proper way to conduct their scientific inquiries, as the next chapter explores further.

91. Gordon 1894, 56; "The Geological Bridgewater Treatise," *Saturday Review*, 27 November 1858, 537–39, on 538.

92. Roscoe 1906, 105–6.

✷ CHAPTER 8 ✷

# *Religion and the Practice of Science*

*Now the peculiar point of view which at present belongs to Natural Philosophy [ . . . ] is that nature, so far as it is an object of scientific research, is a collection of facts governed by* laws *[ . . . ]. And it must therefore here be our aim to shew how this view of the universe falls in with our conception of the Divine Author.*

WILLIAM WHEWELL, Astronomy and General Physics, *1833*[1]

Some of the Bridgewater Treatises were clearly useful to scientific educators and students for the manner in which they offered synthetic accounts of the state of rapidly developing sciences such as paleontology, comparative anatomy, and chemistry. Still more attractive, however, was the manner in which the series addressed fundamental questions concerning the sciences, such as what it meant to know things scientifically and how scientific knowledge related to other kinds of knowledge, including religious knowledge. These were among the great questions of the age and, like other reflective scientific treatises addressed to wide audiences, the Bridgewaters offered their own distinctive "visions of science." Moreover, the fact that they presented a multiplicity of such visions meant that they were well suited to stimulate lively discussion.[2] In doing so, the series not only helped to shape the understanding of the rising generation but also enabled more established men of science to reflect on the religious meaning and legitimacy of what they were doing. Indeed, this was a period in which even practitioners whose religious views were unorthodox found themselves having to wrestle with such questions if for no other reason than that they needed to allay concerns or curry favor in an

1. Whewell 1833b, 3.

2. On differing "visions of science" and the phrase itself, see Secord 2014; see also Yeo 1993, Hull 2003, and Snyder 2006, 2011.

overwhelmingly Christian society, and the Bridgewaters were a valuable means of so doing.

One way in which the visions of science in the Bridgewaters went to the heart of scientific practice related to the scope and limitations of natural laws. Whewell's was the most forthright of the series in making laws the defining feature of modern science, but most of the others echoed this perspective to a greater or lesser extent, insisting that the expanding scope of science's laws was a boon to Christianity rather than a difficulty. Yet in seeking to show that such a view was consistent with "our conception of the Divine Author," they naturally drew attention to the momentous question of whether certain aspects of the history of creation lay beyond the bounds of such laws. In particular, several argued for God's direct action at certain moments in the history of creation, especially in the origination of new species. What was the reader to make of these complications? Just where did the limits of nature's laws lie? This was arguably the most vexed of all the questions concerning the character and connections of scientific knowledge that preoccupied men of science in the 1830s, lying barely hidden beneath the religiously complacent rhetoric of public pronouncements. The Bridgewaters offered an opportunity for scientific readers to evaluate the options, and many asked themselves the question repeatedly as they read. Some did so largely in private, reflecting on the proper way to conduct themselves as men of science while they sought to formulate new theories. Others were much more vocal about their reflections, outlining their disagreements with particular Bridgewaters or developing the emphasis on the uniformity of nature that they found there at the expense of the emphasis on singular acts of creative power.

The Bridgewater Treatises also raised fundamental questions for scientific practice in their emphasis on the design evident in nature. It was one thing to discuss the appearances of design in the created order in a religious context, but the question for scientific practitioners, especially those working in the sciences of life, was whether it should play any part in scientific observation. Some claimed that the perception of design in nature was crucial to scientific advance, but others were sure that, at least in some forms, an interest in identifying design distorted observation and reasoning in ways that were scientifically damaging. This was among the more heated methodological debates of the period, and, in engaging with the Bridgewaters, those at the forefront of research, including Richard Owen and William Carpenter, as well as Charles Darwin, found in the series an array of positions to consider and evaluate. For such readers, the treatises made useful contributions to a key debate about scientific practice that ultimately lay at the heart of the project to understand the history of living beings.

## *A Divine Programmer*

Whewell's emphasis on laws in his characterization of science echoed the account recently offered in his old friend John Herschel's *Preliminary Discourse on the Study of Natural Philosophy*. Herschel had affirmed that the "objects of enquiry" of the natural philosopher were the natural laws that were an expression of the "constant exercise" of God's "direct power." This kind of study, he urged, was ennobling, serving to render doubt of God's existence and principal attributes "absurd" and atheism "ridiculous." There was, therefore, no reason for scientific researchers to be constrained by prior convictions, since scientific knowledge would not undermine divine truth. Just like several of the Bridgewater authors, however, Herschel noted that there were limits to the scope of scientific inquiry in regard to "the origin of things." Speculating on the process of creation was "not the business of the natural philosopher."[3]

Herschel's scientific reputation and profile stood very high, and his inexpensive, six-shilling work carried his message to a wide audience. The Bridgewaters amplified that message, developing it in relation to particular examples. Whewell's treatise, for instance, while arguing that Laplace's lawlike account of the origin of the solar system would not, if true, diminish our sense of divine agency, pointedly referred to the divine intelligence that must have been responsible for the initial conditions on which such laws would have operated. Chalmers's Bridgewater went further. As ever greater numbers of natural phenomena were explained in terms of natural laws, it declared, the natural theologian could rest secure in the knowledge that the original "dispositions" of matter were also designed and that divine "fiat" came into play in the case of the origin of new species and the reestablishment of the living world following each geological catastrophe. Similarly, while Buckland's *Geology and Mineralogy* suggested that the physical history of the globe, including such catastrophic upheavals, could be referred to natural laws, it nevertheless also made much of the role of "Creative Interference" in the earth's repopulation following those catastrophes.[4]

The Bridgewaters thus effectually publicized the religious legitimacy of

3. Herschel [1831], 7–8, 13–14, 37–39.

4. Buckland 1836, 1:49, 586; Chalmers 1833, 1:16–31; Whewell 1833b, 181–91.

science's ambition to expand the explanatory scope of natural laws while at the same time making abundantly evident that such explanations had limitations. For readers actively engaged in scientific researches, the series consequently offered an opportunity to explore the proper limits of natural laws, especially in regard to the history of creation, where it was far from clear where the limits should lie. One such reader was a young Charles Darwin, whose private reading of several of the Bridgewaters following his return from the *Beagle* voyage fed into the theorizing documented in his notebooks and ultimately to the quotation from Whewell's Bridgewater at the opening of *Origin of Species*. But before we examine Darwin's experience, we need to consider that of another reader whose published response to Whewell's Bridgewater itself became important for Darwin, namely, the mathematician Charles Babbage.

Babbage's reading of Whewell's Bridgewater was rooted in personal knowledge of its author. The two had only been separated by a couple of years as Cambridge undergraduates, and Babbage was now the university's Lucasian Professor of Mathematics, although he was largely absent from the university and did not lecture. They were nevertheless poles apart in their religious and political views as well as their social origins. Wealthy banker's son Babbage had from his youth chafed against the rigidity of religious doctrine, unafraid to question or even to poke fun. Moreover, his reformism in both politics and science, combined with his belief in the mechanization of economy and society, was at loggerheads with Whewell's social conservatism. His mindset was epitomized by his wholesale commitment to the analytical mathematics that had been so successfully developed in Paris but which was still considered too mechanistic in Cambridge (they considered it "pure D-ism," Babbage punned, in reference to the Continental differential notation). It was also embodied in his calculating engine, designed to produce mathematical tables, to which much of his time and vast sums of government money were devoted over the course of the 1820s (fig. 8.1).[5]

These differences underpinned Babbage's strong reaction to Whewell's new work. Babbage dined with the author while visiting Cambridge to examine the Smith's prize in Lent 1833, and Whewell probably arranged for him to have a copy of his Bridgewater on its publication in March. But, as one of Britain's most enthusiastic adopters of analytical mathematics, Babbage found Whewell's widely welcomed assessment of the religious tendencies of different scientific habits of mind both offensive and dangerous. No doubt it was with Babbage in mind that Whewell was soon com-

5. Babbage 1864, 21; Schaffer 2003.

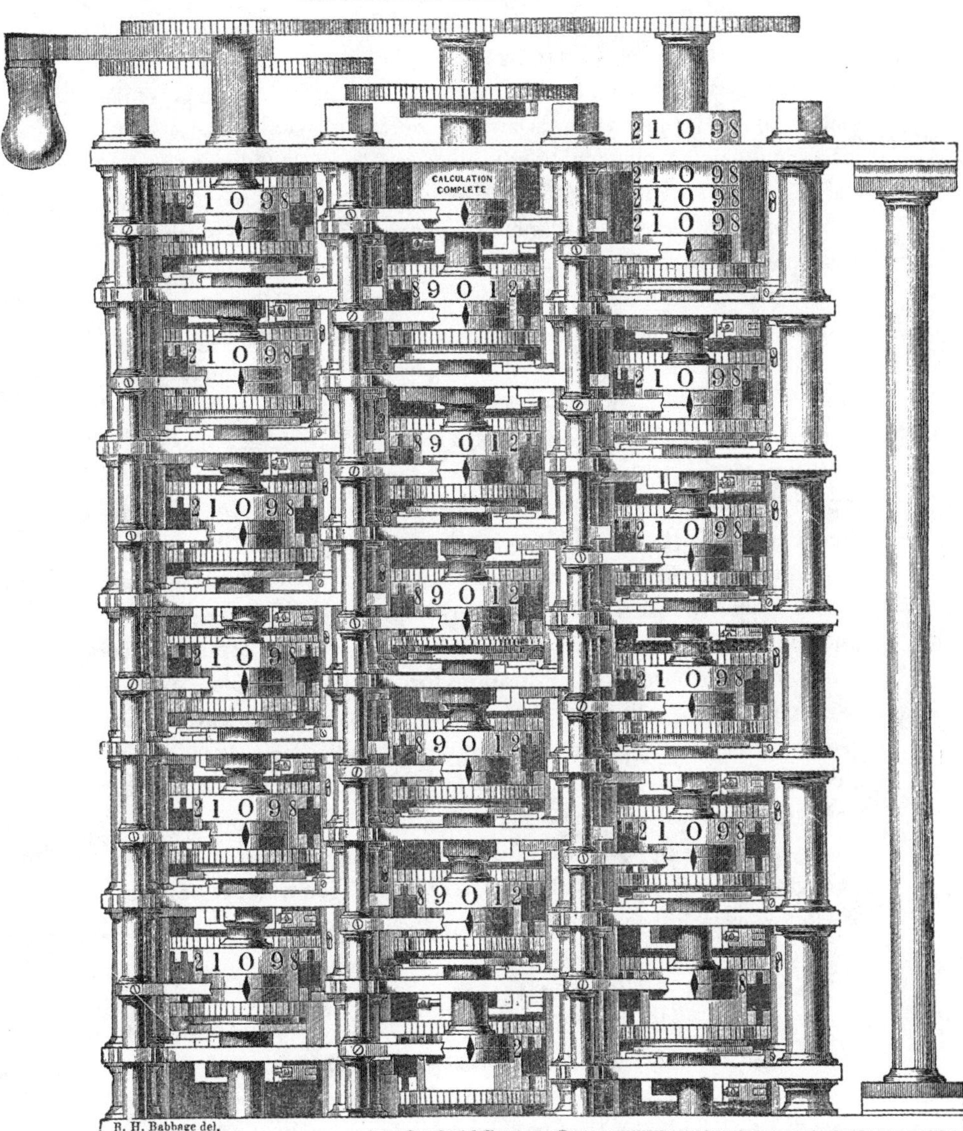


Small Portion of Mr. BABBAGE'S DIFFERENCE ENGINE, No. 1 (CALCULATING MACHINE), the property of Government; in the Museum of King's College, Somerset House (p. 142).

FIGURE 8.1 "Small Portion of Mr. Babbage's Difference Engine, No. 1 (Calculating Machine)." Wood engraving after B. H. Babbage. From John Timbs, *Stories of Inventors and Discoverers in Science and the Useful Arts,* 2nd ed. (London: Lockwood, 1863), opposite 142.

menting on the "complaints" he had received on the subject from "one or two" of his "mathematical friends," although he was still hoping at that point to win them over. According to Whewell it was inductive habits of mind that gave the scientific man an acute sense of the divine presence that contingently upheld natural laws. By contrast, those who engaged primarily in following out deductions from natural laws—notably the analytical devotees of the Laplacian school—too often came to see those laws as necessary. Whewell's treatise consequently asserted that "mechanical philosophers and mathematicians of recent times" had nothing helpful to say about God's "administration of the universe" and might be viewed as positively disadvantaged in this regard. Not surprisingly, Babbage considered this assertion unjust. Whewell's Bridgewater, he later claimed, ranked with Nolan's far different book in offering support to the prejudice "*that the pursuits of science are unfavourable to religion*."[6]

Within weeks Babbage's response was taking shape in the form of a demonstration involving his calculating engine. Part of Babbage's first difference engine had been assembled as a demonstration model the previous year, and in December 1832 it was placed on display in the drawing room of his West End house near Regent's Park. Having already been a topic of much public comment, it now became one of the wonders of fashionable London, with Babbage's demonstrations forming the centerpiece of what Harriet Martineau called his "glorious soirées." These, as Lyell reported, were "very brilliantly attended by fashionable ladies, as well as literary and scientific gents" and were places to meet "with persons high in all professions, and with distinguished foreigners."[7] Almost from the start, Babbage took the opportunity of these fashionable performances with his engine to offer an alternative vision of God's agency in the universe, developing as he did so his answer to Whewell's account of the religious tendency of mechanical habits of thought.

As early as June 1833, Babbage was using his engine to claim that when a creative mind had sufficient foresight, apparent miracles could be made to occur within a visibly lawlike sequence. One of the earliest to hear about these developing notions was Lord Byron's seventeen-year-old daughter, Ada, who visited him for the first time on 21 June. Her mother reported, concerning what she and Ada considered "the *thinking* machine," that Bab-

6. Babbage 1837, x; Whewell 1833b, 334; WW to D Gilbert, 20 May 1833, in Enys 1877, 30–31 (transcription corrected from the original in a private collection); Schaffer 2003, 285.

7. Lyell to Herschel, 1 June 1836, in Lyell 1881, 1:464–69, on 466; Martineau 1877, 1:354–55; Swade 2001, 65–66, 72–87; Hyman 1984, 128–29; Snyder 2011, 189–95.

bage had stated that "it had given him notions with respect to general laws which were never before presented to his mind—For instance, the Machine could go on counting regularly, 1, 2, 3, 4, &c—to 10,000—and then pursue its calculation according to a new ratio." Just as Babbage could set up his calculating machine to vary its law of calculation after an extraordinary number of regular instances, the argument went, so a divine creator might establish natural laws to vary after a similarly vast number of regular instances in ways that would appear miraculous to a scientific observer. According to Babbage's autobiography, he immediately started applying this radical notion to "the successive creations of animal life, as developed by the vast epochs of geological time." Outlining the idea to political economist Thomas Malthus and two visiting Irish natural philosophers he was, he claimed, gratified by the "flash of intellectual light" that it provoked.[8]

Babbage's preordained miracles generated extensive interest, but his decision to commit his reflections to paper in the summer of 1836 seems to have owed much to his notorious sensitivity to what he took to be personal slights. In any case he now conceived of a book that would answer Whewell's charge by drawing together a number of his reflections to show the riches that mathematical reasoning could offer to Christianity. "The Devil," Lyell told Herschel, was to "have his due."[9] By the following January, several of Babbage's friends—notably Lyell, fellow geologist William Henry Fitton, and iron founder James Maclaren—were looking over proofs and concluding that the work was in many ways ill judged.

Fitton was especially blunt about the wisdom of Babbage publishing his "tract" at all. First, Babbage was offering what his subtitle termed "*A Fragment*"—an increasingly disjointed set of chapters and incomplete notes containing more or less developed ideas. He was, he claimed, too busy with "other pursuits" to develop them further. The main text ran to fewer than thirty thousand words, although John Murray padded it out as much as he could using white space and thick paper. Second, Babbage could not be dissuaded from making the treatise a blatant riposte to Whewell's work, designating it a *Ninth Bridgewater Treatise* and offering an epigraph from Whewell on his title page. Some copies of the work even included a variant title page with an additional, outrageous epigraph from the Arabian Nights: "The ninth pedestal redoubled his amazement, for it was covered with a piece of white satin, on which were written these

8. Babbage 1864, 387–91; Toole 1992, 52; N Byron to W King, 21 June 1833, in Moore 1977, 43–44.

9. Lyell to Herschel, 1 June 1836, in Lyell 1881, 1:464–69, on 467. For a sympathetic account of Babbage's character, see Martineau 1877, 1:354–55.

words:—'It cost me much toil to get these eight statues; but there is a ninth in the world, which surpasses them all: that alone is worth more than a thousand such as these.'" Worse still, his introduction provocatively implied that the professional and financial interests of the Bridgewater authors were far more likely to have influenced their judgement than Babbage's deductive habits were to have adversely affected his own.[10]

Babbage represented the Bridgewaters as failing to pursue their argument "to a sufficient extent." It was in the extraordinary foresight of a divine lawgiver who need not intervene in the natural order of things to achieve his purposes, he claimed, that the argument from design found its apotheosis. Many of those not deeply versed in science were inclined to reduce divine action to the measure of human action and to think of God as "perpetually interfering" in the laws he had ordained.[11] Babbage's object was to outline a higher view of divine action in which all was referred to laws, including the apparently miraculous, and his primary example was the party piece with his difference engine. Moreover, he immediately applied this conception to the "vast cycles" of geological change and the origin of new species before defending it from possible charges either of fatalism or of contradicting the scriptural account of the creation. Subsequently, a rather disordered series of chapters explored further his novel perspective and its ramifications, including an analysis of the lawlike character of what were called miracles, an attempt to show that probabilistic thinking could offer an answer to Hume's argument against miracles, and an argument that miracles were, on a priori grounds, more likely to occur than not.

Babbage's account of the uniformity of nature, in which apparent miracles such as the creation of new species were assimilated to higher laws, pushed the perspective of the Bridgewaters into dangerous territory. It was one thing to make such suggestions at fashionable soirees, another to do so in print. From Babbage's pen, the claim that "the toleration of the fullest discussion is most advantageous to truth" generated a certain frisson.[12] However, he fortuitously hit on a powerful expedient for giving his novel perspective authority and respectability. In a masterstroke he quoted

10. Babbage 1837, ix, xvi–xviii; Campbell-Kelly 1989, 9: [iv], 6. On the imprudence of publishing, see Fitton to Babbage, 2 and 4 January [1837], 23 January 1837, and [1837], and Maclaren to Babbage, 2 February 1837; on the blatant book making, see Fitton to Babbage, 11 September 1837 and 18 February 1838, BL, Add.Ms.37189, fols. 261–64, Add.Ms.37190, fols. 19–20, 32, 173–74, 269–74. See also Clark and Hughes 1890, 1:483–84, and Wilson 1972, 461–62.

11. Babbage 1837, vii, 24.

12. Babbage 1837, 28.

a surprising statement from a letter sent to Lyell by Babbage's old friend Herschel to the effect that he also thought that new species had appeared according to natural laws.

Babbage's use of Herschel's authority as the leading Christian theorist of the philosophy of creation came about in an incidental manner. With a view to making his "*Fragment*" into a decently sized volume, Babbage added eighty pages of notes, the relevance of most of which would only have been evident to the well-informed reader. Among these were two introducing Babbage's views concerning the role of a "central heat" in explaining the rise and fall of geological formations, which had first been expounded, through Lyell's encouragement, at the Geological Society in 1834. The knowledgeable reader would see the connection: here was a mechanism, rooted in the laws that governed cooling bodies, that accounted for key phenomena in geological history. Next came a note containing extracts from two related letters—one from Herschel to Lyell, dated February 1836, which described rather similar but independently derived views, and a second, clarifying letter to Roderick Murchison, which Herschel suggested might be read informally at the Geological Society.

Lyell had told Babbage about Herschel's letter in the summer of 1836, but it was not until the spring of 1837 that Babbage saw it as he finalized his text for the press. With Herschel far away undertaking astronomical observations at the Cape of Good Hope, Babbage hurriedly obtained Lyell's and Murchison's permission to publish extracts. In so doing he could not resist quoting the opening comments of Herschel's letter concerning the dignity that Lyell had given in the fourth edition of his *Principles of Geology* to that "mystery of mysteries, the replacement of extinct species by others." Herschel had observed that in this, as in all the creator's other works, analogy led the scientific observer to expect him to act through "intermediate causes" so that the origin of species, "could it ever come under our cognizance, would be found to be a natural in contradistinction to a miraculous process." Herschel considered that there were no known indications of such a process actually in operation at the present time, but Babbage reprinted his statement with glee, exulting in the support his own views received from their "almost perfect coincidence" with those of the renowned philosopher.[13]

The inclusion of such a claim in print was a bold step, and Lyell had taken some persuading to allow it. Telling Herschel the previous summer

13. Babbage 1837, 202–4; Lyell to Babbage, [Spring 1837], BL Add.Ms.37190, fol. 179. The original letter from Herschel to Lyell, 20 February 1836, is reproduced in full in Cannon 1961, 304–11.

how pleased he was to find him positive about the role of "intermediate causes," the geologist had confessed that he had himself "left this rather to be inferred" in his *Principles,* "not thinking it worth while to offend a certain class of persons by embodying in words what would only be a speculation." On seeing Babbage's proofs, Lyell welcomed the "argument of changes of law" for the light it shone on some of his "geological speculations" and, while recognizing that some would not like it, he claimed that Babbage's conception of God was "much higher than theirs." Nevertheless, he was very clear that one could not be "too careful" when publishing a private letter "without asking leave." As Lyell told Herschel, however, Babbage had shown him that he had introduced the quotation "as a counterpart of a passage from Bishop Butler, and that in such company no-one could be otherwise than correct and orthodox." Even Whewell, Lyell suggested, in his new *History of the Inductive Sciences,* seemed almost willing to consider that new species were introduced according to natural laws.[14]

Lyell's reading of Whewell's new book was rather optimistic. In this three-volume work, published just weeks before Babbage's treatise appeared at the end of May 1837, Whewell offered a much fuller and more circumspect account of the province of natural laws than that developed in his Bridgewater, picking up on reservations he had expressed in anonymous reviews following the initial publication of Lyell's *Principles.* In sciences concerned with historical causes—what he termed "palætiological sciences"—the prospect of a thoroughgoing lawlike account seemed like a vain hope. And having delivered himself of this authoritative judgement, Whewell may have considered that he had done enough to counter Babbage's more radical approach. In any event, the short open letter he circulated in response to Babbage's fragmentary treatise was highly measured in tone, restating his analysis of the tendency of deductive habits, denying that it had had any negative effect on the reputation of the sciences, and welcoming Babbage's contribution, with its uniquely "original" approach.[15]

The publication of Babbage's *Ninth Bridgewater Treatise* nevertheless caused a stir. Babbage sent large numbers of presentation copies to dignitaries, including both the dying king and, on her eighteenth birthday,

14. Lyell to Herschel, 1 June 1836 and 24 May 1837, in Lyell 1881, 1:464–69, on 467; 2:11–13, on 11; Lyell to Babbage, [Spring 1837], BL Add.Ms.37190, fols. 180–81, 185–87. For Butler, see Babbage 1837, 175.

15. Whewell 1837b, 7; Whewell 1837a, 3:380–88. Babbage responded in Babbage 1838, iv–vii. For a penetrating exploration of Whewell's changing position, see Hodge 1991.

Princess Victoria. Among the numerous letters of thanks was one from Thomas Chalmers, who considered that the new work left his own argument "grounded on the distinction between the Laws and Dispositions of Matter" untouched. However, even friends found the work to be an extraordinary and eccentric hodgepodge, and to Charles Bell it seemed to lack "the method that belongs to madness." Reviewers agreed with that assessment, and while Babbage's innovative perspective won him praise from Unitarians, a number of more orthodox reviewers were critical of its tendency to undermine belief in special providence, replacing it with a mechanistic vision of the created order. Indeed, Babbage was sufficiently concerned that in a second edition the following year he added a new chapter on "the nature of a superintending providence."[16]

Babbage was by no means a sympathetic reader of the Bridgewaters, but his work makes clear precisely how the perspectives that they offered on the religious significance of natural laws became part of an ongoing and increasingly demanding conversation among scientific practitioners as readers read and responded to them. The liveliness and importance of these conversations is particularly well illustrated by Darwin's experience. We know that a generation later Darwin took the strategic step of quoting at the start of his great work a pithy statement from Whewell's Bridgewater concerning the lawlike character of divine action in the universe. But this comes as little surprise when we realize that reading and being party to discussions of the Bridgewater Treatises represented a significant element of his program of scientific work during the 1830s as he developed his philosophy of creation. Following his return from the *Beagle* voyage, Darwin's reading was extensive, manifesting an impressively voracious capacity for engaging creatively with all manner of publications. Among these, however, the innovative reflective treatises aimed at wider audiences held an especially important place because of the way that they helped him to reach a determination about what constituted an acceptable theory.[17]

Darwin's reading of such books began before he embarked on the *Beagle*. He read Herschel's *Preliminary Discourse* at the start of 1831, just as he approached the end of his university studies, later remembering it as having had a pivotal role in firing his zeal for science. He had evidently

16. Babbage 1838, 141–48; CB to PMR, 2 March 1838, RSL, MC.2.278; TC to Babbage, 30 October 1837, BL, Add.Ms.37190, fol. 295 (this volume of Babbage's manuscript correspondence includes many other responses to the work). Many of the reviews are listed in Topham 1993, 553–615; see also Yule 1976, 245–48.

17. On Darwin's reading as scientific work, see Secord 2000, 426–33, and Topham 2004c, 433–35.

been primed to expect great things of the work, perhaps by Whewell, whom he met at the scientific soirees of professor of botany John Henslow. In any case, it was with Herschel's analysis of the nature and significance of scientific knowledge reverberating in his head that Darwin headed off on the *Beagle* the following December. Moreover, he was soon enriching his thinking further with the three volumes of Lyell's *Principles*, which chimed nicely with the emphasis on causal laws he had learned from Herschel.[18]

Darwin's library on the *Beagle* was necessarily limited, and, while his sisters kept him abreast of news about the Bridgewaters as they were published, he was not in a position to read them until after his return to Britain. Even then, his first encounters with the series were largely through conversation, not infrequently involving the authors themselves. Darwin landed on 2 October 1836, just as Buckland's Bridgewater became the talk of the town. While he was soon very busy with organizing his specimens and beginning to prepare publications, he also found himself feted by scientific worthies and thrust into the midst of their conversations, especially around the Geological Society. Within weeks, Buckland was being called in to look at some of Darwin's fossil specimens. The Oxford man was sufficiently impressed to issue a "supplementary note" on the subject the following spring as a continuation of the second edition of his Bridgewater. By that date Darwin had been present at the Geological Society's February anniversary meeting to hear Lyell as outgoing president praise Buckland's work in warm terms, especially for its invaluable "general view" of paleontology. More generally, mingling with members of the society gave the young man ample opportunity to judge the significance of the new work.[19]

Darwin's new friend Lyell had wanted him present at the Geological Society anniversary meeting to support Whewell's election as incoming president, and he was in turn soon being asked by Whewell to be the society's secretary. Billeted in Cambridge for much of the winter while he sorted his specimens, Darwin had ample opportunities to catch up with a man whose star had risen significantly during his absence, especially following his Bridgewater success. Whewell was likewise impressed by the young man whose work he now discussed at the Cambridge Philosophical Society, becoming part of his patronage network by supporting his appeal for government funding in order to publish illustrations of his *Beagle*

18. Ruse 1975, 2000; Snyder 2006, 185–202; Browne 1995, 126–30; Darwin to WD Fox, [15 February 1831], in *CCD*; Darwin 1958, 67; Herschel to WW, 15 February 1831, TC, Add.Ms.a.207[19].

19. Lyell 1838, 517; Buckland 1837, 603; Darwin to Henslow, [30–31 October 1836], S to C Darwin, [23] May 1834, in *CCD*. See also Rudwick 1982.

specimens. Darwin was, however, also hearing about Whewell's Bridgewater from other sources. Babbage naturally wanted the promising young naturalist at his famous parties, and by February Lyell was urging him to attend with the enticement that there would be "a good mixture of pretty women." He doubtless prevailed, and Darwin thus probably witnessed Babbage's miraculous party trick and perhaps heard first hand of its incorporation into the forthcoming *Ninth Bridgewater Treatise*. Lyell also introduced him to Herschel's letter of the previous year that was shortly to be extracted in Babbage's *Fragment*, and by the summer he was asking Darwin to lend Babbage his copy of Whewell's new *History*.[20]

Thus, while there is no evidence that Darwin read any of the Bridgewaters during his first six months back in Britain, he nevertheless found himself immersed in the lively conversation they fostered concerning nature's laws and the creator's mode of action in the universe. Meanwhile, his thinking about species was beginning to take shape. By late March 1837 he made his first notebook entry concerning the possibility that "one species does change into another." At the start of the summer, with his *Beagle* diary written up, he opened his first notebook on transmutation. The succeeding pages record snatches from his increasingly earnest reading program in which publications were carefully scrutinized for the evidence they might yield. He soon worked his way through Kirby's Bridgewater and was telling himself to "Read Buckland." Notes over the next few years show him reading not only Buckland's but also Bell's Bridgewater (alongside Lamarck's *Philosophie zoologique*) and then Roget's.[21] But Darwin's concern was not merely with factual titbits. There were underlying questions about the character of scientific knowledge and the nature of scientific work that had to be addressed. As discussed later in this chapter, one such question was the role that the perception of design should play in scientific observation, but in relation to the question about the role of laws in the history of creation, it was Whewell and Babbage's recent works that came to the fore.

Darwin was already taking seriously the idea that "the Creator creates by [ . . . ] laws" by the autumn of 1837, apparently before he had read either book. But as he continued to develop this view, his reading program soon encompassed both. By the spring of 1838 he had read Whewell's Bridge-

20. Charles to Caroline Darwin, 27 February 1837, Darwin to Babbage, [June–September 1837], Lyell to Darwin, 13 February 1837, Darwin to WW, [10 March 1837], Darwin to Henslow, 18 [May 1837], in *CCD*. See also Snyder 2011, 189, 217–20.

21. *Notebooks*, B141–43, B149, RN130; CD's Reading Notebooks, *CCD* 4:447, 449, 457, 460, 475, 476.

water, considering it important enough to reread two years later, and in the autumn he read Whewell's *History* while also rereading Herschel's *Preliminary Discourse*. Soon afterward, he read the second edition of Babbage's *Ninth Bridgewater Treatise*, observing with delight, "Herschel calls the appearance of new species. the mystery of mysteries. & has grand passage upon problem.! Hurrah.—'intermediate causes.'" Here was clear confirmation that the notion of creation by law was scientifically respectable—even as applied to new species—and Darwin took mental note. The opening pages of *Origin* included not only Whewell's assertion of the lawlike character of divine action but also the sonorous phrase "mystery of mysteries" that Babbage had extracted from Herschel's letter and which Darwin now attributed to "one of our greatest philosophers." As this implies, Darwin looked to Whewell and Herschel as philosophical heavyweights, and it is widely recognized that their writings did much to shape his developing ideas about scientific method. More particularly, however, their pronouncements clearly helped to inform and confirm his vision of the process of creation as being governed by natural laws—a vision that went to the heart of his theorizing about species.[22]

In soirees, scientific societies, and college halls, the Bridgewater Treatises spawned lively conversation concerning the proper character and limits of scientific inquiry. Above all, they focused attention on natural laws and the extent to which the history of creation should be expected to come within the scope of such laws. However, the confident pronouncements of several of the authors concerning the religious legitimacy of lawlike accounts of aspects of that history invited further augmentation. In the case of the somewhat bellicose Babbage, his provocative comments on the topic at private parties ended up being publicized in startlingly forthright print. But even the cautious Darwin found in his private research notebooks the opportunity to build on the wider scientific conversation as he developed what he trusted would be regarded as a properly scientific theory of species transmutation.

## *The "Higher Law" of Development*

Darwin further codified the notion of creation by law in the outlines of his theory that he penned in the early 1840s, but his engagement with these

22. *Notebooks*, B98, C72, E59; CD's Reading Notebooks, *CCD* 4:456, 459. On creation by law, see Brooke 1985; on Darwin's debt to Herschel and Whewell, see Ruse 1975, 2000.

questions famously continued to be a private matter for many years. There were, however, two readers of the Bridgewaters who did not shy away from commenting in print on how their emphasis on creation by law might be developed to encompass a natural process of species change. Strikingly, one did so from the secure social position of an ordained Oxford don. The other, however, considered it necessary to maintain the protection of anonymity, and his published comments provoked a notable sensation in early Victorian Britain.

The Oxford don was geometry professor Baden Powell, who offered a carefully studied response to the Bridgewaters in February 1838 in his *The Connexion of Natural and Divine Truth; or, The Study of the Inductive Philosophy Considered as Subservient to Theology*. As we have seen, this somewhat embattled cleric's changing sense of the apologetic challenge facing the Church of England had left him convinced that natural theology must form the foundation of any defense of Christianity that was designed to reach the working classes of industrial Britain. That perspective was now set out in a reflective treatise focused explicitly on the relevance of scientific to theological knowledge, which was issued from the seat of Anglican orthodoxy with the assistance of John William Parker, the Society for Promoting Christian Knowledge's official publisher and printer to Cambridge University. In it, Powell urged readers to follow his own progressive reading of Whewell and Babbage's Bridgewaters and to allow species origins to be incorporated within the domain of scientific laws.

On the whole, the Bridgewaters and numerous other works published in their wake had left Powell underwhelmed, he told his readers. Such books were largely preoccupied with amassing "*particular instances* of design" culled from the several sciences rather than with analyzing the underlying "philosophy of the argument." Whewell's Bridgewater and especially Babbage's work were, together with Brougham's *Discourse of Natural Theology*, the least deficient in this regard. What Powell now offered was a thoroughgoing account of how the inductive philosophy worked, how natural theology was based on it, and how revealed religion was founded on that. His account of the inductive philosophy focused on the role of induction in connecting physical causes to universal laws that were grounded in a universal intuitive belief in the uniformity of nature. Moreover, he claimed that it was above all the law-governed character of natural phenomena that enabled the natural theologian to infer the existence and attributes of a divine designer. This assertion was bookended by quotations from Babbage and Whewell's works. The *Ninth Bridgewater* seemed especially suggestive to Powell, who claimed that its importance had been overlooked as a result of its peculiarities. However, Powell also found support for his law-based conception in Whewell's Bridgewater, drawing on some of his Cambridge contemporary's examples.[23]

Powell knew that some of the readers of Babbage's *Fragment* had doubted its religious tendency, but he did not hold back from pushing his view of the subject to a similar extent. Ultimately, questions of the origins of things lay beyond the bounds of inductive science, he asserted. But he just as confidently avowed that the question of species transmutation was full of promise and "fairly open to philosophical discussion." The clear implication was that, ultimately, the appearance of new species would be explained according to law and that natural theology would be the better for it. Unlike Babbage, Powell had promised to avoid "specific controversy" or criticism of recent authors, but statements such as this made perfectly clear where he considered the Bridgewaters to be deficient. Buckland's account of the progressive history of life on earth was excellent so far as it went, Powell suggested, but his ignorance of the natural process by which new species came into being might yet be overcome to the benefit of theists.[24]

The spectacle of an Anglican clergyman in the bosom of the University of Oxford offering such an olive branch to transmutation theorists might have been expected to provoke a storm, but the response was strikingly mild. Many periodicals had recently visited the question of the scientific status of natural theology in reviews of Brougham's *Discourse of Natural Theology*, the author of which had, as lord chancellor between 1830 and 1834, become one of the country's best-known public figures. By contrast, Powell's previous publications and position in Oxford doubtless contributed to his book being viewed as drily theological. Moreover, his former High Church supporters arguably had bigger fish to fry as controversy over Tractarianism intensified. Thus, although his book was welcomed in the liberal weekly *Athenæum*, which praised the author for exhibiting "no ordinary share of moral courage," it was left to the Unitarian *Christian Teacher* to offer more extensive comment. And while Powell was applauded there for being "enlightened, philosophical, and courageous," even the new breed of Unitarians found his approach to Christianity too rationalistic.[25] Yet whatever reviewers might say, Powell's "courageous" public engagement with the Bridgewaters in expanding the philosophy of creation reflected a wider pattern in private among scientific practitioners, as is well exemplified by Darwin's experience.

23. Powell 1838, x–xii, 51–52, 113, 157, 172–76, 202. See Corsi 1988, esp. 178–93.

24. Powell 1838, xii, 148–54.

25. "The Connexion of Natural and Divine Truth," *Christian Teacher*, n.s., 1 (1839): 160–77, on 160; "Our Library Table," *Athenæum*, 2 June 1838, 390.

The response to Baden Powell's work was notable for its subdued character, but another work that engaged publicly with the topic suffered a rather different fate. This was the grand evolutionary synthesis titled *Vestiges of the Natural History of Creation* that rapidly became a cause célèbre following its publication in 1844. Taking to its limit the notion of "creation by law" that the Bridgewaters had rebaptized, the work offered an account of the history of the cosmos in which all aspects, from the stars through to the human species, were to be accounted for in relation to the unbroken operation of natural laws. This depiction, *Vestiges* suggested, offered an altogether more dignified view of God's action in the universe than that ordinarily presented by scientific authors. However, with its highly speculative theory of species origins further tainted by other scientific blunders and limitations, the work prompted as fierce a backlash from men of science as from anyone. For all that its findings were presented as developing themes from the Bridgewaters, *Vestiges*'s captivating but scientifically unreliable account arguably made it more, rather than less difficult for naturalists to advocate publicly the origin of new species according to natural law.[26]

The book's origins lay in the late 1830s, when its anonymous author, Scottish popular journalist and pioneering cheap publisher Robert Chambers, first began to consider how a lawlike account of the entire created order might be constructed. Considering scientific education to be critical in effecting social improvement, Chambers had become increasingly busy in the 1830s writing scientific articles both for *Chambers's Edinburgh Journal* and for a new series of cheap educational works issued by W. & R. Chambers. At the same time, an encounter with phrenology led him to place particular emphasis on the expanding reach of natural laws. He was drawn into phrenology through George Combe, whose recently published *Constitution of Man* argued that the new science enabled people to bring themselves into alignment with the fixed laws of human nature, leading to individual and social progress. Chambers quickly appreciated this vision of human improvement rooted in natural laws, and while remaining cautious concerning phrenology's doctrine of bumps (the "organology"), he actively promoted a "phrenological metaphysics" in relation to mind and morals. By 1835 he had published with his brother a "people's edition" of *Constitution of Man* that helped make it one of the best-selling works of the period, and he had begun to prepare his own work on the mind based on phrenological principles.[27]

26. Secord 2000.

27. Chambers to Combe, 25 November 1835, NLS MS.7234, fols. 140–41; Chambers to A Ireland, [November 1837], NLS Dep. 341/110, fols. 9–10; Gibbon 1878; Cooter 1984; Secord 1989a; Van Wyhe 2004.

It is striking that in seeking to coax Chambers into the phrenological camp, Combe had argued for the superior religious perspective that phrenology's laws brought. In December 1833, as Chambers geared up to offer readers of *Chambers's Journal* their first taste of the philosophy of phrenology, Combe wrote to point out the similarities between his *Constitution of Man* and the intended outcomes of the Bridgewater bequest. These were themes he had just rehearsed in the *Phrenological Journal*, where he regretted Chalmers's failure to respond to his own book on the same subject. Combe now asked Chambers, "Does my book expound the power, wisdom, and goodness of God, as displayed in adapting the constitution of man to external objects, or does it not?" After all, its "leading principles" were "the separate existence and operation of each natural law; the necessity of obeying *all* of them; and the evident adaptation of all to the moral and intellectual advancement of the race." According to Combe, this emphasis on law removed the obscurities from religion and made moral philosophy newly scientific. Chambers saw the matter in much the same light, telling Combe that if he had been the Earl of Bridgewater and had seen his "masterly little volume," he would have given his money to the author and "spared any further writing on the topic."[28]

Like Combe, Chambers found himself increasingly critical of organized, supernatural religion, leaving the ever-more evangelical Church of Scotland for the Scottish Episcopal Church following pulpit criticism of *Chambers's Journal*. Both authors, however, continued to present their views in relation to divinely ordained laws consistently with a somewhat deistic outlook. Moreover, as Chambers's scientific synthesis developed, he found the Bridgewaters and Babbage's *Fragment* useful in making this move seem safely orthodox and far removed from the atheist transmutationism of working-class radicals. Chambers's scientific sources were nevertheless very varied, and he later recalled that a key impetus to his evolutionary work had come from the exciting vision of progressive law-governed change in the universe offered by the nebular hypothesis, which he encountered in the populist *Views of the Architecture of the Heavens* that had been penned in 1837 by Combe's friend, the Glasgow professor of practical astronomy John Pringle Nichol. Chiming with Chambers's commitment to natural law in human science, the new work inspired him

28. Chambers to Combe, 12 and 14 December 1833, NLS, MS.7230, fols. 49–52; Combe to Chambers, 13 and 15 December 1833, in Gibbon 1878, 1:294–97 (originals in NLS, MS.7386, fols. 64–68); *Phrenological Journal* 8 (1834): 338–64; [Chambers] 1834.

to offer a larger vision of law-bound progress in nature that extended to the origin of species.[29]

In developing this evolutionary synthesis, Chambers drew on a wide range of contemporary publications—some more authoritative than others—as well as entering into his own "geology fever." Some measure of his reading is given by the contents of *Chambers's Journal* and the fortnightly serialized encyclopedia *Chambers's Information for the People* (1833–34). Both bore the prominent claim "conducted by William and Robert Chambers," and while Robert did not write either in its entirety, he scrutinized the content assiduously. The two publications soon began to draw content from the Bridgewaters in paraphrases, quotations, and extracts. In the autumn of 1834, an entire number of *Chambers's Information* was devoted to "Natural Theology," drawing heavily on Whewell's and Roget's volumes. Moreover, as Chambers's interest in the history of the creation and the laws of life grew in succeeding years, articles on such themes in *Chambers's Journal* relied significantly on the Bridgewaters, especially those of Roget and Buckland. In 1837 the latter furnished much of the substance for an important six-part series on the four "ages of animal life" that presaged the account in *Vestiges*. However, the series strikingly echoed an earlier critical article on "The Transmutation of Species" when it culminated in a quotation from Buckland concerning the need to refer the origin of the successive systems of life to "the direct agency of creative interference."[30]

Chambers's growing interest in science also found expression in a friendship with Charles Bell, with whom he enjoyed conversations on mental physiology. The publisher viewed Bell as his "father" within the Scottish scientific establishment, and it was Bell who, as a council member of the Royal Society of Edinburgh, nominated him for a fellowship in 1839. By the time Chambers wrote *Vestiges* in the early 1840s, however, he had temporarily moved to St. Andrews in order to recover from the mental strain of overwork. It was in this relaxed family environment that he carefully crafted the text of *Vestiges*, using all his skill in writing for the middle-class home to domesticate his evolutionary synthesis. Like Combe's *Constitution of Man*, the narrative implied, *Vestiges* should be read as a superior

29. [Chambers] 1853, v–vi; Priestley 1908, 30. For Chambers as a "moderate deist" see Secord 2000, 85. On Nichol, see Schaffer 1989.

30. [Chambers?] 1837, 380; Chambers to DR Rankine, 3 September 1837, NLS, Dep.341/109/1. Cf. [Chambers?] 1835. On Chambers's editorial role, see Cooney 1970, 63–64, Secord 1989a, 177, and Secord 2000, 91; see also Chambers 1847 and the marked files at the NLS, Dep.341.

Bridgewater, presenting the unrolling of God's laws in creation as an altogether more ennobling vision than that of a tinkering God. Early chapters outlining the nebular hypothesis and the history of the earth and its inhabitants bore obvious parallels to parts of Whewell's and Buckland's treatises, with Buckland appearing repeatedly as an authority on organic remains.[31] However, the crux came two-fifths of the way into the book, at the point of considering how new species arose.

Reviewing the account hitherto given of geological history, the narrator sought to bring home to readers the superiority of the notion of "creation by law." That God created living creatures was a well-attested fact that the narrator took for granted. As he later pointed out, one of the "most agreeable tasks of modern science" had been to trace creative design in living organisms. Paley's *Natural Theology* and the Bridgewaters had been so conclusive that further proof would be "tiresome." But the higher view of creative design was one that saw the divine will as being expressed through natural laws. And here Buckland's commitment to "creative interference" in the origin of new organic forms was sidestepped as another passage from his Bridgewater was put to powerful use in justifying the superiority of a lawlike account of the history of the material elements of the earth. Having thus cleared the ground, subsequent chapters moved on to introduce the origin of life and the development hypothesis, and it was in the latter that Babbage's *Fragment* came into play. Allowing Babbage's authority full rein in a lengthy quotation concerning the apparently miraculous calculating engine, the anonymous narrator proceeded to apply the underlying "philosophy" to the "higher law" of the development hypothesis.[32] Thus, at pivotal moments in the narrative, the reader of *Vestiges* was invited to see the new approach to the origin of living creatures as a proper extension of a Bridgewater philosophy of creation.

In practice, of course, many readers refused *Vestiges*'s invitation, and the Bridgewater authors were prominent among them. Not surprisingly, it was Whewell who within four months rushed out a robust (if implicit) riposte to *Vestiges* in the form of a fashionable volume of "extracts, bearing upon theology," with the title *Indications of the Creator*. His stated purpose was to document "the Indications of Design in the Creator, and of a Supernatural Origin of the World," as well as the consistency of science and revelation.[33]

31. [Chambers] 1844, 108, 129, 139; Chambers 1994, xiii; Chambers 1872, 277; [Waterston and Shearer] 2006, pt. 1, 174.

32. [Chambers] 1844, 145–58, 206–11, 324.

33. Whewell 1845, viii–ix. On reactions to *Vestiges* see Secord 2000; Whewell's work is discussed on 227–29.

Moreover, while one of the extracts, regarding the compatibility of the nebular hypothesis with divine creation, came from Whewell's Bridgewater, the bulk came from his *History* and *Philosophy*, reiterating his later, much more narrowly proscriptive account of the philosophy of creation, including in regard to the nebular hypothesis.

Chambers was significantly concerned by Whewell's "remarkable reservations" about the reign of law. Whewell was "one of the chief writers of the day" and deserved to be answered in full in the volume of *Explanations* issued in response to critics at the end of the year. Even now, however, the Bridgewaters continued to provide Chambers with rhetorical resources. Puzzling over objections to creation by law, the narrator claimed (in obvious reference to Chalmers) that "one of our own most popular divines has written a Bridgewater Treatise, to show the predominance of natural law over mind, as a proof of the existence and wisdom of God." The problem lay, he continued, with the tendency of devout scientists to invoke miracles in areas of ignorance only to retreat in the face of natural laws. In this context, Whewell's "reservations" could be seen to be weak and demeaning. Elsewhere in *Explanations*, however, Buckland's and Roget's Bridgewaters were wheeled out for support on particular points—much, indeed, to Roget's chagrin.[34]

Nothing shows more clearly than the reaction to *Vestiges* the extent to which the Bridgewater authors intended natural laws to be kept within limits. Yet in the scientific world of the 1830s, the Bridgewaters not only symbolized a rebaptism of the view that the law-based perspective of modern science was religiously sound but also offered scientific readers an opportunity to explore the boundaries of such a view. Babbage's decision to do so flamboyantly, at the heart of London society, while drawing on Herschel's unimpeachable authority, blazed a trail for others to follow. Darwin did so silently, and Powell's more public response remained relatively little regarded. In his anonymous *Vestiges*, however, Chambers's reading had maximum effect, shocking and appalling the Bridgewater authors and leaving Whewell in particular struggling to reclaim the standpoint that he had striven to establish. The Bridgewaters had given new currency and authority to this view of the philosophy of creation, inviting scientific readers to reflect further on how it applied in the several sciences, and some had done so with considerable vigor, especially in regard to the origin of new species.

34. [Chambers] 1845, 88n, 98–99, 106, 126, 128; Sedgwick 1850, xxxvii, 278.

## *Observing Design*

The issue of the lawlike character of scientific knowledge and its relation to theological knowledge was clearly central to scientific practice in the 1830s, but the Bridgewaters also invited scientific readers' consideration of another important topic. Their focus on the evidence of design in nature drew attention to the nature of scientific observation itself and more particularly to the role that the perception of design should have within it. This was a highly topical issue and all the more important at a period when the scientific savants were seeking to establish themselves as cultural leaders who could guide the observational practices of less knowledgeable observers. While reflective works like Herschel's *Preliminary Discourse* were notably thin in their accounts of observation, they nevertheless made clear that the practice required significant discipline and thought. The philosopher must observe without prejudice, Herschel observed, but observation still required training, which should include learning about pertinent theoretical claims.[35] In this context, and in the absence of more focused accounts of the character and qualities of scientific observation, the Bridgewaters were strikingly useful for those who wished to engage with such questions.

The Bridgewater authors were agreed that scientific observation typically entailed a perception of design. However, they considered it to be, to a greater or lesser extent, a result of prior mental training. Roget claimed that the study of final causes or purposes was "forced" on the observer's attention by even "the most superficial survey of nature." Yet while he urged that it was "impossible not to recognize the character of intention" in the structure of living things, he also claimed that the perception was developed by scientific education. The physiologist who had learned to recognize design across the full extent of organic nature was better able to perceive individual instances, some of which were obscure. Likewise, while Bell had a habit of seeing design in everything, his Bridgewater stressed the importance of examining any object in the living world in "all its bearings" as a necessary preliminary to the perception of design. Kirby and Kidd went rather further, suggesting that it was actually a thorough grounding in biblical truth that prepared the mind to perceive divine design in natural phenomena.[36]

For the scientific observer, the key question was what role such perceptions of design should play in the work of science, and this was a subject

35. Herschel [1831], 79–80, 132.

36. Roget 1834, 1:25, 33; Bell 1833a, 1.

on which the Bridgewaters were by no means agreed. The most negative view was expressed by Whewell, who took his cue from the great theorist of the new experimental philosophy of the seventeenth century, Francis Bacon. Bacon had declared that final causes should be excluded from scientific reasoning, since they tended to stand in place of and obscure physical causes. Only when the philosopher subsequently reflected on the religious meaning of his work should they come into view: they were "vestal virgins" that were dedicated to God while being philosophically barren. Whewell summarized Bacon's views in his Bridgewater, recognizing that they had wide currency, not least among French skeptics. His own gloss on the subject was somewhat more positive than Bacon's, but he nevertheless agreed that final causes were "to be excluded *from physical enquiry*." Philosophers should not "assume that we know the objects of the Creator's design, and put this assumed purpose in the place of a physical cause."[37]

This view stood in stark contrast to that elaborated in Roget's opening chapter on final causes. In physiology, he explained, knowledge of relations between phenomena was not limited to cause and effect. The science also encompassed knowledge of relations between means and ends, and the study of such relations was a legitimate aspect of the scientific endeavor to discover new laws. Roget had made a similar claim in his article on "Physiology" in the supplement to the *Encyclopædia Britannica*, but there he had been critical of the extent to which the subject was dominated by study of the "*functions* of life, that is, of the purposes to which the actions constituting life are subservient." There was, he had claimed, a regrettable and damaging "proneness to substitute final for physical causes" in physiology, and his article had focused exclusively on physical causes instead. Ten years later, Roget seemed to have undergone a change of mood, and his new emphasis on the value of final causes persisted when he rewrote his "Physiology" article in 1837.[38] At the same time, however, his Bridgewater also emphasized the limits of the functional interpretation of living organisms, making rather more of the design evident independently of final causes in morphological patterns across taxa (the "unity of plan").

The curious ambivalence concerning the role of final causes in scientific observation that was exhibited by the Bridgewaters as a whole, and by Roget's in particular, strikingly reflects the parameters of a debate that was becoming increasingly heated in the 1830s. Within the sciences of life, the legitimacy of teleological explanation (i.e., explanation in relation to final causes) had become a matter of great consequence. Moreover, it was a

37. Whewell 1833b, 352–56.

38. Roget [1838], 578; Roget 1834, 1:5–7, 21–23, 33; Roget 1824, 180–82.

debate tinged with religious, political, and nationalist overtones, not least because British anatomists and physiologists were keenly aware of the increasingly angry and politicized debate on the question that had taken place in France between Geoffroy Saint-Hilaire and Cuvier.

Geoffroy's transcendental anatomy was firmly rooted in the study of form, and he explicitly argued that final causes had no place in the work of the naturalist. By contrast Cuvier insisted that there was an "essential difference between the general sciences and natural history" that consisted in the necessity of applying the principle of the "conditions of existence," or final causes. As we have seen, these arguments were rehearsed at length in Britain in the 1830s. However, their precise religious significance remained open to question. While some considered the identification of design in the functional adaptedness of living organisms to be useful both in manifesting divine action and in the work of anatomy and physiology, most of those who questioned or doubted the scientific utility of such teleology still nonetheless agreed that living organisms manifested divine action, not least through the "unity of plan." Moreover, few naturalists took really extreme positions, most preferring in practice to blend the two approaches to some extent.[39]

With all their ambivalences, the Bridgewaters were naturally read against the backdrop of this debate. Buckland and Bell's treatises particularly praised Cuvier's work, endorsing the importance of final causes in the work of the anatomist or paleontologist. The naturalist's eye, they made clear, had to be trained to perceive the relevant details in a specimen through an understanding that animal bodies were constructed in relation to particular ends. A man "ignorant of anatomy" who encountered a bone in some unexplored country would learn next to nothing from it, Bell had claimed. By contrast, a trained anatomist would, by observing the new bone in relation to final causes and in the light of what was known of other animal forms, be able to estimate the size, form, diet, and internal economy of the animal from which it came. Such observational practices, replete with inferences about means and ends, were especially important for the paleontologist, often working with limited materials, as Buckland eagerly insisted.[40]

This endorsement of Cuvier's principles contributed significantly to confirming both their religious tendency and their high status. Following

39. Cuvier 1827–1835, 1:4–5; Geoffroy Saint-Hilaire 1830, 66n. See also Appel 1987. For somewhat contrasting accounts of the British debate, see Rehbock 1983 and Desmond 1989b.

40. Buckland 1836, 1:109; Bell 1833b, 80–83.

the Frenchman's death in 1832, some designated Buckland the "English Cuvier" in view of his ability to emulate the latter's practice of fossil reconstruction. Others went further, viewing the entire Bridgewater series as fundamentally rooted in Cuvier's functionalist approach. The Edinburgh anatomist Robert Knox, for instance, saw them as sustaining what one of his acolytes later called an "ultra-teleological school," arguing that the "existence of a general scheme or plan embracing the whole range of the animal kingdom" was, in its fine detail, "quite inconceivable by those who adopt the views and the physiology of the Bridgewater Treatises."[41] Yet both Bell and Buckland also considered the "unity of plan" important in anatomical reconstruction, and both claimed that it offered an alternative source of evidence for design. Moreover, even though Roget emphasized the importance of final causes in physiology, it was his emphasis on the morphological unity of design that predominated.

More than most scientific works, then, the Bridgewaters spelled out a range of ways in which the man of science might expect to find design in nature. In doing so, they made explicit and to some extent codified observational practices that went to the heart of scientific work. Yet the fact that their contribution to the growing debate was mixed encouraged readers to engage with them creatively, navigating their own path through some of the most complex epistemological issues of the age.

One such reader was Richard Owen, at this time still a young and deferential comparative anatomist, whose encounter with the Bridgewaters framed his engagement with these vital questions on which he became an important authority. As assistant conservator at the Hunterian Museum of the Royal College of Surgeons (fig. 8.2), Owen assisted several of the Bridgewater authors as they wrote their treatises. However, as a college servant his was a socially subservient position that reflected the financially straitened circumstances of his once-mercantile family. After a surgical apprenticeship in Lancaster, Owen had received only a brief formal education at Edinburgh University and St. Bartholomew's Hospital in London. However, that was enough to provide him with a thorough exposure to the tradition of functionalist anatomy associated with the great eighteenth-century surgeon John Hunter, whose collections he was now cataloging, and to give him an opportunity to demonstrate the ability that secured him his livelihood. He also became intimately familiar with the Cuvier-Geoffroy debate. Indeed, in 1830, aged just twenty-six, he was deputed to show Cuvier around the Surgeons' museum on what proved to be the

41. Knox 1841, 128; Blake 1870, 334; Dawson 2016, chap. 2; *Church of England Quarterly Review* 2 (1837): 465.

FIGURE 8.2 Royal College of Surgeons, Lincoln's Inn Fields, London. Etching with aquatint by W. O. Geller, 1836, after T. Kearnan. Image: Wellcome Collection. Public Domain Mark.

Frenchman's final trip to England. The following year Owen visited Paris in the midst of the scientific hostilities, hearing and reading much about them and seeing the principals in action.[42]

Owen's first major publication—an 1832 museum memoir describing a recently acquired rare specimen of the pearly nautilus—indicated that his sympathies lay with Cuvier. It was hailed as "an excellent specimen of Hunterian-Cuvierian Natural History" and marked him out as someone who could be useful to Cuvier's English advocates. Those advocates were soon putting him to good use, and prominent among them was Buckland.

42. Owen 1894, 1:48–59; Desmond 1989b, 276–79; Sloan 1992, 39–43.

Owen had sent Buckland an early copy of his memoir with the comment that, since the death of the "lamented Cuvier," he looked to no one more than the Oxford professor for approbation. Buckland was very impressed by Owen's "most able and masterly and satisfactory" investigation, using the memoir in his Bridgewater to make visible the design in fossil forms. He was especially excited by the information it provided regarding the siphuncle or tube passing through the chambers of the nautilus shell. This he used to develop a new account of the structure's function in achieving buoyancy, and he set Owen to work undertaking further observations to verify what he later called "our Theory." Nor was this the only evidence of functional adaptation that Owen's growing body of work offered as Buckland put together his Bridgewater.[43]

Owen commented on Buckland's passages concerning the nautilus in draft, but he subsequently read his presentation copy of the completed Bridgewater twice in quick succession, sending its author various technical "notes and suggestions." The following spring, he incorporated some of Buckland's evidence against transmutation of species into his inaugural course of lectures as Hunterian Professor of Comparative Anatomy, and the two continued to make common cause. The next year Owen was being co-opted by Buckland to defend Cuvierian principles of fossil reconstruction against the attacks of Cuvier's successor as professor in Paris, Henri de Blainville. The debate focused on fossils from the Stonesfield slate near Oxford that Cuvier had adjudged were marsupial, despite their coming from Secondary strata otherwise bereft of terrestrial mammals. Owen now defended both Cuvier's interpretation and his principles in a paper before the Geological Society. Summing up the society's year in February 1839, its president Whewell claimed that this had been a "battle" about "the foundations of our philosophical constitution" and specifically the "great Cuvierian maxim,—that from the fragment of a bone we can reconstruct the skeleton of the animal." This "doctrine of final causes in animal structures" was the foundation of zoology and paleontology, and it was worthy of vigorous defense.[44]

Whewell's emphasis here on the scientific value of final causes contrasts with the critical comments of his earlier Bridgewater, but there his

43. Buckland 1836, 1:73n–75n, 310–32; WB to Owen, 9 March 1838, RCS MS0025/1/5/2; Owen to WB, 28 July 1832, 14 December 1833, WB to W Clift, n.d., and A Carlisle to Owen, [December 1834?], in Owen 1894, 1:64–67, on 64–66.

44. Whewell 1842, 89; Owen to WB, [late 1836–early 1837], TUL, MROX10041; Buckland 1836, 1:[ix]. On the Stonesfield fossils, see Owen 1841, Desmond 1989b, 306–21, Rupke 1994, 143–47, and Dawson 2016, 88–90. For the lectures compare Sloan 1992, 222, and Buckland 1836, 1:294.

mind had been on physics, not on physiology. He had since developed his view in preparing his *History of the Inductive Sciences*, which reported that "those who have studied the structure of animals and plants, have had a conviction forced upon them, that the organs are constructed and combined in subservience to the life and functions of the whole. The parts have a purpose, as well as a law;—we can trace final causes, as well as laws of causation." This principle was, Whewell claimed, "peculiar to physiology," and he conceded that it had recently become highly contested. He offered a lengthy review of the Geoffroy-Cuvier debate, charging the former with rejecting final causes on religious rather than "professional" grounds while himself taking the scientific high ground, clear of "metaphysical and theological" topics. Adaptation, he claimed, was an "irremovable principle of the philosophy of organization" that "constantly forced itself" on the minds of naturalists "not only as an inference, but as a guide whose indications they could not help following." Final causes were a crucial help in "seeing the meaning of the organization." The point was taken further in Whewell's *Philosophy of the Inductive Sciences* (1840), which expressed the view that the "Idea of Final Cause" was an a priori "Fundamental Idea," necessary to render phenomena intelligible.[45]

Somewhat unsure of his ground in physiology, Whewell was anxious to hear Owen's views of his *History*. The naturalist was deferentially appreciative but offered a diplomatic suggestion that "a harmonious theory combining the transcendental and teleological views" was desirable. The anatomist must examine an organ in relation to the "general plan or pattern adhered to on the one hand, and the exigencies of the species to which deviations related on the other—or in other words its transcendental and teleological relations."[46]

According to Owen, the great exemplar of this "comprehensive spirit" in physiology was John Hunter. Owen had outlined this view earlier in the year, near the start of his Hunterian lectures—an annual series at the Royal College of Surgeons that he had been given the honor of relaunching following a hiatus for major building work. Addressing a huge audience of cultural leaders, politicians, and scientific men (including both Buckland and Roget, who sat prominently in the front row), Owen offered an account of the history of comparative anatomy and physiology that climaxed in "the inductive labours of Hunter." Not only had these been far more significant than Cuvier realized, he reported, but Hunter's approach had been more judiciously balanced than that of his successors. While transcenden-

45. Whewell 1840, 2:78–79; Whewell 1837a, 2:456, 463, 467, 469, 471.
46. Owen to WW, 31 October 1837, TC, Add.Ms.a.210[54].

talists like Geoffroy had become obsessed with the unity of composition, allowing "metaphysical dogma" to override the "*Inductive method*," their abuses had led Cuvier to mistakenly underrate the value of the doctrine of analogies, reflecting a dogmatic commitment to functional adaptation. Careful attention to the developmental processes by which creatures were formed, as found in the embryology of Karl Ernst von Baer and others, allowed the naturalist to make the study of homologies properly inductive, and this had been Hunter's approach.[47]

Owen's maturing perspective on the role that function should play in the comparative anatomist's observational practice was thus taking him away from the approaches exhibited in Buckland and Bell's Bridgewaters and closer to the more blended approach found in Roget's. On the surface, Owen appeared to have learned well the lessons of his apprenticeship to Buckland. The geologist viewed him as "destined to be in *Our* country the successor of John Hunter, & to complete what he and our ever lamented friend Cuvier left undone," urging him to make "observation & observation alone" his guide and to "put theory aside." Moreover, Owen continued to develop high-profile instances of Cuvierian reconstruction. These included an improved account of the *Megatherium* that Buckland had made so much of in his Bridgewater. With additional teeth from Darwin's South American specimens, Owen offered a new interpretation of the creature's habits and diet. He told Whewell that his insights had come "teleologically," and he later added his own interpretation to the posthumous third edition of Buckland's Bridgewater. These endeavors climaxed in Owen's reconstruction from a fragment of bone of the extinct flightless New Zealand bird, the Moa, in the autumn of 1839, which became a much-cited Cuvierian tour de force. Likewise, when he assisted Whewell in preparing his *Philosophy of the Inductive Sciences*, he still approved of his powerful ally's "clear statement of the scientific character of teleological reasoning," thanking him on behalf of "the Cuvierian cultivators of Comparative Anatomy."[48]

Nevertheless, Owen's view of the role of final causes in anatomical study was increasingly tempered by his concern with unity of composition, which also began to make its appearance in published research. By 1846, when he prepared a report on "the Archetype and Homologies of the Vertebrate Skeleton" for the British Association, he was very direct in

47. Sloan 1992, 165, 189–94.

48. Owen to WW, 26 March 1840, TC, Add.Ms.a.210[61]; Owen 1840, 100–106; 1842; Buckland 1858, 1:154n, 156n–57n; WB to Owen, n.d., RCS, MS0025/1/5/2; Dawson 2016, 87, 95–132, 158–60.

stating that "Cuvierian principles" were utterly inadequate to account for the correspondence of parts in different animals, referring such homologies instead to God's "higher law of organic conformity." The upshot was even clearer in his famous Royal Institution discourse "On the Nature of Limbs," delivered in February 1849. Bacon's "comparison of 'final causes' to the Vestal Virgins" was, Owen now reflected, entirely apt when seeking to understand the law of conformity to type.[49] In this he was outdoing Whewell in the consistent application of the Baconian proscription, and, as we shall see, he was not the only Bridgewater reader to do so. Yet his complex engagement with the Bridgewaters and their authors during the 1830s nicely illustrates how scientific readers found themselves using the series to navigate around the complex epistemological and methodological issues involved in the observation of design.

While Owen's approach bore some resemblance to that of Roget's Bridgewater, he made little direct reference to it. By contrast, another early advocate of transcendental anatomy, William Carpenter, drew explicitly on his reading of both Roget's and Whewell's Bridgewaters to argue for the limitations of functional adaptation in anatomical observation. As we have already seen, Carpenter was much taken with Roget's emphasis on the unity of creation as found in transcendental anatomy. Like Roget, he was clear that, while "the attainment of every end by the best adapted means" was a fundamental law of creation, it operated alongside a "fundamental unity of structure."[50] However, as he contemplated the implications of this for the practice of physiology and comparative anatomy, he developed and articulated his own sophisticated account of the place that the observation of design occupied in the work of the scientific observer.

Carpenter's account initially appeared in April 1838 in an anonymous review of Whewell's *History of the Inductive Sciences*. Using the review as the springboard for a discussion of the philosophy of physiology, he presented an account of the inductive philosophy grounded in Whewell's work but reflective of his earlier reading of Herschel's *Preliminary Discourse*. According to Carpenter, the main problem that physiology faced lay in the empirical difficulty of accurately ascertaining the fundamental facts of living structure and function from which to ascend inductively to laws. Indeed, what Whewell had taken for physiological laws were, he asserted, in most cases merely facts. Physiologists and naturalists together

49. Owen 1847, 241; 1849, 40. See Ospovat 1981, 19–21, Rehbock 1983, 73–84, and Desmond 1989b, 359–72.

50. Carpenter 1837, 108.

faced a significant observational research program to establish sufficient facts to ground inductive generalizations through a focus on structure, function, and development.[51]

This entailed, Carpenter accepted, a significant commitment to teleology. The study of final causes was "of great value in leading to the *discovery of facts*." The observation of facts was often "suggested," and phenomena were "brought to light," by the perception of "general harmony and adaptation." Carpenter nevertheless roundly criticized the tenor of Whewell's chapter on final causes in physiology. Mistaking facts for laws, Whewell had argued that final causes were useful in the "search after general laws" in physiology and had criticized Geoffroy for refusing to use them in this way. By contrast, Carpenter claimed that the discovery of physiological laws should be independent of final causes. And now he was able to quote at length the assertion in Whewell's Bridgewater that final causes should be "excluded *from physical enquiry*." In Carpenter's view, physics and physiology could not clearly be demarcated, and the same principle should be applied to both sciences in relation to final causes. The upshot, moreover, would be the same in physiology as in physics. Physiological laws would be discovered that would, ultimately, offer stronger evidence of divine design. As Whewell's Bridgewater had claimed of physics, so in physiology the notion of design would be transferred from "the region of facts to that of laws," and an altogether "higher and nobler conception of the Divine Mind" would ensue.[52]

Carpenter offered examples to show how Whewell's more recent conciliatory approach to final causes could lead physiologists astray. The final cause of floral anatomy, he claimed, was no guide to the laws of floral structure, which had to be traced through "all its regular and irregular forms with a total disregard of its function," since the same "rudiments" assumed an endless variety of forms and uses. It was "presumptuous" to think that the "evident purpose" of structures was the sole or even chief design. Similarly, while the "teleologist" would claim that bony processes on vertebrate bones were designed as muscle attachments and make inferences about fossil forms on that basis, the "philosophic anatomist," while "fully acknowledging the adaptation," would temporarily disregard it, seeking to understand the laws of development of the skeletal and muscular systems by extensive observation. Discovering that all such significant bony processes in humans correlated with separate bones in lower animals in

51. [Carpenter] 1838c, 328–31.

52. [Carpenter] 1838c, 338–40, 342.

accordance with a law of specialization, the anatomist would gain a far higher sense of divine wisdom and foresight.[53]

Much of this discussion was repeated in the concluding chapter to Carpenter's 1839 *Principles of General and Comparative Physiology*, where it appeared as a coda that was both metaphysical and epistemological. In restating his position, Carpenter was even clearer in allowing that belief in "Universal Design" in nature often led "to enquiries which would otherwise have been neglected," sometimes putting the scientific investigator "on the right track in the conduct of those enquiries," and he offered William Harvey's researches on the circulation of the blood and Charles Bell's on the functions of the nervous system as examples. The challenge, however, was "not to rest satisfied with the obvious purpose of a particular structure as affording us the supposed reason for which it was created." The "philosophic Physiologist" might find his curiosity awakened by the adaptations he discovered, but he must temporarily disregard the "immediate *purposes* of the adaptations he witnesses" and investigate them as "the *results* or *ends* of the general laws for which he should search." In this enlarged perspective, divine design would appear altogether more impressive, and the physiologist would discover "new objects" for structures that at first seemed "destitute of utility."[54]

Just as with Owen, Carpenter thus found in the Bridgewaters useful materials for thinking through the proper place of the perception of design in the work of the naturalist. Neither wished to deny teleology a place in the practice of scientific observation, but each sought to temper its role in favor of a deeper understanding of design based on the derivation of morphological laws. In this context, it was the caution of Roget's and Whewell's Bridgewaters concerning final causes that provided the obvious points of reference. Moreover, other readers also found themselves championing the Whewell of the Bridgewater against the Whewell of the *History*. Baden Powell, for instance, argued that the later Whewell's prescriptions for teleologically informed observation were not only scientifically damaging in the ways that Bacon and Geoffroy suggested but also undermining of the evidential value of natural theology by dint of making the argument circular.[55] These readers represented the Bridgewaters as ambiguous and multivocal about the role of final causes in scientific

53. [Carpenter] 1838c, 340.
54. Carpenter 1839, 460–62.
55. Powell 1838, 128–33, 176; Corsi 1988, 182–84.

observation while nevertheless seeking to steer the methodological discussion well away from narrow teleology.

## *"Like a Bridgewater Treatise"*

Owen and Carpenter's nuanced readings of the Bridgewater Treatises contrast with the more sweeping assessments of others, such as Robert Knox and Edward Forbes, who saw the series collectively as emblematic of a too-eager desire to find immediate evidence of design in the functional anatomy of living organisms at the expense of the deeper design in morphological laws. Another reader of the Bridgewaters who was impatient of their sometimes facile claims about design in the living world was Charles Darwin (fig. 8.3). However, while his methodological dialogue with the series was critical, it was nevertheless highly productive as he developed his new and strikingly positive perspective on the role of final causes in scientific observation.

Darwin's education as a naturalist had given him a thorough grounding in the teleological perspective. Indeed, he had famously taken delight in Paley's *Natural Theology* while a student at Cambridge, finding the "argument from design in nature" to be "conclusive." For the young Darwin, the expectation was strong that the organisms he observed would be functionally adapted, and as his mind turned in the later 1830s to the question of how new species came into being, he continued to be interested in the question of functional adaptation. Indeed, rather than thinking that the use of final causes in making sense of organic structure was scientifically unproductive, he was preoccupied with how any theory he might suggest would account for the observable adaptedness. Thus, when he opened his first transmutation notebook in the summer of 1837, this question lay at the heart of his theoretical work.[56]

Darwin's novel, transmutationist approach to adaptation, with its emphasis on discovering the creator's "laws of adaptation," undoubtedly made some of the examples of design offered in the Bridgewaters seem risibly parochial. In the autumn of 1837 he noted, "Kirby all through Bridgewater errs greatly in thinking every animal born to consume this or that thing.—There is some much higher generalization in view." A year later he was poking fun at the praise heaped on Whewell's Bridgewater by London surgeon Herbert Mayo. In his *Philosophy of Living* (1837), Mayo had quoted Whewell as "profound," Darwin noted, "because he says length of

56. Ospovat 1981; Kohn 1980; Darwin 1958, 59, 87.

FIGURE 8.3 Charles Darwin, aged forty. From a lithograph. Image: Wellcome Collection. Public Domain Mark.

days adapted to duration of sleep of man!!! whole universe so adapted!!! & not man to Planets.—instance of arrogance!!" In actuality, both Mayo and Whewell had written in terms of mutual adaptation between astronomical and physiological phenomena. Nevertheless, Whewell had discounted the idea that in such cases the astronomical cycle had in any way caused the physiological one as an "arbitrary and baseless assumption" that was "useless for the purposes of explanation."[57] Thus, even if Whewell's explanation

57. *Notebooks*, D49, B141; Mayo 1837, 146–48; Whewell 1833b, 37–39.

of adaptation was not so facilely anthropocentric as Darwin assumed, its particularity suggested a tinkering sort of God quite out of keeping with his own conception. Still, Darwin's beef was with Whewell's narrow conception of divine action rather than with the reality of adaptation.

Darwin's attempt to offer an account of how such adaptation could arise as a result of the operation of natural laws came to fruition in the theory of natural selection in the autumn and winter of 1838–39, especially following his reading of Thomas Malthus's *Principles of Population* in September 1838. With a well-developed and satisfactory theory in hand, his impatience with the common tinkering explanations for adaptedness reached new levels. This became clear when he read part of geologist John MacCulloch's *Proofs and Illustrations of the Attributes of God from the Facts and Laws of the Physical Universe*, probably in March 1839.[58] This was the work whose publication had been held back in anticipation of the Bridgewaters before being published posthumously in the spring of 1837. Darwin's reading notes show that he found it an excellent foil for his new theory, criticizing and recasting many of its examples of design as he read from page to page. This sifting of claims through the filter of his developing theoretical understanding was a common feature of his reading practice, and it brought other readings to mind.

As earlier with the Bridgewaters, Darwin found many of MacCulloch's examples of design ridiculously petty. By contrast, he noted, his new theory made "all organic beings perfectly adapted to all situations where in accordance to certain laws they can live." It was a mistake to think individual species had been specially created by God for particular locations: such a view lowered the creator "to the standard of one of his weak creations." Instead, all adaptations flowed from "some grand & simple laws." A few pages later, Darwin reviewed MacCulloch's description of the woodpecker as an "instance of beautiful adaptation" in the light of the selection theory. Rather than resulting from individual creative acts, he repeated, adaptations were rooted in death and destruction: there was "pain & disease in the world & yet talk of perfection." This brought to mind the Bridgewaters. On his view, Darwin reflected, they were all "reduced simply [to a] statement of productiveness, & laws of adaptation." Turning a couple more of MacCulloch's pages, he noted, "The Final cause of innumerable eggs is explained by Malthus.—" In other words, there was a teleological explanation for the fecundity of species, but it did not relate to God's particular agency in making individual species;

58. Hodge and Kohn 1985, 200.

rather, it was a corollary of the law of natural selection by which God made all species.[59]

Darwin immediately afterward asked himself whether it was anomalous for him to be talking about final causes, adding the injunction, "consider this!" Were they "barren Virgins," he wondered? This was a question that had been brought forcibly home to him by his reading of Whewell's *History* during the preceding autumn. While he had read Geoffroy's account of the debate with Cuvier a year earlier, he was more forthright in his response to Whewell's account of the question, repeatedly blurring the author's carefully constructed boundaries. Darwin marked with uncharacteristic emphasis Whewell's passage asserting the importance of final causes in physiology, placing vertical lines and large question marks in the margins. His marginal comment "mammae in man" indicated his conviction that animal structure—such as male nipples—could never be explained wholly in terms of functional adaptedness. On the next page, however, he marked as "clearly wrong" Geoffroy's reported view that "analogy" and not "fitness" was to be the sole guide in the study of animals. When, a few pages later, Whewell reported Geoffroy's objections to Cuvier's view of animal structure being fitted to "the part which the animal *has to* play in nature," Darwin highlighted Cuvier's view, noting, "This qualified is correct." The modifications necessary in regard to Cuvier's view reflected the operations of the law of natural selection. Adaptedness was not a result of special creation: it was "Owing to external contingencies, & numbers of other allied species & not owing to mandate of God."[60]

As Darwin read Whewell's case for the teleological approach to physiological inquiry, he continued to evaluate it in relation to his new law-based account of adaptedness, which meant that its applicability was limited by the constraints of descent. The point was well exemplified by the case of male nipples, which were evidently not functionally adapted in themselves and could only be accounted for as a by-product of transmutation. Darwin glossed one of Whewell's key claims to reflect how his new theory changed the teleological perspective. It now read, "That the parts of the bodies of animals are ~~made~~ [born, altered] in order to discharge their respective offices [under changing circumstances], is a conviction which we cannot believe to be otherwise than an irremovable principle of the philosophy of organization." On the next page he corrected the narrow, special creation

59. *Notebooks*, MAC54r, MAC57v, MAC58r.

60. For Darwin's reading of Whewell's *History*, see Manier 1978, 52–55, and Di Gregorio and Gill 1990, 866–68. For his reading of Geoffroy, see Di Gregorio and Gill 1990, 300–301, *Notebooks*, B110–14, and Ospovat 1981, 28–29.

teleology of Whewell's claim that the "use of every organ" was always discovered from the "assumption that it must have *some* use," by replacing the word *use* with *relation*. Here again was the transmutationist perspective, reflecting that the process resulted in anatomical peculiarities, and a marginal note contained an interjection concerning the nonfunctional "shrivelled wings of those non-flying Coleoptera?!" At the same time, however, another marginal note stated, "In every animal, final cause or adaptation is applicable to far greatest proportion of structure. For otherwise it would be pressed." Many anatomical features were nonfunctional by-products of descent with modification, but natural selection meant that functional adaptedness predominated.

Darwin considered Kant's argument that all the parts of animals were "mutually ends and means" to be "vitiated" by his new theory (an inherited harelip was not an adaptation in Kant's sense, he noted), and he found Cuvier's principle of the conditions of existence "rather far fetched." Nevertheless, his theory gave new focus to the annotated Cuvierian claim that "As nothing can exist if it do not combine all the conditions which render its existence possible [with innumerable other animals striving to increase], the different parts of each being must be co-ordinated in such a manner as to render the total being possible, not only in itself, but in its relations to <u>those which surround it</u>." A narrow, special-creation-based teleology of the sort implied by Whewell's principles was unproductive and failed to account for the nonfunctional consequences of descent. Adaptedness was not universal, nor did it result from the direct exercise of divine intelligence. Nevertheless, it was a fact of nature that required explanation, and Darwin had an explanation based in the "laws of organization" whose productions seemed analogous to those of intelligence. Thus, his developing theory only redoubled his desire to explain the facts of adaptation beloved of the Bridgewater authors, and it only confirmed the sense that the naturalist should expect to see such adaptedness in living creatures much of the time.

A good example of this comes from Darwin's "Questions & Experiments" notebook, probably begun in the summer of 1839. One question he early marked for attention related to whether bees were guided to flowers by smell or by sight. Darwin noted that botanist Robert Brown had argued that the bee-like appearance of the bee orchid was intended to deter rather than attract insects. An added note raised the question of the "final cause of beauty of flowers," referring to Kirby's Bridgewater, which had noted in passing that, unlike in animals, the reproductive organs of plants were the "most beautiful and admired, and odorous and elevated parts." Understanding such characteristics of plants, Darwin was clear, required attending to their "final cause." Moreover, doing so suggested particular experiments and observations: "cover flower— put artificial flowers—also do with honey." Strikingly, when Darwin pursued this empirical program in detail many years later (following the publication of *Origin*), it led to a book "On the Various Contrivances by which Orchids are Fertilised by Insects" that Darwin described to publisher John Murray as being "like a Bridgewater Treatise" in that its "chief object" was "to show the perfection of the many contrivances in Orchids." But, of course, the book's staggering account of the adaptedness of orchid flowers was now rooted firmly in the causal law of natural selection rather than in individual instances of creative action.[61]

Like other scientific readers of the Bridgewaters, Darwin was far from uncritical of what they had to say about the role of final causes in guiding observation. There were many pitfalls, such as getting fixated on the idea of a tinkering deity thinking adaptations through one at a time or misunderstanding nonfunctional structures that were a consequence of descent with modification. Some of the Bridgewaters were better at avoiding such pitfalls than others. Early in the 1840s, for instance, Darwin jotted a reminder to himself to read Roget's, which was "very good" on "abortive organs." But like other readers, Darwin found in them highly topical explorations of the proper character of scientific inquiry that required his active engagement. And as he elaborated his new theory in the late 1830s, he found in the Bridgewaters a foil for his new view of the role of teleology in biology.[62]

Historians have long recognized that Darwin's teleological theory of natural selection grew out of his exposure to the design literature, and his engagement with the epistemological and methodological issues that the Bridgewaters raised and exemplified clearly formed part of the process by which the theory was developed. These issues were among the most significant of the day, and Darwin was, if anything, on the unfashionable side of the argument in his attachment to teleology. Other rising figures in the sciences of life, such as Owen and Carpenter, were rather more inclined to temper the teleology they found in the Bridgewaters with an emphasis on discovering underlying morphological laws that was rather less evident in the series. Thus, while Buckland was early delighted with how Owen seemed to build on the Cuvierian functionalism of his Bridgewater, the

61. Darwin to Murray, 21 September [1861], in *CCD*; *Notebooks*, QE[5]a; Kirby 1835, 1:139.

62. CUL, DAR119:16v. For a helpful contextualization of Darwin's teleology, see Asma 1996.

younger man came increasingly to prioritize the value of the morphological plan in the practice of anatomical observation. Similarly, Carpenter was from the start taken with Roget's emphasis on the "unity of plan," coming to champion the Baconian Whewell of the Bridgewater against the later Whewell, for whom final causes became the key to physiological observation. For all, however, reading the series offered important materials for exploring the question of how the perception of design should figure in the work of scientific observation.

❋

When Darwin quoted from Whewell's treatise opposite the title page of *Origin* in 1859, the act was highly premeditated. For his generation, the discussions in the Bridgewaters of the meaning and status of scientific knowledge, especially in relation to religious truths, became a key point of reference at a time when natural law was ever-more dominant in the sciences. In the sciences of life in particular, the question of divine design went to the heart of observational practice, and scientific readers used the series to engage with the complexities of a changing understanding of how such work should be done. For many men of science in the 1830s, the Bridgewaters were of real practical importance to their work as well as offering an invaluable token of the Christian tendency of the sciences. That Darwin drew on their emblematic significance at the start of *Origin* indicates that it had not entirely dissipated a quarter of a century later. Yet by then it was clearly a vision of science that was coming under increasingly critical scrutiny.

✷ CONCLUSION ✷

# *"The Fashionable Reign of the Bridgewater Treatises"*

*It is true that a reconciliation of the scientific with the religious views is still possible, but it is not so clear and striking as it was. But it is still a weakness to regret this; and no doubt another generation will find some way of looking at the matter which will satisfy religious men. I should be glad to see my way to this view, and am hoping to do so soon.*

WILLIAM WHEWELL, *4 January 1864*[1]

Darwin's inclusion of an epigraph from Whewell's Bridgewater at the start of *Origin* reflected the extent to which the series had become a standard emblem of the religious safety of the sciences. Widely read and discussed, the Bridgewaters had served to reassure a generation that the rapidly changing disciplinary sciences, grounded in a commitment to the uniformity of nature, would feed rather than undermine Christian faith. Almost from the start, however, some wished to apply this framework to the origin of new species. *Vestiges* had been a seminal moment but *Origin* reopened the question explosively. For Whewell, turning to the first page of his presentation copy, it must have come as a nasty shock to find himself quoted in justification of Darwin's theory. The parcel for Whewell almost certainly arrived at Trinity College in the middle of November 1859, along with that for Sedgwick. But while Sedgwick immediately sent Darwin a friendly and voluble, albeit damning response, it took Whewell seven weeks to return his polite but terse thanks to the author. The book had "interested him very much," but he could not be considered a convert, "yet at least."[2]

The conclusion's title is quoted from Ryall 1844, 197.

1. WW to JD Forbes, 4 January 1864, in Todhunter 1876, 2:435–36. Whewell was commenting on Forbes's recent review of Charles Lyell's *Antiquity of Man* (1863).

2. WW to Darwin, 2 January 1860, in *CCD*. Cf. Sedgwick to Darwin, 24 November 1859, in *CCD*.

There was "much of thought and of fact" in *Origin*, which was "not to be contradicted without careful selection of the ground and manner of the dissent," but that was something for which he did not presently have time.

When, a couple of years later, one of the younger Trinity fellows—the son of the professor of anatomy—proposed that Darwin's book be purchased for the college library, Whewell famously overruled the suggestion. And it was not until 1864, when Cambridge publishers Deighton, Bell issued a ninth edition of his Bridgewater, that Whewell made public his response to *Origin* in a new preface. Even then, while the target was clear and Darwin's book was quoted, neither the book nor its author was named. Whewell's first line of defense looked back to 1833. Theories like Darwin's, he argued, continued not to be a threat even should they prove to be true. Just as he had shown with the nebular hypothesis a quarter century before, a lawlike account of the origin of new species would ultimately need to rest on the supernatural design of initial conditions. In any case, he continued, the theory fell afoul of the principles set out for the palaetiological sciences in his *Philosophy of the Inductive Sciences*. It involved "two enormous assumptions" that made it "a mere work of fancy." There was no evidence that the proposed mechanism actually worked, and the assumption of an "unlimited number of generations" to shore it up was unwarranted and merely rhetorical. Whewell could not see how such a philosophy would "supersede the ancient doctrine of Final Causes" that had held sway since the time of Socrates and been so scientifically productive.[3]

While Whewell's reaction to *Origin* was conservative and critical, there were others raised on the Bridgewaters who felt well able to embrace the new work in the manner suggested by Darwin. Despite being ill, Charles Kingsley replied very promptly to his own presentation copy to tell Darwin that he would greatly prize the new work. He had, he reported, long ago learned to "disbelieve the dogma of the permanence of species" and had also "gradually learnt" to see that it was "just as noble" to believe of God "that he created primal forms capable of self development into all forms needful" as it was to believe that God intervened to "supply the lacunas w[h]. he himself had made." Darwin was naturally delighted and immediately asked if he could quote Kingsley's letter in the conclusion of the second edition of *Origin* that appeared early in January 1860. Moreover,

3. Whewell 1864, xv–xxii; Darwin, ed. 1887, 2:261n. See also Hodge 1991. For Whewell's response to Darwin, see also WW to Brown, 26 October 1863, in Todhunter 1876, 2:433–34. Trinity College archivist Jonathan Smith kindly confirms that the report concerning *Origin* is substantiated by the library accession records.

Kingsley was far from being the only commentator who elaborated on the consistency of Darwin's theory with belief in a divine creator.[4]

Nevertheless, by the 1860s some were looking back to the Bridgewaters with nostalgia. There was, these days, "a good deal of mutual suspicion" between clergy and scientific men, observed High Church cleric John Hannah in September 1867. Even some of the clergy, he noted, were claiming that recent attacks on Darwin reflected a centuries-old pattern of clerical opposition to scientific innovation. Hannah's response was to point to the recent past, when the clergy had been at the heart of the sciences. Claims about clerical obscurantism "must sound harsh in the ears of a generation which remembers Buckland and Chalmers, and owes so much to Sedgwick and Whewell."[5] In the early years of Victoria's reign, the Bridgewaters had helped to codify a dominant vision of the sciences as fundamentally Christian, but by the 1860s that vision was under robust scrutiny. As scientific and clerical identities were remolded, the Bridgewaters looked increasingly outmoded. But they had left a significant legacy. By rebaptism, they had not only eased the passage of the newly disciplinary sciences into the heart of British culture but also made eminently respectable a progressive and largely law-governed account of the history of creation that smoothed the path for Darwin's *Origin* a generation later.

## *A Christian Vision of the Sciences*

The Bridgewaters were a godsend for the sciences at a moment of rapid change in British social and cultural life, helping to secure public confidence that grounding national prosperity, harmony, and progress in natural science would not lead to the abandonment of Christianity. This vision stands out as distinctive within the larger European scene, where scholars have associated the rapid change in the sciences with a fundamental shift "from a 'godly' to a secular activity." British science shared many characteristics with that on the Continent, including the vision of a special role for the scientific class in achieving progress and a commitment to a liberal ideology favoring free inquiry and meritocracy. Yet in its religious character it reflected very particular local circumstances, including Britain's

4. Moore 1979; Roberts 2009; Kingsley to Darwin, 18 November 1859, Darwin to Kingsley, 30 November 1859, in *CCD*; Darwin 1860, 481. On Kingsley, see also Lightman 2007, 71–81.

5. Hannah 1867, 4.

conservative reaction to the French Revolution and the continuing dominance of the aristocracy and the English established church.[6]

Of course, the vision of the sciences that the Bridgewaters fostered was by no means uncontested. Quite apart from working-class radicals, there was no shortage of secularizing and liberalizing reformers, especially beyond the sphere of Anglican influence in London's private medical schools and new secular "university" and in the University of Edinburgh. Nevertheless, there were strong incentives for men of science to maintain a public profile in accord with the rebaptism of the sciences, and even unorthodox practitioners were frequently interested in the religious implications of their work. Thus, while Lyell sought to exclude the Bible from the principles of geological science and Robert Grant strongly resisted a functionalist view of organic design, these two deists both presented the sciences within university lectures as explorations of the divine creation even if much of what they said appeared in preliminary remarks. The question was not *whether* the sciences should be presented within a religious frame but *how* they should be. At the same time, even the Anglican elite were by no means as rigidly opposed to extending the scope of inquiry as they have sometimes been portrayed as being, and some allied themselves with those who shared more reformist attitudes, especially within the British Association.[7]

The development of a newly respectable religious face for the disciplinary sciences depended heavily on the management of the changing print media of the industrial age. The Bridgewater authors, growing increasingly accustomed to the opportunities provided by new forms of publication for addressing wider audiences, took the opportunity to offer sweeping views of the sciences in relation to their wider religious significance. Moreover, while other reflective treatises published during the 1830s also contributed to cultivating a sense of the religious safety of the sciences, the Bridgewaters necessarily placed such questions center stage, together amounting to a grand, symbolic statement.[8]

Taken at the symbolic level, the series offered a vision of the interconnectedness of science and religion that was somewhat ill defined, since the individual treatises were as strikingly different in theological orientation and argument as they were in literary form and style. This, however, served to increase their importance in securing public trust, since readers

6. Cunningham and Williams 1993, 424–26; Morrell and Thackray 1981; Secord 2014, 239–40.

7. Brooke 1983; Secord 1990; Secord 2014, 138–72; Winter 1997.

8. Secord 2014, 243–44.

could select the account of consilience that most appealed. Altogether, they conveyed an overwhelming sense that, notwithstanding any difference of theological party or approach, the disciplinary sciences would be found by orthodox Protestants to be religiously safe and to contribute to an enhanced sense of God's "Power, Wisdom, and Goodness." In an age in which many Christians were skeptical about or frankly alarmed by the claims of a rationally constituted natural theology, the Bridgewaters offered little by way of such formal apologetic argumentation. Yet their frequent references to divine design combined powerfully with their concerted efforts to answer opponents of Christianity, to explain apparent points of disagreement between the sciences and accepted truths, and to suggest that scientific views enhanced the Christian's appreciation of divine action. For some Christians, notably evangelicals and the Anglican High Church, a greater emphasis on the authority of scripture or the distinctive doctrines of Christianity seemed called for. But even for these readers, selective quotation meant that the Bridgewaters could be used to symbolize the religious safety of the sciences. Those who considered the findings of geology irreconcilable with the Bible were almost alone in considering the series fundamentally dangerous.

The sciences that were thus offered up as religiously safe were nevertheless a long way removed from those of a generation before. In Paley's *Natural Theology* the scientific vision had been of God's creation as the static outcome of an original, implicitly miraculous set of events, with the form of living organisms solely determined by functional design. The Bridgewaters offered something altogether different. Taken collectively, and despite some obvious unevenness, they represented creation as a protracted and progressive historical process, many aspects of which took place according to natural laws, with the morphology of living creatures reflecting larger patterns of divine creativity as well as functional adaptation. From a post-Darwinian perspective, and indeed from the perspective of some contemporary critics, the fact that this progressive and law-governed account stopped short of and stood in staunch opposition to the transmutation of species is obviously significant. Such presentism aside, however, what is striking is how novel the perspective offered by the series was. The Christian vision of the sciences embodied in the Bridgewaters was distinctly up to the minute, normalizing for a large audience the idea that the earth had a long and progressive history and that God's creative action was appropriately characterized by the laws discovered by science.

Given their opposition to transmutation, it is no surprise that the role of the Bridgewaters in mobilizing public support for a law-governed and progressionist account of the creation has not been fully recognized. Yet with their remarkable appeal to a wide public, rooted in part in their imprimatur of Christian orthodoxy, they achieved just that. In the pages of, for instance, Buckland's and Whewell's Bridgewaters, readers could find themselves exhilarated like never before by the vision of a creator unfolding an elaborate plan of creation, much of it in accordance with natural laws, and leading progressively to its pinnacle in the human species. According to the Bridgewaters, this vision was entirely in keeping with Christianity. Yet it was also rooted in the claims to cultural authority of the disciplinary sciences, their requirement for freedom in theorizing, and their arrogation of a key role in the progress of the nation. The measured, law-like progress of the creation under God's guidance graduated teleologically into the steady progress of Britain in the age of reform to which the gentlemen of science viewed themselves as contributing. As Buckland's Bridgewater showed, the extensive and easily accessible coal reserves underpinning Britain's industrial and social progress had been laid down as a result of natural laws by a God who foresaw all.

Of course, there were numerous other means by which a progressive and law-governed view of the history of creation came to dominate public discourse in early Victorian Britain. Notable among these was George Combe's *Constitution of Man* (1828), which reached a vast audience and offered a vision of progressive social and moral change rooted in the phrenological laws of human nature. With the extraordinarily low pricing of its "people's edition," Combe's book became a blockbuster that by the end of the 1830s had sold more copies than all the Bridgewaters combined. Yet while it had a broadly religious gloss, Combe's controversial science of mind was seen by many as profoundly undermining of Christianity.[9] By contrast, the Bridgewaters played a distinctive role in rendering a progressive account of the unfolding of God's laws familiar and acceptable to Christian readers at large and in developing the reassuring persona of the Christian "man of science."

This revisionist view of the Bridgewaters contributes to a larger rethink of the Christian character of early Victorian science. As has recently been argued, the "theistic science" that dominated in Britain at this period was rooted in the principle of the "uniformity of nature," with men of science and clergy alike sustaining the notion that the creation was governed and upheld by God through natural laws.[10] Such science had its limits, notably around ultimate origins and around the distinctive moral and spiritual

9. Van Wyhe 2004.
10. Stanley 2015.

qualities of humans, as the Bridgewaters themselves illustrate. Nevertheless, the vision of the sciences that the series publicly baptized in the 1830s was in certain ways more similar to the scientific naturalism that came to dominate in late Victorian Britain than either its advocates or modern historians have been inclined to recognize. As this book has demonstrated, the Bridgewaters not only contributed to securing widespread public support for such a vision of the sciences but also informed discussions among those actively engaged in scientific education or research concerning the practice of science. In particular, they fed into contemporary debate concerning the true extent of the uniformity of nature and the proper role of teleology in the practice of the life sciences.

From a post-Darwinian perspective, the contributions of the Bridgewaters to that debate seem hopelessly outmoded, yet at the time they raised issues that went to the core of the prevailing Christian vision of the sciences. These were not merely reactionary works of "Paleyite natural theology."[11] Rather, in the diversity and (in some cases) sophistication of their engagements with the rapidly changing sciences, they contributed materially to a developing conversation concerning the proper practice of theistic science. The point is well illustrated by the way in which the progressive, law-governed account of creation offered by several of the Bridgewaters was drawn on by authors who wished to develop it further—most especially Robert Chambers and Charles Darwin. Using the same framework to legitimize transmutation and offering quotations from the series to demonstrate the religious orthodoxy of their views, both authors presented their work as conforming to the respectable standards of theistic science. Not, of course, that they were successful in convincing all of their readers—far from it. *Vestiges* in particular, with its anonymous authorship and lack of scientific credentials, was viewed by most savants, including the Bridgewater authors, as betraying the program of theistic science. But in any case, the Christian vision of the sciences broadly symbolized by the Bridgewaters soon came under increasing challenge from other quarters.

As we have seen, the assertions of the incompatibility of science and religion expressed by atheist radicals in the 1820s became more commonplace as Owenism fractured in the early 1840s. In May 1844 George Holyoake's weekly *Movement* carried an article on "Science and Religion" that began with the assertion that the two were "natural enemies." Religionists had, throughout history, either oppressed or enthralled men of

11. See, for example, Desmond 1989b.

science, and the period of "hollow confederacy" between science and religion that had lasted from Bacon to "the fashionable reign of [the] Bridgewater treatises" had only served to constrain science.[12] Such views were further fueled by the *Vestiges* controversy, and by the 1850s Holyoake was proving increasingly successful in selling what he had rebranded as "Secularism" to the respectable middle classes. Moreover, a new generation of liberal intellectuals made common cause both with these secularists and with the growing cadre of young men from middle-class families who were looking to make a living from science, combining to offer a concerted challenge to the prevailing theistic framework for the sciences that was assisted in due course by the publication of *Origin*.[13]

This well-developed story of the rise of Victorian scientific naturalism rightly underpins our understanding of the development of professional science in the second half of the nineteenth century. Nevertheless, in light of the history told here, the transition appears somewhat more gradual and partial than we have been wont to think. Even the scientific naturalists emerged out of the Christian culture of science that the Bridgewaters helped to create, and both their personal experiences and their understanding of the cultural terrain were shaped by that history.

The point is well illustrated by the example of Edward Clodd. The son of a Suffolk maritime pilot, Clodd was given a strict religious upbringing and was intended for the Baptist ministry but chose instead as a teenager in the 1850s to become a banker's clerk in London. As he took the opportunity to develop his scientific interests, he read the Bridgewaters and other books that had as their main theme "The power, wisdom and goodness of God manifested in the Creation." Looking back, he considered that it was these books that "started certain lines of thought, and were a sort of sliding scale towards unorthodox views," but it was a slow process. When *Origin* was published, for instance, a nineteen-year-old Clodd was reading Bell's Bridgewater—just out in a new edition—and he did not find Darwin's work "a greatly disturbing force." Darwin, Clodd later recalled, "after only hinting that his theory 'would throw light on the origin of man and his history,'" had added in good Bridgewater fashion, "there is grandeur in this view of life with its several powers having originally been breathed by the Creator into a few forms or only one." To Clodd, *Origin* itself seemed to be a work of theistic science. His ultimate move to agnosticism and his involvement in the secular movement came much later, following a range

12. Ryall 1844.

13. On the changing "visions of science" at this period, see Secord 2000, chap. 12–14.

of experiences not only during the "*Sturm and Drang* period" of the 1860s but over subsequent decades.[14]

Clodd was not entirely untypical among the scientific naturalists in being conscious of such continuity. Admittedly, when John Tyndall, the recently appointed professor of natural philosophy at the Royal Institution, was asked in 1854 to prepare a new edition of Prout's Bridgewater, he declined, considering it a poor thing, lacking "even common scientific depth, not to speak of religious inspiration." He even doubted Prout's religious sincerity. But while he had a low estimate of what could be known of God through natural theology, he was at the same time developing his own spiritual outlook in which nature was infused with religious meaning. Likewise, when botanist Joseph Hooker delivered his presidential address to the British Association in 1868, he sought to connect the beneficial effect of the Bridgewaters on a previous generation to contemporary efforts to maintain a "mutually considerate and friendly" attitude between "Religion and Science." His prescription involved a very un-Bridgewater-like separation of scientific and religious realms rooted in Herbert Spencer's vision of the inscrutability of God, but the presence of the Bridgewater emblem is nonetheless striking.[15]

This assertion of two independent realms was a characteristic strategy by which the scientific naturalists combined the potential for harmony between science and religion with an insistence that theology and clerical authority had no place in the practice of science. Thus, the grand rhetoric of Thomas Huxley about the "extinguished theologians" who lay round "the cradle of every science" has to be read alongside his assertion of the harmony between science and that religion that was rooted in the "deeps of man's nature." This nevertheless meant jettisoning many aspects of the earlier Christian vision of the sciences, especially in regard to the notion of design, and on that point the Bridgewaters were considered to be singularly outmoded. When, by the late 1880s, Unitarian theologian James Martineau was seeking to root Christianity in natural theology, he was aware that he was going against the latest fashion of "Advanced thought." Many a would-be scientist, eager to act "*comme il faut*" in regard to final causes and to show himself a true Darwinian, now thought it necessary "to have his fling at 'Paley and the Bridgewater Treatises,'" Martineau reported. By 1896 the ridicule that was sometimes "lavished" on the series

14. Clodd 1916, 11–12. On Clodd, see Lightman 2007, 253–66. On the "grandeur" of an "anticipating or prospective intelligence," see Bell 1833a, 221.

15. Hooker 1869, lxxiii–lxxv; Eve and Creasey 1945, 56. On Tyndall, see also Brooke and Cantor 1998, 331–34, and Cantor 2015.

was such that even Andrew Dixon White, in his *History of the Warfare of Science with Theology in Christendom*, felt it was hardly fair.[16]

For the scientific naturalists, the Bridgewaters had thus become signifiers, in many respects, of an outmoded theistic science. It is nevertheless notable that editions of some of the series continued to appear even as late as the 1880s. While the scientific naturalists may have been especially vocal in articulating a vision of science as separated from religion, theirs was only one of a number of competing visions. Not only were there prominent groups of theistic scientists, but many of the most successful scientific popularizers of the second half of the century, such as J. G. Wood and Margaret Gatty, continued to cast the sciences as being in harmony with Christian orthodoxy. Indeed, the great success of the Bridgewaters contributed to establishing a tradition of religiously infused popular science writing that developed and flourished over succeeding decades. The continuing republication of the Bridgewaters is itself significant in this regard. After the initial flurry of sales in the 1830s and early 1840s had ebbed away, most of the series were reissued in the early 1850s at a much cheaper price in Henry Bohn's Scientific Library. Moreover, after Bohn's cheap libraries were taken over by the educational publisher George Bell, new editions continued to appear over the three succeeding decades (fig. C.1).[17]

Reviewers of these later editions of the Bridgewaters continued to point to their value as accessible scientific works that offered a vision of the sciences in harmony with Christian orthodoxy. Reviewing the state of "popular science" at midcentury, Anglican cleric and regular *Quarterly Review* contributor Whitwell Elwin saw the Bridgewaters as being among the historically significant "impulses to popular science." That had not been the earl's intention, but nothing, according to Elwin, had been "less wanted than a work upon Natural Theology, for Paley had left little to add, and little to amend." However, with the exception of Chalmers, the Bridgewater authors had "turned more of their attention to science than theology," and this was the light in which the series "came soon to be regarded." The Bridgewaters thus served as an important reference point for the literature of "popular science" that became, in their wake, one of the staples of nineteenth-century publishing. Like the Bridgewaters, many of the most successful works of "popular science" in the second half of the century featured a significant emphasis on divine design, again offering to readers

16. White 1896, 1:43–44; Martineau 1888, 1:xiii; Lightman 2001; [Huxley] 1864; [Huxley] 1860, 556.

17. Lightman 2001. For details of later Bridgewater editions, see appendix B.

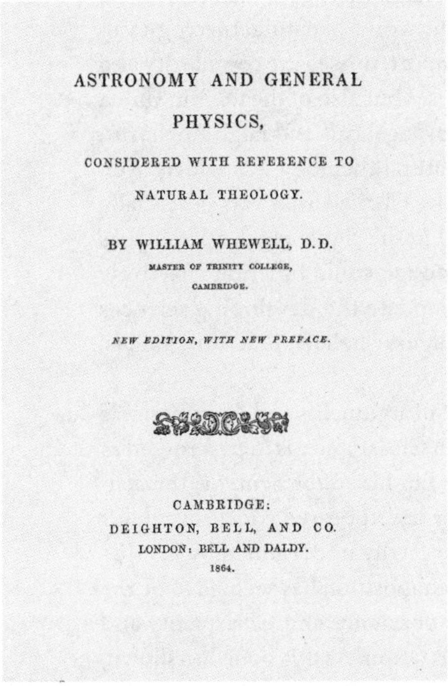

ASTRONOMY AND GENERAL
PHYSICS,
CONSIDERED WITH REFERENCE TO
NATURAL THEOLOGY.

BY WILLIAM WHEWELL, D.D.
MASTER OF TRINITY COLLEGE,
CAMBRIDGE.

*NEW EDITION, WITH NEW PREFACE.*

CAMBRIDGE:
DEIGHTON, BELL, AND CO.
LONDON: BELL AND DALDY.
1864.

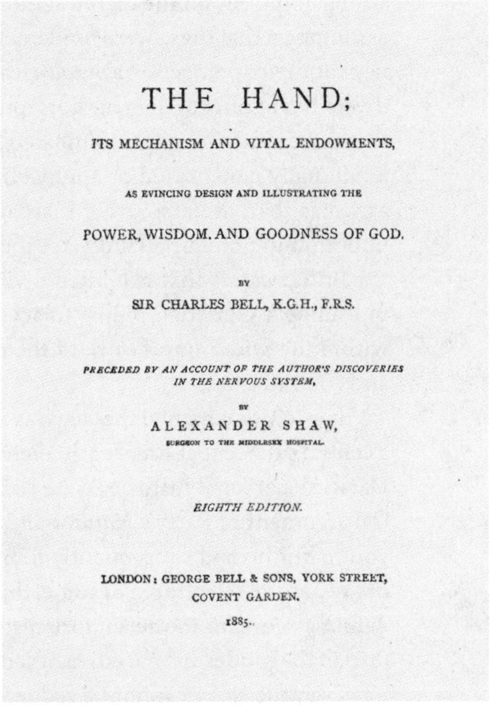

THE HAND;
ITS MECHANISM AND VITAL ENDOWMENTS,
AS EVINCING DESIGN AND ILLUSTRATING THE
POWER, WISDOM, AND GOODNESS OF GOD.
BY
SIR CHARLES BELL, K.G.H., F.R.S.

*PRECEDED BY AN ACCOUNT OF THE AUTHOR'S DISCOVERIES IN THE NERVOUS SYSTEM,*
BY
ALEXANDER SHAW,
SURGEON TO THE MIDDLESEX HOSPITAL.

*EIGHTH EDITION.*

LONDON: GEORGE BELL & SONS, YORK STREET,
COVENT GARDEN.
1885.

FIGURE C.1 Late editions of Whewell's (1864) and Bell's (1885) Bridgewaters Treatises.

opportunities for incorporating the latest scientific findings within quotidian religious experience.[18]

## *Science and Religion: A More Practical View*

Inheriting histories of Victorian science from the victorious scientific naturalists, an earlier generation of historians of science assumed that the Bridgewaters must have "bored the public" with their "dropsical version of the thesis that all of nature shows design."[19] As I have shown in this book, however, that was far from being the case. For many readers, the Bridgewaters offered an inspiring and securely Christian vision of the latest science that was most welcome. That vision certainly incorporated a

18. [Elwin] 1849, 317; Lightman 2007.
19. Young 1985, 32.

strong emphasis on the manifestations of God's design in nature, but the assumption that these were fundamentally works of natural theology, valued for their apologetic or theological import, misses the complexity and diversity not only of their authors' purposes but also of the ways in which they were read. In an age dominated by evangelicals and High Anglicans, a rationally constituted or apologetic natural theology was widely seen as worse than useless: rather than making Christians, it was potentially undermining of Christianity. Yet if used in the right way, works such as the Bridgewaters that exhibited divine design could function effectively in leading a Christian public to accommodate the developing sciences within the wider framework of their religious beliefs, practices, values, and communities.

The view that natural theology, as a set of arguments and doctrines, was a central preoccupation of early Victorian Christianity is in part rooted in Darwin's personal history. As he recalled in his *Autobiography*, the central argument of Paley's *Natural Theology* had appeared conclusive in his youth, but he had subsequently found it fatally undermined by his theory of natural selection. For some, this juxtaposition has seemed to be the defining moment in the Victorian story of science and Christianity and thus of the modernist narrative of secularization. As this book has shown, however, the story cannot be reduced to such a simplistic contrast. Natural theology had been developed far beyond Paley's argument from the functional adaptation of specially created living organisms before Darwin published his theory, as he well knew, and some of the Bridgewaters were among the first to explore these alternative arguments. More fundamentally, a large proportion of Christians in Darwin's youth and afterward were not willing to accord the arguments of natural theology an important role in their religion.

This was clear not only in several of the Bridgewaters themselves but also in the way in which religious reviewers and readers responded to them. Design talk was much more often seen as relevant to the practical and emotional aspects of Christianity than to the production of a rational apologia. Its value lay, for instance, in the ways in which it supported private or public acts of devotion or aided the Christian upbringing of children, particularly by evoking feelings of wonder and love and inculcating mental habits of religious association. Some might call this a natural theology, but if they did they meant something very different from the sort of rational argument offered by Paley and dismissed by Darwin.

This history came to be rewritten in the course of time, including by Christian apologists and philosophical theologians, for many of whom the Bridgewaters became a useful foil for their changing views, especially in the wake of *Origin*. Yet throughout the middle years of the century, religious commentators continued to argue that the value of the series did not depend on any apologetic or philosophical pretensions. When the *Christian Remembrancer* reviewed a wide range of works of natural theology in 1857, for instance, it included a warm appreciation of the Bridgewaters precisely because their purpose was "to illustrate a recognised truth" rather than arguing for a "doubtful conclusion." The winning of converts had not been the purpose of the bequest, but the journal claimed that "many religious minds" had been "strengthened by the striking illustrations" the volumes contained. Moreover, on the eve of the publication of *Origin*, the writer was still anticipating with Whewell that their value would be unabated as scientific discoveries shifted phenomena "from the domain of independent facts to that of general laws."[20]

Taking a book-historical approach to the Bridgewaters has reexposed the full richness of this story. Too often, histories of science and religion have focused primarily on matters of belief, not least because of a pervasive concern to address the question of whether or not the two are locked in an inexorable conflict located in the realm of ideas and cultural authority. Of course Christianity (perhaps especially Protestantism) is notable for its peculiar emphasis on the religious significance of correct beliefs. Yet no religion can be reduced simply to doctrinal matters. Day-to-day religious practice was as important in Victorian Protestantism as in other religious cultures, and by focusing on the practice of reading, this book has begun to reveal the historical significance of such activities. Where historians have typically focused on the doctrinal arguments of largely wealthy men, expressed in acts of authorship, the approach taken here has been to consider the varied concerns of a more diverse range of readers—men, women, and children—seeking to make sense of the rapidly changing sciences within the daily round of lives conditioned by the practical as well as the doctrinal expectations of Protestant Christianity (fig. C.2).

What emerges from this approach should not surprise us. Once we move away from the statements of theological specialists, we find that the daily preoccupations of Christians on the eve of the Victorian age related to a number of deep-seated psychological and social concerns. The lived experience of Christianity was to a significant extent dominated by the ebb and flow of religious feelings, and the quotidian practice of Christianity was heavily structured by a concern to develop and manage those feelings. Whether in acts of private or family devotion, in the management of family relationships and the upbringing of children, or in churchgoing and other

20. "Natural Theology," *Christian Remembrancer* 33 (1857): 70–117, on 80–86.

FIGURE C.2 Mid-Victorian family reading. From the title page of the Religious Tract Society's weekly *Leisure Hour*, 1855.

public activities, a significant preoccupation was with how what was done would foster particular feelings, attitudes, and values that would in turn prompt particular behaviors. To a greater or lesser extent, religious readers of the Bridgewaters read them in relation to these concerns, and one of the most telling reasons why the series became important was that it could be read in this way. Moreover, the concerns of readers with the practical management of religious feelings were echoed by clergy and other theological commentators not only in reviews but also in the kinds of practical religious works that have often been left by historians to gather dust.

This history is quite distinct from one in which the cognitive compatibility of religious and scientific beliefs is taken to be the sole or overriding concern. Of course, ordinary believers as well as theologians have been interested in the implications of changing scientific beliefs for core religious doctrines—the overarching "conceptions of a general order of existence" that underpin religious experience.[21] Yet as modern psychological and social studies of science and religion confirm, people are extraordinarily resourceful in holding together beliefs that others consider should be incom-

21. Geertz 1966, 4.

patible. By offering a history of science and religion from the ground up, this book has shown that the concerns of the readers of the Bridgewaters were not reducible to theology, let alone to the arguments of natural theology, and related in a significant degree to the management of religious experience within the often collective context of quotidian religious practice.

The history of reading is especially valuable in recovering such perspectives. Evidence of the experiences of ordinary people in relation to the interconnectedness of science and religion is naturally more difficult to acquire than it is for those involved in producing published discussions. However, it was very often in the practical activity of reading that ordinary people encountered questions about how the developing sciences related to the religious aspects of their lives. Reading had long held a distinctively important role in the practice of Protestant Christianity, but the dual transformation in print culture and the sciences on the eve of the Victorian age brought the relevance of the sciences to Christianity into focus on a scale never before experienced. How was scientific reading to be made compatible not only with religious beliefs but also with religious practices and the feelings and impulses that they generated? By focusing on such concerns, this book has shown that the history of science and religion in Victorian Britain can be usefully rethought in relation to the transformation that took place in print culture and the associated reconfiguring of religious practices.

To a greater or lesser extent, the Bridgewater authors were conscious of the ways in which the changing character of the sciences had combined with changes in print publication to offer a significantly new context for the practice of Christian faith. Though the task of authorship came to them in an oddly predefined form, they responded in different ways to the challenge of writing books that would allow readers to connect the latest findings of the sciences with their lived religion. Certain of the clerical authors—most notably Whewell and Chalmers—were particularly attuned to the question of how the sciences might relate to the practice of Christianity. More generally, however, the authors applied themselves to blending their sweeping accounts of the sciences with pious reflection in ways that invited a diverse range of Christian readers to look on recent developments "with confidence and pleasure." For all that the financial terms of their recruitment led to jibes about the authors writing "as per contract," they clearly gave significant consideration to the needs of a changing reading public in the age of industrial print.[22] Their efforts were necessarily experimental, as they tried out the possibilities of the

22. [Bunbury] 1890–91, 2:46; Whewell 1833b, vi.

new publishing market, but what they produced were works that readers found they could use in the context of practical religion.

In the process of thus finding their voices as Bridgewater authors, the eight contributed to the construction of a social role and identity for the emerging "man of science." The historic neglect of scientific authorship has begun to be addressed in recent years, but much still remains to be done to understand how the public persona of scientific savants depended on the development and exploitation of different forms of publication. Along with other reflective treatises on the sciences in the 1830s, the Bridgewaters provided a platform that paralleled that of the British Association as a locus in which to perform the new kind of public role and identity that was being forged in these years. In an age distinguished by a characteristically British process of moderate reform, the new public prominence of scientific men was achieved through a rapprochement with traditional forms of power within the church and the aristocracy. Manifesting new authorial purposes and personas, the Bridgewaters became public markers of that alliance and of the capacity of scientific men to behave in accordance with the notions of civility that prevailed within the leading social circles. Their tone of humble piety marked out the men of science as public figures quite as suitable as the clergy to take their place in the cultural leadership of the nation.

These close ties between the practice of authorship and ideas about religious roles, identity, character, and ethics again take us away from the customary emphasis on matters of religious belief. Moreover, similar issues come to the fore when we examine how the Bridgewaters were used by those teaching about the sciences in Britain's universities and beyond. The education of youth was one of the most fraught issues involved in the expanding role of the sciences in British culture during the age of reform, and those involved in scientific education labored hard to make clear how it could foster religious character. Young people engaged in such education were expected to find in the Bridgewaters a more elaborate induction into the virtues as well as the intellectual framework of the Christian "man of science." While student readers not infrequently engaged critically with the series, it is clear that a rising generation found in the Bridgewaters a key point of orientation for their developing ideas about the religious value of the sciences and the religious character of scientific men.

It was not only those learning the sciences for the first time who were touched by the concern about how the Christian man of science was to conduct himself. Once again, the history of reading offers new insights here concerning the more practical, quotidian aspects of the science and religion story. For readers such as Darwin, the Bridgewaters were part of the apparatus of scientific work. Reading, rereading, note taking, and annotating were all aspects of the process by which the mental work of science was carried out, and this included understanding the proper conduct of inquiry and the proper character of scientific theory within a wider religious framework. While Darwin and many of his contemporaries frankly disagreed with the Bridgewaters about aspects of the work of science, they engaged vigorously with the series in considering questions that were practically important in both scientific and religious terms. Moreover, as the processes by which the identity and practice of the man of science were redefined over the succeeding generation come to be better known, it becomes clearer that many of these same concerns about morality and character persisted.[23]

Through its more detailed attention to the production and use of printed matter, then, this book has offered a broader perspective on science and religion in Victorian Britain than has been typical of earlier studies. In particular, it stands as a corrective to the master narrative of conflict, in which the overriding story is of Christian belief forced to retreat in the face of advances in scientific theory. This, of course, was the rhetoric of Huxley and other scientific naturalists. As we have seen, however, even Huxley admitted that there was more to religion than theology. For him, as for several of the scientific naturalists, "religion, like poetry and art, belonged to the realm of feeling and ethics."[24] Of course, whatever religious feelings the likes of Huxley or Tyndall had were far removed from anything like Christianity. Yet the fact that these prophets of theological conflict were aware that the story of science and religion went much deeper than intellectual argumentation gives notice to historians in an altogether more secular age that we should broaden our perspective. The story of science and religion in Victorian Britain is rooted in a much more multifaceted world of belief and practice than we have been accustomed to recognize. The kind of bottom-up history developed in this book, grounded in the material objects of everyday life, offers valuable new avenues of inquiry in pursuit of that richer story.

23. White 2003; Dawson 2007; DeWitt 2013.
24. Lightman 2001, 348.

# *Acknowledgments*

*Of making many books there is no end; and much study is a weariness of the flesh.*

ECCLESIASTES *12:12*

The long journey that has led to this book has been one involving many inspiring and helpful companions to whom I am greatly indebted. My interest in the Bridgewater Treatises and in what they could reveal about science and religion in the quarter century before *Origin* developed almost as soon as I encountered the history and philosophy of science through the inspirational teaching of Simon Schaffer and Andrew Cunningham in the University of Cambridge. That interest led me to the door of John Hedley Brooke at the University of Lancaster, under whose erudite and thoughtful supervision the project blossomed. My researches were assisted by the legendary Eric Korn, who agreed to let me have a set of the Bridgewaters for a knockdown price, and by my parents, who paid for them. I was privileged to meet Geoffrey Cantor of the University of Leeds at an early stage, gaining enormously from his wisdom, understanding, and generous support over many years. Through Geoffrey I was introduced to the wonderful academic community at the Leeds Centre for History and Philosophy of Science, then led by himself, Jon Hodge, and John Christie, who welcomed me with enormous liberality, and to the compelling and always enlightening Jack Morrell. Another crucial early encounter was with Jim and Anne Secord, whose penetrating insight and unfailing helpfulness have since been as inspiring as they have been encouraging.

In the intervening years, I have accrued too many debts of gratitude to enumerate, especially among colleagues and students in Cambridge, Sheffield, and Leeds. While Munby Fellow in Bibliography, I learned a vast amount from colleagues in Cambridge University Library, above all from the brilliant and generous Elisabeth Leedham-Green, who also introduced me to the inspirational Leslie Howsam. The academic communities at Darwin College and the Department of History and Philosophy of Science were wonderful places in which to push the boundaries of knowledge. One former student, the remarkable Aileen Fyfe, has gone on to teach me many things through her researches on topics adjacent to those explored in this book. At the Universities of Leeds and Sheffield, very particular thanks are due to Geoffrey Cantor and Sally Shuttleworth, whose leadership of the Science in the Nineteenth-Century Periodical project was so congenial and productive, with a wonderful team that included Graeme Gooday, Gowan Dawson, Richard Noakes, and Sam Alberti. A succession of wonderful colleagues and students at Leeds have provided a matchless academic community over the last twenty-three years, marked in equal measure by rigor, good humor, and generosity. I am immensely grateful to them all for much help and many kindnesses.

The rebirth of this book project after a lengthy hiatus owes much to the inspirational guidance and practical support of Greg Radick. He and Jon Hodge patiently read the manuscript as it appeared, keeping me supplied with wonderfully insightful comments, words of encouragement, and pints of Yorkshire bitter in a winning combination for which I am extremely thankful. Jo Elcoat and Rosie Alfatlawi deserve my thanks for helping me track down many obscure references to readers' encounters with the Bridgewaters. The Leeds School of Philosophy, Religion, and the History of Science generously supported the work by granting research leave and covering research and publication expenses. A period of secondment to the University of Oxford Constructing Scientific Communities project came at a critical moment, when working with Sally Shuttleworth, Gowan Dawson, and Bernard Lightman provided a much-needed lift. I am enormously indebted to Jim Secord and two anonymous referees who offered invaluable suggestions on the completed manuscript and to Geoffrey Cantor who provided very helpful comments on several portions of it. At the University of Chicago Press, Karen Darling was hugely helpful, and the final book is much better for her astute guidance. I am also much indebted to Tristan Bates, Rebecca Brutus, Elizabeth Ellingboe, and Steven LaRue for their patient and expert management of the manuscript editing and design and for all of the team at Chicago for their professionalism and creativity in making this book.

The research on which this book is based has depended on the knowledgeable assistance of many librarians and archivists across the United Kingdom and beyond and on references and pointers from many colleagues, for which I remain most grateful. I am obliged to Patrick Boylan; the British Geological Survey Archives; the British Library; Bodleian Libraries Oxford; the Syndics of Cambridge University Library; the Chambers family; Christ Church Oxford; Her Majesty Queen Elizabeth II; Andrew Grout; the Linnean Society of London; New College Special Collections, University of Edinburgh; the Library and Archives, Natural History Museum, London; the National Library of Scotland; Oxford University Museum of Natural History; the Royal College of Surgeons of England Archives; the Master and Fellows of Trinity College Cambridge; the Frances Hirtzel Collection of Richard Owen Correspondence, Special Collections Research Center, Temple University Libraries; University College London Library Services, Special Collections; and the Wellcome Library, London, for permission to quote from manuscripts in their possession. I am also grateful to Cambridge Scholars Publishing for permission to reuse some text from my chapter in a volume edited by Neil Spurway titled *Laws of Nature, Laws of God? Proceedings of the Science and Religion Forum Conference*, which was published by them in 2014.

The biggest thanks come last. Roberta Topham has had to live with my interest in the Bridgewaters for more than a third of a century and Rosie and Matthew since birth. They have my undying gratitude for their forbearance, interest, support, and love. I hope that they are able to take much pride in the result and that they will enjoy talking about something else in coming years as much as I will.

APPENDIX A

# Note on Currency and the Value of Money

In nineteenth-century Britain, the pound (£) was divided into twenty shillings (s), each of which was worth twelve pence (d), so that a pound amounted to 240 pence. Coins in common use included the following:

*Copper*
¼d = farthing
½d = ha'penny
1d = penny

*Silver*
3d = threepence
6d = sixpence
1s = shilling
2s 6d = half crown
5s = crown

*Gold*
10s 6d = half guinea
£1. 1s = guinea

Establishing modern equivalents for old currency is fraught with difficulty, not least because the rate of change in the prices of goods and services has differed from the rate of change in mean and median incomes. Using a factor of one hundred can be a useful if crude way of getting a very broad sense of modern equivalents in relation to pounds. Taking decimalization into account, the equivalent factor for pence would be approximately forty.

For those interested in greater accuracy, the relative value calculator at https://www.measuringworth.com/ offers a range of ways of calculating modern equivalents for £1 in 1833. Among these are a 2020 equivalent of £97, where the emphasis is on retail prices, and £880, where the emphasis is on average incomes. The great disparity between the figures reflects the vast increase in personal wealth over the last two centuries.

A more developed sense can, of course, be obtained by paying more attention to detail, such as the sample domestic budgets in Peel (1934, 1:104–8, 126–34). On wealth and poverty, see also Daunton (1995).

APPENDIX B

# British Editions of the Bridgewater Treatises

This list is based on the examination of many library catalogs, publishers' archives, trade publications and advertisements, and numerous original copies. However, it is quite possible that there may be omissions and/or ghosts among the later editions, where the evidence concerning unique editions is more ambiguous. The number of copies printed is derived from the relevant printer or publisher's archives except where otherwise stated.[1]

**William Whewell, *Astronomy and General Physics***

| *Edition* | *Date* | *Format* | *Price* | *No. Printed* |
|---|---|---|---|---|
| **William Pickering, 1833–47** | | | | |
| 1st | March 1833 | Demy 8vo | 9s 6d | 1,000 |
| 2nd | June 1833 | Demy 8vo | 9s 6d | 1,500 |
| 3rd | January 1834 | Demy 8vo | 9s 6d | 2,000 |
| 4th | September 1834 | Demy 8vo | 9s 6d | 1,500 |
| 5th | June 1836 | Demy 8vo | 9s 6d | 1,500 |
| 6th | May 1837 | Fcp 8vo | 6s | 2,000 |
| 7th | June 1839 | Demy 8vo | 9s 6d | 1,250 |
| 8th | May 1847 | Fcp 8vo | 5s | 1,000 |
| **Henry G. Bohn, 1852–62** | | | | |
| "7th" | 1852 | Post 8vo | 3s 6d | |
| "8th" | 1862 | Post 8vo | 3s 6d | |
| **Bell & Daldy / George Bell & Sons, 1864–78** | | | | |
| "New" | 1864 | Fcp 8vo | 5s | |
| "8th" | 1870, 1871, 1878 | Post 8vo | 3s 6d | |

1. For more detail, see also Topham 1993, 282–85, 344–46.

**John Kidd, *Physical Condition of Man*[1]**

| *Edition* | *Date* | *Format* | *Price* | *No. Printed* |
|---|---|---|---|---|
| **William Pickering, 1833–37** | | | | |
| 1st | April 1833 | Demy 8vo | 9s 6d | 1,000 |
| 2nd | June 1833 | Demy 8vo | 9s 6d | 1,000 |
| 3rd | January 1834 | Demy 8vo | 9s 6d | 1,000 |
| 4th | January 1836 | Demy 8vo | 9s 6d | |
| 5th | 1837 | Demy 8vo | 9s 6d | |
| **Henry G. Bohn, 1852** | | | | |
| 6th | 1852 | Post 8vo | 3s 6d | |
| **Bell & Daldy / George Bell & Sons, 1870–87** | | | | |
| | 1870, 1887 | Post 8vo | 3s 6d | |

**Thomas Chalmers, *Moral and Intellectual Constitution of Man*[2]**

| *Edition* | *Date* | *Format* | *Price* | *No. Printed* |
|---|---|---|---|---|
| **William Pickering, 1833–39** | | | | |
| 1st | May 1833 | Demy 8vo | 16s | 1,500 |
| 2nd | July 1833 | Demy 8vo | 16s | 1,500 |
| 3rd | January 1834 | Demy 8vo | 16s | 2,000 |
| 4th | November 1835 | Demy 8vo | 16s | >1,000 |
| 5th | 1839 | Demy 8vo | 16s | >600 |
| **Henry G. Bohn, 1853** | | | | |
| | 1853 | Post 8vo | 5s | |
| **Bell & Daldy / George Bell & Sons, 1869–84** | | | | |
| | 1869, 1870, 1871, 1884 | Post 8vo | 5s | |

**Charles Bell, *The Hand***

| *Edition* | *Date* | *Format* | *Price* | *No. Printed* |
|---|---|---|---|---|
| **William Pickering, 1833–37** | | | | |
| 1st | June 1833 | Demy 8vo | 10s 6d | 1,000 |
| 2nd | September 1833 | Demy 8vo | 10s 6d | 2,000 |
| 3rd | April 1834 | Demy 8vo | 10s 6d | 3,000 |
| 4th | October 1837 | Demy 8vo | 10s 6d | 2,500 |
| **John Murray, 1852–60** | | | | |
| 5th | 1852 | Post 8vo | 7s 6d | 1,500 |
| "6th" | 1854 | Post 8vo | 7s 6d | |
| "6th" | 1860 | Post 8vo | 6s | 1,500 |

| | | | | |
|---|---|---|---|---|
| **Bell & Daldy / George Bell & Sons, 1865–85** | | | | |
| "7th" | 1865, 1870 | Post 8vo | 5s | |
| "8th" | 1872, 1873, 1877, 1885 | Post 8vo | 5s | |
| "9th" | 1874 | Crown 8vo | 9s | |

**William Prout, *Chemistry, Meteorology, and the Function of Digestion***

| *Edition* | *Date* | *Format* | *Price* | *No. Printed* |
|---|---|---|---|---|
| **William Pickering, 1834** | | | | |
| 1st | March 1834 | Demy 8vo | 15s | 2,000 |
| 2nd | June 1834 | Demy 8vo | 15s | 3,000 |
| **John Churchill, 1845** | | | | |
| 3rd | 1845 | Demy 8vo | 15s | 350 |
| **Henry G. Bohn, 1855** | | | | |
| 4th | 1855 | Post 8vo | 5s | |
| **Bell & Daldy, 1870** | | | | |
| "4th" | 1870 | Post 8vo | 5s | |

**Peter Mark Roget, *Animal and Vegetable Physiology***

| *Edition* | *Date* | *Format* | *Price* | *No. Printed* |
|---|---|---|---|---|
| **William Pickering, 1834–40** | | | | |
| 1st | May 1834 | Demy 8vo | £1 10s | 2,000 |
| 2nd | July 1834 | Demy 8vo | £1 10s | 3,000 |
| 3rd | August 1840 | Demy 8vo | £1 10s | 1,500 |
| **Bell & Daldy, 1867–70** | | | | |
| 4th | 1867 | Post 8vo | 12s | |
| 5th | 1870 | Post 8vo | 12s | |

**William Kirby, *History, Habits, and Instincts of Animals***

| *Edition* | *Date* | *Format* | *Price* | *No. Printed* |
|---|---|---|---|---|
| **William Pickering, 1835** | | | | |
| 1st | June 1835 | Demy 8vo | £1 10s | 3,000 |
| 2nd | June 1835 | Demy 8vo | £1 10s | 2,000 |
| **Henry G. Bohn, 1852–53** | | | | |
| "New" | 1852, 1853 | Post 8vo | 10s | |
| **Bell & Daldy, 1870** | | | | |
| "New" | 1870 | Post 8vo | 10s | |

| **William Buckland, *Geology and Mineralogy*[3]** | | | | |
|---|---|---|---|---|
| *Edition* | *Date* | *Format* | *Price* | *No. Printed* |
| **William Pickering, 1836–37** | | | | |
| 1st | September 1836 | Demy 8vo | £1 15s | 5,000 |
| 2nd | April 1837 | Demy 8vo | £1 15s | 5,000 |
| **George Routledge, 1858** | | | | |
| "New" | 1858 | Demy 8vo | 24s | 5,000 |
| **Bell & Daldy, 1869–70** | | | | |
| 4th | 1869, 1870 | Post 8vo | 15s | |

1. Kidd's Bridgewater was printed at Oxford University Press, but the number of copies printed of the first two editions is inferred from the number of labels printed by Whittingham; the number of copies in the third edition is given in Pickering to TC, 26 Nov 1833, NC, CHA 4.212.24.

2. The fourth and fifth editions of Chalmers's Bridgewater were printed by William Collins; the numbers given here refer to the number of title pages printed by Whittingham for Pickering's variant edition of these printings.

3. The size of Buckland's third edition is given in Burgess 1967, 72.

APPENDIX C

# British Reviews of the Bridgewater Treatises, 1833–38

This list of early reviews of the Bridgewaters is arranged chronologically by publication, with the different types of periodicals arranged alphabetically under separate headings. The list is based on systematic manual searching of British and Irish periodicals for the period 1833–38, combined with digital searching of some periodicals and many newspapers. Further details concerning many (though not quite all) of the periodicals searched, including ones in which no reviews were located, are given in Topham (1993), 553–615.[1] The anonymous authors of articles from periodicals included in the *Wellesley Index* (Houghton et al. 1966–89) are identified from that source unless otherwise stated.

## *Literary Weeklies*

### *ATHENÆUM*

[William Cooke Taylor].[2] Review of Whewell. 23 March 1833, 184.
[Perceval B. Lord]. Review of Kidd. 20 April 1833, 247–49.
[William Cooke Taylor]. Review of Chalmers. 22 June 1833, 396–97.
[Perceval B. Lord]. Review of Bell. 6 July 1833, 427–29.
[Perceval B. Lord]. Review of Prout. 10 May 1834, 349–50.
[Perceval B. Lord]. Review of Roget. 12 July 1834, 516–18.
Review of Kirby. 29 August 1835, 663.
Review of Buckland. 4 February 1837, 79–81.

### *EXAMINER*

"Astronomy and General Physics." 9 June 1833, 357a.

1. Note also the direct responses to the Bridgewaters issued in pamphlet form (Best 1837; Brown 1838; [Carter] 1837; Cockburn 1838b; De Johnsone 1838; Ensor 1836; [Morgan] 1834; Roberton 1836).

2. Authors from the *Athenæum* identified using Beaulieu and Holland 2001.

## *LITERARY GAZETTE*

Review of Whewell. 18 May 1833, 306.
Review of Kidd. 1 June 1833, 339–40.
Review of Chalmers. 15 June 1833, 370–71.
Review of Bell. 17 August 1833, 518–20.
Review of Prout. 22 March 1834, 201–2.
Review of Roget. 12 July 1834, 473–74.
Review of Kirby. 4 and 11 July 1835, 417–19, 438–41.
Review of Buckland. 12 November 1836, 721–22; 14 January 1837, 23–25.

## *SPECTATOR*

Review of Whewell. 13 April 1833, 331–32.
Review of Kidd. 20 April 1833, 360.
Review of Bell. 28 September 1833, 909–11.
Review of Roget. 14 June 1834, 566–67.
Review of Kirby. 25 July 1835, 711.
Review of Buckland. 1 October 1836, 946–47.

## LONDON NEWSPAPERS

"The Bridgewater Treatises." *Observer*, 1 July 1833, 4d.
"The Bridgewater Treatises." *Morning Post*, 3 July 1833, 5e.[3]
"Literature—Bridgewater Treatises [Whewell]." *United Kingdom*, 21 July 1833, 3d.

## OTHER NEWSPAPERS

"Sunday's and Tuesday's Posts." *Lincoln, Rutland, and Stamford Mercury*, 19 April 1833, 4c.[4]
"The Literary Review [Kidd]." *Leeds Times*, 25 April 1833, 4c.[5]
"Varieties—Literary and Scientific [Whewell]." *Reading Mercury*, 6 May 1833, 4e.[6]
Notice of Whewell. *Sussex Advertiser*, 15 July 1833, 3c.
"Varieties—The Bridgewater Treatises." *Lancaster Gazette*, 20 July 1833, 4b.[7]
"Literature [Chalmers]." *Scotsman*, 28 August 1833, 1e–f.
"The Bridgewater Treatises [Bell]." *Chester Chronicle*, 6 September 1833, 4b.
"Bridgewater Treatises. No. I [Whewell]." *Oriental Observer*, 23 November 1833, 553a–c.[8]
George Ensor, "The Bridgewater Treatises." *Dublin Evening Post*, 12 July 1834, 4a.

3. Repr. from *Observer*, 1 July 1833, 4d.
4. Repr. from *Spectator*, 13 April 1833, 331.
5. Repr. from *Spectator*, 20 April 1833, 360.
6. Based on *British Magazine* 3 (1833): 586–87.
7. Repr. from *Observer*, 1 July 1833, 4d.
8. Repr. from *Literary Gazette*, 13 April 1833, 331–32.

### RADICAL WEEKLIES

[James E. Smith?]. "Dr Chalmers." *Crisis*, 11 January 1834, 154c–155b.
[James E. Smith]. "Buckland's Bridgewater Treatise on Geology and Mineralogy." *Shepherd*, 15 March 1837, 49–51.

## *Literary Monthlies*

### *DUBLIN UNIVERSITY MAGAZINE*

[John Scouler]. "Geology and Mineralogy Considered." 8 (1836): 692–701.

### *FRASER'S MAGAZINE*

[James A. Heraud with William Maginn]. "The Bridgewater Treatises [Chalmers, Kidd, Whewell]." 8 (1833): 65–80.
[James A. Heraud with William Maginn]. "The Bridgewater Treatises: Dr Chalmers and Sir Charles Bell." 8 (1833): 259–78.
[James A. Heraud with William Maginn]. "The Bridgewater Treatises: Rev. William Kirby and Doctor Roget." 12 (1835): 415–29.
[James A. Heraud with William Maginn]. "The Bridgewater Treatises: Dr Buckland & Dr Prout." 16 (1837): 719–31.

### *GENTLEMAN'S MAGAZINE*

[John Mitford].[9] "Review—Whewell's *Bridgewater Prize Essay*." 103, pt. 1 (1833): 425–26.
[John Mitford]. "Review [Kidd]." 103, pt. 1 (1833): 612–13.
[John Mitford]. "Review—Dr Chalmers's *Bridgewater Treatise*." 103, pt. 2 (1833): 54–56.
[John Mitford]. "Review [Bell]." N.s., 1 (1834): 197–98.
F. S. W. "Remarks on Sir Charles Bell's Objections against the Undulatory Theory of Light in His Bridgewater Treatise." N.s. 2 (1834): 367–69.
[John Mitford]. "Review—Prout's *Bridgewater Treatise*." N.s. 2 (1834): 610–12.
[John Mitford]. "On the History Habits and Instincts of Animals." N.s. 4 (1835): 227–35.
[John Mitford]. "Professor Buckland's Geology." N.s. 7 (1837): 115–32.

### *LADIES' CABINET*

"Bridgewater Treatises. No. I [Chalmers]." 2 (1833): 65–66.[10]
"Bridgewater Treatises, No. IV [Bell]." 2 (1833): 196–201.


9. Authors from the *Gentleman's Magazine* identified using de Montluzin 2003.
10. Both extracted with acknowledgment from the *Literary Gazette*.

### METROPOLITAN MAGAZINE

“Notices of New Works [Whewell].” 7 (1833): 33–37.
“Notices of New Works [Kidd].” 10 (1834): 114–16.
“Notices of New Works [Roget].” 10 (1834): 42–43.

### MONTHLY MAGAZINE

“Amateur Naturalists.” 2nd ser., 15 (1833): 616–19.

### MONTHLY REPOSITORY

H[orne], R[ichard] H[enry]. “Buckland’s Geology.” 2nd ser., 11 (1837): 269–78.

### MONTHLY REVIEW

“Dr Chalmers’s Bridgewater Treatise.” N.s. 2 (1833): 378–87.
“Kidd’s Bridgewater Treatise.” N.s. 2 (1833): 499–509.
“Whewell’s Bridgewater Treatise.” N.s. 2 (1833): 561–68.
“Bell’s Bridgewater Treatise on the Hand.” N.s. 3 (1833): 424–37.
“Prout’s Bridgewater Treatise.” N.s. 1 (1834): 449–64.
“Roget’s Bridgewater Treatise.” N.s. 3 (1834): 219–38.
“Kirby’s Bridgewater Treatise.” N.s. 3 (1835): 1–14.
“Dr Buckland’s Bridgewater Treatise.” N.s. 3 (1836): 330–51.

### SCOTTISH MONTHLY MAGAZINE

“Buckland’s Bridgewater Treatise.” 1 (1836): 676–90.[11]

## *Quarterly Reviews*

### DUBLIN UNIVERSITY REVIEW AND QUARTERLY MAGAZINE

“Chemistry, Meteorology, &c.” N.s. 1 (1834): 230–46.[12]

### EDINBURGH REVIEW

[Brewster, David]. “Whewell’s *Astronomy and General Physics*.” 58 (1833–34): 422–57.
[Brewster, David]. “Dr Roget’s *Bridgewater Treatise—Animal & Vegetable Physiology*.” 60 (1834–35): 142–79.
[Brewster, David]. “Dr Buckland’s *Bridgewater Treatise—Geology and Mineralogy*.” 65 (1837): 1–39.

11. Extracted in *Caledonian Mercury*, 3 November 1836, 3d.
12. Extracted in “Science and Theology,” *Northern Whig*, 15 September 1834, 4a.

### *QUARTERLY REVIEW*

[Holland, Henry].[13] "*The Bridgewater Treatises—The Universe and Its Author* [Whewell, Kidd, Chalmers, Bell]." 50 (1833): 1–34.

[Broderip, William, and G. Poulett Scrope].[14] "Dr Buckland's *Bridgewater Treatise.*" 56 (1836): 31–64.

### *WESTMINSTER REVIEW*

[Thompson, Thomas Perronet]. "Dr Chalmers' *Bridgewater Treatise.*" 20 (1834): 1–21.

## *Religious Journals*

### *BRITISH CRITIC*

[Le Bas, Charles].[15] "Abercrombie *on the Moral Feelings*—Whewell's *Bridgewater Treatise.*" 14 (1833): 72–113.

[Spry, J. H.?].[16] "Chalmers's *Bridgewater Treatise—On the Moral and Intellectual Constitution of Man.*" 14 (1833): 239–82.

[Conybeare, William Daniel]. "Buckland's *Bridgewater Essay on Geology.*" 20 (1836): 295–328.

### *BRITISH MAGAZINE*

"Notices and Reviews [Whewell]." 3 (1833): 586–89.

"Notices and Reviews [Chalmers & Bell]." 4 (1833): 193–95.

"Notices and Reviews [Kidd]." 5 (1834): 69.

"Notices and Reviews [Prout]." 6 (1834): 308–10.

"Notices and Reviews [Roget]." 6 (1834): 666–67.

"Notices and Reviews [Kirby]." 8 (1835): 325.

Winning, W. B. "Correspondence—Dr Buckland's Bridgewater Treatise." 10 (1836): 554–56.

W., "Correspondence—Dr Buckland." 10 (1836): 714.

### *CHRISTIAN EXAMINER AND CHURCH OF IRELAND MAGAZINE*

T. P. K. "On the Moral Constitution of Man." N.s. 2 (1833): 665–76, 737–47.

### *CHRISTIAN REFORMER*

"Review—The Bridgewater Treatises on Natural Theology [Chalmers]." 1 (1834): 146–52.

13. Identification from Patterson 1983, 132–33.

14. See the discussion in chapter 4.

15. Authors from the *British Critic* identified using Curran and Simons 2004–17.

16. Corsi 1988, 180n, attributes this article tentatively to William Sewell.

"Review—The Bridgewater Treatises on Natural Theology [Chalmers & Kidd]." 1 (1834): 234–42.
"Review—The Bridgewater Treatises on Natural Theology [Whewell]." 1 (1834): 391–99.
"Review—The Bridgewater Treatises on Natural Theology [Bell]." 1 (1834): 632–40.
T[urner?], W[illiam?]. "Review—The Bridgewater Treatises on Natural Theology [Prout]." 1 (1834): 853–62.
T[urner?], W[illiam?]. "Review—Buckland's Bridgewater Treatise." 4 (1837): 690–98.

### *CHRISTIAN REMEMBRANCER*

"Whewell's Astronomy and General Physics Considered with Reference to Natural Theology." 15 (1833): 257–61.
"On the Adaptation of External Nature to the Physical Condition of Man [Kidd]." 15 (1833): 329–33.
"Literary Report [Chalmers]." 15 (1833): 402.
"Bell's Bridgewater Treatise—The Hand: Its Mechanism and Vital Endowments." 15 (1833): 466–71.
"Prout's Bridgewater Treatise." 16 (1834): 409–14.
"Animal and Vegetable Physiology." 16 (1834): 723–30.
"Kirby's *Bridgewater Treatise* on the Creation, &c. of Animals." 17 (1835): 707–20.
"Buckland on Geology and Mineralogy." 19 (1837): 1–15, 90–97, 149–58, 272–85, 206–10.

### *CHRISTIAN TEACHER*

"Critical Notices [Kirby]." 1 (1835): 507–9.
L. "Professor Buckland's Bridgewater Treatise." 2 (1836): 664–71.

### *CHURCH OF ENGLAND QUARTERLY REVIEW*

"Infidelity in Disguise—Geology." 2 (1837): 450–91.

### *CONGREGATIONAL MAGAZINE*

"Dr Buckland's Geology and Mineralogy Considered with Reference to Natural Theology." 13 (1837): 42–47.

### *ECLECTIC REVIEW*

"Geology and Natural Theology." 5th ser., 1 (1837): 23–37.

### *EDINBURGH CHRISTIAN INSTRUCTOR*

"Dr Chalmers' Bridgewater Treatise." 2nd ser., 2 (1833): 755–70.

### *INTELLECTUAL REPOSITORY*

Minus. "Remarks on Dr Buckland's Declaration Respecting the Age of the World." 4 (1836–37): 277–82.

### *PRESBYTERIAN REVIEW*

[MacDougall, Patrick C.].[17] "Dr Chalmers's Bridgewater Treatise." 5 (1834): 1–31.
"Dr Kidd's Bridgewater Treatise." 5 (1834): 318–31.
"Whewell's Bridgewater Treatise." 5 (1834): 527–42.
"Prout's Bridgewater Treatise." 6 (1834–35): 1–15.
"Sir C. Bell's Bridgewater Treatise." 6 (1834–35): 470–81.
"Kirby's Bridgewater Treatise." 7 (1835–36): 571–87.
"Roget's Bridgewater Treatise." 8 (1836): 213–30.
[Hamilton, James].[18] "Buckland's Bridgewater Treatise." 9 (1836–37): 222–46.

### *SUNDAY SCHOOL TEACHERS' MAGAZINE*

Esto. "Review—Dr. Chalmers' Bridgewater Treatise." N.s., 6 (1835): 551–58, 617–26.
Esto. "Review—Astronomy and General Physics Considered &c. [Whewell]." N.s., 7 (1836): 237–45, 295–301, 360–66.
Esto. "Review—On the Power Wisdom and Goodness of God [Kirby]." N.s., 8 (1837): 36–42, 118–22, 180–86.

### *YOUTH'S INSTRUCTER*

"Review [Whewell]." N.s., 2 (1838): 61–63.

## *Scientific and Medical Press*

### *EDINBURGH JOURNAL OF NATURAL HISTORY*

[MacGillivray, William?]. "Reviews [Kirby]." 24 October 1835, 4.

### *EDINBURGH MEDICAL AND SURGICAL JOURNAL*

"Dr. Roget's *Bridgewater Treatise*." 43 (1835): 365–409.

### *EDINBURGH NEW PHILOSOPHICAL JOURNAL*

"New Publications—Bridgewater Treatises." 15 (1833): 403–4.

### *ENTOMOLOGICAL MAGAZINE*

"List of Entomological Works [Kirby]." 3 (1835–36): 292–301.

### *LANCET*

"Bell on the Hand." 26 October 1833, 165–69.
Elliotson, John. "Dr. Kidd's Objections to Phrenology." 22 February 1834, 835–37.
Elliotson, John. "Reply of Dr. Elliotson to an Anti-Phrenologist." 1 March 1834, 861–63.
"Kirby's Bridgewater Treatise." 17 October 1835, 105–9.


17. Repr. in MacDougall 1852.
18. Identification from Arnot 1842, 85.

### *LONDON AND EDINBURGH PHILOSOPHICAL MAGAZINE*

Henry, William Charles. "Remarks on the Atomic Constitution of Elastic Fluids [Prout]." 3rd ser., 5 (1834): 33–39.

### *LONDON MEDICAL AND SURGICAL JOURNAL*

[Ryan, Michael?]. "Bridgewater Treatises—The Mechanism of the Hand." 7 September 1833, 179–81.

### *LONDON MEDICAL GAZETTE*

"Sir Charles Bell's Bridgewater Treatise." 16 November 1833, 253–58.

Philip, A.P.W. "Some Remarks on Dr. Prout's Bridgewater Treatise." 15 March 1834, 912–14.

A Stranger to Mr Kirby. "Kirby's Bridgewater Treatise: Strictures on the Criticism in the *Medico-Chirurgical Review*." 28 November 1835, 318-19.

### *MAGAZINE OF BOTANY AND GARDENING*

[Rennie, James?]. "Roget's Animal and Vegetable Physiology." 2 (1834): 165–66.

### *MAGAZINE OF NATURAL HISTORY*

"Review [Kirby]." 8 (1835): 471.

### *MAGAZINE OF POPULAR SCIENCE*

[Powell, Baden?]. "Dr Buckland's Bridgewater Treatise." 2 (1836): 337–46.

### *MECHANICS' MAGAZINE*

"Bridgewater Treatises—Mechanism of the Hand." 30 August 1834, 376–79.

Senex. "The Bridgewater Treatises." 13 September 1834, 412.

### *MEDICAL QUARTERLY REVIEW*

"Sir Charles Bell on the Hand." 1 (1833–34): 358–64.

"Prout on Chemistry, Meteorology, and the Function of Digestion." 2 (1834): 267–72.

### *MEDICO-CHIRURGICAL REVIEW*

"Kirby on Instinct." 23 (1835): 400–413; 24 (1836): 79–93, 358–65.

"Dr Prout on Chemistry, &c." 24 (1836): 385–400; 25 (1836): 95–117.

### *MINING REVIEW*

"Notices of Recent Publications [Buckland]." 4 (1837): 73–104.

### *PHRENOLOGICAL REVIEW*

[Combe, George].[19] "Chalmers's Bridgewater Treatise." 8 (1832–34): 338–64.

### *VETERINARIAN*

K[arkeek William F]. "Review—Bridgewater Treatises No.5 [Roget]." 7 (1834): 618–24; 8 (1835): 283–88.

K[arkeek William F]. "Review—Bridgewater Treatises No.8 [Prout]." 8 (1835): 234–40.

K[arkeek William F] "Review—Bridgewater Treatises No.4 [Bell]." 8 (1835): 408–11, 466–71.

K[arkeek William F] "Review—Bridgewater Treatises No.7 [Kirby]." 8 (1835): 585–90.

K[arkeek William F] "Review—Bridgewater Treatise [Buckland]." 10 (1837): 110–12.

## *Selective List of Newspaper Discussions of Buckland's Bridgewater*

"Dr Buckland's Bridgewater Essay." *Brighton Gazette*, 28 April 1836, 2c–d.

"Geology." *Morning Post*, 7 May 1836, 6f.[20]

"To the Editor." *Brighton Gazette*, 26 May 1836, 3d–e.

"Age of the World." *Belfast Commercial Journal*, 5 September 1836, p. 1b.[21]

"Frightful Blasphemy." *Satirist*, 11 September 1836, 293.

"British Association for the Advancement of Science." *John Bull*, 12 September 1836, 293a–b.

19. Identification from "Combe's Constitution of Man," *Presbyterian Review* 10 (1836): 92–118, on 116n.

20. Repr. in *Standard*, 17 May, 7a; *Waterford Mail*, 25 May, 4c–d; *Manchester Courier*, 28 May, 4b; *Newcastle J.*, 28 May, 4d; *North Wales Chr.*, 31 May, 4b; *Preston Chr.*, 4 June, 4b–c; *Carlisle Patriot*, 4 June, 4e–f.

21. Repr. in *Drogheda J.*, 6 September, 4b; *Standard*, 7 September, 3d; *Newry Examiner*, 7 September, 4c; *St James's Chr.*, 6–8 September, 2e; *Devizes and Wiltshire Gaz.*, 8 September, 2c; *Caledonian Mercury*, 8 September, 2d; *Saunders's News-Letter*, 9 September, 1b; *Athlone Sentinel*, 9 September, 1e; *Dublin Morning Reg.*, 9 September, 1d; *Scotsman*, 10 September, 2b; *Hereford Times*, 10 September, 4c; *Berkshire Chr.*, 10 September, 4a; *Hampshire Advertiser*, 10 September, 4d; *Carlisle Patriot*, 10 September, 3c; *Tipperary Free Press*, 10 September, 4d; *Bell's Life in London*, 11 September, 1e; *Observer*, 11/12 September, 4b; *Reading Mercury*, 12 September, 4f; *Cumberland Pacquet*, 13 September, 4e; *Hereford J.*, 14 September, 4e; *Bradford Observer*, 15 September, 3a; *Perthshire Advertiser*, 15 September, 2d; *Stamford Mercury*, 16 September, 4e; *John o' Groat J.*, 16 September, 4d; *Chester Chr.*, 16 September, 4e; *Preston Chr.*, 17 September, 1d; *York Herald*, 17 September, 4a; *Leicester Chr.*, 17 September, 4d; *Northampton Mercury*, 17 September, 4c; *Windsor and Eton Express*, 17 September, 2c; *Taunton Courier*, 21 September, 8c.

A Reader of the Bible. "To the Editor of the *Standard*." *Standard*, 14 September 1836, 2e.[22]

A. M. "Age of the World." *Devizes and Wiltshire Gazette*, 15 September 1836, 4b.

[Stephenson, George?]. Editorial. *Globe*, 17 September 1836, 2d–e.[23]

Verax. "Dr Buckland's Age of the World." *Morning Chronicle*, 17 September 1836, 3f.

Editorial. *West Kent Guardian*, 17 September 1836, 4c.[24]

Buckland, J. [pseud.]. "Doctor Buckland to the Satirist." *Satirist*, 18 September 1836, 298.

[Giffard, Stanley Lees?]. Editorial. *Standard*, 19 September 1836, 2e.[25]

Editorial. *Scotsman*, 21 September 1836, 2f.[26]

Buckland, William. "Dr. Buckland's Geological Chronology." *Standard*, 22 September 1836, 3a.[27]

G——. "To the Editor." *Standard*, 22 September 1836, 3a.[28]

Oxoniensis. "To the Editor." *Standard*, 24 September 1836, 3a.[29]

[Stephenson, George?]. Editorial. *Globe*, 24 September 1836, 2c.

F.R.S. M.R.I.A. "Mosaic Records and Modern Geology." *Globe*, 24 September 1836, 2e.

A Friend to Truth. "The John Bull versus Dr. Buckland." *Globe*, 24 September 1836, 2e.

"Age of the World." *Berkshire Chronicle*, 24 September 1836, 2f.

Buckland, William. "To John Bull." *John Bull*, 26 September 1836, 313a–c.

[Giffard, Stanley Lees?]. Editorial. *Standard*, 27 September 1836, 2b–d.[30]

G——. "To the Editor." *Standard*, 27 September 1836, 3a.[31]

H. "To the Editor." *Standard*, 27 September 1836, 3a.[32]

Waldron, W. F. G. "Geology Irreconcilable with the Flood." *Standard*, 28 September 1836, 3b.

Fidus. "To the Editor." *Standard*, 29 September 1836, 4d.[33]

"The New Philosophy and Its Dangerous and Fantastic Dogmas." *Bell's Weekly Messenger*, 2 October 1836, 314a–b.[34]

Oxoniensis. "Geological Chronology." *Standard*, 3 October 1836, 3a.[35]

22. Repr. in *St James's Chr.*, 13–15 September, 3d; *West Kent Guardian*, 17 September, 2e; *Blackburn Standard*, 21 September, 6a–b.
23. Repr. in *Scotsman*, 21 September, 4b–c.
24. Repr. in *Blackburn Standard*, 21 September, 4d.
25. Repr. in *St James's Chr.*, 17–20 September, 4b.
26. Repr. in *Globe*, 24 September, 2d.
27. Repr. in *St James's Chr.*, 20–22 September, 2c; *Globe*, 23 September, 4a; *Bell's Weekly Messenger*, 26 September, 309d.
28. Repr. in *St James's Chr.*, 20–22 September, 2c; *Bell's Weekly Messenger*, 26 September, 309d.
29. Repr. in *St James's Chr.*, 22–24 September, 4c.
30. Repr. in *St James's Chr.*, 25–27 September, 4a–b.
31. Repr. in *St James's Chr.*, 27–29 September, 4d.
32. Repr. in *St James's Chr.*, 27–29 September, 4d.
33. Repr. in *St James's Chr.*, 29 September–1 October, 4d.
34. Repr. (in part) in *St James's Chr.*, 1–4 October, 2f.
35. Repr. in *St James's Chr.*, 1–4 October, 2f.

Rusticus. "To the Editor." *St James's Chronicle*, 1–4 October 1836, 2f.
Lee, Samuel. "The Geological Controversy." *Globe*, 4 October 1836, 3d.
Oxoniensis. "To the Editor." *St James's Chronicle*, 4–6 October 1836, 3c.
F. "Dr. Buckland's New Theory of the Creation." *Standard*, 7 October 1836, 2f.[36]
[Giffard, Stanley Lees]. "To Correspondents." *St James's Chronicle*, 8–11 October 1836, 2a.
Observator. "Dr. Buckland's Bridgewater Treatise." *Reading Mercury*, 10 October 1836, 4d.
G——. "To the Editor." *Standard*, 11 October 1836, 3c.[37]
Oxoniensis. "Geology and Scripture." *Standard*, 19 October 1836, 3a.[38]
"Dr Buckland's Geology." *Cheltenham Looker-On*, 22 and 29 October 1836, 267–68, 290–92.
"The Marquis of Northampton and the Recent Meeting of the Philosophical Association." *Bristol Mercury*, 5 November 1836, 4b.
A Correspondent. "Buckland's Bridgewater Treatise." *The Times*, 15 November 1836, 3d.

36. Repr. in *Bell's Weekly Messenger*, 9 October, 325e.
37. Repr. in *St James's Chr.*, 11–13 October, 4c.
38. Repr. in *St James's Chr.*, 18–20 October, 3e.

# *Works Cited*

## *Archives*

| | |
|---|---|
| *AHL* | *Archives of the House of Longman, 1794–1914*. Cambridge: Chadwyck-Healey, 1978. 73 microfilm reels. |
| BGSA | British Geological Survey Archives |
| BL | British Library, London |
| BO | Bodleian Libraries, Oxford |
| CC | Christ Church, Oxford |
| CUL | Cambridge University Library |
| DCY | Dean and Chapter of York Library |
| ESRO | East Sussex Record Office, Brighton |
| EUL | Edinburgh University Library |
| FMC | Fitzwilliam Museum, Cambridge |
| HA | Hertfordshire Archives, Hertford |
| KCL | Kings College London |
| LSL | Linnean Society of London |
| NA | National Archives, London |
| NC | New College Special Collections, University of Edinburgh |
| NHM | Library and Archives, Natural History Museum, London |
| NLS | National Library of Scotland, Edinburgh |
| NPG | National Portrait Gallery, London |
| OUMNH(–BP) | Oxford University Museum of Natural History (–Buckland Papers) |
| OUPA | Oxford University Press Archives |
| RA | Royal Archives, Windsor |
| RCS | Royal College of Surgeons of England |
| RSL | Royal Society, London |
| StAUL | St Andrews University Library |
| TC | Trinity College, Cambridge |
| TUL | Special Collections Research Center, Temple University Libraries |
| UCL | University College London Library Services, Special Collections |
| WL | Wellcome Library, London |
| WYAS-B | West Yorkshire Archive Service, Bradford |

### *Bibliographic Abbreviations*

| | |
|---|---|
| *CCD* | Burkhardt, Frederick, et al., eds. *The Correspondence of Charles Darwin*. 30 vols. Cambridge: Cambridge University Press, 1985–2022. |
| *Notebooks* | Barrett, Paul H., Peter J. Gautrey, Sandra Herbert, David Kohn, and Sydney Smith, eds. *Charles Darwin's Notebooks*. Cambridge: Cambridge University Press, 1987. |
| *ODNB* | *Oxford Dictionary of National Biography*. New ed. 60 vols. Oxford: Oxford University Press, 2004. |

### *Printed Sources*

[Abbott, Jacob]. 1835. *Fire-Side Piety; or, The Duties and Enjoyments of Family Religion*. London: Seeley and Burnside.

Acton, Henry. 1833. *Religious Opinions and Example of Milton, Locke, and Newton: A Lecture*. London: R. Hunter.

Acton, Henry. 1846. *Sermons, by the Late Rev. Henry Acton, of Exeter, with a Memoir of His Life*. Edited by William James and J. Reynell Wreford. London: Chapman, Brothers.

Addinall, Peter. 1991. *Philosophy and Biblical Interpretation: A Study in Nineteenth-Century Conflict*. Cambridge: Cambridge University Press.

Agassiz, Elizabeth Cary. 1885. *Louis Agassiz: His Life and Correspondence*. 2 vols. Boston: Houghton, Mifflin.

Ainslie, Robert. 1836. *The Present State and Claims of London*. London: Seeley.

Ainslie, Robert. 1840a. *An Examination of Socialism: The Last of a Series of Lectures against Socialism, Delivered in the Mechanics' Institution, Southampton Buildings, on the Evening of February 27, 1840, under the Direction of the Committee of the London City Mission*. London: Seeley.

Ainslie, Robert. 1840b. *Is There a God? A Lecture Delivered in the Mechanics' Institution, Southampton Buildings, on the Evening of January, 27, 1840*. London: Seeley.

Alberti, Samuel J. M. M. 2001. "Amateurs and Professionals in One County: Biology and Natural History in Late Victorian Yorkshire." *Journal of the History of Biology* 34:115–47.

Alger, John Goldworth. 1904. *Napoleon's British Visitors and Captives, 1801–1815*. New York: James Pott.

Alison, William Pulteney. 1833. *Outlines of Physiology and Pathology*. Edinburgh: William Blackwood.

Alison, William Pulteney. 1839. *Outlines of Human Physiology*. 3rd ed. Edinburgh: William Blackwood and Sons.

Allen, David Elliston. 2010. *Books and Naturalists*. London: Collins.

Allen, W. O. B., and Edmund McClure. 1898. *Two Hundred Years: The History of the SPCK, 1698–1898*. London: SPCK.

Allibone, S. Austin. 1870–71. *A Critical Dictionary of English Literature and British and American Authors, Living and Deceased, from Earliest Accounts to the Latter Half of the Nineteenth Century*. 3 vols. Philadelphia: J. B. Lippincott.

Allin, Thomas. 1828. *Discourses on the Immateriality and Immortality of the Soul; the*

*Character and Folly of Modern Atheism; and the Necessity of a Divine Revelation.* London: Hurst, Chance.

Allin, Thomas. 1833. *Atheism Refuted; or, Its Falsehood and Folly Demonstrated in Three Discourses. To which Are Added, Remarks on Mr. Carlile's Lecture.* Sheffield: privately printed.

Allison, K. J., ed. 1969. *A History of the County of York, East Riding.* Vol. 1, *The City of Kingston upon Hull.* London: Oxford University Press.

Altholz, Josef L. 1989. *The Religious Periodical Press in Britain, 1760–1900.* New York: Greenwood Press.

Altick, Richard D. 1957. *The English Common Reader: A Social History of the Mass Reading Public, 1800–1900.* Chicago: University of Chicago Press.

Anderson, James S. M. 1835. *Discourses on Elijah, and John the Baptist.* London: Rivington.

Anderson, James S. M. 1837. *Sermons on Various Subjects.* London: Rivington.

Anderson, James S. M. 1839–43. *The Cloud of Witnesses: A Series of Discourses on the Eleventh and Part of the Twelfth Chapters of the Epistle of the Hebrews.* 2 vols. London: Rivington.

Anderson, James S. M. 1842. *Redemption in Christ, the True Jubilee: A Sermon.* London: Rivington.

Anderson, Lyall I., and Michael A. Taylor. 2015–16. "Tennyson and the Geologists." *Tennyson Research Bulletin* 10:340–56, 415–30.

Anderson, Patricia. 1991. *The Printed Image and the Transformation of Popular Culture, 1790–1860.* Oxford: Clarendon.

Anderson, Ronald. 2006. "Augustus De Morgan's Inaugural Lecture of 1828." *Mathematical Intelligencer* 28:16–18.

Anon. 1832. *Report of the Society for Promoting Christian Knowledge.* London: Rivington.

Anon. 1833. *Lithographed Signatures of the Members of the British Association for the Advancement of Science, Who Met at Cambridge, June M.DCCC.XXXIII. with a Report of the Proceedings at the Public Meetings during the Week.* Cambridge: privately printed.

Anon. 1835. *Proceedings of the Fifth Meeting of the British Association for the Advancement of Science, Held in Dublin, During the Week from the 10th to the 15th of August, 1835, Inclusive.* 2nd ed. Dublin: privately printed.

Anon. 1842. *Statistics of Dissent in England and Wales, from Dissenting Authorities; Proving the Inefficiency of the Voluntary Principle to Meet the Spiritual Wants of the Nation.* London: William Edward Painter.

Antommarchi, Francesco. 1825. *The Last Days of the Emperor Napoleon.* 2 vols. London: Henry Colburn.

Appel, Toby. 1987. *The Geoffroy-Cuvier Debate: French Biology in the Decades before Darwin.* Oxford: Oxford University Press.

[Arnold, Thomas]. 1836. "The Oxford Malignants and Dr Hampden." *Edinburgh Review* 63:225–39.

Arnold, Thomas. 1842. *Introductory Lectures on Modern History, Delivered in Lent Term, MDCCCXLII.* Oxford: John Henry Parker.

Arnot, William. 1842. *Memoir of the Late James Halley, A.B., Student of Theology.* 2nd ed. Edinburgh: John Johnstone.

Arnot, William. 1870. *Life of James Hamilton, D.D. F.L.S.* London: James Nisbet.

Arnot, William. 1877. *Autobiography of the Rev. William Arnot* [ . . . ] *and Memoir by his Daughter, Mrs A. Fleming*. London: James Nisbet.

Arnott, Neil. 1827. *Elements of Physics; or, Natural Philosophy, General and Medical, Explained Independently of Technical Mathematics, and Containing New Disquisitions and Practical Suggestions*. 2nd ed. London: Thomas and George Underwood.

Arnott, Neil. 1828. *Elements of Physics; or, Natural Philosophy, General and Medical, Explained Independently of Technical Mathematics, and Containing New Disquisitions and Practical Suggestions*. 3rd ed. London: Longman.

Asma, Stephen T. 1996. *Following Form and Function: A Philosophical Archaeology of Life Science*. Evanston IL: Northwestern University Press.

Astore, William J. 2001. *Observing God: Thomas Dick, Evangelicalism, and Popular Science in Victorian Britain and America*. London: Routledge.

Austen, S. C., ed. 1879. *The Scepticism of the Nineteenth Century. Selections from the Latest Works of the Rev. William Gresley* [ . . . ] *with a Short Account of the Author*. London: J. Masters.

Babbage, Charles. 1830. *Reflections on the Decline of Science in England, and on Some of Its Causes*. London: B. Fellowes.

Babbage, Charles. 1832a. *On the Economy of Machinery and Manufactures*. London: Charles Knight.

Babbage, Charles. 1832b. *On the Economy of Machinery and Manufactures*. 3rd ed. London: Charles Knight.

Babbage, Charles. 1837. *The Ninth Bridgewater Treatise: A Fragment*. London: John Murray.

Babbage, Charles. 1838. *The Ninth Bridgewater Treatise: A Fragment*. 2nd ed. London: John Murray.

Babbage, Charles. 1864. *Passages from the Life of a Philosopher*. London: Longman.

Baines, Edward. 1822. *History, Directory & Gazetteer of the County of York*. 2 vols. Leeds: Edward Baines.

Baker, William J. 1981. *Beyond Port and Prejudice: Charles Lloyd of Oxford, 1784–1829*. Orono: University of Maine at Orono Press.

Baring-Gould, S. 1923. *Early Reminiscences, 1834–1864*. London: John Lane/The Bodley Head.

Barnes, James J. 1964. *Free Trade in Books: A Study of the London Book Trade since 1800*. Oxford: Clarendon.

Barnes, James J. 1976. "John Miller: First Transatlantic Publisher's Agent." *Studies in Bibliography* 29:373–79.

Barr, C. B. L. 1977. "The Minster Library." In *A History of York Minster*, edited by G. E. Aylmer and Reginald Cant, 487–539. Oxford: Clarendon.

Barry, Martin. 1837. "On the Unity of Structure in the Animal Kingdom." *Edinburgh New Philosophical Journal* 22:116–41.

Bates, Alan. 1969. *Directory of Stage Coach Services 1836*. Newton Abbot: David and Charles.

Bathe, Greville, and Dorothy Bathe. 1943. *Jacob Perkins: His Inventions, His Times, & His Contemporaries*. Philadelphia: Historical Society of Pennsylvania.

Baxter, Paul. 1985. "Science and Belief in Scotland, 1805–1868: The Scottish Evangelicals." PhD diss., University of Edinburgh.

Beaulieu, Micheline, and Sue Holland. 2001. *The Athenaeum Index of Reviews and*

*Reviewers: 1830–1870*. London: Information Science, City University. https://athenaeum.city.ac.uk.

Becher, Harvey. 1986. "Voluntary Science in Nineteenth Century Cambridge University to the 1850s." *British Journal for the History of Science* 19:57–87.

Beetham, Margaret. 1996. *A Magazine of Her Own? Domesticity and Desire in the Woman's Magazine, 1800–1914*. London: Routledge.

Bell, Benjamin. 1881. *Lieut. John Irving* [ . . . ] *A Memorial Sketch with Letters*. Edinburgh: David Douglas.

[Bell, Charles]. 1827–29. *Animal Mechanics*. 2 pt. [London]: [Baldwin, Cradock, and Joy].

Bell, Charles. 1829. "Mr Bell's Introductory Lecture, Delivered on the First of October, 1829, at the Opening of the Medical School of the University." *London University Magazine* 1:241–43.

Bell, Charles. 1831. "Mr Bell's Letter to His Pupils of the London University, on Taking Leave of Them." *London Medical Gazette*, 4 December, 308–11.

Bell, Charles. 1833a. *The Hand: Its Mechanism and Vital Endowments as Evincing Design*. London: William Pickering.

Bell, Charles. 1833b. *The Hand: Its Mechanism and Vital Endowments as Evincing Design*. 2nd ed. London: William Pickering.

Bell, Charles. 1833–34. "Lectures on the Hunterian Preparations." *Lancet* 1:279–85, 313–19, 486–92, 912–19, 962–69; 2:216–21, 265–71, 346–52, 410–16, 745–51, 794–806, 824–29, 875–87.

Bell, Charles. 1835. "Observations on the Proper Method of Studying the Nervous System." *Report of the Fourth Meeting of the British Association for the Advancement of Science*, 667–70.

Bell, Charles. 1838. *Institutes of Surgery: Arranged in the Order of the Lectures Delivered in the University of Edinburgh*. 2 vols. Edinburgh: Adam and Charles Black.

Bell, George Joseph, ed. 1870. *Letters of Sir Charles Bell, Selected from His Correspondence with His Brother, George Joseph Bell*. London: John Murray.

Bellon, Richard. 2012. "The Moral Dignity of Inductive Method and the Reconciliation of Science and Faith in Adam Sedgwick's *Discourse*." *Science & Education* 21:937–58.

Bellon, Richard. 2015. *A Sincere and Teachable Heart: Self-Denying Virtue in British Intellectual Life, 1736–1859*. Leiden: Brill.

Bellot, H. Hale. 1929. *University College London, 1826–1926*. London: University of London Press.

Bennett, James R. 1830. *Lecture Introductory to the Course of General Anatomy; Delivered in the University of London, on Wednesday, October 6, 1830*. London: John Taylor.

Bennett, Joshua. 2018. "A History of "Rationalism" in Victorian Britain." *Modern Intellectual History* 15:63–91.

Bennett, Scott. 1982. "Revolutions in Thought: Serial Publication and the Mass Market for Reading." In *The Victorian Periodical Press: Samplings and Soundings*, edited by Joanne Shattock and Michael Wolff, 225–57. Leicester: Leicester University Press.

Berkowitz, Carin. 2011. "The Beauty of Anatomy: Visual Displays and Surgical Education in Early Nineteenth-Century London." *Bulletin of the History of Medicine* 85:248–71.

Berkowitz, Carin. 2015. *Charles Bell and the Anatomy of Reform*. Chicago: University of Chicago Press.

Berman, David. 1988. *A History of Atheism in Britain: From Hobbes to Russell*. London: Croom Helm.

Best, Samuel. 1837. *Afterthoughts on Reading Dr Buckland's Bridgewater Treatise*. London: J. Hatchard and Son.

Bickersteth, E. 1838. *Christian Truth: A Family Guide to the Chief Truths of the Gospel*. 2nd ed. London: Seeley and Burnside.

Bickersteth, E. 1844. *The Christian Student: Designed to Assist Christians in General in Acquiring Religious Knowledge*. 4th ed. London: Seeley, Burnside, and Seeley.

Birks, T. R. 1851. *Memoir of the Rev. Edward Bickersteth*. 2 vols. London: Seeleys.

Blake, C. Carter. 1870. "The Life of Dr Knox." *Journal of Anthropology* 1:332–38.

Blomfield, Charles James. 1831. *The Duty of Combining Religious Instruction with Intellectual Culture. A Sermon Preached in the Chapel of King's College London at the Opening of the Institution on the 8th of October, 1831*. London: B. Fellowes.

Boase, Frederic. 1892–1921. *Modern English Biography: Containing Many Thousand Concise Memoirs of Persons Who Have Died between the Years 1851–1900*. 6 vols. Truro: Netherton and Worth.

Bonner, Thomas Neville. 1995. *Becoming a Physician: Medical Education in Britain, France, Germany, and the United States, 1750–1945*. Baltimore: Johns Hopkins University Press.

[Bowden, John?]. 1834a. "The British Association." *Oxford University Magazine* 1:401–12.

[Bowden, John]. 1834b. "Nolan and Powell." *British Critic* 15:411–34.

[Bowden, John]. 1834c. "Religion in Its Connection with Science." *British Critic* 15:233–34.

Bowler, Peter J. 1977. "Darwinism and the Argument from Design: Suggestions for a Reevaluation." *Journal of the History of Biology* 10:29–43.

Boylan, Patrick John. 1984. "William Buckland, 1784–1856: Scientific Institutions, Vertebrate Palaeontology, and Quaternary Geology." PhD diss., University of Leicester.

Braithwaite, Joseph Bevan. 1854. *Memoirs of Joseph John Gurney*. 2 vols. Norwich: Fletcher and Alexander.

Bray, Edward Atkyns. 1860. *A Selection from the Sermons, General and Occasional*. 2 vols. London: Rivington.

[Brewster, David]. 1830. "Decline of Science in England and Patent Laws." *Quarterly Review* 43:305–42.

Briggs, Asa. 2008. *A History of Longmans and Their Books, 1724–1990: Longevity in Publishing*. London: British Library.

Brightwell, Cecilia Lucy. 1854. *Memorials of the Life of Amelia Opie*. Norwich: Fletcher and Alexander.

Bristed, Charles Astor. 1852. *Five Years in an English University*. 2 vols. New York: G. P. Putnam.

B[ritton], J[ohn]. 1834. "On Wood Engraving and Printing—Puckle's Club." *Arnold's Magazine of the Fine Arts* 3:257–61.

Brock, M. G., and M. C. Curthoys, eds. 1997. *The History of the University of Oxford*. Vol. 6, *Nineteenth-Century Oxford*, pt. 1. Oxford: Clarendon.

Brock, W. H. 1966. "The Selection of the Authors of the *Bridgewater Treatises*." *Notes and Records of the Royal Society* 21:162–79.

Brock, W. H. 1985. *From Protyle to Proton: William Prout and the Nature of Matter, 1785–1985*. Bristol: Adam Hilger.

[Broderip, William John]. 1859. "Buckland's 'Bridgewater Treatise.'" *Fraser's Magazine* 59:227–43.

Brooke, John Hedley. 1977. "Natural Theology and the Plurality of Worlds: Observations on the Brewster-Whewell Debate." *Annals of Science* 34:221–86.

Brooke, John Hedley. 1979. "Nebular Contraction and the Expansion of Naturalism." *British Journal for the History of Science* 12:200–211.

Brooke, John Hedley. 1983. "Middle Positions." *London Review of Books*, 21 July, 11–12.

Brooke, John Hedley. 1985. "The Relations between Darwin's Science and His Religion." In *Darwinism and Divinity: Essays on Evolution and Religious Belief*, edited by John Durant, 40–75. Oxford: Blackwell.

Brooke, John Hedley. 1989. "The Superiority of Nature's Art? Vitalism, Natural Theology and the Rise of Organic Chemistry." In *Science and Religion: Proceedings of the Symposium of the XVIIIth International Congress of History of Science*, edited by Anne Baumer and Manfred Buttner. Bochum: Universitatsverlag Dr. N. Brockmeyer.

Brooke, John Hedley. 1991a. "Indications of a Creator: Whewell as Apologist and Priest." In *William Whewell: A Composite Portrait*, edited by Menachem Fisch and Simon Schaffer, 149–73. Oxford: Clarendon.

Brooke, John Hedley. 1991b. *Science and Religion: Some Historical Perspectives*. Cambridge: Cambridge University Press.

Brooke, John Hedley, and Geoffrey Cantor. 1998. *Reconstructing Nature: The Engagement of Science and Religion*. Glasgow Gifford Lectures. Edinburgh: T and T Clark.

[Brougham, Henry]. 1827. *Objects, Advantages, and Pleasures of Science*. London: Baldwin, Cradock, and Joy.

[Brougham, Henry]. 1828. "London University and King's College." *Edinburgh Review* 48:235–58.

Brougham, Henry. 1835. *A Discourse of Natural Theology, Showing the Nature of the Evidence and the Advantages of the Study*. London: Charles Knight.

Brougham, Henry. 1839. *Dissertations on Subjects of Science Connected with Natural Theology*. 2 vols. London: C. Knight.

Brougham, Henry, and Charles Bell. 1836. *Paley's Natural Theology, with Illustrative Notes*. 2 vols. London: Charles Knight.

Brown, Candy Gunther. 2004. *Word in the World: Evangelical Writing, Publishing, and Reading in America, 1789–1880*. Chapel Hill: University of North Carolina Press.

Brown, Daniel. 2012. *The Poetry of Victorian Scientists: Style, Science and Nonsense*. Cambridge: Cambridge University Press.

Brown, James Mellor. 1838. *Reflections on Geology: Suggested by the Perusal of Dr. Buckland's Bridgewater Treatise*. London: James Nisbet.

Brown, John. 1782. *A Compendious View of Natural and Revealed Religion, in Seven Books*. London: J. Mathews.

[Brown, John]. 1857. "Dr Samuel Brown." *North British Review* 26:376–406.

Brown, Kenneth D. 1988. *A Social History of the Nonconformist Ministry in England and Wales, 1800–1930*. Oxford: Clarendon.

Brown, Niamh. 2016. "Devotional Cosmology: Poetry, Thermodynamics and Popular Astronomy, 1839–1889." PhD diss., University of Glasgow.

[Brown, Samuel]. 1841. *Lay Sermons on the Theory of Christianity, by a Company of Brethren. No. I. The Fidianism of Saint Paul*. London: Smith, Elder.

[Brown, Samuel]. 1842. *Lay Sermons on the Theory of Christianity, by a Company of Brethren. No. II. The Argument of Design Equal to Nothing; or, Nieuentytt and Paley versus David Hume and Saint Paul*. London: Smith, Elder.

[Brown, Samuel]. 1856. *Some Account of Itinerating Libraries and Their Founder*. Edinburgh: privately printed.

Brown, Stewart J. 1982. *Thomas Chalmers and the Godly Commonwealth in Scotland*. Oxford: Oxford University Press.

Brown, William Laurence. 1816. *An Essay on the Existence of a Supreme Creator*. 2 vols. London: T. Hamilton; Aberdeen: A. Brown.

Browne, Janet. 1995. *Charles Darwin*. Vol. 1, *Voyaging*. London: Jonathan Cape.

[Brydges, Egerton]. 1832. "Clavering's Auto-Biography." *Metropolitan Magazine* 5:289–98.

Buckland, Adelene. 2013. *Novel Science: Fiction and the Invention of Nineteenth-Century Geology*. Chicago: University of Chicago Press.

Buckland, William. 1820. *Vindiciæ geologicæ; or, The Connexion of Geology with Religion Explained, in an Inaugural Lecture Delivered before the University of Oxford, May 15, 1819*. Oxford: privately printed.

Buckland, William. 1836. *Geology and Mineralogy Considered with Reference to Natural Theology*. 2 vols. London: William Pickering.

Buckland, William. 1837. *Supplementary Notes to the First and Second Edition of Dr Buckland's Bridgewater Treatise with a Plate of the Fossil Head and Restored Figure of the Dinotherium*. London: William Pickering.

Buckland, William. 1840. *Address Delivered at the Anniversary Meeting of the Geological Society of London, on the 21st of February, 1840*. London: privately printed.

Buckland, William. 1858. *Geology and Mineralogy Considered with Reference to Natural Theology*. 3rd ed. 2 vols. Edited by Francis T. Buckland. London: George Routledge.

Budge, Gavin, et al., eds. 2002. *The Dictionary of Nineteenth-Century British Philosophers*. 2 vols. Bristol: Thoemmes.

[Bunbury, Frances], ed. 1890–91. *Memorials of Sir C. J. F. Bunbury, Bart*. 9 vols. Mildenhall: privately printed.

Burd, Van Akin, ed. 1973. *The Ruskin Family Letters: The Correspondence of John James Ruskin, His Wife, and Their Son, John, 1801–1843*. 2 vols. Ithaca, NY: Cornell University Press.

Burd, Van Akin. 2008. "Ruskin and His 'Good Master,' William Buckland." *Victorian Literature and Culture* 36:299–315.

Burgess, Geoffrey H. O. 1967. *The Curious World of Frank Buckland*. London: John Baker.

Burnett, Gilbert T. 1832a. *A Lecture Delivered in King's College, London, on Tuesday, the 11th of October, 1831, Being Introductory to the First Botanical Course of the Session Opening the Institution*. London: privately printed.

Burnett, Gilbert T. 1832b. *A Lecture Delivered in King's College, London, on the 14th of March, 1832, (Introductory to the Second Course)*. London: privately printed.

Butler, Marilyn. 1993. "Culture's Medium: The Role of the Review." In *The Cambridge Companion to British Romanticism*, edited by Stuart Curran, 120–47. Cambridge: Cambridge University Press.

Butler, Stella V. F. 1981. "Science and the Education of Doctors in the Nineteenth Century: A Study of British Medical Schools with Particular Reference to the Development and Uses of Physiology." PhD diss., University of Manchester.

Bywater, Abel. 1830. *Infidelity Refuted, in a Speech Delivered at Scotland-Street Chapel, at the Second Anniversary of the New Connexion Methodist Tract Society, March 29th, 1830*. Sheffield: privately printed.

Bywater, Abel. 1877. *The Sheffield Dialect*. 3rd ed. Wakefield: W. Nicholson and Sons; Sheffield: Thomas Rogers.

Campbell, John. 1844. *Memoirs of David Nasmith*. London: John Snow.

Campbell, Nancie. 1971. *Tennyson in Lincoln: A Catalogue of the Collections in the Research Centre*. Lincoln: Tennyson Society.

Campbell-Kelly, Martin, ed. 1989. *The Works of Charles Babbage*. Vol. 9, *The Ninth Bridgewater Treatise: A Fragment*. 2nd ed. London: William Pickering.

Cannon, Susan F. 1961. "The Impact of Uniformitarianism: Two Letters from John Herschel to Charles Lyell, 1836–1837." *Proceedings of the American Philosophical Society* 105:301–14.

Cantor, Geoffrey N. 1979. "Revelation and the Cyclical Cosmos of John Hutchinson." In *Images of the Earth: Essays in the History of the Environmental Sciences*, edited by L. J. Jordanova and Roy S. Porter, 3–22. Chalfont St. Giles: British Society for the History of Science.

Cantor, Geoffrey N. 2011. *Religion and the Great Exhibition of 1851*. Oxford: Oxford University Press.

Cantor, Geoffrey N. 2015. "John Tyndall's Religion: A Fragment." *Notes and Records of the Royal Society* 69:419–36.

Cantor, Geoffrey N., et al. 2004. *Science in the Nineteenth-Century Periodical: Reading the Magazine of Nature*. Cambridge: Cambridge University Press.

Cantor, Geoffrey N., et al. 2020. *Science in the Nineteenth-Century Periodical: An Electronic Index*, v. 4.0. http://www.sciper.org.

[Carey, John]. 1795. *Inland Navigation; or, Select Plans of the Several Navigable Canals, throughout Great Britain*. London: J. Carey.

[Carleton, Charlotte], ed. 1909–11. *Recollections of a Long Life, by Lord Broughton (John Cam Hobhouse)*. 6 vols. London: John Murray.

Carlile, Richard. 1821. *An Address to Men of Science*. London: R. Carlile.

Carlile, Richard. 1823. "To Mr William Fitton, of Royton, Lancashire." *Republican* 7:396–411.

Carlile, Richard. 1828. "Phrenology." *Lion* 1:481–82.

Carlile, Richard. 1833a. *A Full Report of the Third Lecture, Delivered in the Sheffield Theatre, on Religion. October 3, 1830*. Sheffield: P. T. Bready.

Carlile, Richard. 1833b. *The Substance of the Two First Lectures, Delivered in the Sheffield Theatre, on Politics and on Morals. September 30, and October 1, 1833*. Sheffield: P. T. Bready.

[Carlyle, Thomas]. 1829. "Signs of the Times." *Edinburgh Review* 49:439–59.

Carpenter, J. Estlin. 1881. *The Life and Work of Mary Carpenter*. 2nd ed. London: Macmillan.

[Carpenter, William Benjamin]. 1835. "On the Structure and Functions of the Organs of Respiration, in the Animal and Vegetable Kingdoms." *West of England Journal of Science and Literature* 1:217–28, 279–87.

Carpenter, William Benjamin. 1837. "On Unity of Function in Organized Beings." *Edinburgh New Philosophical Journal* 23:92–114.

[Carpenter, William Benjamin]. 1838a. "German School of Physiology—English School of Physiology." *British and Foreign Medical Review* 5:75–116.

Carpenter, William Benjamin. 1838b. "On the Differences of the Laws Regulating Vital and Physical Phenomena." *Edinburgh New Philosophical Journal* 24:327–53.

[Carpenter, William Benjamin]. 1838c. "Whewell's *History of the Inductive Sciences*." *British and Foreign Medical Review* 5:317–42.

Carpenter, William Benjamin. 1839. *Principles of General and Comparative Physiology, Intended as an Introduction to the Study of Human Physiology, and as a Guide to the Philosophical Pursuit of Natural History*. London: John Churchill.

Carpenter, William Benjamin. [1840]. *Remarks on Some Passages in the Review of "Principles of General and Comparative Physiology," in the "Edinburgh Medical and Surgical Journal," January, 1840*. [Bristol]: privately printed.

[Carpenter, William Benjamin]. 1846. "Dr Prout's *Bridgewater Treatise*." *British and Foreign Medical Review* 21:109–23.

Carpenter, William Benjamin. 1888. *Nature and Man: Essays Scientific and Philosophical, with an Introductory Memoir by J. Estlin Carpenter, M.A*. London: Kegan Paul.

Carritt, E. F., ed. 1933. *Letters of Courtship between John Torr and Maria Jackson, 1838–43*. London: Oxford University Press.

Carroll, Victoria. 2008. *Science and Eccentricity: Collecting, Writing and Performing Science for Early Nineteenth-Century Audiences*. London: Pickering and Chatto.

Carter, John W. 1932. *Binding Variants in English Publishing, 1820–1900*. London: Constable; New York: Long and Smith.

Carter, John W. 1935. *Publishers' Cloth: An Outline History of Publishers' Binding in England, 1820–1900*. New York: Bowker; London: Constable.

Carter, William George. 1835. "The Deluge: If Universal in the Mosaic Narrative." *British Magazine* 8:48–52.

[Carter, William George]. 1837. *Remarks on Dr Buckland's View of the Mosaic Creation as the Last Fitting Up of the Earth* [ . . . ] *by Eretzsepher*. London: Smallfield and Son.

Carus, William. 1847. *Memoirs of the Life of the Rev. Charles Simeon, M.A*. London: Hatchard and Son.

[Caswall, Edward]. 1837. *Sketches of Young Ladies: In Which These Interesting Members of the Animal Kingdom are Classified, According to Their Several Instincts, Habits, and General Characteristics*. 5th ed. London: Chapman and Hall.

*Census 1851—Religious Worship: Census of Great Britain, 1851. Religious Worship. England and Wales. Report and Tables*. House of Commons Parliamentary Papers, Session 1852–53, 89:1–444.

Chadwick, Owen. 1977. "From 1822 until 1916." In *A History of York Minster*, edited by G. E. Aylmer and Reginald Cant, 272–312. Oxford: Clarendon.

Challinor, John. 1961–63. "Some Correspondence of Thomas Webster, Geologist, 1773–1844." *Annals of Science* 17:175–95; 18:147–75; 19:49–79, 285–97.

Chalmers, Thomas. 1832. *On Political Economy in Connexion with the Moral State and Moral Prospects of Society*. Glasgow: William Collins.

Chalmers, Thomas. 1833. *On the Power, Wisdom and Goodness of God as Manifested in the Adaptation of External Nature to the Moral and Intellectual Constitution of Man*. 2 vols. London: William Pickering.

Chalmers, Thomas. [1836]. *On Natural Theology*. 2 vols. Glasgow: William Collins.

Chalmers, Thomas. 1847–49. *Posthumous Works of the Rev. Thomas Chalmers, D.D., LL.D.* Edited by William Hanna. 9 vols. Edinburgh: Thomas Constable.

[Chambers, Robert]. 1834. "Is Ignorance Bliss?" *Chambers's Edinburgh Journal*, 4 January, 385–86.

[Chambers, Robert?]. 1835. "Transmutation of Species." *Chambers's Edinburgh Journal*, 26 September, 273–74.

[Chambers, Robert?]. 1837. "Fourth Ages of Animal Life." *Chambers's Edinburgh Journal*, 23 December, 379–80.

[Chambers, Robert]. 1844. *Vestiges of the Natural History of Creation*. London: John Churchill.

[Chambers, Robert]. 1845. *Explanations: A Sequel to "Vestiges of the Natural History of Creation."* London: John Churchill.

Chambers, Robert. 1847. *Select Writings of Robert Chambers*. 7 vols. Edinburgh: W. and R. Chambers.

[Chambers, Robert]. 1853. *Vestiges of the Natural History of Creation*. 10th ed. London: John Churchill.

Chambers, Robert. 1994. *"Vestiges of the Natural History of Creation" and Other Evolutionary Writings*. Edited by James A. Secord. Chicago: University of Chicago Press.

Chambers, William. 1832. "The Editor's Address to His Readers." *Chambers's Edinburgh Journal*, 11 February, 1–2.

Chambers, William. 1872. *Memoir of Robert Chambers, with Autobiographical Reminiscences of William Chambers*. 3rd ed. Edinburgh: W. and R. Chambers.

[Chatto, W. A.]. 1839. *A Treatise on Wood Engraving*. London: Charles Knight.

C[hilton], W[illiam]. 1842. "The Cowardice and Dishonesty of Scientific Men." *Oracle of Reason* 1:193–95.

C[hilton], W[illiam]. 1845. "*Vestiges of the Natural History of Creation*. Theory of Regular Gradation." *Movement* 2:9–12.

Chitty, Susan. 1974. *The Beast and the Monk: A Life of Charles Kingsley*. London: Hodder and Stoughton.

Clark, J. F. M. 2006. "History from the Ground Up: Bugs, Political Economy, and God in Kirby and Spence's *Introduction to Entomology* (1815–1856)." *Isis* 97: 28–55.

Clark, J. F. M. 2009. *Bugs and the Victorians*. New Haven, CT: Yale University Press.

Clark, John Willis. 1900. *Old Friends at Cambridge and Elsewhere*. London: Macmillan.

Clark, John Willis, and Thomas McKenny Hughes. 1890. *The Life and Letters of the Reverend Adam Sedgwick*. 2 vols. Cambridge: Cambridge University Press.

Clarke, Adam, ed. 1810. *Reflections on the Works of God in Nature and Providence, for*

*Every Day in the Year. By Christopher Christian Sturm*. New ed. 4 vols. London: Cradock and Joy.

Clarke, W. K. Lowther. 1959. *A History of the S.P.C.K.* London: Society for Promoting Christian Knowledge.

Clarke, William M. 1988. *The Secret Life of Wilkie Collins*. London: Allison and Busby.

Clodd, Edward. 1916. *Memories*. London: Chapman and Hall.

Cobbe, Frances Power. 1894. *Life of Frances Power Cobbe*. 2 vols. London: Richard Bentley and Son.

Cockburn, William. 1838a. *A Letter to Professor Buckland, Concerning the Origin of the World*. 2nd ed. London: J. Hatchard and Son.

Cockburn, William. 1838b. *A Remonstrance, Addressed to His Grace the Duke of Northumberland, upon the Dangers of Peripatetic Philosophy*. London: J. Hatchard and Son.

Cockburn, William. 1844a. *The Bible Defended against the British Association: Being the Substance of a Paper Read in the Geological Section, at York, on the 27th of September, 1844*. London: Whittaker.

Cockburn, William. 1844b. *A Sermon on the Evils of Education without a Religious Basis, Preached in York Minster, on Sunday, 29th of September, 1844*. 2nd ed. London: Whittaker.

Cohen, I. Bernard. 1985. *Revolution in Science*. Cambridge, MA: Belknap Press of Harvard University Press.

Cole, Henry. 1834a. "Mr Sedgwick's Commencement Sermon." *The Times*, 20 February, 1d.

Cole, Henry. 1834b. *Popular Geology Subversive of Divine Revelation! A Letter to the Rev. Adam Sedgwick*. London: Hatchard and Son.

Cole, Henry. 1853. *The Bible a Rule and Test of Religion and of Science*. Cambridge: Hall and Son.

Coleridge, Edith. 1873. *Memoir and Letters of Sara Coleridge*. 2 vols. London: H. S. King.

Combe, George. 1866. *The Constitution of Man Considered in Relation to External Objects*. 9th ed. Edinburgh: MacLachlan and Stewart.

Compton, Philip. 2014. "Through the Looking Glass." *Geoscientist* 24:10–15.

[Compton, Spencer Joshua Alwyne]. 1836. "To the Editor of the Northampton Mercury." *Northampton Mercury*, 8 October, 3f.

Conder, Eustace R. 1857. *Josiah Conder: A Memoir*. London: John Snow.

[Conder, Josiah]. 1835. "Lord Brougham on Natural Theology." *Eclectic Review*, 3rd ser., 14:165–85.

Conolly, John. 1828. *An Introductory Lecture Delivered in the University of London, on Thursday, October 2, 1828*. London: John Taylor.

Conolly, John. 1835. "An Address Delivered at the Second Anniversary of the Association." *Transactions of the Provincial Medical and Surgical Association* 3:13–44.

Conybeare, W. D. 1831. *Inaugural Address on the Application of Classical and Scientific Education to Theology*. London: John Murray.

Conybeare, W. D. 1834. *The Origin and Obligations of Civil and Legal Society Considered*. Oxford: J. H. Parker.

Conybeare, W. D. 1836. *An Elementary Course of Theological Lectures, in Three Parts*. 2nd ed. London: Sherwood, Gilbert, and Piper.

Cook, E. T., and Alexander Wedderburn. 1908. *The Works of John Ruskin*. Vol. 35, *Præterita and Dilecta*. London: George Allen.

Cooney, Sondra Miley. 1970. "Publishers for the People: W. & R, Chambers; The Early Years, 1832–1850." PhD diss., Ohio State University.

Coontz, Stephanie. 2006. *Marriage, A History: How Love Conquered Marriage*. London: Penguin Books.

Cooper, Thomas. 1872. *The Life of Thomas Cooper*. London: Hodder and Stoughton.

Cooter, Roger. 1984. *The Cultural Meaning of Popular Science: Phrenology and the Organization of Consent in Nineteenth-Century Britain*. Cambridge: Cambridge University Press.

C[ope], R[ichard]. 1837. "Conversations at Carringford Lodge." *Youth's Magazine*, 3rd ser. 10:160–65, 193–99, 241–45, 265–69, 296–303, 344–50, 374–79, 410–16.

Copley, [Esther]. [1839?]. *Female Excellence; or, Hints to Daughters*. London: Religious Tract Society.

Corsi, Pietro. 1978. "The Importance of French Transformist Ideas for the Second Volume of Lyell's *Principles of Geology*." *British Journal for the History of Science* 11:221–44.

Corsi, Pietro. 1988. *Science and Religion: Baden Powell and the Anglican Debate, 1800–1860*. Cambridge: Cambridge University Press.

Coult, Douglas. 1980. *A Prospect of Ashridge*. London: Phillimore.

Cox, William Sands. 1873. *Reprint of the Charter; Supplemental Charters; the Warneford Trust Deeds; and the Act of Parliament of the Queen's College, Birmingham*. Birmingham: privately printed.

Croly, George. 1834. *Divine Providence; or, The Three Cycles of Revelation, Showing the Parallelism of the Patriarchal, Jewish, and Christian Dispensations*. London: James Duncan.

[Croly, George]. 1836. "The World We Live In (No. I)." *Blackwood's Edinburgh Magazine* 40:609–26.

[Croly, George]. 1837a. "The World We Live In (No. IV)." *Blackwood's Edinburgh Magazine* 41:163–82.

[Croly, George]. 1837b. "The World We Live In (No. XIII)." *Blackwood's Edinburgh Magazine* 42:673–92.

Crombie, Alexander. 1829. *Natural Theology; or, Essays on the Existence of Deity and of Providence, on the Immateriality of the Soul, and a Future State*. 2 vols. London: R. Hunter and T. Hookham.

Cromwell, Thomas. 1835. *Walks through Islington*. London: Sherwood, Gilbert, and Piper.

Csiszar, Alex. 2018. *The Scientific Journal: Authorship and the Politics of Knowledge in the Nineteenth Century*. Chicago: University of Chicago Press.

Cunningham, Andrew, and Perry Williams. 1993. "De-Centring the 'Big Picture': *The Origins of Modern Science* and the Modern Origins of Science," *British Journal for the History of Science* 26:407–32.

Curran, Eileen, and Gary Simons. 2004–17. *The Curran Index*. Research Society for Victorian Periodicals. http://curranindex.org.

Currie, David Alan. 1990. *The Growth of Evangelicalism in the Church of Scotland, 1793–1843*. PhD diss., University of St. Andrews.

Cuvier, [Georges]. 1827–35. *The Animal Kingdom, Arranged in Conformity with Its*

*Organization*. Edited by Edward Griffiths et al. 16 vols. London: Geo. B. Whittaker.

Dahm, John Johannes. 1969. "Science and Religion in Eighteenth Century England: The Early Boyle Lectures and the Bridgewater Treatises." PhD diss., Case Western Reserve University.

Dale, Antony. 1989. *Brighton Churches*. London: Routledge.

Dale, R. W., ed. 1861. *The Life and Letters of John Angell James*. London: James Nisbet.

Dale, Thomas. 1829. *An Introductory Lecture, Delivered in the University of London, on Friday, October 24, 1828*. 4th ed. London: John Taylor.

Dale, Thomas. 1837. *The Philosopher Entering, as a Child, into the Kingdom of Heaven*. 2nd ed. London: Taylor and Walton.

D'Alton, Eduard. 1821. *Das Riesen-Faulthier Bradypus giganteus, abgebildet, beschrieben und mit den verwandten Geschlechtern verglichen*. Bonn: Eduard Weber.

Daniell, J. F. 1831. *An Introductory Lecture, Delivered in King's College, London, October 11, 1831*. London: B. Fellowes.

Daniell, J. F. 1839. *An Introduction to the Study of Chemical Philosophy*. London: John W. Parker.

Darton, Lawrence. 2004. *The Dartons: An Annotated Check-List of Children's Books Issued by Two Publishing Houses 1787–1876*. London: British Library.

Darwin, Charles. 1859. *On the Origin of Species by Means of Natural Selection, or the Preservation of Favoured Races in the Struggle for Life*. London: John Murray.

Darwin, Charles. 1860. *On the Origin of Species*. 2nd ed. London: John Murray.

Darwin, Charles. 1958. *The Autobiography of Charles Darwin, 1809–1882*. Edited by Nora Barlow. London: Collins.

Darwin, Francis, ed. 1887. *The Life and Letters of Charles Darwin*. 3 vols. London: John Murray.

Daston, Lorraine, and Peter Galison. 2010. *Objectivity*. New York: Zone Books.

Daubeny, Charles. 1823. *Inaugural Lecture on the Study of Chemistry, Read at the Ashmolean Museum, November 2, 1822*. Oxford: privately printed.

Daubeny, Charles. 1831. *An Introduction to the Atomic Theory*. London: John Murray.

[Daubeny, Charles]. 1833. "Apology for British Science." *Literary Gazette*, 7 December, 769–71; 14 December, 789–92.

[Daubeny, Charles]. 1834a. "Dr Nolan's Bampton Lectures." *Literary Gazette*, 11 January, 25.

Daubeny, Charles. 1834b. *An Inaugural Lecture on the Study of Botany, Read in the Library of the Botanic Garden, Oxford, May 1, MDCCCXXXIV*. Oxford: privately printed.

Daubeny, Charles. 1840. *Supplement to the Introduction to the Atomic Theory*. London: J. Murray; Oxford: J. H. Parker.

Daunton, M. J. 1995. *Progress and Poverty: An Economic and Social History of Britain 1700–1850*. Oxford: Oxford University Press.

Davidoff, Leonore, and Catherine Hall. 1987. *Family Fortunes: Men and Women of the English Middle Class, 1780–1850*. London: Hutchinson.

Davies, Richard. 1834. *A Sermon on Natural Theology, Preached at St Mary's, in Brecon, on Sunday, February 11, MDCCCXXXIV*. London: Rivington.

Dawson, Gowan. 2007. *Darwin, Literature and Victorian Respectability*. Cambridge: Cambridge University Press.

Dawson, Gowan. 2016. *Show Me the Bone: Reconstructing Prehistoric Monsters in Nineteenth-Century Britain and America*. Chicago: University of Chicago Press.

Dawson, Gowan, and Jonathan R. Topham. 2020. "Scientific, Medical, and Technical Periodicals in Nineteenth-Century Britain: New Formats for New Readers." In *Science Periodicals in Nineteenth-Century Britain: Constructing Scientific Communities*, edited by Gowan Dawson, Bernard Lightman, Sally Shuttleworth, and Jonathan R. Topham, 35–64. Chicago: University of Chicago Press.

Dean, Dennis R. 1999. *Gideon Mantell and the Discovery of Dinosaurs*. Cambridge: Cambridge University Press.

De Beer, Gavin. 1960. *The Sciences Were Never at War*. London: Thomas Nelson and Sons.

de Flon, Nancy Marie. 2005. *Edward Caswall: Newman's Brother and Friend*. Leominster, UK: Gracewing.

De Johnsone, Fowler. 1838. *Truth, in Defence of the Word of God—Vanquishing Infidelity. A Vindication of the Book of Genesis. Addressed to the Rev. William Buckland*. London: R. Groombridge.

de Montluzin, Emily Lorraine. 2003. *Attributions of Authorship in the "Gentleman's Magazine," 1731–1868: An Electronic Union List*. Charlottesville: Bibliographical Society of the University of Virginia. http://bsuva.org/bsuva/gm2.

De Morgan, Augustus. 1830. *Remarks on Elementary Education in Science. An Introductory Lecture, Delivered* [. . .] *in the University of London, November 2, 1830*. London: John Taylor.

De Morgan, Sophia Elizabeth. 1882. *Memoir of Augustus De Morgan, with Selections from His Letters*. London: Longman.

Desmond, Adrian. 1987. "Artisan Resistance and Evolution in Britain, 1819–1848." *Osiris*, 2nd ser., 3:77–110.

Desmond, Adrian. 1989a. "Lamarckism and Democracy: Corporations, Corruption and Comparative Anatomy in the 1830s." In *History, Humanity and Evolution: Essays for John C. Greene*, edited by James R. Moore, 99–130. Cambridge: Cambridge University Press.

Desmond, Adrian. 1989b. *The Politics of Evolution: Morphology, Medicine, and Reform in Radical London*. Chicago: University of Chicago Press.

Desmond, Adrian, and James Moore. 1992. *Darwin*. London: Penguin.

DeWitt, Anne. 2013. *Moral Authority, Men of Science, and the Victorian Novel*. New York: Cambridge University Press.

[Dibdin, Thomas Frognall]. 1832. *Bibliophobia: Remarks on the Present Languid and Depressed State of Literature and the Book Trade*. London: Henry Bohn.

Dibdin, Thomas Frognall. 1836. *Reminiscences of a Literary Life*. 2 vols. London: John Major.

Dick, Thomas. 1824. *The Christian Philosopher; or, The Connection of Science and Philosophy with Religion*. 2nd ed. Glasgow: Chalmers and Collins.

Di Gregorio, Mario, and N. W. Gill. 1990. *Charles Darwin's Marginalia*. Vol. 1. New York: Garland.

Distad, N. Merrill. 1979. *Guessing at Truth: The Life of Julius Charles Hare (1795–1855)*. Shepherdstown: Patmos Press.

Dixon, Thomas. 2003. *From Passions to Emotions: The Creation of a Secular Psychological Category*. Cambridge: Cambridge University Press.

Doddridge, Philip. 1745. *The Rise and Progress of Religion in the Soul*. London: J. Waugh.

Douglas, [Janet Mary] Stair. 1881. *The Life and Selections from the Correspondence of William Whewell, D.D.* London: Kegan Paul.

Draper, Bourne Hall. 1828a. *Conversations on Some Leading Points in Natural Philosophy*. 1st US ed. Utica, NY: Western Sunday School Union.

Draper, Bourne Hall. [1828b]. *Stories from Scripture, on an Improved Plan. Old Testament*. 2nd ed. London: John Harris.

Draper, Bourne Hall. 1839. *The Juvenile Naturalist*. 2 vols. London Darton and Clark.

Draper, Bourne Hall. [1841]. *Stories of the Animal World*. London: Darton and Clark.

Drew, G. S. 1845. *Eight Sermons Preached in St Pancras Church*. London: Francis and John Rivington.

Duncan, George John C. 1848. *Memoir of the Reverend Henry Duncan*. Edinburgh: William Oliphant and Sons.

Duncan, Henry. 1836–37. *Sacred Philosophy of the Seasons: Illustrating the Perfections of God in the Phenomena of the Year*. 4 vols. Edinburgh: William Oliphant and Son.

[Duncan, P. B.]. 1836. *A Catalogue of the Ashmolean Museum*. Oxford: printed by S. Collingwood.

Duns, John. 1873. *Memoir of Sir James Y. Simpson, Bart*. Edinburgh: Edmonston and Douglas.

Dyson, Anthony. 1984. *Pictures to Print: The Nineteenth-Century Engraving Trade*. London: Farrand Press.

Edmonds, J. M. 1979. "The Founding of the Oxford Readership in Geology, 1818." *Notes and Records of the Royal Society of London* 34:33–51.

Edmonds, J. M., and J. A. Douglas. 1976. "William Buckland, F.R.S. (1784–1856) and an Oxford Lecture, 1823." *Notes and Records of the Royal Society of London* 30:141–67.

Edwards, Edward. 1870. *Lives of the Founders of the British Museum*. London: Trübner.

Edwards, O. C. 2004. *A History of Preaching*. Nashville, TN: Abingdon Press.

Egerton, Francis Henry, ed. 1796. *Εὐριπιδου Ἱππολυτος Στεφανηφορος, cum scholiis, versione Latinâ, variis lectionibus, Valckenarî notis integris, ac selectis aliorum VV. DD*. Oxford: privately printed.

[Egerton, Francis Henry]. 1808. *John Bull*. London: privately printed.

Egerton, Francis Henry. [1809]. *Francis Egerton, Third Duke of Bridgewater*. [London]: privately printed.

Egerton, Francis Henry. [1819–20]. *The First [Second] Part of a Letter, to the Parisians, and, the French Nation, Upon Inland Navigation*. [Paris]: privately printed.

Egerton, Francis Henry. [1821]. *Numbers IX. X. XI. XII. XIII. of Addenda and Corrigenda to the Edition of the Hippolytus Stephanéphoros of Euripides*. [Paris]: privately printed.

Egerton, Francis Henry. [1821?]. *Four Letters from Spa, in May 1819*. [London]: privately printed.

Egerton, Francis Henry. [1823?a]. *Note (c) Indicated at page 113 in the Third Part of the "Letter on Inland Navigation to the Parisians, and French Nation."* [Paris]: privately printed.

Egerton, Francis Henry [1823?b]. *One of the Notes (No. 33) Preparing for Insertion, in the Third Part of the "Letter to the Parisians, and French Nation."* [Paris]: privately printed.

Egerton, Francis Henry. [1828?]. *Catalogue of All the Works of the Right Hon. Francis Henry Egerton, Earl of Bridgewater.* [Paris]: privately printed.

Ehrenberg, Christian Gottfried. 1832. "Beitrage zur Kenntniss der Organisation der Infusorien und ihrer geographischen Verbreitung, besonders in Sibirien." *Abhandlungen der Königlichen Akademie der Wissenschaft Berlin* (1830):1–88.

Eliot, George. 1981. *The Mill on the Floss.* Edited by Gordon S. Haight. Oxford: Oxford University Press.

Elliotson, John, ed. 1828. *The Elements of Physiology, by J. Fred. Blumenbach.* 4th ed. London: Longman.

Elliotson, John. 1832. *Address, Delivered at the Opening of the Medical Session in the University of London, October 1st, 1832.* London: Longman.

Ellis, Heather. 2017. *Masculinity and Science in Britain 1831–1918.* London: Palgrave Macmillan.

Ellis, [Sarah]. [1838]. *The Women of England, Their Social Duties, and Domestic Habits.* London: Fisher, Son.

Ellis, [Sarah]. [1842]. *The Daughters of England, Their Position in Society, Character, & Responsibilities.* London: Fisher, Son.

Ellison, Robert H. 1998. *The Victorian Pulpit: Spoken and Written Sermons in Nineteenth-Century Britain.* Selinsgrove, PA: Susquehanna University Press.

Elshakry, Marwa. 2013. *Reading Darwin in Arabic, 1860–1950.* Chicago: University of Chicago Press.

[Elwin, Whitwell]. 1849. "Popular Science." *Quarterly Review* 84:307–44.

Emblen, D. L. 1970. *Peter Mark Roget: The Word and the Man.* London: Longman.

Engen, Rodney K. 1985. *Dictionary of Victorian Wood Engravers.* Cambridge: Chadwyck-Healey.

Ensor, George. 1836. *Natural Theology: The Arguments of Paley, Brougham, and the Bridgewater Treatises on This Subject Examined.* London: Richard Taylor.

Enys, John D. 1877. *Correspondence Regarding the Appointment of the Writers of the Bridgewater Treatises between Davies Gilbert and Others.* Penryn: privately printed.

Eve, A. S., and C. H. Creasey, 1945. *Life and Work of John Tyndall.* London: Macmillan.

*Evidence on the Universities of Scotland: Evidence, Oral and Documentary, Taken and Received by the Commissioners Appointed [ . . . ] for Visiting the Universities of Scotland.* 1837. House of Commons Parliamentary Papers, Session 1837, vols. 35–38.

[Fairholme, George, and Samuel Wilks]. 1834. "A Layman on Scriptural Geology: With Observations Thereon." *Christian Observer* 34:479–96.

Falk, Bernard. 1942. *The Bridgewater Millions: A Candid Family History.* London: Hutchinson.

[Farrar, Eliza Ware]. 1837. *The Young Lady's Friend.* London: John W. Parker.

Farrar, Frederick W. 1868. "The Attitude of the Clergy towards Science." *Contemporary Review* 9:600–620.

Farrar, George Henry. 1889. *Joseph Farrar, J.P.* Bradford: Wm Byles and Sons.

Farrar, Squire. 1823. "To Mr R. Carlile, Dorchester Gaol." *Republican* 8:177–79.

Farrar, Squire. 1826. "To Mr R. Carlile." *Republican* 14:785–91.

Farrar, Squire. 1832a. "To Mr Richard Carlile." *Isis* 1:28–29.

Farrar, Squire. 1832b. "To Mr Richard Carlile." *Isis* 1:595.

[Fellowes, Robert]. 1834. "Religion and Science Identical." *Examiner*, 2 March, 131.

Fellowes, Robert. 1836. *The Religion of the Universe*. London: Thomas Allman.

[Ferguson, Robert]. 1843. "Sir Charles Bell." *Quarterly Review* 72:192–231.

[Ffoulkes, E. S.]. 1892. *A History of the Church of S. Mary the Virgin, Oxford, the University Church*. London: Longman.

Fleming, George A. 1840. "Progress of Social Reform." *New Moral World* 7:1110–12.

Fletcher, John. 1834–35. "Lectures on the Institutions of Medicine." *London Medical and Surgical Journal* [Ryan's] 6:487–90, 519–21, 551–57, 584–90, 615–19, 649–52, 709–14, 739–43, 777–81; 7:4–10, 37–44, 70–75, 103–9, 131–35, 169–73, 199–204, 225–28, 263–67, 327–32, 356–63, 390–96, 545–51, 577–80, 609–14, 641–45, 673–77, 705–11, 737–42, 769–76; 8:97–103, 129–34, 161–64, 193–97, 225–28, 257–61, 289–95.

Fletcher, John. 1835–37. *Rudiments of Physiology, in Three Parts*. Edinburgh: John Carfrae and Son.

Fletcher, John. 1836–37. "Notes of Lectures on Physiology." *London Medical and Surgical Journal* 10:447–52, 461–65, 493–97, 525–29, 557–62, 589–95, 620–27, 653–59, 685–90, 717–21, 748–53, 781–85, 813–17, 845–48, 877–81, 909–13; 11:1–5, 33–37, 65–70, 91–95, 123–27, 155–59, 187–91, 219–23, 251–55, 283–86, 315–19, 347–51, 359–63, 391–400, 423–27, 455–59, 487–91, 519–23.

Forbes, Duncan. 1952. *The Liberal Anglican Idea of History*. Cambridge: Cambridge University Press.

[Forbes, James David]. 1832. *Testimonials in Favour of James D. Forbes* [ . . . ] *as a Candidate for the Chair of Natural Philosophy in the University of Edinburgh*. [Edinburgh]: privately printed.

Forbes, James David. 1849. *The Danger of Superficial Knowledge: An Introductory Lecture*. London: John W. Parker; Edinburgh: Blackwood and Sons.

Forster, John. 1840. *The Churchman's Guide*. London: John W. Parker.

Fox, Robert, and Graeme Gooday. 2005. *Physics in Oxford, 1839–1939: Laboratories, Learning, and College Life*. Oxford: Oxford University Press.

Francis, Keith A. 2012. "Paley to Darwin: Natural Theology versus Science in Victorian Sermons." In *Oxford Handbook of the British Sermon, 1698–1901*, edited by Keith A. Francis and William Gibson, 444–62. Oxford: Oxford University Press.

Francis, Keith A., and William Gibson, eds. 2012. *Oxford Handbook of the British Sermon, 1698–1901*. Oxford: Oxford University Press.

Fraser, Lucy A. 1905. *Memoirs of Daniel Fraser, M.A., LL.D.* London: Percy Lund, Humphries.

Freeman, John. 1852. *Life of the Rev. William Kirby, M.A. F.R.S. F.L.S.* London: Longman.

Fyfe, Aileen. 1997. "The Reception of William Paley's *Natural Theology* in the University of Cambridge." *British Journal for the History of Science* 30:321–35.

Fyfe, Aileen. 2000. "Reading Children's Books in Eighteenth-Century Dissenting Families." *Historical Journal* 43:453–74.

Fyfe, Aileen. 2002. "Publishing and the Classics: Paley's *Natural Theology* and the Nineteenth-Century Scientific Canon." *Studies in History and Philosophy of Science* 33:729–51.

Fyfe, Aileen. 2004. *Science and Salvation: Evangelical Popular Science Publishing in Victorian Britain*. Chicago: University of Chicago Press.

Fyfe, Aileen. 2012. *Steam-Powered Knowledge: William Chambers and the Business of Publishing, 1820–1860*. Chicago: University of Chicago Press.

Gaëde, Henri-Maurice. 1828. *Deux nouveaux discours développant le but de l'étude de l'histoire naturelle*. Liège: P. J. Collardin.

Gascoigne, John. 1988. "From Bentley to the Victorians: The Rise and Fall of British Newtonian Natural Theology." *Science in Context* 2:219–56.

Gaskell, Philip. 1974. *A New Introduction to Bibliography*. Oxford: Clarendon.

Gay, Hannah, and John W. Gay. 1997. "Brothers in Science: Science and Fraternal Culture in Nineteenth-Century Britain." *History of Science* 35:425–53.

Geertz, Clifford. 1966. "Religion as a Cultural System." In *Anthropological Approaches to the Study of Religion*, edited by Michael Banton, 1–46. London: Tavistock.

Geoffroy Saint-Hilaire, [Étienne]. 1830. *Principes de philosophie zoologique*. Paris: Pichon et Didier.

George, M. Dorothy. 1952. *Catalogue of the Political and Personal Satires in the British Museum*. Vol. 10, *1820–1827*. London: British Museum.

Gettman, Royal A. 1960. *A Victorian Publisher: A Study of the Bentley Papers*. Cambridge: Cambridge University Press.

Gibbon, Charles. 1878. *The Life of George Combe, Author of "The Constitution of Man."* 2 vols. London: Macmillan.

Gibson, William. 2012. "The British Sermon, 1689–1901: Quantities, Performance, and Culture." In *Oxford Handbook of the British Sermon, 1698–1901*, edited by Keith A. Francis and William Gibson, 3–30. Oxford: Oxford University Press.

Gilbert, Davies. 1831. "Statement Respecting the Legacy Left by the Late Earl of Bridgewater, for Rewarding the Authors of Works, to be Published in Pursuance of His Will, and Demonstrative of the Divine Attributes, as Manifested in the Creation." *Philosophical Magazine*, 2nd ser., 9:200–202.

Gillespie, Neal C. 1987. "Natural History, Natural Theology, and Social Order: John Ray and the 'Newtonian ideology.'" *Journal of the History of Biology* 20:1–49.

Gillispie, Charles Coulston. 1959. *Genesis and Geology: The Impact of Scientific Discoveries upon Religious Believers in the Decades before Darwin*. New York: Harper and Row.

Gisborne, Thomas. 1794. *An Enquiry into the Duties of Men in the Higher and Middle Classes of Society in Great Britain*. London: B. and J. White.

Gisborne, Thomas. 1797. *An Enquiry into the Duties of the Female Sex*. London: Cadell and Davies.

Gisborne, Thomas. 1818. *The Testimony of Natural Theology to Christianity*. London: Cadell and Davies.

Glaser, Sholem. 1995. *The Spirit of Enquiry: Caleb Hillier Parry*. Stroud: Alan Sutton.

Gliserman, Susan. 1975. "Early Science Writers and Tennyson's *In Memoriam*: A Study in Cultural Exchange." *Victorian Studies* 18:275–308, 437–59.

Godwin, Benjamin. 1834. *Lectures on the Atheistic Controversy*. London: Jackson and Walford.

Godwin, Benjamin. 1839–55. *Reminiscences of Three Score Years and Ten*. Typescript, Bradford Local Studies Library.

Golden, Catherine J. 2003. *Images of the Woman Reader in Victorian British and American Fiction*. Gainesville: University Press of Florida.

Goodfield-Toulmin, June. 1969. "Some Aspects of English Physiology: 1780–1840." *Journal of the History of Biology* 2:283–320.

Goodsir, Joseph Taylor. 1868. *The Divine Rule Proceeds by Law: An Old Sermon*. London: Williams and Norgate.

Gordon, [Elizabeth Oke]. 1894. *The Life and Correspondence of William Buckland, D.D., F.R.S.* London: John Murray.

Gordon, [Margaret Maria]. 1869. *The Home Life of Sir David Brewster*. Edinburgh: Edmonston and Douglas.

Gornall, Thomas, ed. 1981. *The Letters and Diaries of John Henry Newman*. Vol. 5, *Liberalism in Oxford, January 1835 to December 1836*. Oxford: Clarendon.

Grant, James. 1836. *The Great Metropolis*. 2 vols. London: Saunders and Otley.

Grant, James. 1837. *The Great Metropolis*. 2nd ser. 2 vols. London: Saunders and Otley.

Grant, James. 1839. *The Metropolitan Pulpit*. 2 vols. London: George Virtue.

Grant, Robert E. 1828. *An Essay on the Study of the Animal Kingdom: Being an Introductory Lecture Delivered in the University of London, on the 23rd of October, 1828*. London: John Taylor.

Grant, Robert E. 1834. "Lectures on Comparative Anatomy and Animal Physiology. Lecture XII." *Lancet* 1:537–46.

Grant, Robert E. 1846. "Dr Roget's Bridgewater Treatise." *Lancet* 1:445–46.

Gray, Jane Loring. 1893. *Letters of Asa Gray*. 2 vols. Boston: Houghton, Mifflin.

Green, Joseph Henry. 1832. *An Address Delivered in King's College, London, at the Commencement of the Medical Session, October 1, 1832*. London: B. Fellowes.

[Green, Joseph Henry]. 1836. "The British Association: Bristol Meeting." *Fraser's Magazine* 14:582–94.

Gresley, William. 1835. *Ecclesiastes Anglicanus*. London: Rivington.

Grinnell, George James. 1985. "The Rise and Fall of Darwin's Second Theory." *Journal of the History of Biology* 18:51–70.

[Grote, George, and Jeremy Bentham]. 1822. *Analysis of the Influence of Natural Religion on the Temporal Happiness of Mankind. By Philip Beauchamp*. London: R. Carlile.

Gunther, R. T. 1923–67. *Early Science in Oxford*. 15 vols. Oxford: Clarendon.

Gurney, Joseph John. 1832. *Hints on the Portable Evidence of Christianity*. London: J. and A. Arch.

Gurney, Joseph John. 1834. *Essay on the Habitual Exercise of Love to God, Considered as a Preparation for Heaven*. London: Seeley and Burnside.

Gurney, Joseph John. 1853. *Chalmeriana; or, Colloquies with Dr Chalmers*. London: Richard Bentley.

Haight, Gordon S., ed. 1954–78. *The George Eliot Letters*. 9 vols. New Haven, CT: Yale University Press.

Hall, Marie Boas. 1984. *All Scientists Now: The Royal Society in the Nineteenth Century*. Cambridge: Cambridge University Press.

Hall, Sophy. 1910. *Dr Duncan of Ruthwell, Founder of Savings Banks*. Edinburgh: Oliphant, Anderson, and Ferrier.

[Hamilton, William]. 1831. "Universities of England—Oxford." *Edinburgh Review* 53:384–427.

Hampden, Renn Dickson. 1835. *A Course of Lectures Introductory to the Study of Moral Philosophy, Delivered in the University of Oxford, in Lent Term, MDCCCXXXV*. London: B. Fellowes.

Hampden, Renn Dickson. 1837. *The Scholastic Philosophy Considered in Its Relation to Christian Theology*. 2nd ed. London: B. Fellowes.

Hanna, William. 1854. *Memoirs of Thomas Chalmers*. 2 vols. Edinburgh: Thomas Constable.

Hannah, J[ohn]. 1867. "The Attitude of the Clergy towards Science." *Contemporary Review* 6:1–17.

Hannah, J[ohn]. 1868. "A Few More Words on the Relation of the Clergy to Science." *Contemporary Review* 9:395–404.

Hannah, J[ohn]. 1869. "One Word More on the Clergy and Science." *Contemporary Review* 10:74–80.

Hansard, T. C. 1825. *Typographia: An Historical Sketch of the Origin and Progress of the Art of Printing*. London: Baldwin, Cradock, and Joy.

Hardin, Jeff, Ronald L. Numbers, and Ronald A. Binzley. 2018. *The Warfare between Science and Religion: The Idea That Wouldn't Die*. Baltimore: Johns Hopkins University Press.

Harper, John. 1823. "To the Republicans of the Island of Albion." *Republican* 8: 635–39.

Harrison J. F. C. 1961. *Learning and Living, 1790–1960: A Study in the History of the English Adult Education Movement*. London: Routledge and Kegan Paul.

Harrison J. F. C. 1969. *Robert Owen and the Owenites in Britain and America: The Quest for the New Moral World*. London: Routledge and Kegan Paul.

Harrison, Peter. 2015. *The Territories of Science and Religion*. Chicago: University of Chicago Press.

Harte, Negley. 1986. *The University of London, 1836–1986: An Illustrated History*. London: Athlone.

Hartley, J. 1840. *Continental Sermons*. London: James Nisbet.

Hearnshaw, F. J. C. 1929. *The Centenary History of King's College London, 1828–1928*. London: George G. Harrap and Company.

Helmstadter, Richard. 2004. "Condescending Harmony: John Pye Smith's Mosaic Geology." In *Science and Dissent in England, 1688–1945*, edited by Paul Wood, 167–95. Aldershot: Ashgate.

Herschel, John Frederick William. [1831] (1830 on title page). *Preliminary Discourse on the Study of Natural Philosophy*. London: Longman.

Heyworth-Dunne, J. 1940. "Printing and Translations under Muḥmmad 'Alī of Egypt: The Foundation of Modern Arabic." *Journal of the Royal Asiatic Society* 72:325–49.

Hill, George. 1833. *Lectures in Divinity*. Edited by Alexander Hill. 3rd ed. 2 vols. Edinburgh: Waugh and Innes.

Hilton, Boyd. 1988. *The Age of Atonement: The Influence of Evangelicalism on Social and Economic Thought 1785–1865*. Oxford: Clarendon.

Hinton, D. A. 1979. "Popular Science in England, 1830–1870." PhD diss., University of Bath.

Hodge, M. J. S. 1991. "The History of the Earth, Life, and Man: Whewell and Palaetiological Science." In *William Whewell: A Composite Portrait*, edited by Menachem Fisch and Simon Schaffer, 255–88. Oxford: Clarendon.

Hodge, M. J. S., and David Kohn. 1985. "The Immediate Origins of Natural Selection." In *The Darwinian Heritage*, edited by David Kohn, 185–206. Princeton, NJ: Princeton University Press.

Holland, Susan, and Steven Miller. 1997. "Science in the Early *Athenaeum*: A Mirror of Crystallization." *Public Understanding of Science* 6:111–30.

Hollingworth, John Banks. 1825. *Heads of Lectures in Divinity, Delivered in the University of Cambridge*. London: Geo. B. Whitaker.

Hollis, Patricia. 1970. *The Pauper Press: A Study in the Working-Class Radicalism of the 1830s*. Oxford: Oxford University Press.

Holyoake, George Jacob. [1843?]. *Paley Refuted in His Own Words*. London: Hetherington.

Holyoake, George Jacob. 1892. *Sixty Years of an Agitator's Life*. 2 vols. London: T. Fisher Unwin.

Hooker, Joseph Dalton. 1869. "Address." In *Report of the Thirty-Eighth Meeting of the British Association for the Advancement of Science*, lviii–lxxv. London: John Murray.

Hoppus, John. 1830. *On the Study of the Philosophy of the Mind and Logic: An Introductory Lecture, Delivered in the University of London, on Monday, Nov. 8, 1830*. London: John Taylor.

Hoppus, John. 1840. *The Province of Reason, in Reference to Religion: Considered in a Lecture against Socialism*. London: L. and G. Seeley.

[Horne, Richard Henry]. 1835a. "Lord Brougham's *Natural Theology*." *Westminster Review* 23:547–48.

[Horne, Richard Henry]. 1835b. "Remarks on Lord Brougham's *Discourse of Natural Theology*." *Tait's Edinburgh Magazine* 6:806–19.

Houghton, Esther Rhoades, and Josef L. Altholz, 1991. "The *British Critic*, 1824–1843." *Victorian Periodicals Review* 24:111–18.

Houghton, Walter E., Josef L. Altholz, Eileen Curran, Harold E. Dailey, Esther Rhoads Houghton, John A. Lester, Damian McElrath, and Jean Harris Slingerland, eds. 1966–89. *The Wellesley Index to Victorian Periodicals, 1824–1900*. 5 vols. Toronto: Toronto University Press.

Hull, David L. 2003. "Darwin's Science and Victorian Philosophy of Science." In *The Cambridge Companion to Darwin*, edited by Jonathan Hodge and Gregory Radick, 168–91. Cambridge: Cambridge University Press.

Hullmandel, Charles, trans. 1821. *A Manual of Lithography*. 2nd ed. London: Rodwell and Martin.

Hullmandel, Charles. [1824]. *The Art of Drawing on Stone*. London: C. Hullmandel.

Hulme, Samuel. 1881. *Memoir of the Rev. Thomas Allin*. London: Hamilton, Adams.

[Huxley, Thomas Henry]. 1860. "Darwin on the Origin of Species." *Westminster Review* 17:541–70.

[Huxley, Thomas Henry]. 1864. "Science and 'Church Policy.'" *Reader*, 31 December 1864, 821.

Hyman, Anthony. 1984. *Charles Babbage: Pioneer of the Computer*. Oxford: Oxford University Press.

Ince, William. 1878. *The Past History and Present Duties of the Faculty of Theology in Oxford*. Oxford: James Parker.

Inkster, Ian. 1975. "Science and the Mechanics' Institutes, 1820–1850: The Case of Sheffield." *Annals of Science* 32:451–74.

Irons, William J. 1836. *On the Whole Doctrine of Final Causes*. London: Rivington.

Jackson, Thomas. 1834. *Memoirs of the Life and Writings of the Rev. Richard Watson, Late Secretary to the Wesleyan Methodist Missionary Society*. London: John Mason.

Jacob, W. M. 2007. *The Clerical Profession in the Long Eighteenth Century, 1680–1840*. Oxford: Oxford University Press.

Jacyna, L. S. 1983. "Immanence or Transcendence: Theories of Life and Organization in Britain, 1790–1835." *Isis* 74:311–29.

Jacyna, L. S. 1994. *Philosophic Whigs: Medicine, Science and Citizenship in Edinburgh 1789–1848*. London: Routledge.

James, Frank. 2004. "Reporting Royal Institution Lectures, 1826 to 1867." In *Science Serialized: Representations of the Sciences in Nineteenth-Century Periodicals*, edited by Sally Shuttleworth and Geoffrey Cantor, 67–79. Cambridge, MA: MIT Press.

James, John. 1841. *The History and Topography of Bradford*. London: Longman.

James, John Angell. 1825. *The Christian Father's Present to His Children*. 3rd ed. London: Francis Westley.

James, John Angell. 1838. *The Anxious Inquirer after Salvation Directed and Encouraged*. London: Religious Tract Society.

James, John Angell. 1844. *The Flower Faded: A Short Memoir of Clementine Cuvier*. 6th ed. London: Hamilton, Adams.

[Jameson, Robert]. [between 1830 and 1839–?]. *Syllabus of Lectures on Natural History*. Edinburgh: privately printed.

Jamieson, John. 1959. *The History of the Royal Belfast Academical Institution, 1810–1960*. Belfast: William Mulland and Son.

Jenkins, Bill. 2019. *Evolution before Darwin: Theories of Transmutation of Species in Edinburgh, 1804–1834*. Edinburgh: Edinburgh University Press.

Jenyns, Leonard. 1862. *Memoir of the Rev. John Stevens Henslow*. London: John Van Voorst.

Jerdan, William. 1852–53. *The Autobiography of William Jerdan*. 4 vols. London: Arthur Hall, Virtue.

Johnson, Richard. 1979. "'Really Useful Knowledge': Radical Education and Working-Class Culture, 1790–1848." In *Working Class Culture: Studies in History and Theory*, edited by John Clarke, Chas Critcher, and Richard Johnson, 75–102. London: Hutchinson.

[Johnstone, Christian]. 1834. "*Johnstone's Edinburgh Magazine*: The Cheap and Dear Periodicals." *Tait's Edinburgh Magazine* 4:490–500.

Jones, William. 1850. *The Jubilee Memorial of the Religious Tract Society*. London: Religious Tract Society.

Kaser, David. 1963. *The Cost Book of Carey & Lea, 1825–1838*. Philadelphia: University of Pennsylvania Press.

[Keble, John]. 1827. *The Christian Year*. 2 vols. Oxford: J. Parker.

Keir, David. 1952. *The House of Collins: The Story of a Scottish Family of Publishers from 1789 to the Present Day*. London: Collins.

Kendall, Joshua. 2008. *The Man who Made Lists: Love, Death, Madness, and the Creation of "Roget's Thesaurus."* New York: G. P. Putnam's Sons.

Kennedy, Meegan. 2017. "Discriminating the 'Minuter Beauties of Nature': Botany as Natural Theology in a Victorian Medical School." In *Strange Science: Investigating the Limits of Knowledge in the Victorian Age*, edited by Lara Karpenko and Shalyn Claggett, 40–61. Ann Arbor: University of Michigan Press.

Keynes, Geoffrey. 1969. *William Pickering, Publisher: A Memoir and a Check-List of His Publications*. Rev. ed. London: Galahad.

Kidd, John. 1809. *Outlines of Mineralogy*. 2 vols. Oxford: J. Parker.

Kidd, John. 1818. *An Answer to a Charge against the English Universities Contained in the "Supplement to the Edinburgh Encyclopædia."* Oxford: privately printed.

Kidd, John. 1824. *An Introductory Lecture to a Course in Comparative Anatomy, Illustrative of Paley's Natural Theology*. Oxford: privately printed.

Kidd, John. 1833. *On the Adaptation of External Nature to the Physical Condition of Man, Principally with Reference to the Supply of His Wants and the Exercise of His Intellectual Faculties*. London: William Pickering.

King, John. 1833. *Sermons Preached in Christ's Church, Sculcoates*. London: L. B. Seeley and Sons.

King's College London. 1845. *Student's Hand-Book for the Medical Department of King's College, London*. London: John William Parker.

[Kingsley, Frances Eliza]. 1877. *Charles Kingsley: His Letters and Memories of His Life*. 3rd ed. 2 vols. London: Henry S. King.

Kirby, William. 1835. *On the Power, Wisdom and Goodness of God as Manifested in the Creation of Animals and in Their History, Habits and Instincts*. 2 vols. London: William Pickering.

Kirby, William, and William Spence. 1815–26. *An Introduction to Entomology; or, Elements of the Natural History of Insects*. 4 vols. London: Longman.

Knickerbocker, Driss Richard. 1981. "The Popular Religious Tract in England, 1790–1830." DPhil diss., University of Oxford.

[Knight, Charles]. 1834. "The Market for Literature." *Printing Machine*, 15 February, 1–5.

Knight, Charles. 1854. *The Old Printer and the Modern Press*. London: John Murray.

Knight, Charles. 1864–65. *Passages of a Working Life during Half a Century*. 3 vols. London: Bradbury and Evans.

Knight, David. 1978. *The Transcendental Part of Chemistry*. Folkestone: Dawson.

Knox, Robert. 1831. "Observations on the Stomach of the Peruvian Lama." *Transactions of the Royal Society of Edinburgh* 11:479–98.

Knox, Robert. 1841. "Contributions to Anatomy and Physiology: Communicated at Various Times to the Anatomical and Physiological Society in Edinburgh." *Edinburgh Medical and Surgical Journal* 56:125–39.

Knox, Robert. 1850. *The Races of Men: A Fragment*. London: Henry Renshaw.

Koditschek, Theodore. 1990. *Class Formation and Urban-Industrial Society: Bradford, 1750–1850*. Cambridge: Cambridge University Press.

Kohn, David. 1980. "Theories to Work By: Rejected Theories, Reproduction, and Darwin's Path to Natural Selection." *Studies in the History of Biology* 4:67–170.

Kolb, Jack, ed. 1981. *The Letters of Arthur Henry Hallam*. Columbus: Ohio State University Press.

Kramer, Jack. 1996. *Women of Flowers: A Tribute to Victorian Women Illustrators*. New York: Stewart, Tabori, and Chang.

Landon, Whittington H. 1835. *Ten Sermons, Preached in the Parish Church of Tavistock*. London: Rivington.

Lang, Cecil Y., and Edgar F. Shannon, eds. 1982. *The Letters of Alfred Lord Tennyson*. 3 vols. Oxford: Clarendon.

Lardner, Dionysius. 1829. *A Discourse on the Advantages of Natural Philosophy and Astronomy, as Part of a General and Professional Education. Being an Introductory*

*Lecture Delivered in the University of London, on the 28th October, 1828*. London: John Taylor.

Larsen, Timothy. 2006. *Crisis of Doubt: Honest Faith in Nineteenth-Century England*. Oxford: Oxford University Press.

Latimer, John. 1887. *The Annals of Bristol in the Nineteenth Century*. Bristol: W. and F. Morgan.

Law, James Thomas. 1873. "An Address Delivered at the First Anniversary Meeting, 1835, of the School of Medicine and Surgery, Birmingham." In *Annals of the Queen's College, Birmingham. Reprint of Addresses*, 4 vols., edited by William Sands Cox, 2:63–76. London: privately printed.

Ledger-Lomas, Michael. 2009. "Mass Markets: Religion." In *The Cambridge History of the Book in Britain*, vol. 6, *1830–1914*, edited by David McKitterick, 324–58. Cambridge: Cambridge University Press.

Lewis, Samuel. 1842. *The History and Topography of the Parish of Saint Mary, Islington, in the County of Middlesex*. London: J. H. Jackson.

Liddon, Henry Parry. 1893–97. *Life of Edward Bouverie Pusey*. 4 vols. London: Longman.

Lightman, Bernard. 2001. "Victorian Sciences and Religions: Discordant Harmonies." *Osiris*, 2nd ser., 16:343–66.

Lightman, Bernard. 2007. *Victorian Popularizers of Science: Designing Nature for New Audiences*. Chicago: University of Chicago Press.

Lindley, John. 1829. *An Introductory Lecture Delivered in the University of London, on Thursday, April 30, 1829*. London: John Taylor.

Lindley, John. 1830a. *An Introduction to the Natural System of Botany*. London: Longman.

Lindley, John. 1830b. *An Outline of the First Principles of Botany*. London: Longman.

Lindley, John. 1835. *An Introduction to Botany*. 2nd ed. London: Longman.

*London University Parliamentary Returns: London University. A Copy of the First and of the Second Charter of the University of London; of the Minutes of the Senate of the University, and of All Committees Appointed by the Senate [etc.]*. 1840. House of Commons Parliamentary Papers, Session 1840, 40:29–357.

Lonsdale, Henry. 1870. *A Sketch of the Life and Writings of Robert Knox*. London: Macmillan.

[Lord, Percival B.]. 1834. Review of *Discourse on the Studies of the University*, by Adam Sedgwick. *Athenaeum*, 12 April, 267–68.

Loudon, Irvine. 1986. *Medical Care and the General Practitioner, 1750–1850*. Oxford: Clarendon.

[Lyell, Charles]. 1827. "State of the Universities." *Quarterly Review* 36:216–68.

Lyell, Charles. 1838. "Address to the Geological Society, Delivered at the Anniversary, on the 17th of February, 1837." *Proceedings of the Geological Society of London* 2:479–523.

Lyell, [Katherine M.]. 1881. *Life, Letters and Journals of Sir Charles Lyell, Bart*. 2 vols. London: John Murray.

[Macaulay, Thomas Babington]. 1826. "The London University." *Edinburgh Review* 43:315–41.

MacDougall, Patrick C. 1852. *Papers on Literary and Philosophical Subjects*. Edinburgh: Johnstone and Hunter.

Macfarlane, Alan. 1986. *Marriage and Love in England: Modes of Reproduction, 1300–1840*. Oxford: Basil Blackwell.

MacGillivray, William. 1834. *Lives of Eminent Zoologists, from Aristotle to Linnæus*. Edinburgh: Oliver and Boyd.

MacGillivray, William. 1836. *Descriptions of the Rapacious Birds of Great Britain*. Edinburgh: MacLachlan and Steward.

MacGillivray, William, and J. Arthur Thomson. 1910. *Life of William MacGillivray M.A., LL.D., F.R.S.E.* London: John Murray.

MacGregor, Arthur, and Abigail Headon. 2010. "Re-Inventing the Ashmolean: Natural History and Natural Theology at Oxford in the 1820s to 1850s." *Archives of Natural History* 27:369–406.

Mackintosh, D. 1843. *Supplement to the Bridgewater Treatises: The Highest Generalizations in Geology and Astronomy, Viewed as Illustrating the Greatness of the Creator*. 2nd ed. London: Smith, Elder.

Mackintosh, T. Simmons. 1837–38. "Mackintosh's Electrical Theory of the Universe." *New Moral World* 3:239, 255–56, 283–84, 300–301, 334–35, 342–44, 374–75, 378–80, 386–87, 402; 4:6–7, 16, 23–24, 26–28, 34–35, 143–44, 151–52, 154, 162–63, 170–71, 178–79, 186–87, 219–20, 227–28, 243–44.

Mackintosh, T. Simmons. 1842. *A Dissertation on the Being and Attributes of God*. Leeds: Joshua Hobson.

Mackintosh, T. Simmons. [1845?]. *The "Electrical Theory," of the Universe; or, The Elements of Physical and Moral Philosophy*. London: H. Hetherington.

Mackintosh, T. Simmons. [1846]. *The "Electrical Theory" of the Universe; or, The Elements of Physical and Moral Philosophy*. [1st US ed., from the 1st UK ed. (1838)]. Boston: Josiah P. Mendum.

Macleay, W. S. 1819–21. *Horæ entomologicæ; or, Essays on the Annulose Animals*. London: S. Bagster.

MacLeod, Roy. 1983. "Whigs and Savants: Reflections on the Reform Movement in the Royal Society, 1830–48." In *Metropolis and Province, Science and British Culture, 1780–1850*, edited by Ian Inkster and Jack Morrell, 55–90. London: Hutchinson.

Malet, Hugh. 1977. *Bridgewater: The Canal Duke, 1736–1803*. Nelson: Hendon.

Manier, Edward. 1978. *The Young Darwin and His Cultural Circle: A Study of Influences Which Helped Shape the Language and Logic of the First Drafts of the Theory of Natural Selection*. Dordrecht: D. Reidel.

Mantell, Gideon. 1838. *The Wonders of Geology*. 2 vols. London: Relfe and Fletcher.

Marchand, Leslie A. 1941. *The "Athenaeum": A Mirror of Victorian Culture*. Chapel Hill: North Carolina University Press.

Marcou, Jules. 1896. *Life, Letters, and Works of Louis Agassiz*. 2 vols. New York: Macmillan.

M[artineau], H[arriet]. 1839. "Literary Lionism." *London and Westminster Review* 32:261–81.

Martineau, Harriet. 1877. *Harriet Martineau's Autobiography, with Memorials by Maria Weston Chapman*. 3 vols. London: Smith, Elde.

Martineau, James. 1888. *A Study of Religion: Its Sources and Contents*. 2 vols. Oxford: Clarendon.

[Mason, David Mather]. 1864. "Dead I Have Known; or, Recollections of Three

Cities—Reminiscences of Edinburgh University—Professors and Debating Societies." *Macmillan's Magazine* 11:123–40.

A Master of Arts. 1832. *A Short Criticism of a Lecture Published by the Savilian Professor of Geometry*. Oxford: W. Baxter.

Mayo, Herbert. 1834. *An Introductory Lecture to the Medical Classes in King's College, London*. London: Burgess and Hill.

Mayo, Herbert. 1837. *The Philosophy of Living*. London: John W. Parker.

[McCulloch, John Ramsay]. 1831. "Taxes on Literature." *Edinburgh Review* 53: 427–37.

McDannell, Colleen. 1994. *The Christian Home in Victorian America, 1840–1900*. Bloomington: Indiana University Press.

McDonnell, James Martin. 1983. "William Pickering (1796–1854), Antiquarian Bookseller, Publisher, and Book Designer: A Study in the Early Nineteenth Century Book Trade." PhD diss., Polytechnic of North London.

Medway, John. 1853. *Memoirs of the Life and Writings of John Pye Smith, D.D., LL.D., F.R.S., F.G.S.* London: Jackson and Walford.

Melvill, Henry. 1837. *Four Sermons, Preached before the University of Cambridge, During the Month of February, MDCCCXXXVII*. Cambridge: J. and J. J. Deighton.

Melvill, Henry. 1838. *Sermons*. 2 vols. London: Rivington.

[Mill, John Stuart]. 1835. "Professor Sedgwick's *Discourse*—State of Philosophy in England." *London Review* 1:94–135.

Miller, David Philip. 1981. "The Royal Society of London, 1800–1835: A Study in the Cultural Politics of Scientific Organization." PhD diss., University of Pennsylvania.

Mills, William. 1832. *The Duty of Christian Humility as Opposed to the Pride of Science: A Discourse*. Oxford: J. H. Parker.

[Mitchell, John]. 1838. "Captain Orlando Sabertash to Oliver Yorke, Esq., on Manners, Fashion, and Things in General." *Fraser's Magazine* 17:291–309.

Mitchell, Sally. 2004. *Frances Power Cobbe: Victorian Feminist, Journalist, Reformer*. Charlottesville: University of Virginia Press.

Moon, Marjorie. 1976. *John Harris's Books for Youth, 1801–1843*. Cambridge: Marjorie Moon and Alan Spilman.

Moon, Norman. 1979. *Education for Ministry: Bristol Baptist College, 1679–1979*. Bristol: Bristol Baptist College.

Moore, Doris Langley. 1977. *Ada, Countess of Lovelace: Byron's Legitimate Daughter*. London: John Murray.

Moore, James R. 1979. *The Post-Darwinian Controversies: A Study of the Protestant Struggles to Come to Terms with Darwin in Great Britain and America, 1870–1900*. Cambridge: Cambridge University Press.

[Moore, Thomas]. 1836. "Anticipated Meeting of the British Association in the Year 2836." *Morning Chronicle*, 8 September, 3a.

More, Hannah. 1811. *Practical Piety; or, The Influence of the Religion of the Heart on the Conduct of the Life*. 2nd ed. 2 vols. London: Cadell and Davies.

More, Hannah. 2007. *Cœlebs in Search of a Wife: Comprehending Observations on Domestic Habits and Manners, Religion and Morals*. Edited by Patricia Demers. Peterborough, Canada: Broadview Editions.

[Morgan, John Minter]. 1834. *The Critics Criticized, With Remarks on a Passage in Dr. Chalmers's "Bridgewater Treatise."* London: Edward Moxon.

Morrell, J. B. 1971. "Professors Robison and Playfair, and the *Theophobia Gallica*: Natural Philosophy, Religion and Politics in Edinburgh, 1789–1815." *Notes and Records of the Royal Society of London* 26:43–63.

Morrell, J. B. 1972. "Science and Scottish University Reform: Edinburgh in 1826." *British Journal for the History of Science* 6:39–56.

Morrell, J. B. 1975. "The Leslie Affair: Careers, Kirk, and Politics in Edinburgh in 1805." *Scottish Historical Review* 54:62–82.

Morrell, J. B. 1985. "Wissenschaft in Worstedopolis: Public Science in Bradford, 1800–1850." *British Journal for the History of Science* 18:1–23.

Morrell, J. B. 2005. *John Phillips and the Business of Victorian Science*. Aldershot: Ashgate.

Morrell, J. B., and Arnold Thackray. 1981. *Gentlemen of Science: Early Years of the British Association for the Advancement of Science*. Oxford: Clarendon.

Morrell, J. B., and Arnold Thackray, eds. 1984. *Gentlemen of Science: Early Correspondence of the British Association for the Advancement of Science*. London: Royal Historical Society.

Morren, Édouard. 1865. "Prologue à la mémoire de Henri-Maurice Gaede, 1795–1834." *La Belgique horticole* 15:v–xv.

Morrison-Low, A. D., and J. R. R. Christie, eds. 1984. *"Martyr of Science": Sir David Brewster, 1781–1868*. Edinburgh: Royal Scottish Museum.

Mortenson, Terry. 2004. *The Great Turning Point: The Church's Catastrophic Mistake on Geology—Before Darwin*. Green Forest: Master Books.

Morus, Iwan Rhys. 1998. *Frankenstein's Children: Electricity, Exhibition, and Experiment in Early-Nineteenth-Century London*. Princeton, NJ: Princeton University Press.

Moseley, H. 1839. *Lectures on Astronomy, Delivered at King's College, London*. London: John W. Parker.

Moseley, H. 1847. *Astro-Theology*. London: Arthur Varnham.

[Mozley, T.]. 1882a. *Letter to the Rev. Canon Bull, Ch. Ch. Oxford* [ . . . ] *on Two Passages* [ . . . ] *Relating to Dr Buckland, Dean of Westminster*. London: Longman.

Mozley, T. 1882b. *Reminiscences, Chiefly of Oriel College and the Oxford Movement*. 2 vols. London: Longman.

Munk, William. 1878. *The Roll of the Royal College of Physicians of London*. 2nd ed. 3 vols. London: Royal College of Physicians.

Murphy, P. 1834. *The Anatomy of the Seasons, Weather Guide Book, and Perpetual Companion to the Almanac*. London: J. R. Bailliere.

Musgrove, F. 1959. "Middle-Class Education and Employment in the Nineteenth Century." *Economic History Review*, 2nd ser., 12:99–111.

Newman, John Henry. 1843. *Sermons, Chiefly on the Theory of Religious Belief, Preached before the University of Oxford*. London: Rivington.

Newman, John Henry. 1872. *Fifteen Sermons Preached before the University of Oxford, between A.D. 1826 and 1843*. 3rd ed. London: Rivington.

Noel, Baptist Wriothesley. 1835. *The State of the Metropolis Considered*. London: James Nisbet.

Nolan, Frederick. 1833a. *The Analogy of Revelation and Science Established in a Series of Lectures Delivered before the University of Oxford, in the year MDCCCXXXIII.* Oxford: J. H. Parker.

Nolan, Frederick. 1833b. "Strictures on 'The Apology for British Science.'" *Standard,* 12 December, 4d–e; 24 December 1833, 4d.

Nolan, Frederick. 1834. "To the Editor of the *Standard.*" *Standard,* 15 January, 3b.

Numbers, Ronald L. 1977. *Creation by Natural Law: Laplace's Nebular Hypothesis in American Thought.* Seattle: University of Washington Press.

O'Connor, Ralph. 2007a. *The Earth on Show: Fossils and the Poetics of Popular Science, 1802–1856.* Chicago: University of Chicago Press.

O'Connor, Ralph. 2007b. "Young-Earth Creationists in Early Nineteenth-Century Britain? Towards a Reassessment of 'Scriptural Geology.'" *History of Science* 45: 357–403.

Oldroyd, D. R., and D. W. Hutchings. 1979. "The Chemical Lectures at Oxford (1822–1854) of Charles Daubeny, M.D., F.R.S." *Notes and Records of the Royal Society* 33:217–59.

Orange, Derek. 1981. "Science in Early Nineteenth-Century York: The Yorkshire Philosophical Society and the British Association." In *York, 1831–1981: 150 Years of Scientific Endeavour and Social Change,* edited by Charles Feinstein, 1–29. York: Ebor.

Ospovat, Dov. 1981. *The Development of Darwin's Thought: Natural History, Natural Theology, and Natural Selection.* Cambridge: Cambridge University Press.

Ostrander, Rick. 2000. *The Life of Prayer in a World of Science: Protestants, Prayer, and American Culture, 1870–1930.* Oxford: Oxford University Press.

[Otter, William]. 1824. *The Life and Remains of the Rev. Edward Daniel Clarke, LL.D.* London: George Cowie.

Outram, Dorinda. 1984. *Georges Cuvier: Vocation, Science, and Authority in Post-Revolutionary France.* Manchester: Manchester University Press.

Owen, Richard. 1840. "Fossil Mammalia." In *The Zoology of the Voyage of H.M.S. Beagle, under the Command of Captain Fitzroy, R.N., during the Years 1832 to 1836,* 5 vols., edited by Charles Darwin, 1:13–111. London: Smith, Elder.

Owen, Richard. 1841. "Observations on the Fossils Representing the *Thylacotherium Provostii,* Valenciennes, with Reference to the Doubts of Its Mammalian and Marsupial Nature Recently Promulgated; and on the *Phascolotherium Bucklandi.*" *Transactions of the Geological Society of London,* 2nd ser., 6:47–65.

Owen, Richard. 1842. *Description of the Skeleton of an Extinct Gigantic Sloth.* London: John Van Voorst.

Owen, Richard. 1847. "Report on the Archetype and Homologies of the Vertebrate Skeleton." *Report of the Sixteenth Meeting of the British Association for the Advancement of Science,* 169–340.

Owen, Richard. 1849. *On the Nature of Limbs: A Discourse.* London: John Van Voorst.

Owen, Richard. 1894. *The Life of Richard Owen by His Grandson.* 2 vols. London: John Murray.

[Owen, Robert]. 1835. "The Religion of the Millennium." *New Moral World* 2:33.

[Owen, Robert]. 1836. "The Religion of the New Moral World." *New Moral World* 2:237.

Owen, Robert. 1839. "Mr Owen to the Social Missionaries." *New Moral World* 6:[593]–97.

Owen, Robert. 1840. *Socialism; or, The Rational System of Society.* [ . . . ] *First Lecture*. London: Effingham Wilson.

Owen, Robert. 2005. "Social Tracts [1838–1839]." In *Owenite Socialism: Pamphlets and Correspondence*, vol. 5, *1838–1839*, edited by Gregory Claeys, 54–99. London: Routledge.

Page, Frederick G. 2008. "James Rennie (1787–1867), Author, Naturalist and Lecturer." *Archives of Natural History* 35:128–42.

Paley, William. 1802. *Natural Theology; or, Evidences of the Existence and Attributes of the Deity, Collected from the Appearances of Nature*. 2nd ed. London: R. Faulder.

ΠΑΝ [pseud.]. 1825. "On Cheap Periodical Literature." *Gentleman's Magazine* 95, pt. 1:483–86.

Parker, Charles Stuart, ed. 1891–99. *Sir Robert Peel from His Private Papers*. 3 vols. London: John Murray.

[Parker, John Henry]. 1851. *Bibliotheca parva theologica: A Catalogue of Books Recommended to Students in Divinity*. Oxford: John Henry Parker.

Parkinson, Richard. 1838. *Rationalism and Revelation; or, The Testimony of Moral Philosophy, the System of Nature, and the Constitution of Man, to the Truth of the Doctrines of Scripture; in Eight Discourses*. London: Rivington.

Parr, Victor G. 1900. *List of Lecturers and Lectures at the Royal College of Surgeons of England, 1810–1900*. London: privately printed.

[Parry, Charles Henry]. [1846?]. *Ellen Parry*. [Bath]: privately printed.

P[aterson], T[homas]. 1843. "Harmony of the Godlies." *Oracle of Reason* 2:217–18, 235–38.

Patterson, Elizabeth Chambers. 1983. *Mary Somerville and the Cultivation of Science, 1815–1840*. The Hague: Martinus Nijhoff.

Peacock, George. 1855. *Life of Thomas Young, M.D., F.R.S., &c*. London: John Murray.

Peacock, Sandra J. 2002. *The Theological and Ethical Writings of Frances Power Cobbe, 1822–1904*. Lewiston, NY: Edwin Mellen.

Peddie, Robert Alexander, and Quintin Waddington, eds. 1914. *The English Catalogue of Books* [ . . . ] *1801–1836*. London: Sampson Low, Marston.

Peel, C. S. 1934. "Homes and Habits." In *Early Victorian England, 1830–1865*, 2 vols., edited by G. M. Young, 1:79–151. London: Oxford University Press.

Perron, A. 1843. "Lettre sur les écoles et l'imprimerie du pacha d'Égypte." *Journal asiatique*, 4th ser., 2:5–61.

Plant, Marjorie. 1939. *The English Book Trade: An Economic History of the Making and Sale of Books*. London: George Allen and Unwin.

Pollard, Graham. 1978. "The English Market for Printed Books: The Sandars Lectures, 1959." *Publishing History* 4:7–48.

[Post, Jacob]. 1838. *Extracts from the Diary and Other Manuscripts of the late Frederic James Post, of Islington*. London: privately printed.

Potts, Robert. 1855. *Liber Cantabrigiensis*. Cambridge: Cambridge University Press.

Powell, Baden. 1826a. *The Advance of Knowledge in the Present Times, Considered; Especially in Regard to Religion. A Sermon Delivered in the Parish Church of Dartford, Kent, on Thursday, April 27*. London: Rivington.

Powell, Baden. 1826b. *Rational Religion Examined; or, Remarks on the Pretensions of Unitarianism*. London: Rivington.

Powell, Baden. 1832. *The Present State and Future Prospects of Mathematical and Physical Studies in the University of Oxford, Considered in a Public Lecture.* Oxford: privately printed.

Powell, Baden. 1833. *Revelation and Science: The Substance of a Discourse, Delivered before the University of Oxford, at St Mary's, March VIII, MDCCCXXIX.* Oxford: J. H. Parker.

Powell, Baden. 1838. *The Connexion of Natural and Divine Truth; or, The Study of the Inductive Philosophy Considered as Subservient to Theology.* London: John W. Parker.

[Prestwich, Grace Anne]. 1899. *Life and Letters of Sir Joseph Prestwich.* Edinburgh: William Blackwood.

Priestley, Eliza. 1908. *The Story of a Lifetime.* London: Kegan Paul.

Priestman, Martin. 2000. *Romantic Atheism: Poetry and Freethought, 1780–1830.* Cambridge: Cambridge University Press.

Prothero, Iowerth. 1979. *Artisans and Politics in Early Nineteenth Century London: John Gast and His Times.* London: Methuen.

Prout, William. 1821. *An Inquiry into the Nature and Treatment of Gravel, Calculus, and Other Diseases Connected with a Deranged Operation of the Urinary Organs.* London: Baldwin, Cradock, and Joy.

Prout, William. 1834a. *Chemistry, Meteorology and the Function of Digestion Considered with Reference to Natural Theology.* London: William Pickering.

Prout, William. 1834b. *Chemistry, Meteorology and the Function of Digestion Considered with Reference to Natural Theology.* 2nd ed. London: William Pickering.

Prout, William. 1834c. "Reply to Dr. W. Charles Henry." *London and Edinburgh Philosophical Magazine,* 3rd ser., 5:132–33.

Prout, William. 1840. *On the Nature and Treatment of Stomach and Urinary Diseases.* 3rd ed. London: John Churchill.

Prout, William. 1845. *Chemistry, Meteorology and the Function of Digestion Considered with Reference to Natural Theology.* 3rd ed. London: John Churchill.

Pym, Horace N. 1882. *Memories of Old Friends, Being Extracts from the Journals and Letters of Caroline Fox, of Penjerrick, Cornwall from 1835 to 1871.* London: Smith, Elder.

Pyne, Thomas. [1855]. *A Memoir of the Rev. Robert Francis Walker, M.A.* [London]: Nisbet.

Quain, Jones. 1831. *University of London: Lecture Introductory to the Course of Anatomy and Physiology, Delivered at the Opening of Session, 1831–32.* London: John Taylor.

Raven, James. 2007. *The Business of Books: Booksellers and the English Book Trade, 1450–1850.* New Haven, CT: Yale University Press.

Reade, John Edmund. 1838. *Italy: A Poem, in Six Parts.* London: Saunders and Otley.

Rees, Thomas. 1896. *Reminiscences of Literary London from 1779 to 1853.* London: Suckling and Galloway.

Rehbock, Philip F. 1983. *The Philosophical Naturalists: Themes in Early Nineteenth-Century British Biology.* Madison: University of Wisconsin Press.

Reid, Wemyss. 1899. *Memoirs and Correspondence of Lyon Playfair.* London: Cassell.

Rennell, Thomas. 1819. *Remarks on Scepticism, Especially as It Is Connected with the Subjects of Organization and Life.* London: Rivington.

Rennie, James. 1834. *Alphabet of Natural Theology, for the Use of Beginners*. London: Orr and Smith.

*Report into Ecclesiastical Revenues: Report of the Commissioners Appointed by His Majesty to Inquire into the Ecclesiastical Revenues of England and Wales*. 1835. House of Commons Parliamentary Papers, Session 1835, 22:15–1060.

*Report into the Universities of Scotland: Report Made to His Majesty by a Royal Commission of Inquiry into the State of the Universities of Scotland*. 1831. House of Commons Parliamentary Papers, Session 1831, 12:111–546.

*Report into the University of Cambridge: Report of Her Majesty's Commissioners Appointed to Inquire into the State, Discipline, Studies, and Revenues of the University and Colleges of Cambridge*. 1852–53. House of Commons Parliamentary Papers, Session 1852–53, 44:2–701.

*Report into the University of Oxford: Report of Her Majesty's Commissioners Appointed to Inquire into the State, Discipline, Studies, and Revenues of the University and Colleges of Oxford: Together with the Evidence, and an Appendix*. 1852. House of Commons Parliamentary Papers, Session 1852, 22:1–773.

Rice, Adrian. 1997. "Inspiration or Desperation? Augustus De Morgan's Appointment to the Chair of Mathematics at London University in 1828." *British Journal for the History of Science* 30:257–74.

Rice, Adrian. 1999. "What Makes a Great Mathematics Teacher? The Case of Augustus De Morgan." *American Mathematical Monthly* 106:534–52.

Rice, Daniel F. 1971. "Natural Theology and the Scottish Philosophy in the Thought of Thomas Chalmers." *Scottish Journal of Theology* 24:23–46.

Richards, Joan. 1997. "The Probable and the Possible in Early Victorian England." In *Contexts of Victorian Science*, edited by Bernard Lightman, 51–71. Chicago: University of Chicago Press.

Richards, Joan. 2002. "'In a Rational World All Radicals Would Be Exterminated': Mathematics, Logic, and Secular Thinking in Augustus De Morgan's England." *Science in Context* 15:137–64.

Richards, Joan. 2007. "In Search of the 'Sea-Something': Reason and Transcendence in the Frend/De Morgan Family." *Science in Context* 20:509–36.

Richards, Robert J. 1981. "Instinct and Intelligence in British Natural Theology: Some Contributions to Darwin's Theory of the Evolution of Behaviour." *Journal of the History of Biology* 14:193–230.

Richardson, Ruth. 2008. *The Making of Mr Gray's Anatomy*. Oxford: Oxford University Press.

Ricks, Christopher, ed. 1987. *The Poems of Tennyson*. 2nd ed. 3 vols. London: Longman.

Rivers, Isabel. 1982. "Dissenting and Methodist Books of Practical Divinity." In *Books and Their Readers in Eighteenth-Century England*, 127–64. Leicester: Leicester University Press.

Rivers, Isabel. 2018. *Vanity Fair and the Celestial City: Dissenting, Methodist, and Evangelical Literary Culture in England, 1720–1800*. Oxford: Oxford University Press.

Roberton, John. 1836. *Critical Remarks on Certain Recently Published Opinions Concerning Life and Mind*. London: Longman.

Roberts, Jon H. 2009. "That Darwin Destroyed Natural Theology." In *Galileo Goes to Jail and Other Myths about Science and Religion*, edited by Ronald L. Numbers, 161–69. Cambridge, MA: Harvard University Press.

Robertson, John M. 1929. *A History of Freethought in the Nineteenth Century*. 2 vols. London: Watts.

Robson, John M. 1990. "The Fiat and Finger of God: The Bridgewater Treatises." In *Victorian Faith in Crisis: Essays on Continuity and Change in Nineteenth-Century Religious Belief*, edited by Richard J. Helmstadter and Bernard Lightman, 71–125. Stanford, CA: Stanford University Press.

[Roget, Peter Mark]. 1824. "Physiology." In *Supplement to the Fourth, Fifth, and Sixth Editions of the Encyclopædia Britannica, with Preliminary Dissertations on the History of the Sciences*, vol. 6, [edited by Macvey Napier], 180–97. Edinburgh: Archibald Constable; London: Hurst, Robinson.

Roget, Peter Mark. 1826. *An Introductory Lecture on Human and Comparative Physiology, Delivered at the New Medical School in Aldersgate Street*. London: Longman.

Roget, Peter Mark. 1832. *Treatises on Electricity, Galvanism, Magnetism, and Electro-Magnetism*. London: Baldwin and Cradock.

Roget, Peter Mark. 1834. *Animal and Vegetable Physiology Considered with Reference to Natural Theology*. 2 vols. London: William Pickering.

Roget, Peter Mark. [1838]. "Physiology." In *Encyclopædia Britannica; or, Dictionary of Arts, Sciences, and General Literature*, 7th ed., vol. 17, [edited by Macvey Napier], 577–733. Edinburgh: Adam and Charles Black.

Roget, Peter Mark. 1846a. "Dr Roget's Bridgewater Treatise." *Lancet* 1:482–83.

———. 1846b. "The Proceedings of the Royal Society." *Lancet* 1:420.

Roscoe, Henry Enfield. 1906. *The Life & Experiences of Sir Henry Enfield Roscoe*. London: Macmillan.

Rose, Hugh James. 1826. *The Tendency of Prevalent Opinions about Knowledge Considered: A Sermon Preached before the University of Cambridge on Commencement Sunday, July 2, 1826*. Cambridge: J. Deighton and Sons.

Rose, Hugh James. 1831. *Eight Sermons Preached before the University of Cambridge at Great St. Mary's, in the Years 1830 and 1831. To Which Is Added a Reprint of a Sermon Preached before the University on Commencement Sunday, 1826*. Cambridge: J. and J. J. Deighton.

[Rose, Hugh James]. 1832. "Address." *British Magazine* 1:1–10.

Rose, Hugh James. 1834a. *An Apology for the Study of Divinity*. London: Rivington.

Rose, Hugh James. 1834b. *The Duty of Maintaining the Truth: A Sermon Preached before the University of Cambridge, on Sunday, May 18, 1834*. Cambridge: J. and J. J. Deighton.

Rose, Hugh James. 1834c. *The Study of Church History Recommended*. London: Rivington.

Ross, Sydney. 1962. "Scientist: The Story of a Word." *Annals of Science* 18:85.

Royle, Edward. 1971. *Radical Politics 1790–1900: Religion and Unbelief*. London: Longman.

Royle, Edward. 1974. *Victorian Infidels: The Origins of the British Secularist Movement, 1791–1866*. Manchester: Manchester University Press.

Rudwick, Martin J. S. 1974. "The Emergence of a Visual Language for Geological Science, 1760–1860." *History of Science* 14:149–95.

Rudwick, Martin J. S. 1982. "Charles Darwin in London: The Integration of Public and Private Science." *Isis* 73:186–206.

Rudwick, Martin J. S. 1992. *Scenes from Deep Time: Early Pictorial Representations of the Prehistoric World*. Chicago: University of Chicago Press.

Rudwick, Martin J. S. 2008. *Worlds before Adam: The Reconstruction of Geohistory in the Age of Reform*. Chicago: University of Chicago Press.

Rupke, Nicolaas A. 1983. *The Great Chain of History: William Buckland and the English School of Geology (1814–1849)*. Oxford: Clarendon.

Rupke, Nicolaas A. 1994. *Richard Owen: Victorian Naturalist*. New Haven, CT: Yale University Press.

Rupke, Nicolaas A. 1997. "Oxford's Geological Awakening and the Role of Geology." In *The History of the University of Oxford*, vol. 6, *Nineteenth-Century Oxford*, pt. 1, edited by M. G. Brock and M. C. Curthoys, 543–62. Oxford: Clarendon.

Ruse, Michael. 1975. "Darwin's Debt to Philosophy: An Examination of the Influence of the Philosophical Ideas of John F. W. Herschel and William Whewell on the Development of Charles Darwin's Theory of Evolution." *Studies in the History and Philosophy of Science* 6:159–81.

Ruse, Michael. 2000. "Darwin and the Philosophers: Epistemological Factors in the Development and Reception of the Theory of the *Origin of Species*." In *Biology and Epistemology*, edited by Richard Creath and Jane Maienschein, 3–26. Cambridge: Cambridge University Press.

Russell, John. 1853–56. *Memoirs, Journal, and Correspondence of Thomas Moore*. 8 vols. London: Longman.

Ryall, Maltus Questell. 1844. "Science and Religion." *Movement* 1:196–97.

Ryle, J. C. 1852. *How Readest Thou? A Question for 1853*. 4th ed. Ipswich: Hunt and Son; London: Wertheim and Macintosh and Nisbet.

Sadleir, Michael. 1930. *The Evolution of Publishers' Binding Styles, 1770–1900*. London: Constable; New York: Smith.

Salmon, Richard. 2013. *The Formation of the Victorian Literary Profession*. Cambridge: Cambridge University Press.

Sanders, Charles Richard et al., eds. 1970–2021. *The Collected Letters of Thomas and Jane Welsh Carlyle*. 47 vols. Durham, NC: Duke University Press.

[Saunders, Frederick]. 1839. *The Author's Printing and Publishing Assistant*. London: Saunders and Otley.

Savage, William. 1841. *A Dictionary of the Art of Printing*. London: Longman.

Schaffer, Simon. 1986. "Scientific Discoveries and the End of Natural Philosophy." *Social Studies of Science* 16:387–420.

Schaffer, Simon. 1989. "The Nebular Hypothesis and the Science of Progress." In *History, Humanity and Evolution: Essays for John C. Greene*, edited by James R. Moore, 131–64. Cambridge: Cambridge University Press.

Schaffer, Simon. 2003. "Paper and Brass: The Lucasian Professorship, 1820–39." In *From Newton to Hawking: A History of Cambridge University's Lucasian Professors of Mathematics*, edited by Kevin C. Knox and Richard Noakes, 241–93. Cambridge: Cambridge University Press.

Schilke, Paul, ed. 2012. *Charles Dickens, "Sketches of Young Gentlemen and Young Couples" with "Sketches of Young Ladies" by Edward Caswall, Illustrated by Phiz*. Oxford: Oxford University Press.

Schlesinger, Elizabeth Bancroft. 1965. "Two Early Harvard Wives: Eliza Farrar and Eliza Follen." *New England Quarterly* 38:147–67.

Searby, Peter. 1997. *A History of the University of Cambridge*. Vol. 3, *1750–1870*. Cambridge: Cambridge University Press.

Sebastiani, Silvia. 2019. "A 'Monster with Human Visage': The Orangutan, Savagery, and the Borders of Humanity in the Global Enlightenment." *History of the Human Sciences* 32:80–99.

Secord, Anne. 2002. "Botany on a Plate: Pleasure and the Power of Pictures in Promoting Early Nineteenth-Century Scientific Knowledge." *Isis* 93:28–57.

Secord, Anne, ed. 2013. *The Natural History of Selborne, by Gilbert White*. Oxford: Oxford University Press.

Secord, James A. 1989a. "Behind the Veil: Robert Chambers and *Vestiges*." In *History, Humanity and Evolution: Essays for John C. Greene*, edited by James R. Moore, 165–94. Cambridge: Cambridge University Press.

Secord, James A. 1989b. "Extraordinary Experiment: Electricity and the Creation of Life in Victorian England." In *The Uses of Experiment: Studies in the Natural Sciences*, edited by David Gooding, Trevor Pinch, and Simon Schaffer, 337–83. Cambridge: Cambridge University Press.

Secord, James A. 1990. "Darwin in Another Light." *Times Literary Supplement*, 13–19 July, 751.

Secord, James A. 1991. "The Discovery of a Vocation: Darwin's Early Geology." *British Journal for the History of Science* 24:133–57.

Secord, James A., ed. 1997. *Principles of Geology, by Charles Lyell*. London: Penguin Books.

Secord, James A. 2000. *Victorian Sensation: The Extraordinary Publication, Reception, and Secret Authorship of "Vestiges of the Natural History of Creation."* Chicago: University of Chicago Press.

Secord, James A. 2009. "Science, Technology and Mathematics." In *The Cambridge History of the Book in Britain*, vol. 6, *1830–1914*, edited by David McKitterick, 443–74. Cambridge: Cambridge University Press.

Secord, James A. 2014. *Visions of Science: Books and Readers at the Dawn of the Victorian Age*. Oxford: Oxford University Press.

Sedgwick, Adam. 1833. *A Discourse on the Studies of the University*. London: John W. Parker; Cambridge: Deightons.

Sedgwick, Adam. 1850. *A Discourse on the Studies of the University of Cambridge*. 5th ed. Cambridge: John Deighton.

Sedra, Paul. 2011. *From Mission to Modernity: Evangelicals, Reformers and Education in Nineteenth Century Egypt*. London: I. B. Taurus.

Sell, Alan. 2004. *Philosophy, Dissent and Nonconformity*. Cambridge: James Clarke.

Sell, Jeremy Michael. 2016. "Victorian Sermonic Discourse: The Sermon in Nineteenth-Century British Literature and Society." PhD diss., University of California, Riverside.

Seville, Catherine. 2006. *The Internationalisation of Copyright Law: Books, Buccaneers and the Black Flag in the Nineteenth Century*. Cambridge: Cambridge University Press.

Sewell, William. 1834. *The Attack upon the University of Oxford, in a Letter to Earl Grey*. London: James Bohn.

[Sewell, William]. 1838. "Animal Magnetism." *British Critic* 24:301–32.

Shairp, John Campbell, Peter Guthrie Tait, and A. Adams-Reilly, eds. 1873. *Life and Letters of James David Forbes, F.R.S.* London: Macmillan.

Shatto, Susan, and Marion Shaw. 1982. *Tennyson: "In Memoriam."* Oxford: Clarendon.

Shattock, Joanne. 1989. *Politics and Reviewers: The "Edinburgh" and the "Quarterly" in the Early Victorian Age*. Leicester: Leicester University Press.

Shaylor, Joseph. 1912. *The Fascination of Books with Other Papers on Books & Bookselling*. London: Simpkin Marshall.

Shteir, Ann B. 2004a. "Green-Stocking or Blue? Science in Three Women's Magazines, 1800–50." In *Culture and Science in the Nineteenth-Century Media*, edited by Louise Henson, Geoffrey Cantor, Gowan Dawson, Richard Noakes, Sally Shuttleworth, and Jonathan R. Topham, 1–13. Aldershot: Ashgate.

———. 2004b. "'Let Us Examine the Flower': Botany in Women's Magazines, 1800–1830." In *Science Serialized: Representations of the Sciences in Nineteenth-Century Periodicals*, edited by Geoffrey Cantor and Sally Shuttleworth, 17–36. Cambridge MA: MIT Press.

Shteir, Ann B., and Bernard Lightman. 2006. *Figuring It Out: Science, Gender, and Visual Culture*. Hanover, NH: Dartmouth College Press and University Press of New England.

Shuttleworth, Sally. 1984. *George Eliot and Nineteenth-Century Science: The Make-Believe of a Beginning*. Cambridge: Cambridge University Press, 1984.

Simeon, Charles. 1838. *Claude's Essay on the Composition of a Sermon; With Notes and Illustrations, Together with One Hundred Skeletons*. New ed. London: S. Cornish.

Simeon, Charles. 1853. *Claude's Essay on the Composition of a Sermon; With Notes and Illustrations, and One Hundred Skeletons*. New ed. London: James Cornish.

Sloan, Phillip Reid, ed. 1992. *The Hunterian Lectures in Comparative Anatomy, May–June, 1837*. Chicago: University of Chicago Press.

Smiles, Samuel. 1891. *A Publisher and His Friends: Memoir and Correspondence of the Late John Murray, with an Account of the Origin and Progress of the House, 1768–1843*. 2 vols. London: John Murray.

[Smith, James Elishama]. 1835. "Revolution of Philosophy." *Shepherd*, 28 March 1835, 243–45.

Smith, John Pye. 1834. "Dr J. Pye Smith on the Geological Opinions of Professor Sedgwick." *Congregational Magazine* 11:469–71.

[Smith, John Pye]. 1836. "Geology and the Holy Scriptures." *Magazine of Popular Science* 2 (1836):465–68.

Smith, John Pye. 1837. "Suggestions on the Science of Geology, in Answer to the Question of T.K." *Congregational Magazine* 13:765–76.

Smith, John Pye. 1839a. *On the Relation between the Holy Scriptures and Some Parts of Geological Science*. London: Jackson and Walford.

Smith, John Pye. 1839b. "Pulpit Echo, No. 4. Lecture to Young Men and Others; Delivered at the Weigh-House Chapel, Fish-Street-Hill. The Mosaic Account of the Creation and the Deluge Supported by the Discoveries of Modern Science." *Christian's Penny Magazine*, 27 January, 27–34; 3 February, 35–40.

Smith, Jonathan. 1991. "The 'Wonderful Geological Story': Uniformitarianism and *The Mill on the Floss*." *Papers on Language and Literature* 27:430–52.

Smith, W. Anderson. 1892. *"Shepherd" Smith the Universalist: The Story of a Mind.* London: Sampson Low, Marston.

Smyth, John, et al. 1842. *A Course of Lectures on Infidelity: By Ministers of the Church of Scotland in Glasgow and Neighbourhood.* Glasgow: William Collins.

Snyder, Laura J. 2006. *Reforming Philosophy: A Victorian Debate on Science and Society.* Chicago: University of Chicago Press.

Snyder, Laura J. 2011. *The Philosophical Breakfast Club: Four Remarkable Friends Who Transformed Science and Changed the World.* New York: Broadway Paperbacks.

Southwell, Charles. 1842. "A Voice from Bristol Gaol." *Oracle of Reason* 1:78–79.

Southwell, Charles. 1844. "Sentimental Theists, and Natural Theology." *Movement* 1:34–36, 43–45, 52–55, 74–75, 81–83.

Southwell, Charles. [1845]. *Confessions of a Free-Thinker.* London: privately printed.

Stanley, Arthur Penrhyn. 1844. *The Life and Correspondence of Thomas Arnold, D.D.* 2 vols. London: B. Fellowes.

Stanley, Matthew. 2015. *Huxley's Church and Maxwell's Demon: From Theistic Science to Naturalistic Science.* Chicago: University of Chicago Press.

St Clair, William. 2004. *The Reading Nation in the Romantic Period.* Cambridge: Cambridge University Press.

Steadman, Thomas. 1838. *Memoir of the Rev. William Steadman, D.D.* London: Thomas Ward.

Stearn, William T., ed. 1998. *John Lindley, 1799–1865: Gardener, Botanist and Pioneer Orchidologist; Bicentenary Celebration Volume.* Woodbridge: Antique Collectors' Club.

Steel, T. H. 1882. *Sermons Preached in the Chapel of Harrow School and Elsewhere.* London: Macmillan.

Stein, Dorothy. 1985. *Ada: A Life and a Legacy.* Cambridge, MA: MIT Press.

Stevens, Henry. 1884. *Who Spoils Our New English Books Asked and Answered.* London: Henry Newton Stevens.

Stock, Eugene. 1899. *The History of the Church Missionary Society: Its Environment, Its Men and Its Work.* 3 vols. London: Church Missionary Society.

Stoddard, Roger E. 1987. "Morphology and the Book Form in American Perspective." *Printing History* 9:2–14.

Storrs, G. W., and M. A. Taylor. 1996. "Cranial Anatomy of a New Plesiosaur Genus from the Lowermost Lias (Rhaetian/Hettangian) of Street, Somerset, England." *Journal of Vertebrate Paleontology* 16:403–20.

Swade, Doron. 2001. *The Difference Engine: Charles Babbage and the Quest to Build the First Computer.* New York: Viking.

Swann, Elsie. 1934. *Christopher North (John Wilson).* Edinburgh: Oliver and Boyd.

[Talfourd, T. Noon]. 1820. "Modern Periodical Literature." *New Monthly Magazine* 14:304–10.

Talfourd, T. Noon. 1842. *Critical and Miscellaneous Writings of T. Noon Talfourd.* Philadelphia: Carey and Hart.

Tennyson, G. B. 1981. *Victorian Devotional Poetry: The Tractarian Mode.* Cambridge, MA: Harvard University Press.

Tennyson, Hallam. 1897. *Alfred Lord Tennyson: A Memoir.* 2 vols. New York: Macmillan.

Terrey, Henry. 1937. "Edward Turner, M.D., F.R.S. (1798–1837)." *Annals of Science* 2:137–52.

[Thom, John Hamilton?]. 1835. "Lord Brougham's *Natural Theology*." *Christian Teacher* 1:411–17.

Thomas, Vaughan. 1843. *The Educational and Subsidiary Provisions of the Birmingham Royal School of Medicine and Surgery*. Oxford: printed privately.

Thomas, Vaughan. 1855. *Christian Philanthropy Exemplified in a Memoir of the Rev. Samuel Wilson Warneford, LL.D.* Oxford: privately printed.

Thompson, David M. 2008. *Cambridge Theology in the Nineteenth Century: Enquiry, Controversy, and Truth*. Aldershot: Ashgate.

Thompson, Henry. 1838. *The Life of Hannah More*. London: T. Cadell.

Thompson, Malcolm Caldwell. 1949. "Bridgewater Treatises: Their Theological Significance." PhD diss., University of Edinburgh.

[Thomson, John]. 1824. *Hints Respecting the Improvement of the Literary & Scientific Education of Candidates for the Degree of Doctor of Medicine in the University of Edinburgh*. Edinburgh: David Brown.

[Thomson, John]. 1826a. *Additional Hints Respecting the Improvement of the System of Medical Instruction Followed in the University of Edinburgh*. [Edinburgh]: privately printed.

[Thomson, John]. 1826b. *Observations on the Preparatory Education of Candidates for the Degree of Doctor of Medicine, in the Scottish Universities*. [Edinburgh]: privately printed.

[Thomson, Richard, Edward William Brayley, and William Upcott]. 1835–52. *A Catalogue of the Library of the London Institution*. [London]: privately printed.

Timbs, John. 1866. *English Eccentrics and Eccentricities*. 2 vols. London: John Bentley.

Timpson, Thomas. 1841. *Memoirs of British Female Missionaries*. London: William Smith.

Todd, A. C. 1967. *Beyond the Blaze: A Biography of Davies Gilbert*. Truro: D. Bradford Barton.

Todd, Robert Bentley. 1837. "Education of Medical Students." *British Magazine* 11:335–38, 460–63; 12:95–100, 337–341.

Todhunter, Isaac. 1876. *William Whewell, D.D., Master of Trinity College, Cambridge: An Account of His Writings with Selections from His Literary and Scientific Correspondence*. 2 vols. London: Macmillan.

Toole, Betty A. 1992. *Ada, the Enchantress of Numbers: A Selection from the Letters of Lord Byron's Daughter and Her Description of the First Computer*. Mill Valley, CA: Strawberry Press.

Topham, Jonathan R. 1992. "Science and Popular Education in the 1830s: The Role of the Bridgewater Treatises." *British Journal for the History of Science* 25: 397–430.

Topham, Jonathan R. 1993. "'An Infinite Variety of Arguments': The *Bridgewater Treatises* and British Natural Theology in the 1830s." PhD diss., University of Lancaster.

Topham, Jonathan R. 1998. "Beyond the Common Context: The Production and Reading of the Bridgewater Treatises." *Isis* 89:233–62.

Topham, Jonathan R. 1999. "Evangelicals, Science, and Natural Theology in Early Nineteenth-Century Britain: Thomas Chalmers and the *Evidence* Controversy."

In *Evangelicals and Science in Historical Perspective*, edited by David N. Livingstone, D. G. Hart, and Mark A. Noll, 142–74. New York: Oxford University Press.

Topham, Jonathan R. 2000a. "Scientific Publishing and the Reading of Science in Nineteenth-Century Britain: A Historiographical Survey and Guide to Sources." *Studies in History and Philosophy of Science* 31:559–612.

Topham, Jonathan R. 2000b. "A Textbook Revolution." In *Books and the Sciences in History*, edited by Marina Frasca-Spada and Nicholas Jardine, 317–37. Cambridge: Cambridge University Press.

Topham, Jonathan R. 2004a. "Periodicals and the Making of Reading Audiences for Science in Early Nineteenth-Century Britain: The *Youth's Magazine*, 1828–37." In *Culture and Science in the Nineteenth-Century Media*, edited by Louise Henson, Geoffrey Cantor, Gowan Dawson, Richard Noakes, Sally Shuttleworth, and Jonathan R. Topham, 57–69. Aldershot: Ashgate.

Topham, Jonathan R. 2004b. "Science, Natural Theology, and the Practice of Christian Piety in Early Nineteenth-Century Religious Magazines." In *Science Serialized: Representations of the Sciences in Nineteenth-Century Periodicals*, edited by Geoffrey Cantor and Sally Shuttleworth, 37–66. Cambridge, MA: MIT Press.

Topham, Jonathan R. 2004c. "Scientific Readers: A View from the Industrial Age." *Isis* 95:431–42.

Topham, Jonathan R. 2004d. "The *Wesleyan-Methodist Magazine* and Religious Monthlies in Early Nineteenth-Century Britain." In *Science in the Nineteenth-Century Periodical: Reading the Magazine of Nature*, by Geoffrey Cantor et al., 67–90. Cambridge: Cambridge University Press.

Topham, Jonathan R. 2007. "Publishing 'Popular Science' in Early Nineteenth-Century Britain." In *Science in the Marketplace: Nineteenth-Century Sites and Experiences*, edited by Aileen Fyfe and Bernard Lightman, 135–68. Chicago: University of Chicago Press.

Topham, Jonathan R. 2010. "Biology in the Service of Natural Theology: Darwin, Paley, and the Bridgewater Treatises." In *Biology and Ideology from Descartes to Dawkins*, edited by Denis R. Alexander and Ronald L. Numbers, 88–113. Chicago: University of Chicago Press.

Topham, Jonathan R. 2011. "Science, Print, and Crossing Borders: Importing French Science Books into Britain, 1789–1815." In *Geographies of Nineteenth-Century Science*, edited by David N. Livingstone and Charles W. J. Withers, 311–44. Chicago: University of Chicago Press.

Topham, Jonathan R. 2013. "Science, Mathematics, and Medicine." In *The History of Oxford University Press*, vol. 2, *1780–1896*, edited by Simon Eliot, 512–57. Oxford: Oxford University Press.

Topham, Jonathan R. 2020. "Redrawing the Image of Science: Technologies of Illustration and the Audiences for Scientific Periodicals in Britain, 1790–1840." In *Science Periodicals in Nineteenth-Century Britain: Constructing Scientific Communities*, edited by Gowan Dawson, Bernard Lightman, Sally Shuttleworth, and Jonathan R. Topham, 65–102. Chicago: University of Chicago Press.

Torr, Cecil. 1918–23. *Small Talk at Wreyland*. 3 vols. Cambridge: Cambridge University Press.

Turner, Frank M. 1993. *Contesting Cultural Authority: Essays in Victorian Intellectual Life*. Cambridge: Cambridge University Press.

Turner, William, ed. 1868. *The Anatomical Memoirs of John Goodsir [ . . . ] with a Biographical Memoir by Henry Lonsdale*. 2 vols. Edinburgh: Adam and Charles Black.

Turton, Thomas. 1836. *Natural Theology Considered with Reference to Lord Brougham's Discourse on That Subject*. London: John William Parker.

Twyman, Michael. 1970. *Lithography, 1800–1850: The Techniques of Drawing on Stone in England and France and Their Application in Works of Topography*. London: Oxford University Press.

Twyman, Michael. 1976. *A Directory of London Lithographic Printers 1800–1850*. London: Printing Historical Society.

Twyman, Michael. 2009. "The Illustration Revolution." In *Cambridge History of the Book in Britain*, vol. 6, *1830–1914*, edited by David McKitterick, 117–43. Cambridge: Cambridge University Press.

University of London. 1827. *Statement by the Council of the University of London, Explanatory of the Nature and Objects of the Institution*. London: Longman.

University of London. 1828. *Second Statement by the Council of the University of London, Explanatory of the Plan of Instruction*. 2nd ed. London: John Taylor.

Valone, David A. 2001. "Hugh James Rose's Anglican Critique of Cambridge: Science, Antirationalism, and Coleridgean Idealism in Late Georgian England." *Albion* 33:218–42.

Van Wyhe, John. 2004. *Phrenology and the Origins of Victorian Scientific Naturalism*. Aldershot: Ashgate.

[Vieusseux, André]. 1836–37. *Serious Thoughts, Generated by Perusing Lord Brougham's "Discourse of Natural Theology;" with a Few Broad Hints on Education and Politics*. 4 parts. London: John Brooks.

[Walker, Robert]. 1832. *A Few Words in Favour of Professor Powell, and the Sciences, as Connected with Certain Educational Remarks*. Oxford: privately printed.

Warren, Arthur. 1896. *The Charles Whittinghams, Printers*. New York: Grolier Club.

Warrington, Bernard. 1985. "William Pickering and the Book Trade in the Early Nineteenth Century." *Bulletin of the John Rylands University Library of Manchester* 68:247–66.

Warrington, Bernard. 1987. "William Pickering, His Authors and Interests: A Publisher and the Literary Scene in the Early Nineteenth Century." *Bulletin of the John Rylands University Library of Manchester* 69:572–628.

Warrington, Bernard. 1989. "William Pickering, Bookseller and Book Collector." *Bulletin of the John Rylands University Library of Manchester* 71:121–38.

Warrington, Bernard. 1990. "The Bankruptcy of William Pickering in 1853: The Hazards of Publishing and Bookselling in the First Half of the Nineteenth Century." *Publishing History* 27:5–25.

Warrington, Bernard. 1993. "William Pickering and the Development of Publisher's Binding in the Early Nineteenth Century." *Publishing History* 33:59–76.

[Waterston, C. D., and A. Macmillan Shearer]. 2006. *Former Fellows of the Royal Society of Edinburgh, 1783–2002: Biographical Index*. 2 pts. Edinburgh: Royal Society of Edinburgh.

[Watkins, John, and Frederick Shoberl]. 1816. *A Biographical Dictionary of the Living Authors of Great Britain and Ireland*. London: Henry Colburn.

[Watson, Richard]. 1824. "Review of Dick's 'Christian Philosopher.'" *Wesleyan-Methodist Magazine*, 3rd ser., 3:33–40.

Watson, Richard. 1827. *An Address Delivered at the Ordination of the Reverend John Bell, Jonathan Crowther, and Others, at the Conference of Wesleyan-Methodist Ministers, Held in Manchester, August, MDCCCXXVII*. London: John Mason.

Watt, Hugh. 1943. *The Published Writings of Dr Thomas Chalmers (1780–1847): A Descriptive List*. Edinburgh: privately printed.

Weatherall, Mark. 2000. *Gentlemen, Scientists and Doctors: Medicine at Cambridge, 1800–1940*. Woodbridge, Suffolk: Boydell.

Webb, R. K. 1990. "The Faith of Nineteenth-Century Unitarians: A Curious Incident." In *Victorian Faith in Crisis: Essays on Continuity and Change in Nineteenth-Century Religious Belief*, edited by Richard. J. Helmstadter and Bernard Lightman, 126–49. Stanford, CA: Stanford University Press.

Weld, Charles R. 1848. *A History of the Royal Society, with Memoirs of the Presidents, Compiled from Authentic Documents*. 2 vols. London: John W. Parker.

Weylland, John Matthias. [1884]. *These Fifty Years: Being the Jubilee Volume of the London City Mission*. London: S. W. Partridge.

Whalley, Robert. 1825a. "The Secrets of Crystamancy Divulged." *Republican* 11: 179–84.

Whalley, Robert. 1825b. "To the Editor of the *Republican*." *Republican* 11:334–43.

Whalley, Robert. 1835. *Revolution of Philosophy, Containing a Concise Analysis and Synthesis of the Universe. Part First. Developing the Fundamental Principles*. Manchester: A. Heywood.

Whalley, Robert. 1840. *A Philosophical Refutation of the Theories of Robert Owen, and His Followers*. Manchester: privately printed.

Wheeler, Joseph Mazzini. 1889. *A Biographical Dictionary of Freethinkers of All Ages and All Nations*. London: Progressive.

Wheeler, Joseph Mazzini. 1980. "Sixty Years of Freethought." In *An Anthology of Atheism and Rationalism*, edited by Gordon Stein, 334–51. New York: Prometheus Books.

[Whewell, William]. 1831a. "Cambridge Transactions: Science of the English Universities." *British Critic* 9:71–90.

[Whewell, William]. 1831b. "Herschel's *Preliminary Discourse*: Modern Science—Inductive Philosophy." *Quarterly Review* 45:374–407.

[Whewell, William]. 1831c. "Lyell—*Principles of Geology*." *British Critic* 9:180–206.

Whewell, William. 1833a. *Address Delivered in the Senate-House at Cambridge, June XXV, M.DCCC.XXXIII. On the Occasion of the Opening of the Third General Meeting of the British Association for the Advancement of Science*. Cambridge: privately printed.

Whewell, William. 1833b. *Astronomy and General Physics Considered with Reference to Natural Theology*. London: William Pickering.

Whewell, William. 1834a. "Bridgewater Treatises. Mr Whewell's Reply to the *Edinburgh Review*." *British Magazine* 5:263–68.

[Whewell, William]. 1834b. "Mrs Somerville *On the Connection of the Sciences*." *Quarterly Review* 51:54–68.

Whewell, William. 1837a. *History of the Inductive Sciences, from the Earliest to the Present Times*. 3 vols. London: John W. Parker; Cambridge: J. and J. J. Deighton.

Whewell, William. 1837b. *Letter to Charles Babbage, Esq, A.M., Lucasian Professor of Mathematics in the University of Cambridge*. [Cambridge]: privately printed.

Whewell, William. 1840. *The Philosophy of the Inductive Sciences, Founded upon Their History*. 2 vols. London: John W. Parker; Cambridge: J. and J.J. Deighton.

Whewell, William. 1842. "Address to the Geological Society, Delivered at the Anniversary, on the 15th of February, 1839." *Proceedings of the Geological Society of London* 3:61–98.

Whewell, William. 1845. *Indications of the Creator: Extracts, Bearing upon Theology from the History and the Philosophy of the Inductive Sciences*. London: John W. Parker.

Whewell, William 1864. *Astronomy and General Physics, Considered with Reference to Natural Theology*. New ed. Cambridge: Deighton Bell.

White, Andrew Dickson. 1896. *A History of the Warfare of Science with Theology in Christendom*. 2 vols. London: Macmillan.

White, Paul. 2003. *Thomas Huxley: Making the "Man of Science."* Cambridge: Cambridge University Press.

White, Paul. 2014. "The Conduct of Belief: Agnosticism, the Metaphysical Society, and the Formation of Intellectual Communities." In *Victorian Scientific Naturalism: Community, Identity, Continuity*, edited by Gowan Dawson and Bernard Lightman, 220–41. Chicago: University of Chicago Press.

Wiener, Joel H. 1969. *The War of the Unstamped: The Movement to Repeal the Newspaper Tax, 1830–1836*. Ithaca, NY: Cornell University Press.

Wiener, Joel H. 1983. *Radicalism and Freethought in Nineteenth-Century Britain: The Life of Richard Carlile*. Westport, CT: Greenwood.

Wilberforce, Robert, and Samuel Wilberforce. 1838. *The Life of William Wilberforce*. 5 vols. London: John Murray.

[Wilkinson, Christopher, and Squire Farrar]. 1835. *An Examination of the Arguments for the Existence of a Deity, Being an Answer to Mr Godwin's Lectures on the Atheistic Controversy*. London: Simpkin, Marshall.

[Wilkinson, Christopher, and Squire Farrar]. 1853. *An Examination of the Arguments for the Existence of a Deity; Being an Answer to Dr Godwin's "Philosophy of Atheism Examined and Compared with Christianity."* 2nd ed. London: J. Watson.

W[ilks], M[ark]. 1828. "Extract of a Letter on the Death of Mademoiselle Cuvier, Daughter of Baron Cuvier." *Evangelical Magazine*, n.s., 6:69–72.

[Wilks, Samuel Charles]. 1834. "Review of Cole's Letter to Sedgwick on Geology." *Christian Observer* 34:369–87.

Williamson, George C. 1903–5. *Bryan's Dictionary of Painters and Engravers*. Rev. ed. 5 vols. New York: Macmillan.

[Willmott, Robert Eldridge Aris]. 1837. "A Scourging Soliloquy about the Annuals." *Fraser's Magazine* 15:33–48.

Wilson, George. 1840. "Narrative of the Operations of the London City Mission in Westminster. Paper III." *London City Mission Magazine* 5:178–84.

Wilson, George. 1862. *Religio Chemici: Essays*. London: Macmillan.

Wilson, George, and Archibald Geikie. 1861. *Memoir of Edward Forbes, F.R.S.* London: Macmillan.

Wilson, Jessie Aitken. 1860. *Memoir of George Wilson, M.D., F.R.S.E.* Edinburgh: Edmonston and Douglas.

Wilson, Leonard G. 1972. *Charles Lyell: The Years to 1841; The Revolution in Geology*. New Haven, CT: Yale University Press.

Winter, Alison. 1997. "The Construction of Orthodoxies and Heterodoxies in the

Early Victorian Life Sciences." In *Victorian Science in Context*, edited by Bernard Lightman, 24–50. Chicago: University of Chicago Press.

Winter, Alison. 1998. *Mesmerized: Powers of Mind in Victorian Britain*. Chicago: University of Chicago Press.

Wright, Thomas. 1838. "Observations on Dr Buckland's Theory of the Action of the Siphuncle in the Pearly Nautilus." *London and Edinburgh Philosophical Magazine*, 3rd ser., 12:503–8.

Yeo, Richard. 1986. "The Principle of Plenitude and Natural Theology in Nineteenth Century Britain." *British Journal of the History of Science* 19:263–82.

Yeo, Richard. 1993. *Defining Science: William Whewell, Natural Knowledge, and Public Debate in Early Victorian Britain*. Cambridge: Cambridge University Press.

Yeo, Richard. 2001. *Encyclopaedic Visions: Scientific Dictionaries and Enlightenment Culture*. Cambridge: Cambridge University Press.

Young, Robert M. 1985. *Darwin's Metaphor: Nature's Place in Victorian Culture*. Cambridge: Cambridge University Press.

Yule, John David. 1976. "The Impact of Science on British Religious Thought in the Second Quarter of the Nineteenth Century." PhD diss., University of Cambridge.

Zincke, F. Barham. 1866. *On the Duty and the Discipline of Extemporary Preaching*. London: Rivington.

# Index

Page numbers in italics refer to illustrations.

Abbott, Jacob, *Fire-Side Piety*, 302, 303
Abercrombie, John, 383; *Philosophy of the Moral Feelings*, 249
Acton, Henry, 348–49
Adelaide, Queen, 344
Adelaide Gallery, 266
Affleck, James, 358
Agassiz, Louis, 210n63
*Age* (newspaper), 43
Ainslie, Robert, 366–68
Airy, George Biddell, 39
à Kempis, Thomas, *Imitation of Christ*, 44, 314
Aldersgate Street medical school (London), 36
Aldus Manutius, 182
Alison, William Pulteney, 425, 426; *Outlines of Physiology*, 422
Allen, David, 198
Allin, Thomas, 357–58, 362
All Souls College, Oxford, 25
*Analyst* (periodical), 265
Anderson, James, 344–45
Anderson, John, 216
*Annales des arts et manufactures*, 24
*Annals of Philosophy*, 99
*Archives du Christianisme* (periodical), 322n57
Argyle Square medical school (Edinburgh), 422
Aristotle, 81; *History of Animals*, 129–30
Arnold, Thomas, 317, 404–5
Arnott, Neil, *Elements of Physics*, 85
Ashmolean Museum, 76, 96, 202, 203, 211
Association of All Classes of All Nations, 362, 364, 365
Astronomical Society, 37
atheism: Bridgewaters seen as promoting, 279, 280, 284, 370, 371–72; Bridgewater's views on, 27–29, 30, 126; Chalmers's account of, 122; history of, 355n; Laplace's reputation for, 2, 112–15, 120, 158, 261, 271, 352, 421; Leslie's reputation for, 421; at London University, 407; medical students, reputation for, 130; natural laws and, 107; natural theology ineffectual against, 130, 238, 250–51, 252, 272, 341, 370, 371–72, 373; Owenite views regarding, 362, 363–65, 368–72; and the prevalence of science, 6, 395; protection against, 125–26, 251; and working-class radicals, 26, 354–62, 448, 477–78. *See also* infidelity
*Athenæum* (periodical), 48, 232–35, *232*, 446, 499n2
*Atlas* (newspaper), 278n82
atomism, Greek: Bridgewater and, 27–28, 29, 30; Kidd and, 126, 129, 130; Lord Holland and, 387
Augustus Frederick. *See* Sussex, Duke of (Augustus Frederick)

authorship: generic fluidity, 80–81; income from, 56–58, 60–68; practicalities of, 91–105, 179; the rise of the professional author, 57; and scientific credit, 68
*Author's Printing and Publishing Assistant*, 181

BAAS. *See* British Association for the Advancement of Science (BAAS)
Babbage, Charles: calculating engine, 250, 434–37, *435*; *On the Economy of Manufactures*, 215; on payment for men of science, 56–57, 58; reading of Whewell's Bridgewater, 434–36; *Reflections on the Decline of Science in England*, 20, 33, 39, 375
Babbage, Charles, *Ninth Bridgewater Treatise*: mimics Bridgewaters, 221; publication of, 437, 440–41; reading of, 301, 444, 445–46, 448, 450; reviews of, 441n16; Whewell's response to, 440; writing of, 437–40, 443, 451
Bacon, Francis: on active powers, 154; Baconian induction, 116; on final causes, 154, 453, 460, 462, 469; on God's two books, 10; patronized by Bridgewater family, 20; on science and religion, 164, 259, 478
Baer, Karl Ernst von, 459
Baily, Francis, 37
Bakewell, Frederick Collier, *Natural Evidence of a Future Life*, 221
Baldwin, Charles, 278
Baldwin and Cradock (publishers), 172
Banks, Joseph, 19–20, 22, 24, 28, 31, 66
*Baptist Magazine*, 247
Barclay, John, 423
Baring-Gould, Edward, 319–20
Baring-Gould, Sophia, 319–20
Barry, Martin, 425
Baxter, Richard, *Saints Everlasting Rest*, 293–94
*Belfast Commercial Chronicle*, 276n81
Bell, Charles, 73
Bell, Charles, as a Bridgewater author: appointment, 19, 38; attending meetings of authors, 42, 174, 177; concerns about public criticism, 47; dictating text, 92–93; drawing on experience as SDUK author, 84–86; drawing on own lectures, 92–93, 269; line of demarcation from Kidd, 99; on Pickering's charges, 223; reported to be a "competitor," 42; reputation attracts attention to Bridgewaters, 218; response to Paley's *Natural Theology*, 72–74; stretched by theological task, 72
Bell, Charles, Bridgewater Treatise, 130–37, *481*; and comparative anatomy, 132–36, 269; excerpted, 327; functionalism of, 133n44, 146, 409, 454; illustrations, 198–201, *200*; in libraries, 412n61, 415–16; opposing materialism, 109, 130–31, 135, 136–37; opposing transmutation, 109, 131, 133, 134–37, 270, 272; on philosophical anatomy, 133, 134, 137, 269, 423, 455; reading of, 297, 336, 368, 416, 423, 426, 443, 478; reviews of, 92–93, 235, 239, 243, 265–66, 268–70; theologically limited, 72, 108, 131–32, 137; use of by preachers, 345n27
Bell, Charles, life and career: as an artist, 66, 72, 93, 198–99, *200*; on Babbage's *Ninth Bridgewater*, 441; British Association address, 383; discoveries in nervous system, 462; Edinburgh lectures of, 422; friendship with Chambers, 449; and functional adaptation, 424; habit of seeing design in nature, 72, 452; London University lectures of, 408–9; ownership of Great Windmill Street medical school, 40; sincerity of, 92, 390; sources of income, 65–67
Bell, Charles, other publications: *Anatomy of Expression*, 66, 67, 173, 199; *Anatomy of the Human Body*, 66, 361; *Animal Mechanics*, 72, 84–86, 92, 135, 181, 361; *Exposition of the Natural System of the Nerves*, 67; *Illustra-*

*tions of the Great Operations of Surgery*, 67; *Paley's Natural Theology Illustrated*, 71–74, 411; *System of Dissections*, 66; *System of Operative Surgery*, 66
Bell, George, 480, 481
Bell, John, 65–66; *Anatomy of the Human Body*, 66, 361
Bell, Marion, 92–93
*Bell's Weekly Messenger*, 278n82, 280
Bennett, James, 409
Bentham, Jeremy, 407
Benthamites, 45, 85, 244, 407
Berzelius, Jöns, 82, 148
Bewick, Thomas, 190, 200; *General History of Quadrupeds*, 191; *History of British Birds*, 191
bible and geology: the BAAS on, 383–85; biblical literalism, 156, 274; Buckland on, 109–10, 158–59, 164, 165, 276; concerns regarding, 14, 158, 381, 394, 414–15; flood geology, 158, 421; reception of Buckland's views, 266–67, 273–74, 278–84, 299, 319, 350–52, 353–54, 388–89, 403; reinterpretation of Genesis needed, 3, 10; separation of, 266, 474
bible and science: Bridgewaters on, 13; Kirby on, 109, 153–58, 165, 452; Powell on, 400–401; reception of Kirby's views, 264, 265, 268, 271, 298, 422
Bickersteth, Edward, 307–8, 309; *Christian Psalmody*, 307, 308n30; *Christian Truth*, 307–8
Birmingham School of Medicine and Surgery, 417
*Blackwood's Edinburgh Magazine*, 237, 238, 241, 246, 283–84, 420
*Blackwood's Lady's Magazine*, 311n36
Blainville, Henri de, 138, 457
Blomfield, Charles James (bishop of London): agreement to delay in Bridgewaters, 177; criticized over Bridgewater bequest, 44; High Churchmanship, 56, 116; involvement in King's College London, 40, 414; management of Bridgewater bequest, 32–41, 58, 111, 152; on science and religion, 414–15; support of Whewell for Mastership of Trinity, 389; symbolic significance of his patronage, 158
Bohn, Henry G., Scientific Library, 222, 480
Bond, Juliana, 319–20
Bonnet, Charles, 140
book manufacture: binding, 187–89; of Bridgewaters, 179–89; format, 180–81, 185–87; mechanization of, 4; steam printing, 180, 183; stereotyping, 180, 187, 198; type, 181–83
Bowden, John, 400–401
Boyle, Robert, 144, 164
Bradford Mechanics' Institute, 359, 360
*Bradford Observer*, 360
Bray, Edward Atkyns, 345–46
Brewster, David: candidate as Bridgewater author, 39; critical of Royal Society, 39; on the dangers of scientific authorship, 60, 68; on the decline of science in England, 244; and Henry Duncan, 304; editor of *Edinburgh Encyclopædia*, 39, 47, 63, 118; *The Life of Sir Isaac Newton*, 17; on payment for men of science, 57–58, 60; "Portraits of Eminent Philosophers," 238; as a reviewer, 242; reviews of Bridgewaters, 34, 46–47, 48, 49, 53, 243–44, 283; *Treatise on Optics*, 39
Bridgewater, 3rd Duke of (Francis Egerton), 19, 23–24, 25, 42
Bridgewater, 7th Earl of (John William Egerton), 24
Bridgewater, 8th Earl of (Francis Henry Egerton), 22
Bridgewater, 8th Earl of (Francis Henry Egerton), bequest, 4, 10, 19–21, 26–31; appearing reactionary, 15; ineffectiveness of, 418; objectives applauded, 270, 272; speculation regarding, 44, 254, 448, 480; suggested topics, 28–30, 36, 38, 40, 47, 239; symbolic significance of, 179; will, 25–26

Bridgewater, 8th Earl of (Francis Henry Egerton), life and career, 21–31; eccentricity, 21–22, 25; education for the church, 23; fellow of the Royal Society, 19–21, 23; immorality, 24, 45; inheritance unexpected, 20, 24; life in Paris, 21, 24–25; wealth at death, 25

Bridgewater, 8th Earl of (Francis Henry Egerton), publications: account of the inclined plane, 24; addenda to the edition of *Hippolytus*, 27, 29; *Hippolytus* of Euripides, 23, 27; *John Bull*, 24, 26, 28; *Letter to the Parisians*, 30; life of Lord Chancellor Egerton, 23; Milton's *Comus*, 23; supposed work of natural theology, 27

Bridgewater, 8th Earl of (Francis Henry Egerton), views: on atheism, 27–29, 126; on equality, 26, 28; on natural theology, 26–30

Bridgewater Treatises, authors, *12*; concerns about payment, 59; correspondence between, 33n28, 54, 176; diversity of, 15–16, 54–56; largely known to each other, 53; meetings of, 42, 59, 172, 174, 223; requesting extension, 177, 213. *See also individual authors*

Bridgewater Treatises, distribution of, 16, 213–25; advertisements for, 214–15; cheaper editions, 186–87, 222; commercial success, 4, 170, 213–14, 217–20, 224; edition sizes, 214, 218–20, 495–98; high price, 4, 16, 169–70, 180, 213, 255, 264, 272; in libraries, 16, 225, 302n21, 350, 360, 412, 415–16; "literary replication," 225, 227–28, 263, 273, 285; serialization, 215; trade sales, 215–17

Bridgewater Treatises, form of, 179–89; binding, *17*, 187–89; compared with cheap publications, 181, 183–84; criticism of, 169–70, 181, 185–86, 224; dependent on technicians of print, 170; format, 180–81, 185–87; hybrid, 16; separate treatises, 42; size due to dignity of bequest, 172, 178, 181, 224; stereotype, 187; type, 181–83

Bridgewater Treatises, illustration of, 95, *160–61*, 189–212, *184*, *193*, *200*, *207*, *211*; charge of plagiarism regarding, 101; costs of, 198, 201, 204; technologies of illustration, choice of, 170, 190–91, 212–13; use of casts in overseas editions, 223

Bridgewater Treatises, management of bequest, 31–33, 59, 414; advice to authors, 68; criticism of, 19, 42–49, 53, 56, 233, 239, 244, 246, 253–54, 255, 265–66; selection of authors and subjects, 33–41, 54–56

Bridgewater Treatises, natural theology: not primarily valued for, 228, 233–34, 241, 243, 246, 250–51, 253–54, 263, 272, 285, 394, 480–83; not primarily written as works of, 14, 68–80, 108–10, 111, 122–23, 125, 130, 136–37, 138, 159–60, 162, 164, 165–66

Bridgewater Treatises, publication of: charges, dispute over, 222–23; choice of publisher, 42, 170–79; copyright, 177, 221, 223–24; delays in, 212–13; editions, 218–20, 222, 495–98; overseas editions, 223, 224n; profits from, 58–59, 222; republication by Bell, 480, *481*; republication by Bohn, 222; republication by Collins, 222, 498n2; republication by Deighton, Bell, 472, *481*; serialization suggested, 176

Bridgewater Treatises, reading of: and anti-infidel campaigns, 360–61, 368, 369n68, 372; and Christian missions, 373; and commonplace books, 298n12; and courtship, 311–18; and diversity of, 13; and the education of children, 320–29; and family devotion, 303–9; fictional accounts of, 313–14, 326, 327–29; and the identity of "men of science," 376–77, 379–90; key to understanding impact, 16; and marriage, 318–20; and pamphleteering, 350–52, 499n1; and preaching, 331, 335–49, 392–93; and private

devotion, 291–300; and radical politics, 363, 364n61, 370–72; and scientific education, 376–77, 390–429, 486–87; and scientific practice, 431–69, 486–87; and theological education, 335–36, 396, 416; and versification, 297–302; widespread, 4, 9
Bridgewater Treatises, reviews and excerpts, 227–85; administration of bequest criticized, 44–49; in cheap miscellanies, 258–63; in devotional works, 306, 308; in evangelical periodicals, 251–57, 282–83; excerpted widely, 225, 227, 285; in High Church periodicals, 246–51, 280–82, 284; list of periodical reviews, 499–509; in literary monthlies, 236–41, 275–76; in literary weeklies, 229–36; in medical journals, 268–73; in newspapers, 273–74, 276–78, 280; in quarterly reviews, 241–45, 274–76, 283; in scientific periodicals, 263–68; timeline of periodical reviews, 230–31; in Unitarian periodicals, 257, 283; in women's magazines, 311; in youth's magazines, 256, 327–29
Bridgewater Treatises, symbolic status of: imitated by other publications, 220–21; importance of, for BAAS, 11, 235–36, 376, 379–90; importance of, for "men of science," 18, 375–77, 379–90; role in rebaptizing the sciences, 3, 16, 229, 241, 245–46, 251, 254, 263, 273, 275, 283–84, 372, 429, 451, 469, 471, 473–76
Bridgewater Treatises, writing of: diversity in, 15–16, 108–66, 477, 485–86; practicalities of, 91–105, 179; as reflective treatises, 80–91; in relation to Paley's *Natural Theology*, 68–80
*Brighton Gazette*, 388
Bristol Baptist Academy, 325
Bristol College, 346, 406, 413
Bristol Literary and Scientific Institution, 293
Bristol Society of Artists, 386
British Association for the Advancement of Science (BAAS), 377–86; Bristol meeting of (1836), 276, 277, 278, 279, 384–86; Cambridge meeting of (1833), 235, 377, 379, 381–83, 392; and Christianity, 11–12, 376–86, 474; Dublin meeting of (1835), 383–84; Edinburgh meeting of (1834), 383, 384; founding of, 11, 350, 379; meetings of, reported in press, 235, 276, 277; members of, 233, 235, 341; Newcastle meeting of (1838), 351; Norwich meeting of (1868), 336, 478; Oxford meeting of (1832), 96, 234, 235, 379–80, 383, 398–99, 401; platform for "men of science," 11, 486; presidential addresses, 331, 479; reports for, 94, 459; York meeting of (1844), 351–53, *352*, 378
*British Critic*: censuring Powell, 400; identification of contributors to, 503n15; reviews of Bridgewaters, 246, 249–51, 281, 346, 401; Whewell in, 60
*British Magazine*, 48, 244, 246, 247–49, 281
British Museum: Bridgewater's bequests to, 25; location of, 177; Mantell's sale of collections to, 388; Pickering as a bookseller for, 178; specimens at, 102, 203; trustees of, 41, 42
*British Pulpit* (periodical), 340n19
Broderip, William, 275
Brougham and Vaux, 1st Baron (Henry Brougham): and Bell, 218; Buckland quoting, 164; *Discourse of Natural Theology*, 71, 240–41, 251, 254–55, 257, 396–97, 445, 446; and Henry Duncan, 304; *Paley's Natural Theology Illustrated*, 71–74, 262, 411; and SDUK, 71, 84, 262
Brown, Robert, 275, 467
Brown, Samuel, 427–28; *Argument from Design Equal to Nothing*, 372; *Lay Sermons on the Theory of Christianity*, 427
Brown, Thomas, 120, 122
Brydges, Egerton, 22
Buckland, Frank, *89*, *104*, 206n57
Buckland, John, 404

Buckland, Mary, *89*; as an amanuensis, 93–94, 96; family illness and death, 105; as an illustrator, 94, 209, 403; on the importance of illustrations, 201; on "men of science," 375–76, 389; as a mentor, 88, 279; on persecution of her husband, 105

Buckland, William, *89*, *104*, *202*

Buckland, William, as a Bridgewater author: appointment, 19, 38; attending meetings of authors, 42, 59, 172, 174, 177; concerns about format, 181; concerns about interest, 59; concerns about Murray as publisher, 176; criticism alarming, 43; dictating text, 93–94, 96; drawing on own lectures, 88–89, 94–95, 159, 160–61, 164, 402; original research, 95–96; Owen assisting, 457; profits, 222; reported to be a "competitor," 42; reputation attracts attention to Bridgewaters, 218; requesting extension, 95, 102, 213; response to Paley's *Natural Theology*, 76–77, 95, 160

Buckland, William, Bridgewater Treatise, 97, 158–65; benefiting from coming last, 220, 274; establishing high reputation for series, 387; excerpted, 328; functionalism of, 454; on Genesis and geology, 109–10, 158–59, 164, 165, 266–67, 273–74, 279, 280–84, 299, 319, 350–52, 353, 383–85, 388–89, 403–4; illustrations in, 95, *160–61*, 161, 162, 189, 198, 201–12, *207*, *211*, 265, 403; on law-like changes in geology, 162–63, 164, 393, 433, 476; in libraries, 350, 415; making additions to natural theology, 76–77, 159–60, 164; opposing transmutation, 159, 162–63, 164, 165, 446, 457; "as per contract," 390, 485; prompting press storm, 229, 273–74, 276–80; reading of, 293, 299, 300, 301–2, 313–14, 315–18, 319, 329, 350–52, 353, 370, 388–89, 390, 393–94, 442, 443, 446, 449, 450, 451, 457, 459, 468; reviews of, 241, 255, 257, 266–67, 268, 274–76, 280–85, 336, 346, 388, 428–29; revised edition of, 459; set text for students, 415, 428–29; use of by preachers, 344

Buckland, William, life and career: active in British Association, 379, 381; advocate of science education at Oxford, 76–77, 160, 397, 405; attending Owen's lectures, 458; British Association address, 276, 277, 379–80, 384; clerical status, 387; Dean of Westminster, 389; the "English Cuvier," 455; family illness and death, 105, 351; flood geology of, 62, 158, 266, 274, 401, 405, 421; Harcourt's teacher, 350; jocular manner of, 381, 384; lectures on Bridgewater, 313, 386, 402–3; on the *Megatherium*, 234, 379–80; reactions to Bristol announcement, 276–80, 384–85; reactions to Oxford lectures, 62n15, 403–5; reputation of, 238, 473; on Royal Society council, 53; sermon of, 403; sources of income, 62; supporter of Charles Daubeny, 401; taught by Kidd, 53; use of illustrations in teaching, 201–2, *202*, 403; visits by, 305, 386–87; on young ladies as ichthyosauri, 311

Buckland, William, other publications: *Reliquiæ diluvianæ*, 62, 88, 94, 170, 201, 210, 331, 403; *Vindiciæ geologicæ*, 159, 164

Buffon, comte de, 136

Burgon, Miss S. C. (Sarah Caroline?), 204

Burke, William, 423

Burnett, Charles Mountford, *The Power, Wisdom, and Goodness of God*, 221

Bushnan, John Stevenson, *Introduction to the Study of Nature*, 220

Butler, Joseph: *Analogy of Religion*, 74, 78, 116, 121, 235, 395, 404, 418, 440; and Chalmers's Bridgewater, 118–21; *Fifteen Sermons*, 120; and Kidd's Bridgewater, 130

Byfield, John, 198, 206, 209

Byron, Ada (Ada Lovelace), 315, *316*, 436

Byron, Lord, 173, 299, 315, 436; *Childe Harold's Pilgrimage*, 300
Bywater, Abel, 358; *The Sheffield Dialect*, 358

Cabinet Cyclopædia, *17*; and Brewster, 39, 58; format of, 184, 186; and Herschel, 36, 58, 82–83; reading of, 296; and Roget, 68, 87, 100, 174, 176
*Cambridge Chronicle*, 244
Cambridge Philosophical Society, 442; *Transactions*, 39
Cambridge Platonists, 154
Campbell, Thomas, 237
Carey, Lea and Blanchard (publishers), 223
Carlile, Richard: *Address to Men of Science*, *8*, 356; critical of Bridgewater, 21; editor of the *Deist*, 356, 359, 360; editor of the *Lion*, 357n47; editor of the *Republican*, 356; followers of, 356–62, 364; as "infidel" leader, 356–57; infidel mission of, 357, 358; and scientific materialism, 7, 127, 356, 362
Carlyle, Thomas, 239, 317
Carpenter, Lant, 348
Carpenter, William Benjamin: on authorship, 65; family of, 387; on form and function, 432, 460–62; *Principles of Physiology*, 413, 425, 426, 462; reading of Bridgewaters, 412–13, 425–26, 460–62, 468–69; as a student at Edinburgh, 412, 425–26; as a student at London University, 412–13
Carter, William George, 281n91
Carus, Carl Gustav, 138
Caswall, Edward, *Sketches of Young Ladies*, 311
*Catholic Magazine*, 247
Catlow, Agnes, 197–98
Catlow, Samuel, 198
Chalmers, Thomas, *117*, 343
Chalmers, Thomas, as a Bridgewater author: account with Pickering, 222; appointment, 40–41, 56; attending meeting of authors, 59; changing publisher, 222; concerns about interest, 59; concerns with practice of Christianity, 485; drawing on Butler, 118–21; drawing on own lectures, 79, 98, 118–19, 121, 122, 420; on edition sizes, 219; line of demarcation from Kidd, 53, 127, 129; profits, 222; reputation attracts attention to Bridgewaters, 218; response to criticism, 185; response to Paley's *Natural Theology*, 78; Royal Society, not a fellow of, 50; seeking larger readership, 187
Chalmers, Thomas, Bridgewater Treatise, 116–23; on the argument from conscience, 119–21, 326, 342, 344, 345, 347, 396; excerpted, 252, 261; excessively prolix, 79, 335; in libraries, 415; on the limitations of natural theology, 108, 122–23; on natural laws, 120, 271, 433, 441; on political economy, 119–22; reading of, 145n70, 293, 299n15, 301, 326, 335–36, 363, 396, 448, 451; reviews of, 234, 236, 239, 244–45, 249, 250–52, 270–71, 278, 448, 480; similarities to Whewell, 113, 116, 120; unillustrated, 192; use of by preachers, 342, 344, 345
Chalmers, Thomas, life and career: admired by Gurney, 297; as an author, 92, 174; on Babbage's *Ninth Bridgewater*, 441; changing views of the grounds of belief, 116–19; on clerical career and science, 54; discourses on astronomy, 348; Edinburgh lectures of, 78–80, 420–21; an exemplary preacher, 118, 338, 340, 342, 343; Glasgow visitation scheme of, 366; leader in Church of Scotland, 304; reputation of, 473; sermons plagiarized, 429; sources of income, 63; students of starting the *Presbyterian Review*, 253
Chalmers, Thomas, other publications: *Astronomical Discourses*, 63; *Christian and Civic Economy*, 41; *Evidence and Authority of the Christian Revelation*, 63, 118; *Natural Theology*, 79–80, 98, 187, 251, 253n37; *On Political Economy*, 121, 245

Chambers, Robert: editor of *Chambers's Edinburgh Journal*, 262, 447–49; editor of *Chambers's Information for the People*, 449; *Explanations*, 451; reading of Bridgewaters, 448–51, 477; *Vestiges of the Natural History of Creation*, 172, 215, 219, 447–51, 471, 478
Chambers, William, 261–62, 449
*Chambers's Edinburgh Journal*, 258, 261–62, 447–49
Channing, William Ellery, 257
Chantrey, Francis, 208, 212, 276
*Chemist* (periodical), 429
Children, John George, 41, 178
*Children's Friend* (periodical), 327n66
*Child's Companion* (periodical), 327n66
*Child's Magazine*, 327n66
Chilton, William, 370
Christ Church, Oxford, 38, 61, 62, 158, 385, 404
*Christian Examiner*, 252
*Christian Guardian*, 247
Christian Instruction Society, 353
*Christian Lady's Magazine*, 311n36
*Christian Observer*, 254–55, 282, 322n57
*Christian Reformer* (periodical), 257
*Christian Remembrancer* (periodical), 56, 246, 249, 281–82, 483
*Christian Teacher* (periodical), 257, 446
churches (in London unless otherwise stated): Belgrave Chapel (Pimlico), 351; Camden Chapel (Camberwell), 340; Cardiff parish church, 346; Christ Church (Oxford), 403; Collegiate Church of St Mary, St Denys and St George (Manchester), 395; Durham Cathedral, 25; George's Meeting (Exeter), 348; Great St. Mary's (Cambridge), 110, 111, 394; High Street Chapel (Huddersfield), 358; National Scotch Church (Regent Square), 336, 342; Old Gravel Pit Chapel, 353; Scotland Street Chapel (Sheffield), 358; Sion Chapel (Bradford), 359; St. Bride's Fleet Street, 338; St. George's Chapel (Brighton), 344; St. Mary le Strand, 338; St. Mary's (Oxford), 105, 380–81, 400; St. Pancras (Bloomsbury), 341; St. Paul's Cathedral, 299; Weigh House Chapel, 353; York Minster, 350, 352–53
Churchill, John, 172, 215
Church Missionary Society, 307, 342
*Church of England Quarterly Review*, 284
*Church of Scotland Magazine*, 247
Clapham Sect, 254, 322, 324
Clark, William, 394
Clarke, Adam, 305
Clarke, Samuel, 98
Clarmont, Jean-Charles, 34
Claude, Jean, *Essay on the Composition of a Sermon*, 333–34, 335, 336
Clift, William, 102
Clift, William Home, 93n72
Clodd, Edward, 478–79
Cobbe, Charles, 324, 325
Cobbe, Frances, 324
Cobbe, Frances Power, 324–25
Cockburn, William, 349–53, 354; *The Bible Defended against the British Association*, 351
Cole, Henry, 282, 394; *Popular Geology Subversive of Divine Revelation!*, 394
Coleridge, Samuel Taylor, 221n81, 239, 417, 427
Collins, William, 63, 79, 222, 498n2
Combe, George: *Constitution of Man*, 224–25, 270, 447–48, 449–50, 476; friend of Nichol, 448; reading of Ellen Parry's head, 322; review of Chalmers's Bridgewater, 270–71, 448
"commonsense" philosophy, 120, 122, 249
Conder, Josiah, 255
*Congregational Magazine*, 255, 282–83, 353
Congregational Union, 322, 353
Conolly, John, 197, 409, 410
Constable, Archibald, 87, 175
Conybeare, William Daniel, 205, 280–81, 346–47
Cooper, Thomas, 372–73
Cope, Richard, 328–29
Copleston, Edward, 414

Copley, Esther, 260; *Female Excellence,* 309
Corpus Christi College, Oxford, 62
*Court Magazine,* 311n36
Cox, Elizabeth, 172
Crabbe, George, *Outline of a System of Natural Theology,* 221
Craigie, David, 426
*Crisis* (periodical), 363
Croly, George, 283–84
Crombie, Alexander, *Natural Theology,* 17, 243
Crosse, Andrew, 279
Cruickshank, George, 355
Curtis, Charles Morgan, 193, 194–95
Curtis, John, 194, 195; *British Entomology,* 195
Cuvier, Clémentine, 322–24, 323
Cuvier, Georges: author of learned synopses, 82; on comparative anatomy, 132–36, 140, 163–64, 409, 424, 454–55, 456–57, 460; debate over form and function, 109, 454, 455, 458–59, 466–67; family, 324; on geology, 421; *Le règne animal,* 129–30; a model naturalist, 134, 379–80; passages reprinted by Kidd, 81; public reputation of, 311; *Recherches sur les ossemens fossiles,* 206; religious views, 324; on transmutation, 142

Dale, Thomas, 410–11
D'Alton, Eduard, *Die vergleichende Osteologie* (with Christian Pander), 211
Dalton, John, 146, 364; *New System of Chemical Philosophy,* 196
Daniell, John Frederick, *Meteorological Essays,* 102
Darton, William, 326
Darwin, Charles, 464; *Autobiography,* 482; on form and function, 432, 463–69; on natural laws, 1, 443–44, 451, 465; *Orchids,* 468; *Origin of Species,* 1, 2, 14, 444, 468, 471–73, 478, 483; quoting Bacon, 10; quoting Whewell, 1, 2, 3, 10, 14, 18, 434, 441, 444, 469, 471; as a reader, 441–44, 463–68; reading of Bridgewaters, 434, 441, 443–44, 463–64, 465, 467, 468, 477, 486–87; as a student in Edinburgh, 425; talks with Ruskin, 403
Darwin, Erasmus, 3, 280
Daubeny, Charles, 401–2; *Introduction to the Atomic Theory,* 401
Daubuisson, Jean-François, 390
Davies, Richard, 338–39
Davy, Humphry, 31
Dawkins, Boyd, 429
Day, William, 206, 209
De Candolle, Augustin, *Théorie élémentaire de la botanique,* 402
Deighton, Bell (publishers), 472, 481
De la Beche, Henry, 319; *Sections and Views,* 205
De Morgan, Augustus, 410, 411
Denham, Frederick, 338
Denny, Henry, 195
Desmond, Adrian, 269n68
D'Holbach, Baron (Paul-Henry Thiry): materialism of, 356, 358; *Système de la nature,* 116, 356, 357, 359, 360, 361, 364
Dibdin, Thomas Frognall: *Bibliophobia,* 175; *Reminiscences of a Literary Life,* 213
Dick, Thomas: *Christian Philosopher,* 302, 336–37; *On the Improvement of Society,* 253; on science and Christianity, 9
Dickens, Charles, 311
Dinorben, Lord (William Hughes), 386
Disraeli, Benjamin, 175
Doddridge, Philip, *Rise and Progress of Religion in the Soul,* 293–94, 299
D'Oyly, George, 32
Draper, Bourne Hall, 325–27; *Bible Story Book,* 325; *Conversations on Some Leading Points in Natural Philosophy,* 325–26; *Juvenile Naturalist,* 326; *Stories from Scripture,* 325; *Stories of the Animal World,* 326–27
Drew, George Smith, 339, 341
*Dublin University Magazine,* 237n14, 240, 241, 281
Duncan, Henry, 303–7, 304, 309; *Sacred Philosophy of the Seasons,* 256, 306–7

Duncan, John Shute, *Botano-Theology*, 76
Duncan, Philip, 76, *104*
Dunfermline Mechanics Institute, 220
Duns, John, 313n40

East India Company Army, 319
East India Company College (Haileybury), 249
*Eclectic Review*, 255, 282–83
*Edinburgh Christian Instructor*, 246, 252–53, 254, 306–7
*Edinburgh Encyclopædia*, 39, 47, 63, 118
*Edinburgh Journal of Natural History and of the Physical Sciences*, 264
*Edinburgh Medical and Surgical Journal*, 426
*Edinburgh New Philosophical Journal*, 263–64, 422
*Edinburgh Review*: advertisements in, 214; coterie of, 244; as a forum for scientific practitioners, 245; inception of, 241; on the "Oxford malignants," 404; publicizing SDUK, 84–85; reviews of Bridgewaters, 46–47, 53, 243–44, 283; rivalry with *Quarterly Review*, 276; Roget invited to contribute to, 68; sales of, 241, 246
Edinburgh Zetetic Society, 358
Egerton, Francis. *See* Bridgewater, 3rd Duke of (Francis Egerton)
Egerton, Francis Henry. *See* Bridgewater, 8th Earl of (Francis Henry Egerton)
Egerton, John William. *See* Bridgewater, 7th Earl of (John William Egerton)
Egerton, Thomas (1st Viscount Brackley), 23, 25
Egremont, 3rd Earl of (George Wyndham), 388
Ehrenberg, Christian Gottfried, 195
Eliot, George, *The Mill on the Floss*, 313–14
Elliotson, John, 99, 271, 410
Ellis, Sarah: *Daughters of England*, 310–11; *Women of England*, 318
Elwin, Whitwell, 480
*Encyclopædia Britannica*: Napier's supplement to (1815–24), 36, 65, 67, 87, 139, 453; secularity of, 36; seventh edition of (1830–42), 87, 453
*Encyclopædia Metropolitana*, 36, 196
English, Henry, 267
Ensor, George, *Natural Theology*, 371
Entomological Club, 264
*Entomological Magazine*, 264–65
Epicureanism. *See* atomism, Greek
Euclid, 398
Euripides, *Hippolytus*, 23, 27
*Evangelical Magazine*, 246, 247, 322
Evelyn, John, *Silva*, 29
evolution. *See* transmutation
*Examiner* (newspaper), 45, 278
Exeter Hall, 366

Fairholme, George, 282, 284; *General View of the Geology of Scripture*, 274
Farnborough, 1st Baron (Charles Long), 31, 34, 41–42, 177, 213
Farrar, Eliza Ware, *Young Lady's Friend*, 310
Farrar, Squire, 359, 361
Featherstonehaugh, George, 223
Fellowes, Robert, 411–12; *Religion of the Universe*, 412
Fergus, Henry, *The Testimony of Nature and Revelation*, 220, 238
Fisher, Joseph, 205, 206, 207–8
Fitton, William Henry, 222
Fletcher, John, 422–23, 425, 426
Fonblanque, Albany, 45
Forbes, Edward, 426–28, 463
Forbes, James David, 46, 58, 420–21
Forster, John, *Churchman's Guide*, 335
Fox, Caroline, 313
Fox, Robert Were, 313, 386
*Fraser's Magazine*, 237, 238–40, 246, 284, 315
French Revolution, 6, 26, 69, 107, 474
Froude, Hurrell, 405

Gaëde, Henri-Maurice, 153n84
Gainsborough, Thomas, 194
Gaisford, Thomas, 404
Galen, 29, 99, 346
Galileo, 283
Gall, Franz, 126, 128

Gardner, James, 205
Gatty, Margaret, 480
*Gentleman's Magazine*, 27, 236–37, 501n9
Geoffroy Saint-Hilaire, Étienne: debate over form and function, 109, 454, 455, 458–59, 461, 462, 466; as a philosophical anatomist, 10, 100–101, 131, 134, 137–38, 140–43, 269, 270, 409, 423; as a transmutationist, 128, 130, 163
Geological Society: fellows of, 353; proceedings at, 439, 442, 457; specimens at, 203; *Transactions*, illustrations in, 191, 201, 205, 210; Wollaston fund, 208
George IV (king), 31, 355
Gérard, Baron, 21, 22
Gibson, Edmund, *Family Devotion*, 303
Gibson, William Sidney, *The Certainties of Geology*, 221
Gilbert, Davies (president of the Royal Society): advising Bridgewater authors, 68, 81, 83; agreement to delay in Bridgewaters, 102; management of Bridgewater bequest, 15, 31–41, 46, 50, 54; response to criticism, 43; symbolic significance of his patronage, 179, 290
*Globe* (newspaper), 279
Godwin, Benjamin, 359–62
Goodsir, John, 427, 428
Goodsir, Joseph Taylor, 336
*Gospel Magazine*, 247
Gould, John, 195
Grant, James, 219n78, 278n82, 340; *Metropolitan Pulpit*, 338
Grant, Robert Edmond: lectures attended by Carpenter, 412; lectures used by Roget, 100–101, 138, 197, 413; referring to divine design in lectures, 409, 474
Graves, John Thomas, 33
Gray, Charles, 200
Gray, Henry, *Anatomy Descriptive and Surgical*, 197
Great Windmill Street medical school, 40, 67, 199
Green, Joseph Henry, 138, 417
Greenwich Hospital (London), 34
Greg, Mary, 319
Greg, Robert, 319
Grenfell, Fanny, 315–18
Gresley, William, 334, 335
Grey, 2nd Earl (Charles Grey), 401
Grindlay, Jessie, 312–13
Grote, George, 407
Gurney, Joseph John: admirer of Chalmers and Whewell, 297; *Essay on the Habitual Exercise of Love to God*, 297
Guy's Hospital (London), 54

Hackney Phalanx, 151, 155, 249, 417
Hallam, Arthur, 301
Halley, James, 336
Hamilton, James, 336
Hampden, Renn Dickson, 404
Hannah, John, 473
Hannam, Thomas, *Pulpit Assistant*, 333
Harcourt, William Vernon, 350, 383
Hare, William, 423
Harris, John, 326
Harrow School, 336
Hartley, John, 342
Harvey, William, 462
Hatchard, John, 171
Hawkins, Thomas, 203, 207
Heath, William, 22
Henslow, John Stevens, 394, 442
Heraud, John Abraham, 239
Herschel, John: on astronomy, 404; and Babbage's *Ninth Bridgewater*, 437; candidate for Royal Society presidency, 33; critical of management of Bridgewater bequest, 20; friend of Whewell, 81; independently wealthy, 58; invited to become Bridgewater author, 36, 56; letter on species, 439–40, 443, 444, 451; *Preliminary Discourse*, 36, 82–83, 146, 172, 192, 413, 433, 441–42, 443, 444, 452, 460; quoted by Buckland, 164; *Treatise on Astronomy*, 36, 361
Hervey, James, *Meditations and Contemplations*, 305
Heywood, Abel, 365

Highley, Samuel, 172
Hill, George, *Lectures in Divinity*, 78
Hobhouse, John Cam, 385, 386
Holland, 3rd Baron (Henry Richard Vassall-Fox), 387
Holland, Henry, 242–43
Hollingworth, John Banks, 75
Holyoake, George Jacob, 371, 477–78; editor of the *Movement*, 477; *Paley Refuted in His Own Words*, 371
Homer, 125
Homerton Academy, 353
Hone, William, *The Man in the Moon*, 355
Hooker, Joseph Dalton, 331, 333, 336, 479
Hoppus, John, 368, 407
Horne, Richard Henry, 241n20
Horton Academy (Bradford), 359
Howley, William (archbishop of Canterbury): agreeing to delay in Bridgewaters, 177; criticized over Bridgewater bequest, 44; High Churchmanship, 56; involvement in *Encyclopædia Metropolitana*, 36; involvement in King's College London, 40, 414, 417; management of Bridgewater bequest, 32–41, 50; Nolan's patron, 381; symbolic significance of his patronage, 158, 179, 285, 290; warned about Buckland's "blasphemy," 278
Hullmandel, Charles, 194, 202, 209, 210
Humboldt, Alexander von, 196
Hume, David: on causation, 119; critique of cosmological argument, 98; critique of teleological argument, 115, 145, 371; *Dialogues Concerning Natural Religion*, 69, 356, 360; on miracles, 438; problem of induction, 122
Hunter, Alexander, 29
Hunter, John, 455, 456, 458, 459
Huntley, Miss, 321
Hutchinson, John, 155
Hutchinsonianism, 154–56, 254, 271
Hutton, James, *Theory of the Earth*, 300n17
Huxley, Thomas, 479, 487
illustration: in Bridgewater Treatises, 16, *160–61*, *184*, 189–212, *193*, *200*, *207*, *211*; importance of, in science, 191–93, 197, 198–99, 201–2; intaglio, 190–91, 193, 204, 205–9; lithography, 190–91, 194, 196, 204, 209–12; technological revolution in, 190–91, 201, 212; transfer lithography, 205–6, 209–10; use in lectures, 197, 201–2, *202*, 403; use of casts for overseas editions, 223; wood engraving, 190–91, 196–201, 204
infidelity: anti-infidel campaigns, 332, 357–62, 365–69, 372–73; Carlilean movement, 356–57; in France, 6; and geology, 105, 284; and London University, 407; and natural theology, 240; Owenite atheists, 362, 369–72; and the Reform Bill, 152; rise of, in Britain, 332, 354–57; and science, 9, 14, 370. *See also* atheism; materialism
Institut de France, 27
*Investigator!* (periodical), 371
Irons, William Josiah, 405
Irving, John, 340–41
Islington Literary and Scientific Society, 296, 297

Jackson, John, 200
Jackson, Maria, 312
James, James Angell, 322–24; *The Anxious Inquirer after Salvation*, 324; *Christian Father's Present*, 324; *The Flower Faded*, 322–24, *323*
Jameson, Robert, 421–22
Jaques, John, 206n58
Jenks, Benjamin, *Prayers and Offices of Devotion*, 303
*John Bull* (newspaper), 278, 279
Johnson, James, 271
Johnstone, Christian, 239–40
Jones, Richard, 103
Jones, William (of Nayland), 155
Joséphine, Empress, 24
*Journal of Natural Philosophy*, 24

Kant, Immanuel, 115, 239, 250, 467
Karkeek, William Floyd, 267–68

Keble, John: and Buckland's geology, 405; *Christian Year*, 297, 314
Kidd, John, *124*
Kidd, John, as a Bridgewater author: appointment, 40; attending meeting of authors, 177; drawing on earlier publications, 81, 99–100, 126, 127; line of demarcation from other authors, 53, 99, 127, 129; popular exposition, 81; response to Paley's *Natural Theology*, 76–77
Kidd, John, Bridgewater Treatise, 124–30; design rooted in revelation, 452; excerpted, 261; intended to protect students against atheism, 125–26; in libraries, 412n61, 415–16; opposing materialism, 109, 126–27, 128; opposing transmutation, 109, 127–30; reading of, 293, 361, 416; reviews of, 234, 236, 243, 249, 270, 271; theologically limited, 77, 108, 124–25, 137; unillustrated, 192; use of by preachers, 345n27, 346
Kidd, John, life and career: active in British Association, 379; advocate of science education at Oxford, 76–77, 124–26, 397; Buckland's teacher, 53; friend of Richard Davies, 338; Harcourt's teacher, 350; sources of income, 61–62; supporter of Daubeny, 401
Kidd, John, other publications: *Introductory Lecture*, 99–100, 126; *Outlines of Mineralogy*, 62
King, John, 345
King, William, 315
King's College, Cambridge, 333
King's College London: Bridgewaters read at, 407, 416, 418; founding of, 40, 406, 414; professors at, 264; Ruskin educated at, 403; scientific education at, 414–18; theological education at, 414
Kingsley, Charles, 315–18, 372, 472–73
Kirby, William, 55
Kirby, William, as a Bridgewater author: appointment, 41; attending meetings of authors, 42, 174; clashes with Pickering over second edition, 220; drawing on own sermons, 90, 155; noninvolvement in education, 50, 151; not responding to Paley's *Natural Theology*, 78; requesting extension, 102, 213; seeking help from others, 90, 102; undertaking research, 101–2
Kirby, William, Bridgewater Treatise, 151–58; excerpted, 260, 308; illustrations, 192–96, *193*; in libraries, 412n61, 415; making no contribution to natural theology, 152–53; opposing transmutation, 153–54, 155, 157, 165, 195, 271, 272; and quinarianism, 157; reading of, 298–99, 301, 308, 326, 388, 422, 423, 443, 463, 467; reviews of, 240, 256, 264–65, 268, 271–73; on the scientific value of the Bible, 109–10, 153–58, 165, 264, 265, 268, 271, 298, 388, 422, 452; on vicarious suffering, 157, 308, 326
Kirby, William, life and career: academic aspirations of, 55; cousin of Sarah Trimmer, 184, 321; experience with illustrations, 193–95; located far from London, 54; sources of income, 63–64
Kirby, William, other publications: *Introduction to Entomology* (with William Spence), 41, 63, 64, 90–91, 102, 174, 192, 194, 195; *Monographia apum Angliæ*, 90
Knapp, Leonard, 41; *The Journal of a Naturalist*, 41
Knight, Charles: Babbage's publisher, 215; editor of *Penny Magazine*, 262; editor of *Plain Englishman*, 34; on literary output, 227; pioneer of cheap publishing, 172, 213; SDUK publisher, 84–85
Knox, Robert, 423–25, *424*, 428, 455, 463

*Ladies' Cabinet*, 311
*Lady's Magazine*, 311n36
Lamarck, Jean-Baptiste: Bridgewaters on, 131, 134–35, 142–43, 153–54, 155,

Lamarck, Jean-Baptiste (*continued*) 156, 157–58, 163, 165, 195, 270, 271; followers of, 100; *Histoire naturelle des animaux sans vertèbres*, 102; *Lancet* on, 272–73; Lyell's critique of, 127, 154, 162n99, 361; *Philosophie zoologique*, 153, 443; reception of, 127n32, 153–54; transmutation theory of, 11
*Lancet*: on Bell, 72, 85, 270; on phrenology, 271; publication of Grant's lectures, 101; reviews of Bridgewaters, 92–93, 269, 272–73; transformation of medical journalism, 268
Landells, Ebenezer, 200
Landon, Laetitia, 345
Landon, Whittington H., 345
Lansdowne, 3rd Marquis of (Henry Petty-Fitzmaurice), 385, 386
Laplace, Pierre-Simon: Laplacian physics, 112n9, 392, 436; on natural laws, 10, 112, 120, 153, 245, 433; religious skepticism, reputation for, 2, 112–15, 120, 158, 261, 271, 352, 421
Lardner, Dionysius: editor of Cabinet Cyclopædia, 36, 39, 58, 68, 82, 87, 100, 296; professor at London University, 410; *Treatise on Hydrostatics and Pneumatics*, 17
Latreille, Pierre André, 138
Lavater, Johann Kaspar, 299
Law, William, *Serious Call to a Devout and Holy Life*, 295
Lawrence, William, 7, 126, 127, 131, 150, 356
Le Bas, Charles, 249–50
Lee, Samuel, 279
Leighton, Archibald, 188–89
Leslie, John, 421
libraries and book clubs: Bradford Amicable Book Society, 360; Bradford Mechanics' Institute, 360; and the Bridgewater authors, 102; in churches, 302; circulating libraries, 311; Frances Cobbe's library, 324; Congregational Library, 353; Darwin's *Beagle* library, 442; King's College London, 415–16; literary and philosophical societies, 16; literary weeklies in, 232; London University, 412; at mechanics' institutes, 16; Miles Platting Reading Society, 364; ministers' libraries, 308, 335; proliferating variety, 225; Royal Society, 178, 223; St. Thomas's Hospital, 412; Tennyson's library, 301; Trinity College Cambridge, 472; Webb Street Anatomy School, 412; York Minster library, 350
Lieder, Alice, 373
Lindley, John, 411
Linnean Society: conversazione at, 42; John Curtis employed by, 195; *Transactions*, technologies of illustration in, 191, 194, 210
*Literary Gazette*: Daubeny in, 401; reports on Bridgewater bequest, 42; reviews of Bridgewaters, 234–36; reviews reprinted by *Ladies' Cabinet*, 501n10; Roget's lectures reported in, 100n85; on the state of literature, 217; as a weekly literary journal, 232
Lloyd, Bartholomew, 383–84
Lloyd, Charles, 75, 76
Lloyd, William Freeman, 256
Locke, John, 164, 259, 348
Locker, Edward Hawke, 34
Lockhart, John Gibson, 242, 244
London City Mission, 365–69, 367
London Institution, 87, 100n85
London Mechanics' Institution, 366, 369
*London Medical and Surgical Journal*, 269
*London Medical Gazette*, 85, 269, 272
London Phrenological Society, 271
*London Review*, 394
London University (University College London), *408*; Bell one of founding professors, 38, 408–9; Bridgewaters read at, 391–92, 407, 412–13, 418; and Christianity, 6, 130, 407, 410–11, 413; criticized for being anticompetitive, 270; founding of, 6, 391, 406, 416; professors at, 368; scientific education at, 376–77, 407–14; secularizing

reformers at, 474; theological education at, 407, 410
Long, Charles. *See* Farnborough, 1st Baron (Charles Long)
Longman, Thomas Norton, 173
Longmans (publishers): Bell's publisher, 65; considered for Bridgewaters, 172–74, 176; and dispute concerning underselling, 217; mimicking Bridgewaters, 220, 221; publication of Cabinet Cyclopædia, 174, 176, 186; publication of Roget's *Introductory Lecture*, 35; at trade sales, 216
Lord, Perceval, 234, 235; *Popular Physiology*, 234
Loveday, A., *104*
Lovelace, Ada. *See* Byron, Ada (Ada Lovelace)
Lucretius, 126, 127, 129; *De rerum natura*, 387
*lusus naturæ*, 99–100, 128, 130, 142
Luther, Martin, 384
Lyell, Charles: as an author, 58, 88, 92, 96; on Babbage's *Ninth Bridgewater*, 437, 439–40; on Babbage's soirees, 436; on Buckland, 62, 158, 387, 390; and Darwin, 442–43; on English universities, 418; and Herschel's letter on species, 439–40, 443; on Lamarck, 127, 154, 162n99, 361; *Principles of Geology*, 82, 83, 88, 127, 154, 162, 172, 173, 186, 301, 313, 361, 370, 387, 439, 440, 442; as professor at King's College London, 414–15, 474; on the recent origin of humans, 360; as a reviewer, 242

MacCulloch, John: candidate as Bridgewater author, 35–36, 38; *Proofs and Illustrations of the Attributes of God*, 221, 465–66; *System of Geology*, 35–36
MacDougall, Patrick C., 253
MacGillivray, William, 264, 422
Mackintosh, Daniel, *Supplement to the Bridgewater Treatises*, 221, 298
Mackintosh, Thomas Simmons, 364
Maclaren, James, 437
MacLeay, William, 102; *Horæ entomologicæ*, 138, 153, 157; quinarian classification, 140, 157
*Magazine of Botany and Gardening*, 264, 415
*Magazine of Natural History*, 181
*Magazine of Popular Science*, 266–67
Maginn, William, 239
Malcolm, William, 340–41
Maltby, Edward, 351, 407
Malthus, Thomas, 121, 437; *Principles of Population*, 465
Manchester Mechanics' Institute, 319
Mantell, Gideon, 156, 388–89; *Wonders of Geology*, 388–89
"march of intellect," 4, 5
Marryat, Frederick, 237
Martineau, Harriet: on authorship, 57, 91–92, 96, 98; on Babbage, 436, 437n9; on Bell's sincerity, 92, 390; shocked by sales of Bridgewaters, 220
Martineau, James, 479
materialism: Bridgewaters oppose, 109, 126–27, 128, 131, 135, 136–37, 143, 150, 154, 166, 251; and Carpenter, 426; in cheap publications, 258; and D'Holbach, 116, 356, 357–58; in Edinburgh, 425; and Elliotson, 410; and geology, 284; and Grant, 100; and Lawrence, 7, 126, 127, 131, 150; and medical education, 72, 126, 130–31; and phrenology, 126, 128, 269, 356; Tyndall on "Scientific Materialism," 336; in working-class radicalism, 7, 127, 356, 357–58, 359, 362, 364–65, 370, 372
Matfin, John, 359
Maund, Benjamin, 264n58
Maurice, Frederick, 317
Mayo, Herbert, 40; *Outlines of Physiology*, 40; *Philosophy of Living*, 463
McCulloch, John Ramsay, 219
*Mechanics' Magazine*, 258, 265–66
Meckel, Johann Friedrich, 138, 141, 272, 409
*Medico-Chirurgical Review*, 47–48, 169–70, 182, 271–72

*Megatherium*: in Buckland's Bridgewater, 163–64, 210, 211–12, *211*, 234; Buckland's lecture on, 379–80; Owen on, 459; at Royal College of Surgeons, 96
Melvill, Henry, 339–41
"men of science:" as authors, 56–58, 60–68, 233, 486; changing identity, 1–3, 11, 83, 238, 375–79, 486; as educators, 390–92, 486; moral character, 376–86, 392–93, 409; and national progress, 6, 11, 384–85; religious character, 11, 18, 31, 130, 134, 373, 375–90, 476, 486; religious heterodoxy, 7, 8, 12–13, 30–31, 114–15, 123, 134, 152–54, 390, 421, 426, 474; as reviewers, 242, 244, 245; sources of income, 20–21, 56–58, 60–68; and the term "scientist," 83, 377
Merthyr Rising (1831), 346
*Metropolitan Magazine*, 45, 237
Mill, James, 407
Miller, John, 223
Mills, William, 380
Milton, John, 348; *Comus*, 23
Milton, Viscount (Charles William Wentworth Fitzwilliam), 380
*Mining Journal*, 267
*Mining Review*, 267
*Mirror of Literature*, 191, 227
Mitford, John, 237
Monboddo, Lord (James Burnett), 278
*Moniteur universel* (newspaper), 25
*Monthly Magazine*, 43, 237n14, 238
*Monthly Repository*, 237n14
*Monthly Review*, 19, 43, 236–37
Moore, Thomas, 385–86
More, Hannah, 309; *Cœlebs in Search of a Wife*, 302; *Practical Piety*, 294
Morgan, John Minter, *Hampden in the Nineteenth Century*, 363
*Morning Chronicle*, 279, 386
*Morning Post*, 240
Moseley, Henry, 416n67
*Movement* (periodical), 371, 477–78
Muhammad Ali (Pasha of Egypt), 373
Mulgrave, 2nd Earl of (Constantine Henry Phipps), 189
Munster, 1st Earl of (George FitzClarence), 388
Murchison, Roderick, 351, 384, 439
Murphy, Patrick: *Rudiments of the Primary Forces of Gravity*, 49; seeking a share of the Bridgewater bequest, 49
Murray, John: advertising Bridgewaters, 43–44, 214; appointed publisher of Bridgewaters, 42, 172–76; arbiter for Bridgewater authors, 223; Bell's publisher, 67; on Buckland's Bridgewater, 390; Buckland's publisher, 62, 94; buying woodblocks for Bell's Bridgewater, 201; and cheap print, 175, 180; in danger of bankruptcy, 176; and dispute concerning underselling, 217; Family Library, 17, 58, 82, 175, 176, 184; lapse of judgement concerning Bridgewaters, 214; Lyell's publisher, 186; mimicking Bridgewaters, 220; organizing review of Buckland's Bridgewater, 274–75; publication of Babbage's *Ninth Bridgewater*, 221, 437; publication of Darwin's *Orchids*, 468; publication of Darwin's *Origin*, 1n2; publication of Knapp's *Journal of a Naturalist*, 41; on scientific writers, 88; as a sole publisher, 216; suggesting size of type, 181

Napier, Macvey, 67
Napoleon I, 24
Nasmith, David, 366
natural laws: the Bridgewaters on, 3, 14, 165–66, 473, 475–77 (*see also under individual authors*); creation by law, 438–40, 443–51, 471–72, 477; Darwin on, 1, 441–42, 443–44, 451, 463–66, 471–72; and deism, 363, 411–12; distinctive of modern science, 2, 10–11, 14, 83–84, 107, 112, 166, 431, 432, 433, 445, 471, 476–77; explanatory scope of, 432–51; as God's mode of action, 112, 154, 155, 340, 341, 349, 426, 433; as implying a divine lawmaker, 114, 382, 392–93, 400; importance of initial conditions, 120,

441, 472; and irreligion, 2, 14, 107, 112, 114, 123, 142–43, 152–53, 160–61, 248, 341; Laplace on, 10, 112, 120, 153, 245, 433; and miracles, 120, 162, 433, 436–37, 438–39; in palaetiological sciences, 440, 472; philosophical anatomy and, 10, 139–40, 409, 413, 425–26, 460–62; phrenology and, 11, 271, 447–48

natural theology: Bridgewater's views on, 26–30; Chalmers's changing views on, 116–19; contested, 13–14, 70, 228, 240–41, 243–44, 246, 252, 256, 285, 397, 427; definition, 13–14; evangelical criticism, 70, 74–75, 251–52, 253–54, 255, 256, 306–7, 394, 397, 421, 482; High Church criticism, 70, 74–75, 125, 152, 239–40, 250–51, 395, 397, 400, 401, 405, 416, 417–18, 482; ineffectiveness of, 130, 238, 250–51, 252, 272, 332, 341, 344, 370, 371–72, 373, 418, 423; Moderate endorsement of, 70, 116, 118, 304; and moral philosophy, 404; in Paley's *Natural Theology*, 69; and politics, 26, 240–41, 346–47; and preaching, 338–39, 344, 347–48; requirements for a classic work of, 47; as a science, 71, 397–98, 445–46; and Unitarianism, 257; in university education, 74–80, 118–19, 125–26, 393–97, 400–405, 407–10, 411–13, 414–16, 417–18, 419–23, 427–28; and working-class atheism, 360–62, 368–69. *See also* Bridgewater Treatises, natural theology; Paley, William, *Natural Theology*

nebular hypothesis: Chambers and, 448, 450; Laplace and, 2, 10, 153; Whewell and, 113, 162, 261, 451, 472; and working-class radicals, 364

Newman, Edward, 264–65

Newman, John Henry, 400, 405

*New Monthly Magazine*, 236, 237n14

*New Moral World* (periodical), 362, 363, 365, 368

Newton, Isaac: on active powers, 154; anti-Newtonianism, 112n9, 356, 364; on instability of solar system, 113; Newtonian natural theology, 36, 112; as a religious exemplar, 113, 130, 143, 259, 348

Nichol, John Pringle, *Views of the Architecture of the Heavens*, 448

Nolan, Frederick: *Analogy of Revelation and Science*, 235, 381, 436; attacks on British Association, 235–36, 383; Bampton Lecturer (1833), 105, 107, 279, 348, 350, 380–81, 400, 401–2; and geology, 158

Northampton, 2nd Marquis of (Spencer Compton), 385

Northern Dispensary (London), 67

Northumberland, 3rd Duke of (Hugh Percy), 351

*Nouveau dictionnaire d'histoire naturelle*, 102

*Observer* (newspaper), 44, 278n82

*Of the Power, Wisdom, and Goodness of God* (anon.), 221

Oliver and Boyd (publishers), 220

Opie, Amelia, 299n15

*Oracle of Reason* (periodical), 369–71

orangutan, 29

Ordnance Survey, 205

Oriel College, Oxford, 398, 400, 404, 405

*Orthodox Presbyterian* (periodical), 247, 252

Otter, William, 416

Owen, Richard: assisting Bridgewater authors, 90, 102, 195, 203, 457; on form and function, 432, 455–60; Hunterian lectures, 457, 458–59; reading of Bridgewaters, 457, 459–60, 468; revising Buckland's Bridgewater, 459

Owen, Robert, 362–64, 368–69

Owenites, 121, 362–72, 477

Owens College (Manchester), 429

Oxford Lunatic Asylum, 417

*Oxford University Magazine*, 381

Oxford University Museum of Natural History, 96

Oxford University Press, 325, 498n1

Page, David, 427
Paine, Thomas, 359; *Age of Reason*, 354, 356
Paley, William: emblematic figure, 44; *Evidences of Christianity*, 70, 74–75, 395; *Moral and Political Philosophy*, 70; *Reasons for Contentment*, 26
Paley, William, *Natural Theology*: and the Bridgewaters, 45, 56, 68–80, 95, 108, 130, 133, 137, 139, 143–44, 146–47, 150, 157, 160, 163–64, 248, 269, 475, 477, 479, 480; Brougham and Bell's edition of, 71–74, 262, 411; composition of, 70; conclusive as to design, 450; dangers of, 427; and functional adaptation, 109, 112, 133, 139, 151, 157, 424; inconclusiveness of, 372; Paxton's edition of, 76; and preaching, 338; rationalism of, 157; reading of, 294, 323, 325, 371, 482; reception of, 69–70, 254; on religious associations, 326; on the religious value of the book of nature, 10; social views in, 26; use of, in scientific education, 75–77, 126, 393, 421, 422, 463; use of, in theological education, 74–75, 78
Palmer, Elihu, *Principles of Nature*, 356, 360
Pander, Christian, *Die vergleichende Osteologie* (with Eduard D'Alton), 211
Panizzi, Antonio, 178
Parish, Woodbine, 211
Parker, John William, 258, 310, 335, 445
Parkinson, John, 59
Parkinson, Richard, 395–96
Parry, Charles, 291–92, 293n, 295, 321–22
Parry, Edward, 292
Parry, Ellen, 291–96, *291*, 302, 321–22
Parry, Emma, 292, 321–22
Paterson, Thomas, 370
Paxton, James, edition of Paley's *Natural Theology*, 76
Peacock, George, 85
Peel, Robert, 350, 389
Penn, Granville, 284
*Penny Pulpit* (periodical), 340
Perkins, Jacob, 266
Peterhouse, Cambridge, 340
Peterloo Massacre (1819), 169, 355, 356
Phillips, John, 415
Phillips, Richard, 356
philosophical anatomy: Bell on, 133, 134, 137, 269, 423, 455; Carpenter on, 413, 425–26, 460–62; Daubeny on, 402; Grant on, 100–101, 138, 409; at London University, 409–10, 422; and natural laws, 10, 409, 461; Owen on, 458–60; Roget on, 109, 137–38, 139–43, 151, 157, 402, 423, 453, 455, 469; at University of Edinburgh, 422–26. *See also* Geoffroy Saint-Hilaire, Étienne
*Philosophical Magazine*, 43, 258, 263–64
Phiz (Hablot Knight Browne), 311
*Phrenological Journal*, 270–71, 448
phrenology: defended from the Bridgewaters, 270–71, 273; Draper and, 325n62; and education, 321–22; and environmental determinism, 364; and materialism, 126, 128, 269, 356; and natural laws, 11, 271, 447–48
Pickering, William, *178*; Aldine Poets, 184, *185*, 187; *Booksellers' Monopoly*, 217; charges of angering Bridgewater authors, 222–23; cheap publishing, innovator in, 184–85, 189, 217; defending copyright of Bridgewaters, 221; dispute with London trade, 179; employment of Byfield as engraver, 206; and form of Bridgewaters, 180–89; innovator in case binding, 188–89; mimicking Bridgewaters, 221; publication of Bridgewaters, 177–79, 213–20; publication of *Gentleman's Magazine*, 237; and sale of Bridgewaters, 215–17, 222; selling casts of some Bridgewater woodblocks, 223; selling woodblocks from Bell's Bridgewater, 201; typographical taste, 182–83
*Plain Englishman* (periodical), 34
Playfair, Lyon, 427
Plinian Natural History Society (Edinburgh), 425

*Poor Man's Guardian* (periodical), 169, 365n63
Pope, Alexander, 90
*Popular Illustrations of Natural History* (anon.), 221
Post, Frederic, 296–97
Powell, Baden: *The Connexion of Natural and Divine Truth*, 445–47, 451; editor of *Magazine of Popular Science*, 266; and the Oriel Noetics, 404, 405; reading of Bridgewaters, 266–67, 445–46, 451; seeking to reform Oxford curriculum, 398–401, *399*; on Whewell's *History*, 462
*Preacher* (periodical), 340n19
preaching: Chalmers exemplary in, 118, 338, 340, 342, 343; guides to, 333–38; importance of, in Victorian Britain, 331–32, 333; reviews of, 338; and science, 331–32, 335–48, 352–53, 373, 392–93; and scientific authorship, 90; sermon periodicals, 340n19; sermons printed to appear as manuscripts, 339n17; Whewell exemplary in, 338
*Presbyterian Review*, 48, 253–54, 282, 336
Prichard, James Cowles, 406
Priestley, Joseph, 6, 144, 257
*Primitive Methodist Magazine*, 247
print: communication revolution in, 4–5, 15, 18, 180, 289, 474, 485–86; politics of, 169
Prout, William, *144*
Prout, William, as a Bridgewater author: appointment, 39; attending meetings of authors, 42, 174, 177; drawing on manuscript work, 98–99, 143; drawing on own lectures, 88; on Pickering's charges, 223; response to Paley's *Natural Theology*, 74, 146–47, 150
Prout, William, Bridgewater Treatise, 143–51; on the design argument, 143–45; excerpted, 328; illustrations, 196–97; in libraries, 412n61, 415; on matter theory, 146, 147–48; on natural laws, 109, 143, 145–51; reading of, 293, 410, 413, 423, 425, 426, 479; reviews of, 48, 253–54, 267–68, 272, 426; revised edition of, 479; on vitalism, 143, 149–50, 423, 425
Prout, William, life and career: active in British Association, 379; sources of income, 64–65
Prout, William, other publications: *An Inquiry into the Nature and Treatment of Gravel*, 64; *Observations on the Functions of the Digestive Organs* (unpublished), 98–99
publishing: advertising, 214–15; agreements, forms of, 174; of Bridgewaters, 170–79; cheap, 2, 4–5, 175–76, 179–80, 183–84, 186–87, 189, 395; circulation of periodicals, 227–28, 237, 241, 246, 258, 263n56, 278; copyright, 100–101, 177, 221, 223–24; depression in book trade, 175, 217–18; edition sizes, 177, 214, 219; international agents, 223; medical specialism in, 64; "people's editions," 224–25; "popular science," 2; reduction in newspaper taxes, 278; trade sales, 215–17; two main hubs in London, 171; underselling, dispute regarding, 179, 217. *See also* book manufacture; illustration
*Pulpit* (periodical), 340n19
Pusey, Edward, 159, 385, 405

Quain, Jones, 410
*Quarterly Review*: advertisements in, 214; on the decline of science in England, 39, 57, 244; as a forum for scientific practitioners, 242, 245; reviews of Bridgewaters, 45, 185–86, 241–43, 275–76, 281, 283, 284, 388, 480; source for newspaper comment, 276, 388; support for Murray's literary coterie, 173; thriving state of, 175
Queen's Hospital (Birmingham), 417

Ramsay, Andrew, 427
Reade, John Edmund, 299–300, 301; *Cain the Wanderer*, 299; *Italy*, 300

reading, 484; and commonplace books, 298n12; and courtship, 309–18; Darwin and, 441–44, 463–68; and the education of children, 320–29; and family devotion, 303–9; fictional accounts of, 313–14, 326, 327–29; and the historiography of science and religion, 485; and marriage, 318–20; and private devotion, 291–300; religious advice concerning, 290, 294, 295; and versification, 297–302; Whewell and, 102–3. *See also* Bridgewater Treatises, reading of
Rees, Abraham, *Cyclopædia*, 67
Reform Act (1832): defining moment in "age of reform," 6; difficulties in passing, 42, 175; effect on book trade, 217; extension of franchise by, 213; political debate regarding, 21; and political unrest, 152, 346, 414
religious devotion: advice regarding, 290, 293–95, 297, 302–3, 324; in families, 303–9; petitionary prayer, 303n23; and poetry, 297–300; in private, 291–300, 302–3
Religious Tract Society (RTS): *Domestic Visitor*, 260; publication of devotional works, 302; publications, 296n, 309, 324; reaction to "infidel" movement, 357; response to SDUK publications, 169, 260; *Weekly Visitor*, 259, 260–61
Rennie, James, 264, 415; *Alphabet of Natural Theology*, 415; *The Faculties of Birds*, 17
*Representative* (newspaper), 175
Richeraud, Antholme, *Elements of Physiology*, 361
Ritchie, William, 37
Rivingtons (publishers), 171
Robison, John, 116
Rogers, Samuel, 387
Roget, Peter Mark, 35
Roget, Peter Mark, as a Bridgewater author: account with Pickering, 215; acting as unofficial secretary, 42, 54, 176–77; appointment, 19, 35–36; attending meetings of authors, 42, 174, 177; charge of plagiarism, 100–101, 413; concerns about Murray as publisher, 176; drawing on experience as SDUK author, 86–87; drawing on own encyclopedia writings, 87, 100, 138; drawing on own lectures, 87, 100; profits, 222; reported to be a "competitor," 42, 43; requesting extension, 177, 213; response to Paley's *Natural Theology*, 74, 139
Roget, Peter Mark, Bridgewater Treatise, 17, 137–43; illustrations, 101, *184*, 197–98; in libraries, 412n61, 415; opposing transmutation, 141–43; on philosophical anatomy, 109, 137–38, 139–43, 151, 157, 402, 423, 453, 455, 469; reading of, 145n70, 293, 301, 361, 413, 423, 425, 426, 443, 449, 451, 460, 462, 468, 469; reviews of, 240, 264, 266, 267–68, 415; on teleological reasoning, 138–39, 452–53, 455; theologically limited, 138
Roget, Peter Mark, life and career: attending Owen's lectures, 458; contributions to *Encyclopædia Britannica*, 453; death of wife, 104–5, 177, 213; at Laycock Abbey, 386; negotiating contribution to Cabinet Cyclopædia, 174, 176; on Royal Society council, 53; Royal Society secretary, 35, 43; sources of income, 67–68; support for Carpenter, 426; writing for *Quarterly Review*, 173
Roget, Peter Mark, other publications: *Electricity*, 86; *Electro-Magnetism*, 86; *Galvanism*, 86; *Introductory Lecture on Human and Comparative Physiology*, 35; *Magnetism*, 86
Romilly, Samuel, 67
Roscoe, Henry, 429
Rose, Hugh James: editor of the *British Magazine*, 244, 247, *248*, 281; principal of King's College London, 416–18; on the religious tendency of science, 110, 349, 381, 395, 396, 400; Whewell's response to views of, 110–11, 112–15, 116, 381
Rosse, 2nd Earl of (Lawrence Parsons),

388; *An Argument to Prove the Truth of the Christian Revelation*, 220
Royal Academy, 66
Royal Belfast Academical Institution, 406
Royal College of Physicians (London), 88
Royal College of Surgeons (London), 457; Bell's lectures at, 92–93, 269; Green's lectures at, 138; Hunterian Museum, 455; Lawrence's lectures at, 7, 356; Owen as assistant conservator at, 90; Owen's lectures at, 457, 458–59; specimens at, 96, 102, 195, 197, 200, 203
Royal College of Surgeons of Edinburgh, 422
Royal Institution, 37, 87, 100n85, 137, 460, 479
Royal Irish Academy, 279
*Royal Lady's Magazine*, 311n36
Royal Manchester Institution, 319
Royal Medical Society (Edinburgh), 425, 426
Royal Pavilion (Brighton), 344
Royal Physical Society (Edinburgh), 425
Royal Society, 32; Bridgewater authors meeting at, 59, 223; changing character, 19–21, 22, 28; Copley Medal, 38, 39, 40, 98; council, 53; criticized by members of the Astronomical Society, 37; discussion of Bridgewaters at, 386–87; fellows, 279, 381; library, 178; *Philosophical Transactions*, 98, 201, 203, 206; president administering Bridgewater bequest, 4, 26, 27, 172, 213, 414; Royal Medals, 31, 38, 39; and scientific patronage, 19–21, 31–33, 43–44, 46–49, 385
Royal Society of Arts, 24
Royal Society of Edinburgh, 424, 449
RTS. *See* Religious Tract Society (RTS)
Rugby School, 404
Ruskin, John, 403, 404
Russell, Lord John, 385
Russell Institution (London), 87
Ryan, Michael, 269
Ryland, John, 325
Ryle, John Charles, 403–4

Sabine, Edward, 319
*Satirist* (newspaper), 278
*Saturday Review*, 428
Scharf, George Johann, 205n56, 210–12, *211*
Schelling, Friedrich, 239
science and Christianity: historiography of, 9n13, 16–17, 18, 290, 373, 476, 481–82; perception of conflict between, 7–9, 11, 283, 284, 477–78, 487; reconsideration of relationship between, 3, 11, 477–78; "theistic science," 14, 476–78, 480
scientific education: at home, 321–29; at university, 60–62, 75–77, 124–26, 376–77, 390–429, 486. *See also under individual universities and colleges*
scientific patronage: and artisans, 265–66; in France, 48; and the identity of "men of science," 376, 384–90; and the Royal Society, 19–21, 31–33, 43–44, 46–49, 385
scientific practice, 486–87; lawlike explanation in, 432–51; observation in, 432–33, 452–69
scientific practitioners. *See* "men of science"; women in science
*Scotsman* (newspaper), 278, 280
Scott, Walter, 173; Waverley Novels, 184
*Scottish Missionary Register*, 246
*Scottish Pulpit* (periodical), 246, 340n19
Scouler, John, 241n21
Scrope, George Poulett, 275
SDUK. *See* Society for the Diffusion of Useful Knowledge (SDUK)
Sedgwick, Adam: and the admission of dissenters, 394, 401; on Buckland's buffoonery, 384; as a clerical geologist, 388; correspondence with Whewell, 179; declining to review Buckland's Bridgewater, 275; defense of geology from literalists, 351, 383; *Discourse on the Studies at the University*, 347, 383, 392–94, 401; on natural theology, 77, 393–94; on *Origin*,

Sedgwick, Adam (*continued*) 471; president of the British Association, 382; quoted by Buckland, 164; reading of Bridgewaters, 382, 392–94; on the recent origin of humans, 360; reputation of, 473; on scientific education, 392–94, 396; teaching Kingsley, 317
Serres, Étienne, 142; *Anatomie comparée du cerveau*, 128
Sewell, William, 401, 503n16
Seymour, Robert, 5
*Shepherd* (periodical), 365
Sherwood, William, 223
Simeon, Charles, 333, 383
Simpkin and Marshall (wholesalers), 216
Simpson, James Young, 312–13
Smith, Elder (publishers), 221
Smith, James Elimalet, 363, 365
Smith, John Pye, 283, 353–54
Society for Promoting Christian Knowledge (SPCK): Anti-Infidel Committee of, 357; Parker publishes for, 445; publication of scientific works, 234; response to SDUK publications, 169, 258, 289; *Saturday Magazine*, 258–60, 259, 302
Society for Promoting Female Education in the East, 373
Society for the Diffusion of Useful Knowledge (SDUK): attacked as irreligious, 107, 289; Bridgewater authors active in, 15, 36, 38, 67, 72, 84–86, 87; Brougham active in, 71, 84–85; case binding, innovative use of, 188; cheap publications of, 4–5, 169, 172, 175, 212, 289; Henry Duncan supports, 304; eschewing religious subjects, 262; Library of Entertaining Knowledge, 17, 415; Library of Useful Knowledge, 67, 84–86, 181, 191, 197, 199, 361; *Penny Cyclopædia*, 181; *Penny Magazine*, 180, 183, 191, 258, 262–63; production of wall charts, 197; *Quarterly Journal of Education*, 399; steam presses, innovative use of, 180, 183; wood engraving, innovative use of, 183, 191, 212
Society of Antiquaries, 23, 177
Socrates, 472
Somerville, Mary: as an author, 92; *Connexion of the Physical Sciences*, 83
South, James, 37
Southey, Robert, *Life of Cowper*, 314
Southwell, Charles, 369–72
Sowerby, James De Carle, 207, 208–9; *Mineral Conchology*, 208
SPCK. *See* Society for Promoting Christian Knowledge (SPCK)
*Spectator* (periodical), 44, 45, 220, 233, 238
Spence, William: drawing of Kirby, 55; *Introduction to Entomology* (with William Kirby), 41, 63, 64, 90
Spencer, Herbert, 479
Spineto, Elizabeth, Marchesa di, 103
Spring-Rice, Thomas, 385
Spurzheim, Johann Gaspar, 322
*Standard* (newspaper), 278–80, 282
St. Bartholomew's Hospital (London), 455
Steel, Thomas Henry, 336
Stevens, Henry, 182
St. George's Hospital (London), 67
*St James's Chronicle*, 278–80, 282
Stokes, Charles, 275
St. Thomas's Hospital (London), 172
Sturm, Christoph Christian, *Reflections on the Works of God*, 112, 305, 306
Sturtevant, Saunderson Turner, *Preacher's Manual*, 333
*Sunday School Teachers' Magazine*, 256
Sunday School Union, 256, 327
*Sunday Times*, 278n82
Surgeon's Square medical school (Edinburgh), 423
Sussex, Duke of (Augustus Frederick): agreeing to delay in publication of Bridgewaters, 177; announcing completion of Bridgewaters, 49, 386–87; not involved in administering Bridgewater bequest, 41; president

of the Royal Society, 33, 43, 385; visit to Kinmel Park, 386
Sussex Scientific and Literary Institution, 388
Sutton, Charles, 90

*Tait's Edinburgh Magazine*, 237n14, 240–41, 246
Talbot, (William) Henry Fox, 386
Tate, James, 299
Taylor, Jeremy, *Holy Living*, 317
Taylor, Robert, 357, 358
Taylor, William Cooke, 233, 235
*Teacher's Offering* (periodical), 327n66
Tegg, Thomas, 183
Tennyson, Alfred, 300–302; *In Memoriam*, 301
Tennyson, Emily, 301n18
Thomas, Vaughan, 417
Thompson, Thomas Perronet, 245
Thomson, Allen, 420
Thomson, John, 420
Thomson, Thomas, *History of Chemistry*, 361
Timbs, John, *English Eccentrics and Eccentricities*, 21
*Times, The*, 44, 277, 280, 284, 394
Todd, Robert Bentley, 416
Torr, John, 312
Tractarianism: and Blomfield, 389; and the *British Critic*, 281, 401; and Buckland, 158; controversy regarding, 446; and Grenfell, 316; and Hampden, 404; and Pusey, 159; and Rose, 247, 395; and science, 405
transmutation: Bridgewaters opposing, 14, 109, 127–30, 131, 133, 134–37, 141–43, 151, 153–54, 155, 157, 159, 162–63, 164, 165, 166, 195, 251, 270, 271, 272–73, 423, 446, 457, 475–76; Bridgewaters seen as supporting, 280; British naturalists repudiating, 10–11, 349; Chambers on, 448–51; in conversation, 387; and creation by law, 443–51, 471, 472, 477; Darwin on, 14, 443–44, 463–68; geology, role in undermining, 95; Grant on, 100, 409; as legitimate subject for consideration, 446; pulpit opposition to, 349; and working-class radicalism, 361, 370, 448. *See also under* Lamarck, Jean-Baptiste
Trimmer, Mary, *Natural History*, 183, *185*
Trimmer, Sarah, 184, 321; *An Easy Introduction to the Knowledge of Nature*, 64, 321
Trinity College, Cambridge: Bridgewaters read at, 336; British Association visit to, 382; natural theology at, 75, 77; *Origin* at, 471–72; Sedgwick's preaching at, 392; students at, 340–41; Whewell as a fellow at, 61; Whewell's becoming master of, 389
Trinity College, Dublin, 281, 410
Turner, Edward, 410–11
Turner, Sharon, *Sacred History of the World*, 274
Turton, Thomas (Carlilean), 358
Turton, Thomas (professor), 396–97
Tyndall, John, 336, 479, 487

*Unitarian Magazine*, 247
*United Kingdom* (newspaper), 45
Universal Brotherhood of Friends of Truth (Maga Club), 426–28
Universal Community Society of Rational Religionists, 369
University College London. *See* London University (University College London)
University of Cambridge: Bridgewaters read at, 336, 391–92; Carlile's infidel mission at, 357; Christian advocate at, 131, 350; Hulsean lecturer at, 395; professorial salaries at, 60; scientific education at, 77–78, 376, 391–97; Select Preacher at, 340; theological education at, 70, 74–75
University of Durham, 395, 416
University of Edinburgh: Bridgewaters read at, 391–92; Owen as a student at, 455; reformers at, 474; scientific education at, 376–77, 391–92, 418–28; theological education at, 78–80, 118, 419–21

University of Glasgow, 336, 415, 420
University of London, 406, 416. *See also* King's College London; London University (University College London)
University of Oxford: Bridgewaters read at, 336, 391–92, 415; scientific education at, 40, 75–77, 376, 391–92, 397–405, *399*; theological education at, 74–75, 404
University of St. Andrews, 54, 78, 116, 118, 122
utilitarianism, 121–22, 245

*Vestiges of the Natural History of Creation. See* Chambers, Robert
*Veterinarian* (periodical), 267–68
Victoria, Queen, 1, 344, 389, 441, 473
Vieusseux, André, 371n73
Vincent, Joseph, 76
Vincent, William, 61
Virey, Julien-Joseph, 127, 138, 156; *Histoire des mœurs et de l'instinct des animaux*, 141
Volney, François de, 358; *Ruins of Empires*, 356

Wakley, Thomas, 101, 268, 271
Walker, Robert (Oxford), 401
Walker, Robert (Purleigh), 298–99, 300, 301
Warneford, Samuel Wilson, 417–18
Watson, Joshua, 417–18
Watson, Richard, 9, 337–38; *Theological Institutes*, 337
Webster, Thomas, 205, 210
Wesley, John, 295
Wesleyan Methodist conference, 335
*Wesleyan-Methodist Magazine*, 247, 255–56, 282, 337
*Westminster Review*, 45, 241n20, 244–45
Westminster School, 61
Whalley, Robert, 364–65; *Philosophical Refutation of the Theories of Robert Owen*, 365; *Revolution of Philosophy*, 364–65
Whewell, William, 37
Whewell, William, as a Bridgewater author: appointment, 37–38; attending meeting of authors, 177; concerns about Murray as publisher, 176; concerns with practice of Christianity, 485; consulted regarding Airy, 39; drawing on own sermons, 111, 113, 339, 393; original research, 102–3; pleased with form, 186; popular exposition, 81–84; response to criticism, 48–49, 244, 440; response to Paley's *Natural Theology*, 77; response to Rose, 110–11, 112–15, 381; seeking clarification, 58, 68; seeking larger readership, 181, 186; writing "hieroglyphics," 179
Whewell, William, Bridgewater Treatise, *17*, 110–16, *185*, *481*; Babbage's response to, 434–36, 437–38; cheaper edition of, 186–87, 222; on deductive and inductive habits, 114–15, 248–49, 338–39, 341, 369n68, 382, 421, 434–36; establishing high reputation for series, 218, 387; excerpted, 261, 262; on final causes, 453, 457–58, 460; financial history of, 186–87; in libraries, 412n61, 415, 469; on the limitations of natural theology, 108, 111, 114–16, 123, 248–49; on natural laws, 1, 83–84, 112–16, 123, 239, 248, 340, 349, 382, 392–93, 400, 426, 431, 432, 433, 440, 441, 443–44, 446, 461, 472, 476, 483; on the nebular hypothesis, 113, 162, 261, 451, 472; quoted in Darwin's *Origin*, 1, 2, 10, 14, 434, 441, 444, 469, 471; reading of, 299–300, 301–2, 312, 313, 315, 319–20, 341, 360–61, 364n61, 369n68, 382, 383, 392–93, 394, 400, 416, 421, 426, 434–36, 441, 443–44, 445–46, 449, 450, 451, 460, 462, 463–64, 469; on reconciliation of Christians with science, 107, 111, 485; reviews of, 49, 233–34, 235–36, 237–38, 239, 243–44, 247–50, 256, 276–78; revised edition of, 472; similarities to Chalmers, 113, 116, 120; unillustrated, 192; use of

by preachers, 338–39, 340, 341–42, 349, 373
Whewell, William, life and career: active in British Association, 235, 379, 381–82; admired by Gurney, 297; and the admission of dissenters, 394, 405; British Association address, 382; busy in term time, 104; changing views on laws, 440, 451, 457–58; clerical status, 387; on form and function, 457, 459, 466–67; on geology and natural theology, 76, 162; at Laycock Abbey, 386; mentor to Darwin, 1, 442–43; on *Origin of Species*, 471–72; as a preacher, 111, 338, 349, 392; as president of the Geological Society, 442, 457; on professorial salaries, 60; reputation of, 473; as a reviewer, 7n9, 81–84, 242; sources of income, 61; Trinity College master, 389; writing for *Quarterly Review*, 173
Whewell, William, other publications: *Elementary Treatise on Mechanics*, 61; *History of the Inductive Sciences*, 81, 440, 443, 444, 451, 458, 460–61, 462, 466–67; *Indications of the Creator*, 450–51; *Philosophy of the Inductive Sciences*, 81, 451, 458, 459, 472; "Recent Progress and Present State of Mineralogy," 94
White, Andrew Dixon, *History of the Warfare of Science with Theology*, 480
White, Gilbert, *Natural History of Selborne*, 41, 63
Whittingham, Charles, *178*, 182–85, 198, 214, 219, 498nn1–2; Whittingham's Cabinet Library, 183, *185*
*Whole Duty of Man, The*, 324
Wilberforce, William, 254, 342
Wilkin, Simon, 194, 195
Wilkinson, Christopher, 361
Wilks, Samuel Charles, 254–55
Williams, Edward, *Christian Preacher*, 333
William IV (king), 344, 440
Wilson, George, 427, 428
Wilson, John, 420, 425
Winning, William Balfour, 281
Wix, Samuel, 34
women in science: as amanuenses, 56, 92–94, 96, 103; as authors, 83, 92, 183–84; at British Association meetings, 377, 379, 381; as collectors, 204; as illustrators, 94, 197–98, 204, 206, 209, 403; as learners, 293, 315, 321; as manuscript readers, 103; in private intercourse with men, 88, 279, 310–20; as role models, 322–24; in social intercourse with men, 436, 443
Wood, J. G., 480
Wordsworth, William, 292

Yorkshire Philosophical Society, 350
Young, Robert M., 240
Young, Thomas, 65
*Youth's Instructer* (periodical), 256
*Youth's Magazine*, 256n44, 327–29

Zeitter, John Christian, 207
Zoological Society, 102, 195, 210